Catalysis in Asymmetric Synthesis
Second Edition

Postgraduate Chemistry Series

A series designed to provide a broad understanding of selected growth areas of chemistry at postgraduate student and research level. Volumes concentrate on material in advance of a normal undergraduate text, although the relevant background to a subject is included. Key discoveries and trends in current research are highlighted, and volumes are extensively referenced and cross-referenced. Detailed and effective indexes are an important feature of the series. In some universities, the series will also serve as a valuable reference for final year honours students.

Editorial Board

Professor James Coxon (Editor-in-Chief), Department of Chemistry, University of Canterbury, New Zealand.
Professor Pat Bailey, Department of Chemistry, University of Manchester, UK.
Professor Les Field, School of Chemistry, University of New South Wales, Sydney, Australia.
Professor Dr. John Gladysz, Institut für Organische Chemie, Universität Erlangen-Nürnberg, Germany.
Professor Philip Parsons, School of Chemistry, Physics and Environmental Science, University of Sussex, UK.
Professor Peter Stang, Department of Chemistry, University of Utah, Utah, USA.

Titles in the Series

Protecting Groups in Organic Synthesis
James R. Hanson

Organic Synthesis with Carbohydrates
Geert-Jan Boons and Karl J. Hale

Organic Synthesis using Transition Metals
Roderick Bates

Stoichiometric Asymmetric Synthesis
Mark Rizzacasa and Michael Perkins

Catalysis in Asymmetric Synthesis (Second Edition)
Vittorio Caprio and Jonathan M. J. Williams

Reaction Mechanisms in Organic Synthesis
Rakesh Kumar Parashar

Forthcoming

Photochemistry of Organic Compounds: From Concepts to Practice
Petr Klán and Jakob Wirz

Practical Biotransformations
Gideon Grogan

Catalysis in Asymmetric Synthesis

Second Edition

Vittorio Caprio

Department of Chemistry
University of Auckland, Auckland, New Zealand

and

Jonathan M. J. Williams

Department of Chemistry
University of Bath, Bath, UK

A John Wiley and Sons, Ltd., Publication

This edition first published 2009
© Vittorio Caprio and Jonathan M.J. Williams

This text is based on "Catalysis in Asymmetric Synthesis" by Jonathan M.J. Williams, published by Sheffield Academic Press in 1999 and acquired by Blackwell Publishing in 2002.

Blackwell Publishing was acquired by John Wiley & Sons in February 2007. Blackwell's publishing programme has been merged with Wiley's global Scientific, Technical, and Medical business to form Wiley-Blackwell.

Registered office
John Wiley & Sons Ltd, The Atrium, Southern Gate, Chichester, West Sussex, PO19 8SQ, United Kingdom

Editorial offices
9600 Garsington Road, Oxford, OX4 2DQ, United Kingdom
2121 State Avenue, Ames, Iowa 50014-8300, USA

For details of our global editorial offices, for customer services and for information about how to apply for permission to reuse the copyright material in this book please see our website at www.wiley.com/wiley-blackwell.

The right of the author to be identified as the author of this work has been asserted in accordance with the Copyright, Designs and Patents Act 1988.

Library of Congress Cataloging-in-Publication Data
Caprio, Vittorio.
  Catalysis in asymmetric synthesis / Vittorio Caprio and Jonathan M.J. Williams. – 2nd ed.
    p. cm.
  Includes bibliographical references and index.
  ISBN 978-1-4051-9091-6 (cloth) – ISBN 978-1-4051-7519-7 (pbk. : alk. paper)
  1. Asymmetric synthesis.   2. Catalysis.   I. Williams, Jonathan M. J.   II. Title.
  QD262.C28 2008
  541'.39–dc22   2008045023

A catalogue record for this book is available from the British Library.
Set in 10/12 pt Minion by Aptara Inc., New Delhi, India
Printed and bound in Great Britain by TJ International, Ltd, Padstow, Cornwall
1   2008

To Jenny

Contents

Preface to the Second Edition

Since publication of the first edition of this book in 1998 the area of asymmetric catalysis has grown apace. Major advances in organo- and bifunctional catalysis have occurred in this millennium. Thus, the release of a second, updated edition is timely.

The book aims to capture the material published up to the present. As in the first edition the emphasis is on nonenzymatic methods although references to reviews on catalysis with enzymes are included.

We have strived to include as many examples of catalysts and their scope as possible within the confines of a book of this size, while maintaining readability of this text and its use as a teaching aid. Further coverage of specific catalysts can be obtained from the many comprehensive reviews referenced.

We are grateful to Amanda Heapy for all her helpful feedback and to Dr Jenny Gibson for proof reading and editorial assistance.

V. Caprio
University of Auckland

J. M. J. Williams
University of Bath

Preface to the First Edition

Asymmetric synthesis has become a major aspect of modern organic chemistry. The importance of stereochemical purity in pharmaceutical products has been one driving force in the quest for improved control over the stereochemical output of organic reactions. The fact is, like it or not, stereochemistry is hard to avoid.

Asymmetric catalysis is a very important aspect of asymmetric synthesis, and one that has seen tremendous activity during the 1990s. This book aims to capture the latest results in asymmetric catalysis, and covers the literature up until June/July 1998. A few references as late as October 1998 have also been included. The emphasis in this book has been on nonenzymatic methods for asymmetric catalysis, although key references to enzyme-catalysed reactions have been incorporated where appropriate.

It cannot be possible to be comprehensive in a book of this size, although hopefully there are no major omissions. I apologise if I have omitted any 'favourite' reactions, or if a topic has not been treated with the depth that it may deserve. The emphasis, in general, has been on asymmetric catalytic reactions which are as current as possible.

I am grateful to my former mentors, Professor S. G. Davies (Oxford) and Professor D. A. Evans (Harvard), who introduced me to asymmetric synthesis and asymmetric catalysis. I am also indebted to my current coworkers and students who help to keep organic chemistry alive for me with their enthusiasm for the subject.

My thanks also go to Mrs J. W. Curtis, who has typed the bulk of the manuscript, and Miss H. L. Haughton, who helped with many of the chemical structures.

Finally, I am especially grateful to my wife, Cathy, who has not complained about the extra time I have needed to work during the preparation of this book.

J. M. J. Williams
University of Bath

Chapter 1
Introduction

There are several ways of producing compounds as single enantiomers. The resolution of a mixture of enantiomers can often be the cheapest, most practical way of obtaining enantiomerically pure material. The conversion of a cheap enantiomerically pure starting material into another derivative represents another useful technique in some cases.

However, asymmetric synthesis can provide a more general approach to the preparation of enantiomerically enriched compounds. Asymmetric synthesis is limited by the range and scope of available methodology. Fortunately, with so much research in the area, there are many methods now available, with further asymmetric reactions being developed each year. There may still be cases where asymmetric synthesis doesn't provide the best method for the preparation of a particular enantiomerically pure compound, but it certainly allows for the preparation of a more diverse range of structures.

Asymmetric catalysis is an especially appealing aspect of asymmetric synthesis. Small amounts of a catalyst (frequently less than 1 mol%) can be used to control the stereochemistry of the bulk reaction. The use of a catalyst often makes the isolation of a product easier, since there is less unwanted material to remove at the end of a reaction.

1.1 Reactions Amenable to Asymmetric Catalysis

From a synthetic viewpoint, the individual chapter titles in the book give a fair impression of the scope of synthetic processes involved. The types of reaction amenable to asymmetric catalysis can also be considered from a more stereochemical viewpoint.

Most reactions involving asymmetric catalysis are based around the conversion of a planar sp^2 carbon atom into a tetrahedral sp^3 carbon atom. This category of reactions includes asymmetric hydrogenation of alkenes and ketones, as well as the addition of other reagents to these groups as identified in Figure 1.1.

Substrates containing enantiotopic groups can be converted into enantiomerically enriched compounds using asymmetric catalysis with representative examples

Catalysis in Asymmetric Synthesis 2e   © 2009 Vittorio Caprio and Jonathan M.J. Williams

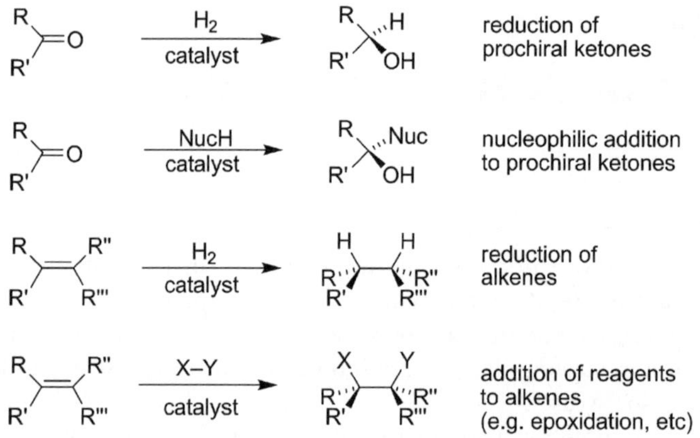

Figure 1.1 Asymmetric catalytic reactions involving sp^2 to sp^3 conversion

outlined in Figure 1.2. These reactions break the symmetry (e.g. meso or achiral) of the starting material.

Asymmetric catalysis may also be achieved by the kinetic resolution of racemic substrates, where one enantiomer of starting material is selectively converted into product, leaving the other enantiomer unreacted. In some instances, both enantiomers of a starting material are converted into the same enantiomer of product, i.e. dynamic kinetic resolution. Most reactions fit into one of the categories identified in Figures 1.1–1.3, even if the exact structure is not represented.

1.2 Assignment of (*R*) and (*S*) Stereochemical Descriptors

The Cahn–Ingold–Prelog (CIP) system for describing the stereochemistry of chiral molecules is universally accepted.[1,2] Simple molecules containing one chiral centre

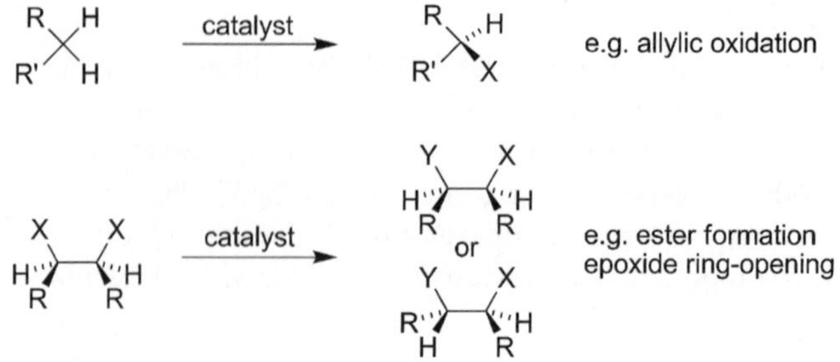

Figure 1.2 Asymmetric catalytic reactions involving symmetry breaking

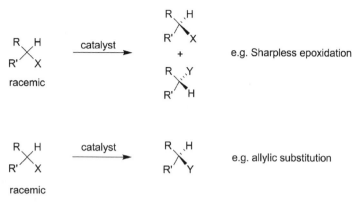

Figure 1.3 Kinetic and dynamic resolution of racemic substrates

are described as (*R*) (from Latin: rectus) or (*S*) (from Latin: sinister). A short account of how to distinguish between (*R*) and (*S*) follows, although more detailed information is available elsewhere.[3]

For a tetrahedral carbon-based chiral centre, the priority of the four attached groups must be determined according to the sequence rules. The two enantiomers in Figure 1.4 have groups attached where a>b>c>d.

The structure must be viewed from the side opposite the lowest priority group. In these cases, the group d must therefore point away from us. If the priority of the remaining groups a–b–c is in an anticlockwise sense, the stereochemical descriptor is (*S*). When the a–b–c sequence is clockwise, the stereochemical descriptor is (*R*).

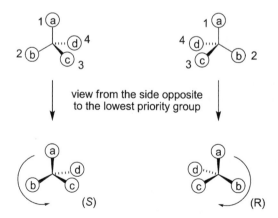

Figure 1.4 (*R*) and (*S*) assignments to tetrahedral chiral centres

Of course, for real molecules, we need to be able to decide which groups have priority over other groups. In general, the most important sequence rule is that groups of higher atomic number precede groups of lower atomic number. Hence, the molecule (**1.01**) has the (*R*)-configuration.

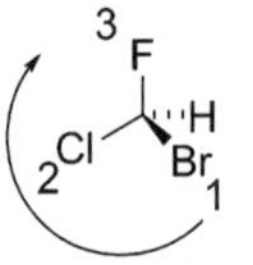

- lowest priority group pointing away
- Br > Cl > F > H
- clockwise priority, hence (*R*)

(**1.01**)

However, not all molecules are so simple! The groups may need to be considered in more detail, when the first atoms attached to the chiral centre have the same priority.

In the case of molecule (**1.02**) we can quickly identify the lowest priority group, the H atom. However, the three remaining groups each contain a carbon attached to the chiral centre. We have to consider the next "sphere" of atoms. In the second sphere, the Cl atom takes priority over the O, which takes priority over the H. The compound therefore has the (*S*)-configuration.

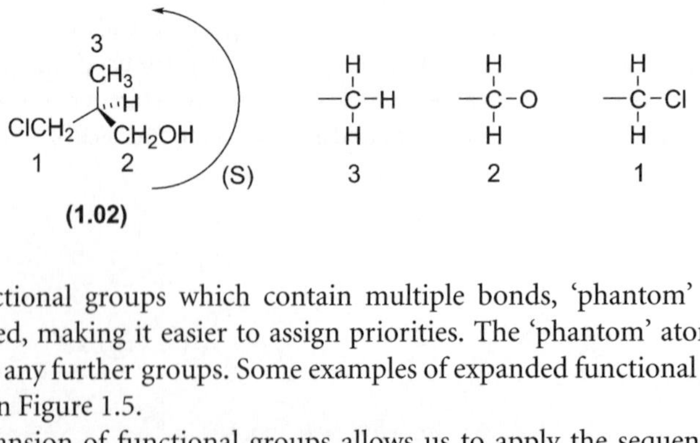

(**1.02**)

For functional groups which contain multiple bonds, 'phantom' atoms are incorporated, making it easier to assign priorities. The 'phantom' atoms are not attached to any further groups. Some examples of expanded functional groups are identified in Figure 1.5.

The expansion of functional groups allows us to apply the sequence rules to more complex structures, (**1.03–1.08**). A few comments may be helpful for some of these assignments. In structure (**1.04**), the -CH_2Cl group takes priority over the CF_3 group. The chloride takes precedence over the fluoride, even though there is only one chloride, because the fluorides are not added together. In the phosphine (**1.07**), the lone pair of electrons is considered to have the lowest priority (lower than any atom). In structure (**1.08**), the -CH_2Cl group takes priority over the -CH_2CH_2I group. This is because the Cl atom comes in the second sphere, whereas the I atom does not appear until the third sphere.

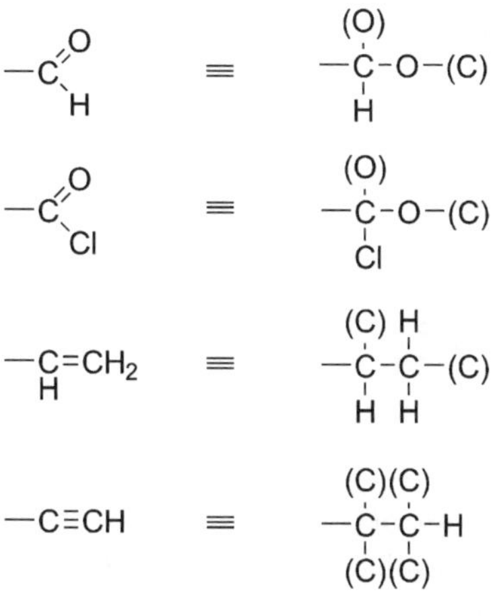

Figure 1.5 Expanded functional groups with phantom atoms

Another aspect of the (R,S)-nomenclature involves assigning structures which possess axial chirality or planar chirality. Because of the number of ligands which possess axial chirality (BINOL, BINAP and related ligands), the assignment of (R) and (S) to these structures is briefly considered. An additional sequence rule is helpful here: the nearer end of an axis (or plane) precedes the further end. For the axially chiral molecule, BINOL (**1.09**), a simple way to assign stereochemistry is by looking down the chiral axis. The nearer groups take priority over the further groups, and hence the assignment of sequence is as shown in Figure 1.6. Again, clockwise corresponds to the (R)-configuration, and anticlockwise

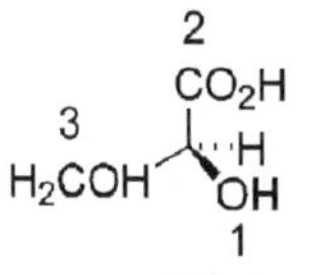

(S)-(**1.03**) 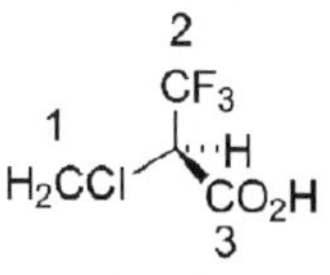 (R)-(**1.04**) 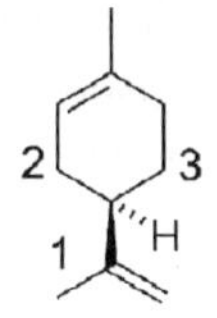 (R)-(**1.05**)

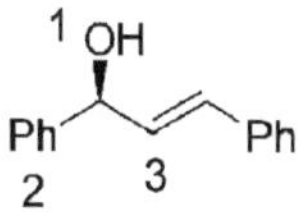

(S)-(**1.06**) 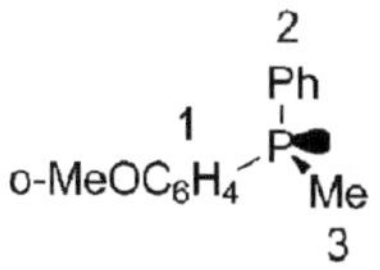 (R)-(**1.07**) 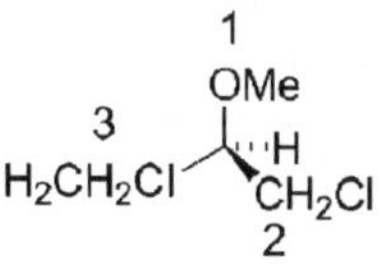 (R)-(**1.08**)

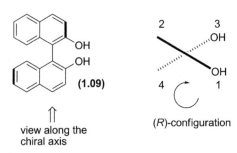

Figure 1.6   Assignment of axial chirality

corresponds to the (S)-configuration. This is like a helix spiralling away from the viewpoint. The assignment will be made by looking along the chiral axis from the other end.

Axial chirality can also be assigned using (M,P)-nomenclature (see reference 2). For axial chirality, (R) is equivalent to (M), and (S) is equivalent to (P).

Further reading

There are several other textbooks available, dealing with various aspects of stereo chemistry, asymmetric synthesis and catalysis. The following titles may be useful.

E.L. Eliel, S.H. Wilen and L.N. Mander *Stereochemistry of Organic Compounds*, John Wiley & Sons Inc., New York, **1994**.

R.E. Gawley and J. Aubé, *Principles of Asymmetric Synthesis*, Tetrahedron Organic Chemistry Series, Volume *14*, Pergamon, Oxford, **1996**.

R.A. Sheldon, *Chirotechnology. Industrial Synthesis of Optically Active Compounds*. Marcel Dekker, New York, **1993**.

Stereoselective Synthesis, ed. G. Helmchen, R.W. Hoffman, J. Mulzer and E. Schaumann, Georg Thieme, Stuttgart, **1995**.

Catalytic Asymmetric Synthesis, ed. I. Ojima, VCH, New York, **1993**.

R. Noyori, *Asymmetric Catalysis in Organic Synthesis*, John Wiley & Sons, Inc., New York, **1994**.

M. Nógrádi, *Stereoselective Synthesis*, 2nd edition, VCH, New York, **1995**.

G. Procter, *Asymmetric Synthesis*, Oxford University Press, Oxford, **1996**.

Comprehensive Asymmetric Catalysis, eds. E. N. Jacobsen, A. Pfaltz and H. Yamamoto, Springer-Verlag, Berlin, **1999**.

New Frontiers in Asymmetric Catalysis, eds. K. Mikami and M. Lautens, John Wiley & Sons, Ltd, Chichester, **2007**.

Asymmetric Synthesis: The Essentials, eds, M. Christmann and S. Bräse, Wiley-VCH, Weinheim, **2007**.

Asymmetric Organocatalysis, A. Berkessel and H. Gröger, Wiley-VCH, Weinheim, **2005**.

References

1. R.S. Cahn, C. Ingold and V. Prelog, *Angew. Chem., Int. Ed. Engl.*, **1966**, *5*, 385.
2. V. Prelog and G. Helmchen, *Angew. Chem., Int. Ed. Engl.*, **1982**, *21*, 567.
3. E.L. Eliel, S.H. Wilen and L.N. Mander, Stereochemistry of Organic Compounds, John Wiley & Sons, Inc., New York, **1994**, Chapter 5.

Chapter 2
Reduction of Alkenes

The reduction of alkenes was the earliest catalytic reaction to be exposed to a substantial research effort directed at achieving an asymmetric variant. The reduction of an alkene to an alkane is a particularly important synthetic transformation, since construction of a C=C bond is often straightforward, and it can subsequently be converted into the corresponding C–C bond. The ability to achieve this transformation with asymmetric induction has been the most widely studied of all transition-metal catalysed asymmetric reactions, and the area has been the focus of many reviews.[1] Enantiomerically pure rhodium and ruthenium-based complexes are the most commonly used catalysts. While the former display great utility in the asymmetric hydrogenation of dehydroamino acids, the latter show much wider scope. Both these systems require the presence of chelating functionality in the substrate and the asymmetric hydrogenation of unfunctionalised alkenes has proved challenging. A major breakthrough in this area has occurred with the advent of iridium-based hydrogenation catalysts that do not require the presence of coordinating groups and the asymmetric reduction of alkyl-substituted alkenes can now be achieved. Metal free organocatalyst-mediated transfer hydrogenation has been recently investigated and while enantioselectivities achieved are high, this method is so far limited to the reduction of α,β-enals and enones.

Related processes such as hydrosilylation, hydroboration and hydroamination are considered in this chapter. Hydroformylation is also considered, as well as hydroacylation and hydrocyanation reactions of alkenes.

2.1 Asymmetric Hydrogenation with Rhodium Complexes

Achiral Wilkinson's catalyst, $RhCl(PPh_3)_3$, is an effective homogeneous catalyst for the hydrogenation of alkenes. The first examples of homogenous asymmetric hydrogenation were reported independently by Horner[2] and Knowles.[3] These were variants of Wilkinson's catalyst using enantiomerically pure monodentate phosphine ligands, and provided definite, but low, asymmetric induction. In the early 1970s, Kagan prepared the bidentate ligand DIOP (2.01) which possesses $C2$-symmetry, and gave good enantioselectivity in the rhodium-catalysed hydrogenation of α-(arylamino)acrylic acids, such as substrate (2.02), which affords the α-amidoacid product (2.03).[4]

Initial research in this area focussed on the development of enantiomerically pure bidenate bisphosphines, often possessing C_2-symmetry, as ligands in the hydrogenation of alkenes, and the α-(acylamino)acrylic acids have remained popular substrates. Many of these ligands have provided high enantioselectivity in the reduction of enamides, and a representative set of such structures (2.04–2.15) is shown, all of which have given over 90% ee (often higher).[5]

The high asymmetric induction obtained with these ligands is attributed to the formation of a rigid chelate ring in the rhodium complexes and for some time the use of bidenate phosphorus ligands was considered a prerequisite for the achievement of highly enantioselective hydrogenation. This view no longer holds and the use of monodentate ligands has seen a recent resurgence.[6] In particular the phosphoramidites (2.16),[7] phosphites (2.17)[8] and phosphonites (2.18)[9] developed by the groups of Feringa, Reetz and Pringle respectively form highly efficient catalyst systems that are comparable or superior to the best bidentate systems in terms of both enantioselectivity and reactivity. Furthermore, these monodentate ligands are generally much less expensive and more readily accessible than bidentate phosphines. The ease of preparation of these ligands, in two steps from BINOL, has led to the synthesis and testing of a variety of analogues. These results indicate that replacement of the dimethylamino group of MONOPHOS with other symmetrical amino moieties such as a diethylamino group,[10] a piperidinyl or a morpholino group[7e] can lead to an increase in enantioselectivity. In contrast, replacement with an unsymmetrically substituted amine functionality often leads to decreases in ee. However, replacement of the dimethylamino group with an α-methylbenzylamine moiety[7b] or ferrocinyl-derived functionality,[11] to give ligand (2.16b) or (2.19), respectively, does lead to improvements in enantioselectivity and/or rate[7c] in some cases. Two monodentate ligands bind to the rhodium centre to give the active catalyst and it has been discovered that hetero-complexes prepared using mixtures of two enantiopure phosphoramidites[7d] or phosphites[8c] can be more effective than

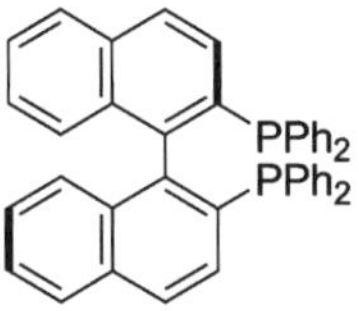

(2.04) BINAP

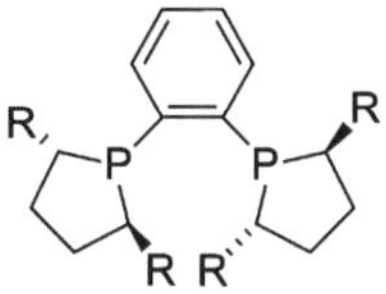

(2.05) DUPHOS

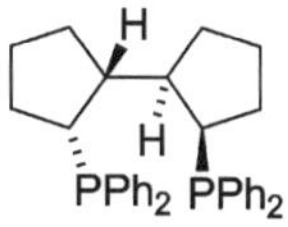

(2.06) BICP

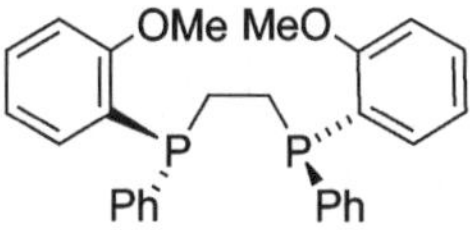

(2.07) DIPAMP

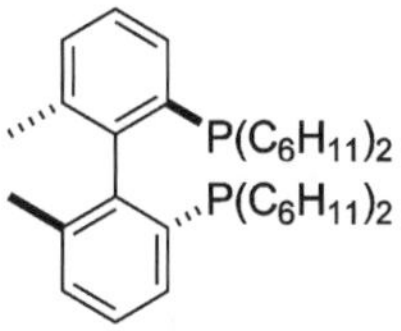

(2.08) (*R*)-BICHEP

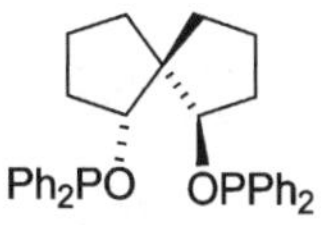

(2.09) (*R*)-SIROP

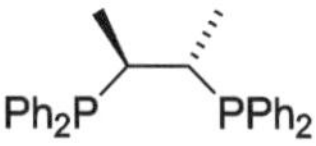

(2.10) CHIRAPHOS

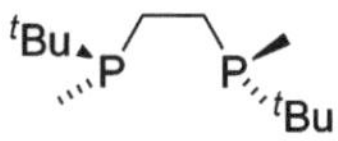

(2.11)

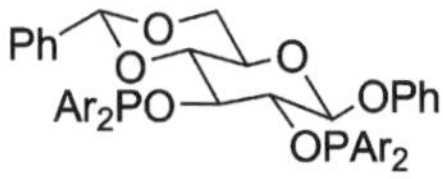

(2.12) Ar = 3,5-(CH$_3$)$_2$C$_6$H$_3$-

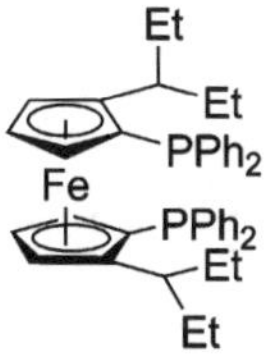

(2.13) (*S,S*)-FERROPHOS

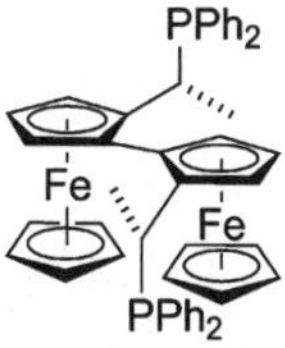

(2.14) (*R*)-TRAP

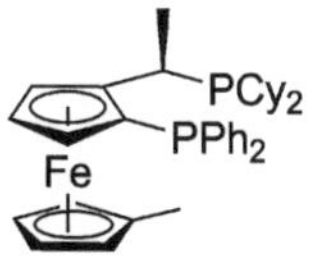

(2.15) JOSIPHOS

the homocombinations. Furthermore, heterocombinations prepared from enantiomerically pure ligands and cheap achiral ligands such as triphenylphosphine can function as highly effective catalysts.[7f, 12]

Other phosphorus-based ligands include the phosphine-phosphoramidite ligand Me-AnilaPhos (**2.20**) that effects hydrogenation of acetamidocinnamates in 98% ee[13] and biphenol-based monodentate phosphites of general structure (**2.21**) that are readily fine-tunable and show ees comparable to those obtained using BINOL-based phosphites (**2.17**).[14] Some research has also been directed towards the generation of mixed donor ligands. In particular, mixed phosphorus/sulfur ligands, such as (**2.22**) are highly effective catalysts in the hydrogenation of a variety of enamides including tetrasubstituted derivatives.[15]

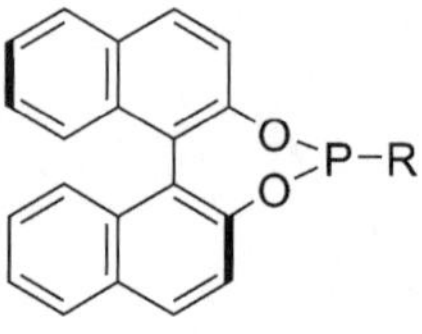

(2.16a) R = NMe$_2$, (*S*)-MONOPHOS

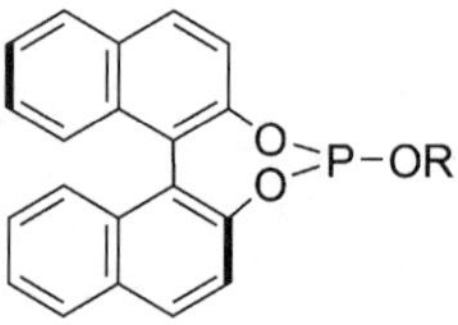

(2.17)

(2.16b) R = HN⁀Ph

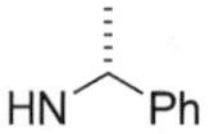

(2.16c) R =  N⬡ , PipPhos

(2.16d) R = 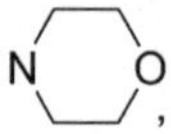 N⬡O , MorfPhos

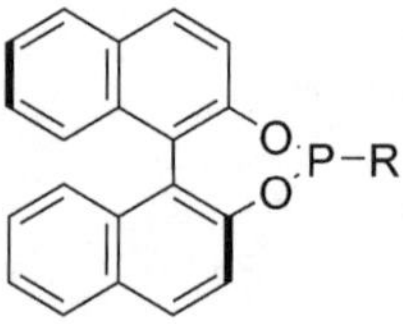

(2.18)

R = Me, Ph, ^{t}Bu

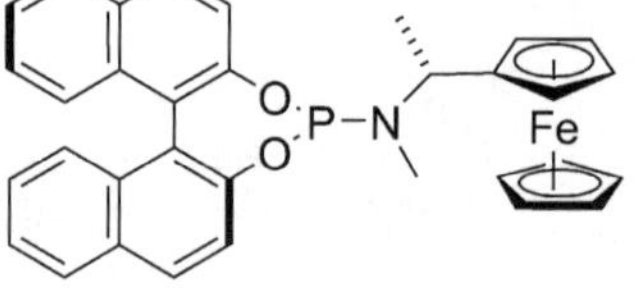

(2.19)

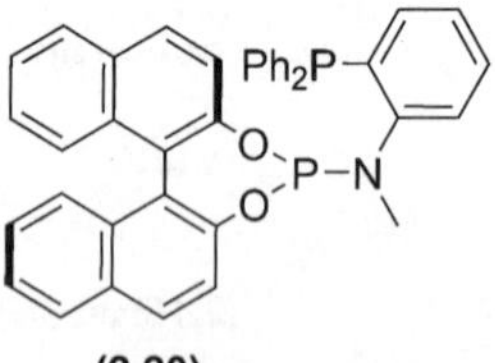

(2.20) 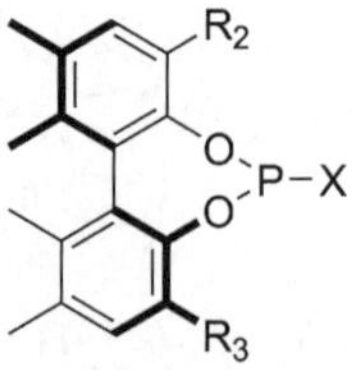   **(2.21)**   **(2.22)**

X = alcohol, secondary amine

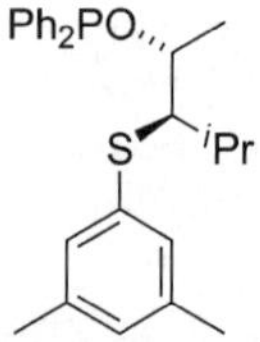

The catalytic species developed to date are generally substrate specific and the ligands may require some fine tuning before optimum performance is achieved with new classes of substrate. The discovery of the best catalyst system for a particular substrate (ligand optimisation) often requires a degree of trial and error. This process could be accelerated by the automated preparation and screening of libraries of enantiomerically pure phosphine ligands, prepared by combinatorial methods.[16] Monodentate ligand–metal complexes are especially amenable to fine-tuning by such a process. The availability of diverse monodentate phosphoramidites (**2.16**) and phosphites (**2.17**) coupled with the potential for preparing mixed ligand catalysts allows the preparation of relatively large libraries of homo and hetero-complexes. This combinatorial approach has led to the discovery of effective catalysts for the hydrogenation of α−[8c,14b] and β−dehydroamino acid derivatives [7d,12,17] and dimethyl itaconate.[14a] For example, hydrogenation of β−dehydroamino acid derivative (**2.23**) using phosphoramidite (**2.24**) proceeded with only 54% ee. However, automated parallel screening of mixtures of ligand (**2.16b**) and six other phosphoramidites in four different solvents, using a 44-well set up, revealed that this transformation could be achieved in up to 91% ee using a 1:1 mixture of (**2.16b**) and (**2.24**) in dichloromethane.[7d]

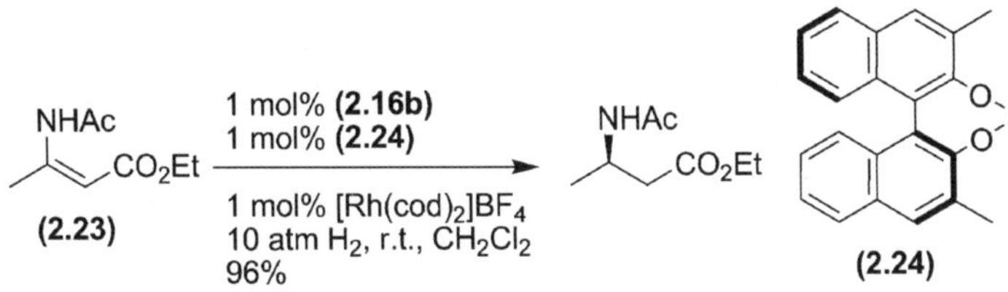

The standard laboratory scale asymmetric alkene hydrogenations of this type are carried out in an autoclave in the presence of the catalytic species preformed from [Rh(cod)$_2$]BF$_4$ and one equivalent of bidentate ligand or two equivalents of monodentate ligand. The reactions may be carried out in a wide variety of organic solvents. However, the use of supercritical carbon dioxide as the reaction medium was found to give the highest enantioselectivities in the reduction of α-enamides using the Et-DUPHOS ligands (**Et-2.05**).[18] The efficacy of these complexes in the catalytic hydrogenation is generally evaluated using alkenes with chelating functionality, commonly acylamino groups. The most common test substrates are α− and β−dehydroamino acids. The reduction of α−dehydroamino acids is certainly a useful process, since it leads to α-amino acid derivatives. Monsanto synthesised the drug L-DOPA (**2.25**), used in the treatment of Parkinson's disease, using rhodium catalysed asymmetric hydrogenation.[19] The α-(acylamino)acrylic acid (**2.26**) was reduced by a Rh/DIPAMP combination with high enantioselectivity to give the L-DOPA precursor (**2.27**). This was the first commercial application of a transition-metal catalysed asymmetric reaction.

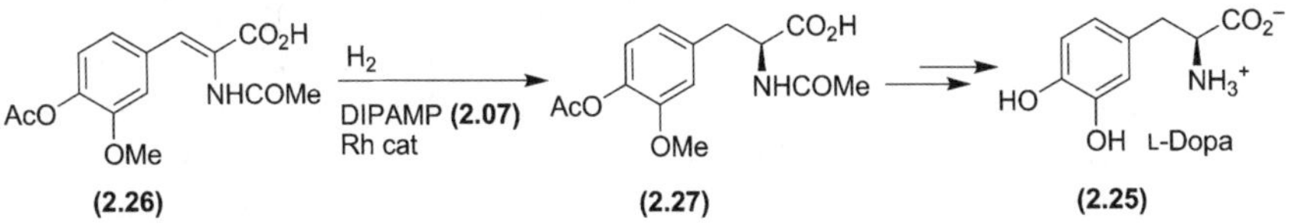

The particular susceptibility that substrates such as α-dehydroamino acids have for highly enantioselective asymmetric reduction has been attributed to their two-point binding. The mechanism deduced for the reduction of (Z)-α−acetamidocinnamates with catalysts formed from bidentate ligands with backbone chirality, such as CHIRAPHOS (2.10) involves the initial formation of two diastereomeric intermediates (2.28) and (2.29) followed by oxidative addition of H_2 to the rhodium centre and subsequent hydrometallation and reductive elimination to release the reduced product. Enantioselectivity is not dependant on the thermodynamic stability of these diastereomers. The minor diastereomer is much more reactive to hydrogen and the reaction therefore proceeds via the minor diastereomer to give the enantioselectivity observed in what is known as an anti lock-and-key process.[20,21] The same principles hold true for DUPHOS ligands (2.05), where the higher reactivity of these ligands is attributed to a low binding constant between the rhodium and enamide, which facilitates rapid ligand exchange.[22]

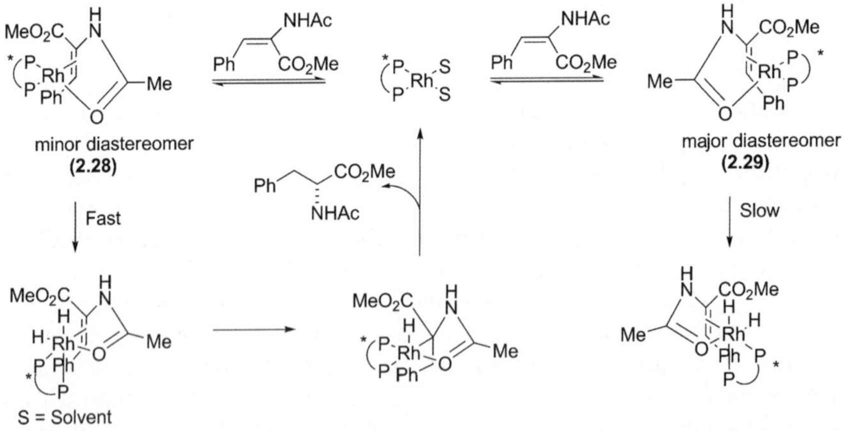

minor diastereomer
(2.28)

major diastereomer
(2.29)

S = Solvent

Recent research has shown that this mechanism does not hold for all catalyst and substrate combinations.[23] For instance, the rhodium-catalysed hydrogenation with electron-rich bidentate phosphine (2.11) is thought to proceed via initial oxidative addition of hydrogen to a catalyst–substrate complex.[24] Furthermore, enantioselectivity arises from reaction of the major catalyst–ligand complex, in a lock-and-key process, when using monodentate phosphites as ligands.[25]

[Rh(BINAP)]ClO_4 was amongst the first catalysts for the asymmetric reduction of enamides, and is still one of the best catalysts available. The geometry of the enamide has been shown to be important.[26] While the (Z)-alkene substrate (2.30) is converted into the α−amidoacid (2.31) with the (R)-enantiomer predominating, asymmetric hydrogenation of the corresponding (E)-isomer provides access to the (S)-enantiomer.

The diene (**2.32**) undergoes selective hydrogenation of the enamide double bond to give an amino acid derivative (**2.33**) containing an alkene functionality.[27] The most enantioselective catalyst for this reaction was found to be the cationic rhodium complex of DUPHOS (**2.05**), which afforded less than 1% of the fully hydrogenated product as a by-product.

β,β-Disubstituted enamides are also substrates for enantioselective hydrogenation reactions. In these cases, the Me-BPE catalyst (**2.34**) generally gave better results than the normal DUPHOS ligands.[28] The enamide (**2.35**) is converted into the β-branched α-amido ester (**2.36**) with high enantioselectivity. By use of the correct geometry of the enamide starting material, the stereochemistry in the β-position can also be controlled. Thus, the (Z)-enamide (**2.37**) affords the $(2R,3S)$-amido ester (**2.38**), whereas the (E)-enamide affords the $(2R,3R)$-diastereomer. The mixed phosphorus/sulfur ligands such as (**2.22**) developed by Evans and coworkers also function as effective ligands in the hydrogenation of β,β−disubstituted acetamidoacrylates.[15]

TRAP ligands (**2.14**) have also been used to control the stereochemistry in the β-position, including the reduction of oxygenated enamides.[29] Reduction of the (Z)-alkene (**2.39**) affords the *anti*-diastereomer of amido ether product (**2.40**), whereas the (E)-alkene affords the *syn*-diastereomer. Electron-rich P-chirogenic phosphines such as (**2.11**) form effective catalysts for the hydrogenation of β,β−disubstituted enamides such as (**2.41**) to provide the (R)-amide (**2.42**) in quantitative yield.[30] Monodentate ligands such as phosphoramidites exert poor stereocontrol during the hydrogenation of tetrasubstituted enamines. However, ligands such as PipPhos (**2.16c**) have been used to good effect in the asymmetric hydrogenation of cyclic enamides such as (**2.43**).

Rhodium-catalysed asymmetric hydrogenation is not restricted to the use of enamides as substrates and while a coordinating group other than the alkene is needed in the substrate for high enantioselectivities, recent research has revealed that the presence of an *N*-acyl group is not necessary for high enantioselectivity. Nevertheless the rhodium-catalysed hydrogenation of unfunctionalised styrene derivatives has been studied and shown to proceed with relatively poor ee.[31] One disadvantage associated with the use of acyl enamides is the difficulty involved

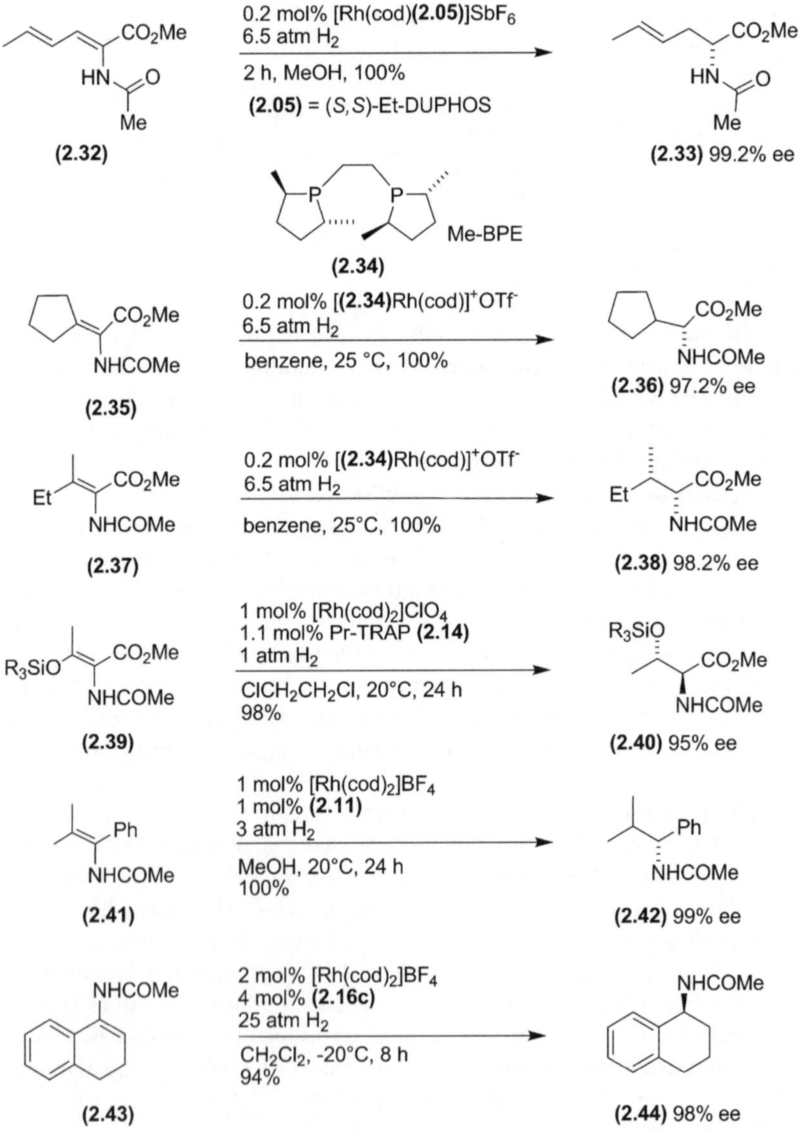

(2.32)

0.2 mol% [Rh(cod)(**2.05**)]SbF$_6$
6.5 atm H$_2$

2 h, MeOH, 100%

(**2.05**) = (S,S)-Et-DUPHOS

(2.33) 99.2% ee

Me-BPE

(2.34)

(2.35)

0.2 mol% [(**2.34**)Rh(cod)]$^+$OTf$^-$
6.5 atm H$_2$

benzene, 25 °C, 100%

(2.36) 97.2% ee

(2.37)

0.2 mol% [(**2.34**)Rh(cod)]$^+$OTf$^-$
6.5 atm H$_2$

benzene, 25°C, 100%

(2.38) 98.2% ee

(2.39)

1 mol% [Rh(cod)$_2$]ClO$_4$
1.1 mol% Pr-TRAP (**2.14**)
1 atm H$_2$

ClCH$_2$CH$_2$Cl, 20°C, 24 h
98%

(2.40) 95% ee

(2.41)

1 mol% [Rh(cod)$_2$]BF$_4$
1 mol% (**2.11**)
3 atm H$_2$

MeOH, 20°C, 24 h
100%

(2.42) 99% ee

(2.43)

2 mol% [Rh(cod)$_2$]BF$_4$
4 mol% (**2.16c**)
25 atm H$_2$

CH$_2$Cl$_2$, -20°C, 8 h
94%

(2.44) 98% ee

in removal of the acyl group. This can be overcome by using *N*-formyl protected amino acid esters as substrates. The formyl group functions as an effective chelating group and enantioselectivities up to 99% ee can be achieved during hydrogenation of these alkenes using PipPhos (**2.16c**) as ligand. Moreover, the formyl group is readily removed using mild acidic conditions.[32] The rhodium-catalysed hydrogenation also tolerates the presence of aryl groups on nitrogen. *N*-aryl β-amino acid derivatives such as **(2.45)** undergo asymmetric hydrogenation with high ee using the bisphosphine TangPhos **(2.47)** as ligand.[33]

(2.45)   **(2.46)** 96.3% ee   **(2.47)** (*S, S, R, R*)-TangPhos

The protecting group can be dispensed with in some cases. A range of simple, unprotected enamino-esters and amides have been shown to undergo enantiose-lective hydrogenation with up to 97% ee using JOSIPHOS-type ligands.[34] In this case, the nitrogen atom of an imine tautomer may function as the necessary chelat-ing species. Furthermore, enantioselective hydrogenation of pyrrolidine enamines bearing a 1,2-diaryl substitution pattern can be effected, via an undetermined mechanism, using phosphonite ligands with a 1,1′-spirobisindane skeleton devel-oped by Zhou and coworkers.[35]

The reduction of α-(aryloxy)acrylates has been reported with the Rh/DUPHOS **(2.05)** catalyst.[36] Interestingly, the substrate **(2.48)** was used as a 3:1 (*E*)/(*Z*) mixture, but the enantioselectivity is still very high in the product **(2.49)**. When this reaction was attempted using benzene as solvent, no product was formed, even though benzene is a suitable solvent for enamide reduction. This was attributed to coordination of the benzene to the cationic rhodium complex. The enol ester is not able to displace the benzene, whilst an enamide can do so.

Other nonenamide substrates which have been reduced enantioselectively with rhodium catalysts include dimethyl itaconate **(2.50)**,[37] and this alkene has recently become a popular test substrate for the evaluation of new ligands.[7a, 7e, 8a, 8c, 12,14] Trisubstituted acrylic acids, such as alkene **(2.52)**, where the carboxylic acid is believed to bind to the amino groups of the ligand **(2.54)**, also undergo enantios-elective rhodium-catalysed hydrogenation.[38] α-Methylcinnamic acid is reduced with 99% ee using the rhodium catalyst prepared from a 2:1 mixture of PipPhos **(2.16c)** and triphenylphosphine.[7f]

The enantioselective hydrogenation of *O*-substituted enol derivatives is a po-tential route to optically active alcohols. Aryl, alkenyl and vinyl-substituted enol acetates undergo highly enantioselective hydrogenation using bidentate ligands, but alkyl-substituted derivatives have been shown to be more challenging sub-strates. Boaz has shown that whilst aliphatic enol esters only react with moderate selectivity using DUPHOS **(2.05)**, alkynyl enol esters are good substrates.[39] Thus, the enol ester **(2.55)** affords the acetate product **(2.56)** with only 64% ee, whilst the corresponding alkynyl substrate **(2.57)** provides the (*Z*)-alkene **(2.58)** with 98.5% ee. Enol acetates bearing aromatic substituents can be reduced to acetates with up to 99% ee using TangPhos as ligand[40] and similar ee values have been achieved using PipPhos (**2.16c**) as ligand.[41] This latter procedure can also be extended to the reduction of enol carbamates. Alkyl-substituted enol acetates can be hydrogenated in high ee using monodentate ligands although enantioselectivity is highly dependant on the nature of the carboxylate moiety.[42] For example, 2-furanylcarboxylate (**2.59**) is hydrogenated to give (**2.60**) with 94% ee using carbohydrate-derived phosphate (**2.61**) while reduction of the corresponding acetate leads to an ee of only 73.6%. The ester group of enol esters such as (**2.48**) can be replaced with a phosphonate group with little deleterious effect on enantioselectivity. Enol benzoate phosphonates such as (**2.62**) can be hydrogenated in up to 96% ee using Et-DUPHOS (**Et-2.05**). Some enamido phosphonates can also be reduced with high enantioselectivity using this catalyst system.

Asymmetric hydrogenation of vinylboronic esters such as (**2.64**) is an attractive goal giving rise to difficult-to-access enantiomerically enriched secondary organoboranes. This transformation can now be achieved in excellent yield and good to high ee using the ferrocine-based WALPHOS ligand (**2.65**).[43] Clearly the boronate moiety is an important stereocontrol element in this process, but the exact role of this functionality in the reaction is yet to be determined.

2.2　Asymmetric Hydrogenation with Ruthenium Catalysts

The ruthenium catalysts used most frequently for alkene hydrogenation are BINAP derivatives, such as complex (**2.67**).[44] Alternative BINAP complexes, including RuBr$_2$(BINAP), cationic BINAP complexes (e.g. [RuCl(arene)(BINAP)]Cl) can also be used. Ligands other than BINAP have been used, including TolBINAP (**2.68**) and H$_8$-BINAP (**2.69**)[45] which sometimes provide a small increase in enantioselectivity. Ruthenium-catalysed hydrogenation reactions have also been reported using a polymer-supported BINAP ligand, with only slightly lower enantioselectivities than in the free homogeneous system.[46]

Genêt and coworkers have screened a series of ligands in ruthenium-catalysed alkene and ketone reductions.[47] The ligands which provided the highest selectivities were those which possessed axial chirality, such as BINAP, but also other axially chiral ligands (**2.70**). Other chelating diphosphines generally provided lower enantioselectivity, although a Ru/Me-DUPHOS catalyst afforded up to 80% ee in the hydrogenation of tiglic acid and has been deployed to effect the 2.3 tonne scale preparation of a key intermediate in the synthesis of candoxatril.[48] Other, more recently developed diphosphine ligands that show some promise in ruthenium-catalysed hydrogenations include the axially chiral dipyridyl-based P-Phos ligands (**2.71**)[49] and the bridged PQ-Phos ligands (**2.72**).[50]

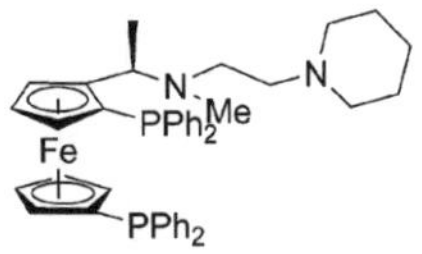

(2.54)

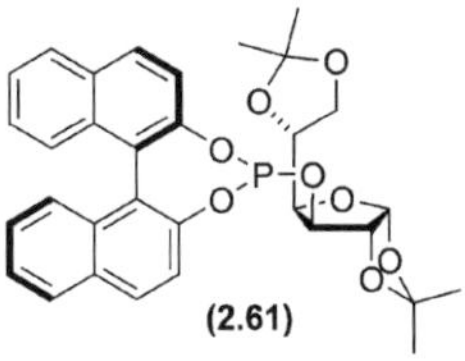

(2.61)

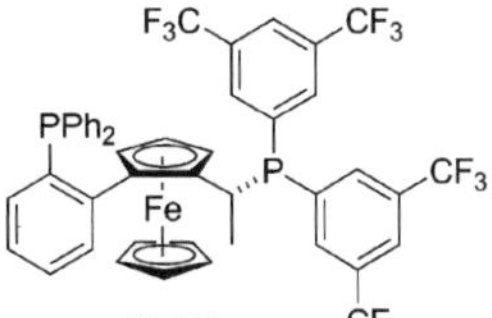

(2.65)

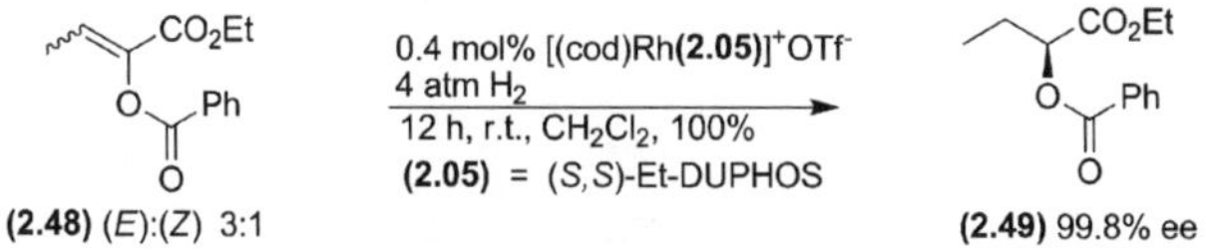

(2.48) (*E*):(*Z*) 3:1	0.4 mol% [(cod)Rh**(2.05)**]⁺OTf⁻ 4 atm H₂ —————————————— 12 h, r.t., CH₂Cl₂, 100% **(2.05)** = (*S,S*)-Et-DUPHOS	**(2.49)** 99.8% ee

(2.50)	1.1 mol% **(2.08)** Rh⁺ 5 atm H₂ —————————— EtOH **(2.08)** = (*R*)-BICHEP	**(2.51)** 99% ee

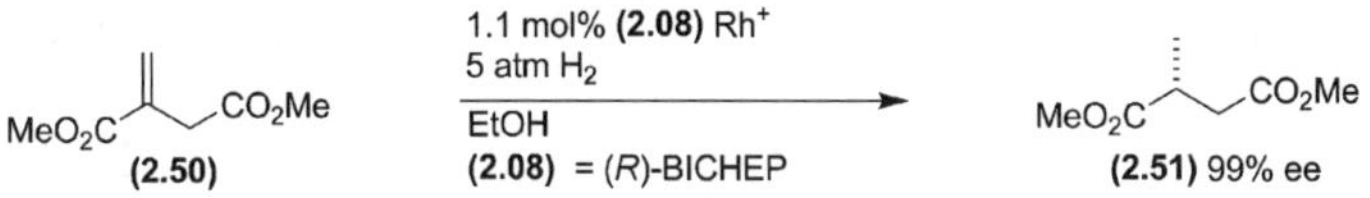

(2.52)	0.5 mol% **(2.54)** 0.5 mol% AgBF₄ 0.25 mol% [RhCl(nbd)]₂ —————————————— 50 atm H₂, 5 mol% Et₃N r.t., THF/MeOH	**(2.53)**

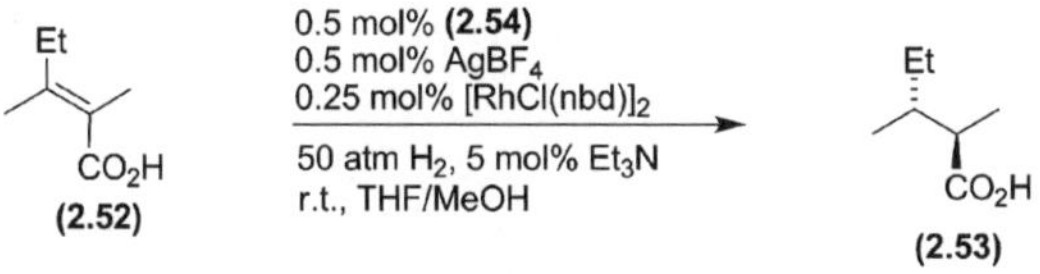

(2.55)	2.4 mol% (*R,R*)-**(2.05)** 2 mol% [Rh(cod)₂]BF₄ —————————— 2 atm H₂ r.t., THF or MeOH >95%	**(2.56)** 64% ee

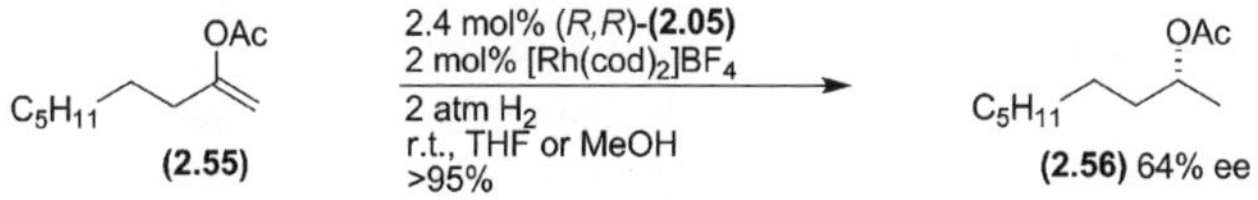

(2.57)	2.4 mol% (*R,R*)-**(2.05)** 2 mol% [Rh(cod)₂]BF₄ —————————— 2 atm H₂ r.t., THF or MeOH >97%	**(2.58)** 98.5% ee

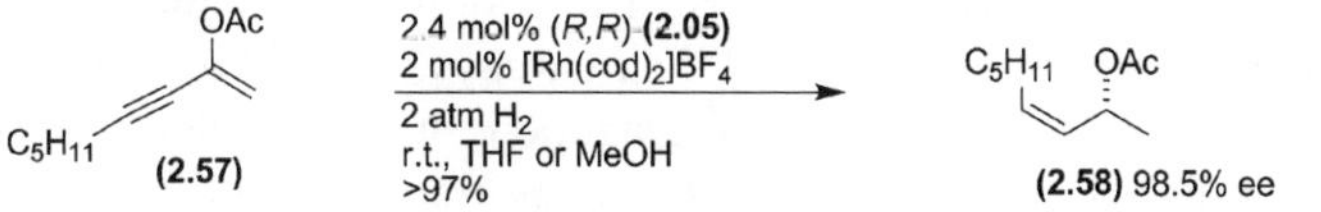

(2.59)	1 mol% **(2.61)** 0.05 mol% [Rh(cod)₂]BF₄ —————————— 60 atm H₂ r.t., CH₂Cl₂ 100%	**(2.60)** 94% ee

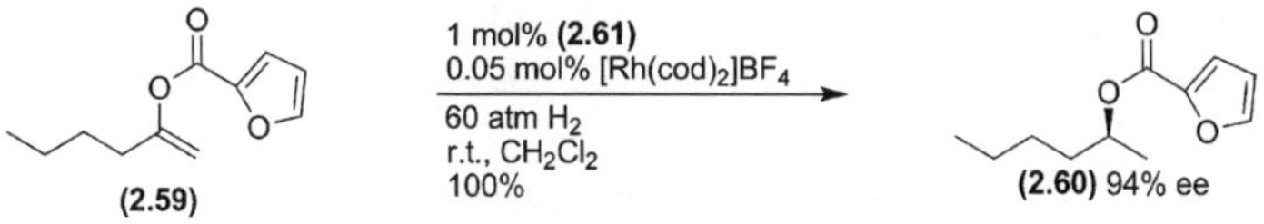

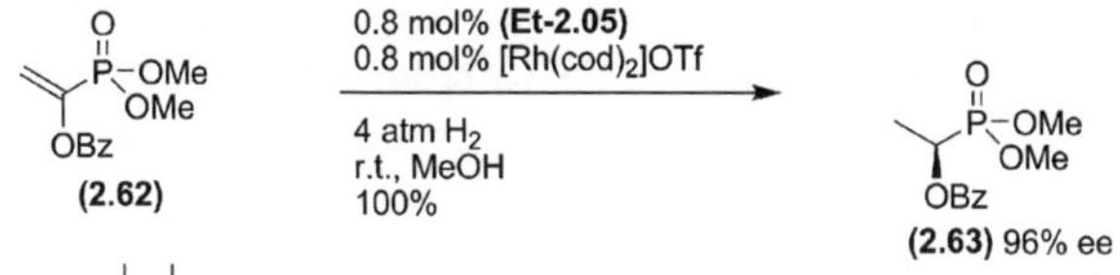

(2.62)	0.8 mol% **(Et-2.05)** 0.8 mol% [Rh(cod)₂]OTf —————————— 4 atm H₂ r.t., MeOH 100%	**(2.63)** 96% ee

(2.64)	8 mol% **(2.65)** 5 mol% [Rh(nbd)₂]BF₄ —————————— 35 atm H₂ -35°C, PhMe >95%	**(2.66)** 94% ee

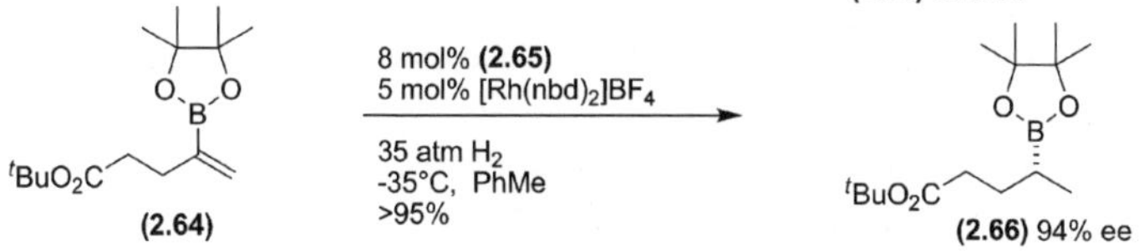

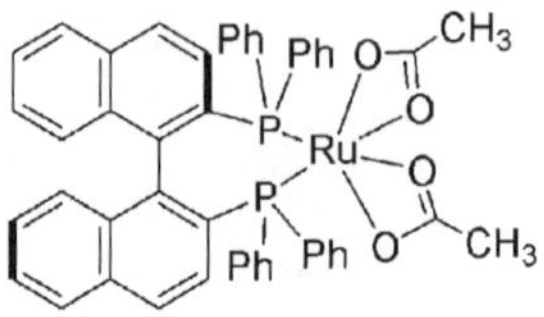

Ru[(*R*)-BINAP](OCOCH$_3$)$_2$ **(2.67)** 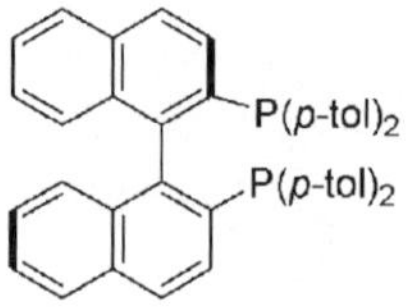   (*R*)-TolBINAP **(2.68)** 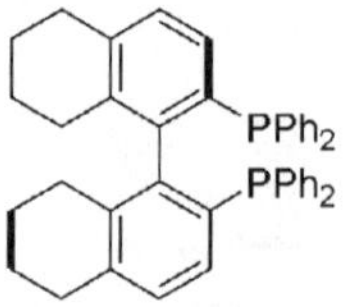   H$_8$-BINAP **(2.69)**

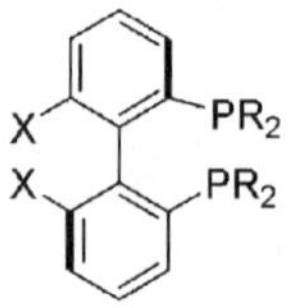

(2.70) 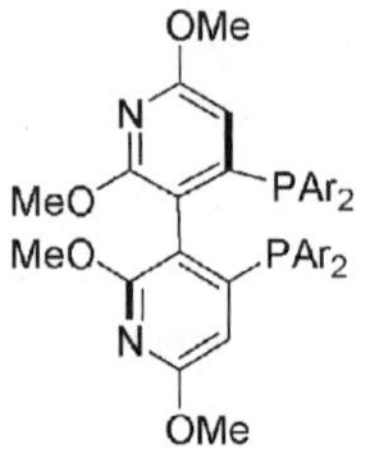   P-Phos **(2.71)**   PQ-Phos **(2.72)**

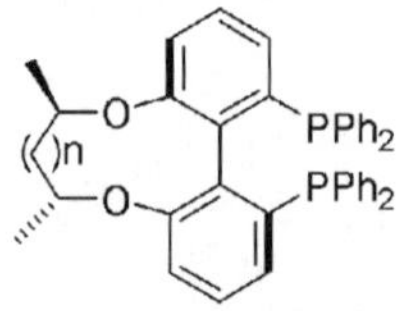

Ruthenium/BINAP complexes can be used in the asymmetric reduction of enamides, although often enantioselectivity is worse than for the corresponding rhodium/BINAP complexes. Interestingly, using the same enantiomer of BINAP affords the opposite enantiomer of product depending on whether rhodium or ruthenium is employed as catalyst.[51] The ruthenium/BINAP catalysts were found to be very effective in the reduction of the enamide (**2.73**).[52] The product tetrahydroisoquinolines (**2.74**) are useful in the synthesis of isoquinoline alkaloids. Simple, β-(acylamino)acrylates such as (**2.75**) can be transformed into β—amino acid derivatives in up to 99.9% ee using PQ-Phos ligands (**2.72**).[50] The asymmetric hydrogenation of (*E*)- β-(acylamino)acrylates can be effected with similar ees using P-Phos catalysts (**2.71**).[49]

Ruthenium/BINAP complexes have been successfully used in the asymmetric reduction of acrylic acids.[53] This methodology has been used to prepare the antiinflammatory drug (*S*)-Naproxen (**2.78**) by reduction of the acrylic acid (**2.77**). Ruthenium/PQ-Phos species catalyse the same transformation with comparable ee.[50]

The reaction has also been used in the preparation of α-fluorocarboxylic acids with good enantioselectivity, including the conversion of the alkenyl fluoride (**2.79**) into α-fluorohexanoic acid (**2.80**).[54]

Allylic alcohols make good substrates for ruthenium/BINAP-catalysed hydrogenation.[55] Geraniol (**2.81**) and nerol (**2.82**) are (*E*) and (*Z*) isomers, and these substrates afforded opposite enantiomers of the product citronellol (**2.83**).

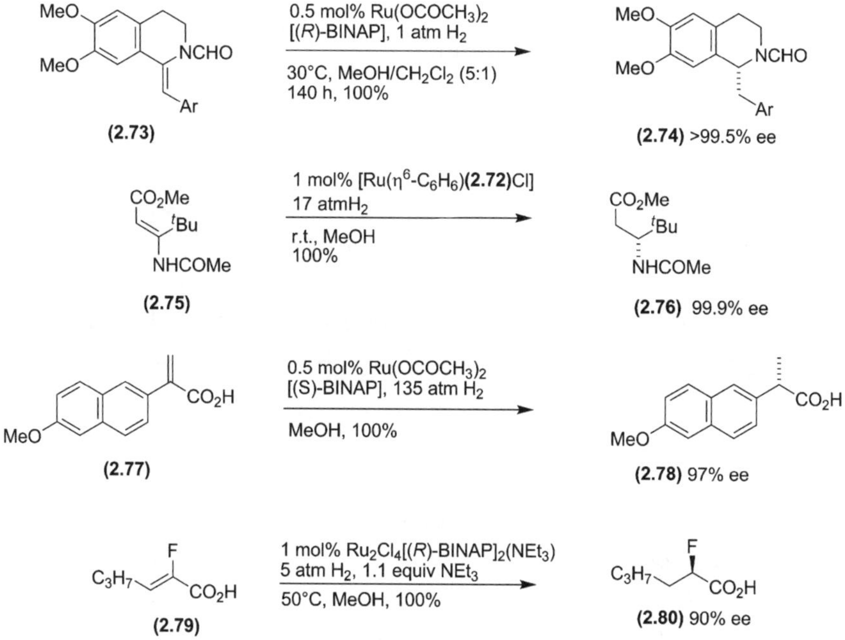
MeO
MeO
NCHO
Ar
(2.73)
0.5 mol% Ru(OCOCH3)2
[(R)-BINAP], 1 atm H2
30°C, MeOH/CH2Cl2 (5:1)
140 h, 100%
MeO
MeO
NCHO
Ar
(2.74) >99.5% ee
CO2Me
tBu
NHCOMe
(2.75)
1 mol% [Ru(η6-C6H6)(2.72)Cl]
17 atmH2
r.t., MeOH
100%
CO2Me
tBu
NHCOMe
(2.76) 99.9% ee
CO2H
MeO
(2.77)
0.5 mol% Ru(OCOCH3)2
[(S)-BINAP], 135 atm H2
MeOH, 100%
CO2H
MeO
(2.78) 97% ee
F
C3H7
CO2H
(2.79)
1 mol% Ru2Cl4[(R)-BINAP]2(NEt3)
5 atm H2, 1.1 equiv NEt3
50°C, MeOH, 100%
F
C3H7
CO2H
(2.80) 90% ee

The allyl alcohol (**2.84**) could be hydrogenated with exceptionally high diastereoselectivity using ruthenium/(*R*)-TolBINAP complexes.[56] The substrate diastereoselectivity and catalyst selectivity represent a matched pair, since, when the enantiomeric (*S*)-TolBINAP ligand was used, the opposite diastereomer was formed with only 56% de.

Kinetic resolution of racemic allylic alcohols has been achieved with a high selectivity factor.[57] Racemic allylic alcohol (**2.86**) undergoes kinetic resolution when undergoing hydrogenation with a ruthenium/BINAP complex. The product (**2.87**) and recovered starting material (**2.86**) can both be obtained with high enantioselectivity. Faller and Tokunaga demonstrated chiral poisoning in the kinetic resolution of allylic alcohol (**2.88**).[58] Using a ruthenium catalyst made from racemic BINAP, and adding an enantiomerically pure poison, (1*R*,2*S*)-ephedrine (**2.89**) still gives good selectivity. The ephedrine is assumed to deactivate one enantiomer of the ruthenium/BINAP complex preferentially, leaving the other enantiomer available to perform the kinetic resolution.

Dixneuf, Bruneau and coworkers have reported an interesting reduction of the unsaturated acyl oxazolidinone (**2.90**).[59] The reduction works with high yield and asymmetric induction, and the product (**2.91**) is effectively propionic acid with a chiral auxiliary attached. The chiral auxiliary was then used to induce asymmetry in a subsequent step. 2,3-Substituted *N*-Boc indoles undergo hydrogenation to

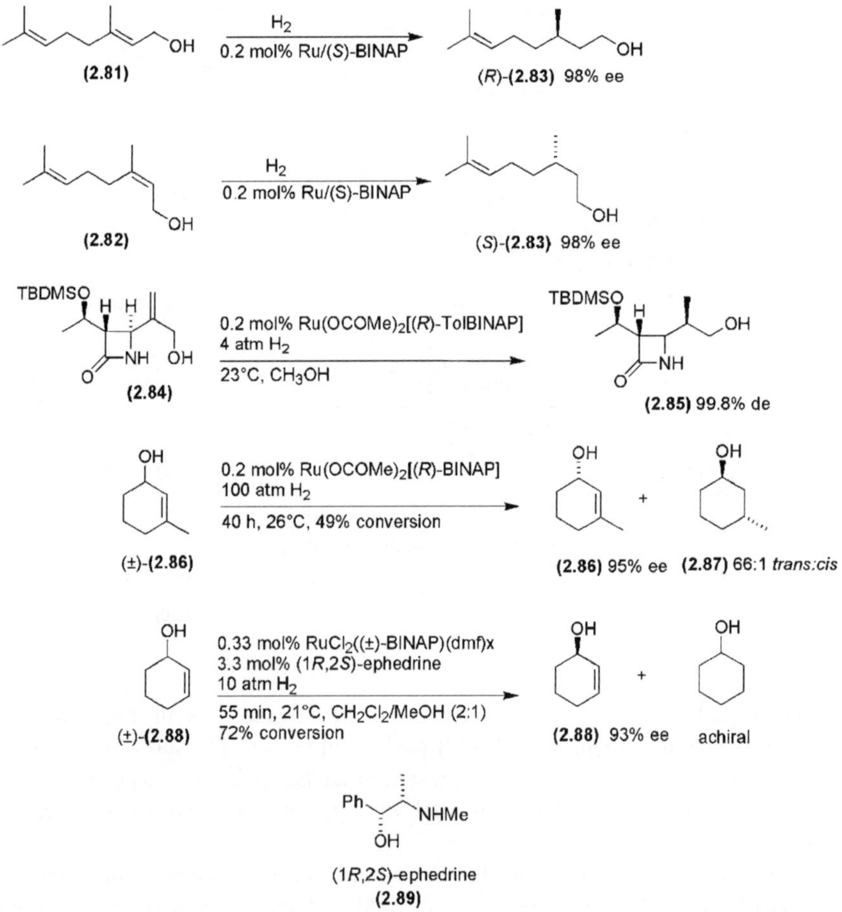

(2.81)   H₂ / 0.2 mol% Ru/(S)-BINAP   (R)-**(2.83)** 98% ee

(2.82)   H₂ / 0.2 mol% Ru/(S)-BINAP   (S)-**(2.83)** 98% ee

(2.84)   0.2 mol% Ru(OCOMe)₂[(R)-TolBINAP] / 4 atm H₂ / 23°C, CH₃OH   **(2.85)** 99.8% de

(±)-(2.86)   0.2 mol% Ru(OCOMe)₂[(R)-BINAP] / 100 atm H₂ / 40 h, 26°C, 49% conversion   **(2.86)** 95% ee   **(2.87)** 66:1 *trans:cis*

(±)-(2.88)   0.33 mol% RuCl₂((±)-BINAP)(dmf)x / 3.3 mol% (1R,2S)-ephedrine / 10 atm H₂ / 55 min, 21°C, CH₂Cl₂/MeOH (2:1) / 72% conversion   **(2.88)** 93% ee   achiral

(1R,2S)-ephedrine
(2.89)

give indolines with up to 95% ee using a ruthenium complex with Ph-TRAP **(2.14)** ligands.[60] Use of the catalyst prepared from [RuCl₂(*p*-cymene)]₂ leads to formation of indoline **(2.92)** with 95% ee from 2-methylindole **(2.93)**. While this catalyst performs poorly during reduction of 2,3-disubstituted indoles the catalyst prepared by reaction of Ph-TRAP with Ru(η³-2-methylallyl)₂(cod) can be used to effect the hydrogenation of 2,3-dimethylindole to give the *cis*-indoline with 72% ee.

The scope of the ruthenium-catalysed asymmetric hydrogenation is relatively wide and a diverse array of alkenes has been reduced with high enantioselectivity. For example, diketene **(2.94)**,[61] and unsaturated lactone **(2.96)**[62] undergo highly enantioselective hydrogenation using BINAP as ligand. The enantioselective hydrogenation of 2-pyrones has been studied in some detail.[63] 3,6-Dialkyl-4-hydroxypyrones such as (**2.98**) can be hydrogenated with high ee and chemoselectively across the 5,6-double bond using catalysts derived from MeO-(**2.70**). The chemoselectivity is influenced by the substituent at position 3 and the absence of functionality at this position results in complete hydrogenation. 4,6-Dialkylpyrones also undergo complete hydrogenation.

(2.90) **(2.91)**

(2.93) **(2.92)**

(2.94) **(2.95)** 92% ee + 3% butyric acid

(2.96) **(2.97)** 95% ee

(2.98) **(2.99)** 95.5% ee

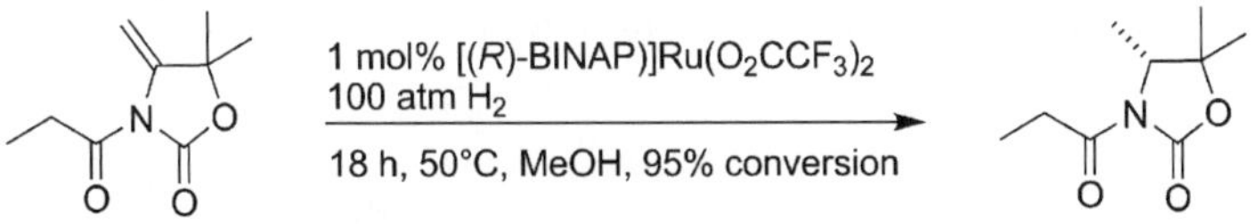

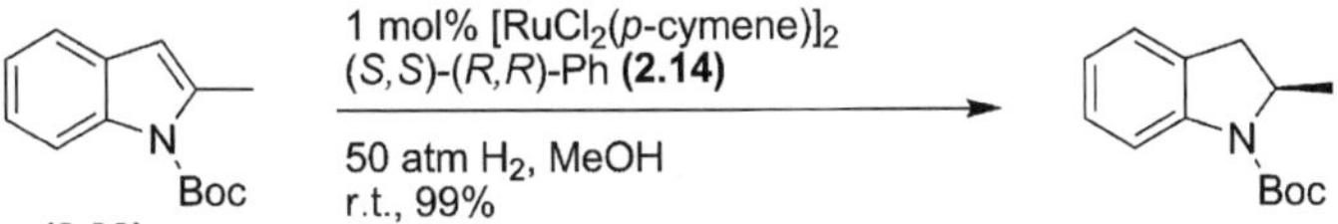

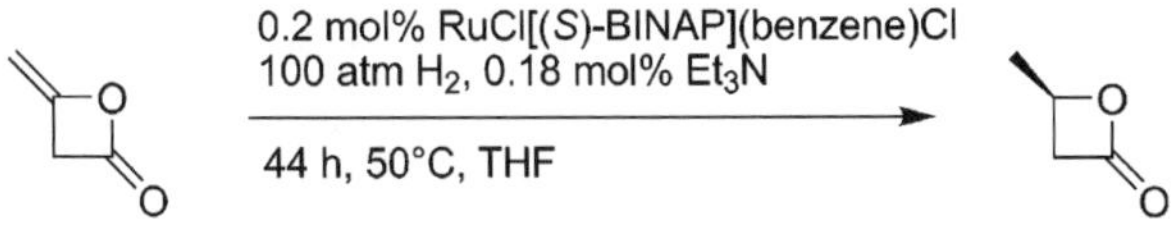

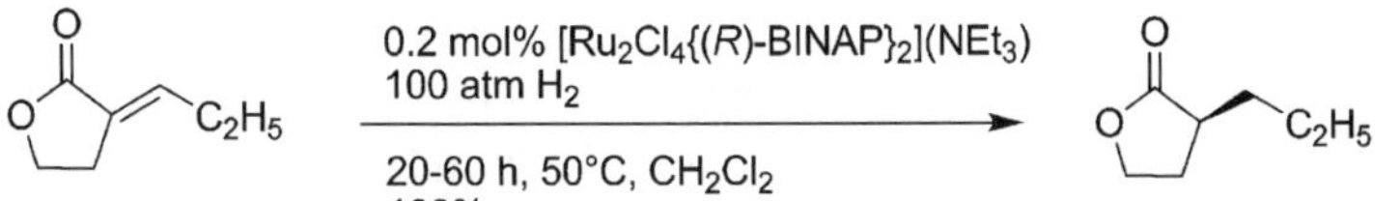

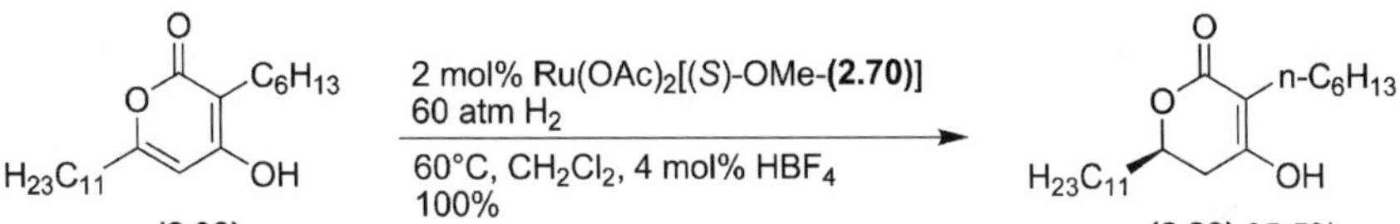

Detailed studies into the mechanism of the ruthenium-catalysed hydrogenation of enamides using catalyst (**2.67**) in methanol reveals that, while there are similarities to the rhodium-catalysed hydrogenation there are also some fundamental differences.[64] The ruthenium-catalysed process involves the initial formation of a ruthenium monohydride that complexes with the alkene in a similar manner to rhodium catalysts to give two catalyst–ligand diastereomers. In this case the structure of the major diastereomer determines the absolute configuration of the reduced product as both are equally reactive in subsequent steps. This factor explains the reversal in the sense of asymmetric induction that occurs in moving from rhodium to ruthenium-based BINAP catalysts. This major diastereomer then undergoes a migratory insertion to give a ruthenacycle that is cleaved with hydrogen, in the rate-determining step, to give the product.

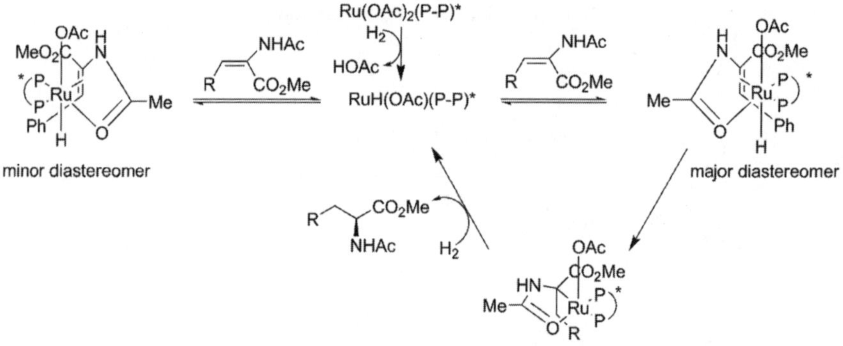

2.3　Alkene Hydrogenation with Titanium and Zirconium Catalysts

All the titanium and zirconium-based hydrogenation catalysts investigated to date are Ziegler-type metallocene systems. Unlike rhodium and ruthenium-based catalysts, enantiomerically pure titanocenes and zirconocenes do not require the presence of chelating functionality on the olefin substrate and have shown some use in the asymmetric hydrogenation of tri- and tetrasubstituted unfunctionalised alkenes. Unfortunately the often high pressures of hydrogen required in these procedures does somewhat limit the utility of these catalysts.[31] The use of enantiomerically pure titanocene-derived catalysts for the asymmetric hydrogenation of alkenes was first reported by Kagan in 1979.[65] Alternative titanocene derivatives were reported by Vollhardt and coworkers, which afforded up to 96% ee.[66]

Buchwald and coworkers have used the titanium version of the Brintzinger catalyst (**2.100**)[67] in the asymmetric reduction of trisubstituted alkenes.[68] The catalyst is reduced *in situ* to a titanium(III) hydride species (see Section 3.6). The reduction is achieved with ^{n}butyllithium and hydrogen, whilst the silane serves to stabilise the catalyst. The catalyst works well for trisubstituted alkenes, including substrates (**2.101**) and (**2.102**). A transition state model assumes that the alkene approaches the titanium hydride 'front-on', and explains the selectivities seen. Earlier work by Waymouth and Pino with a related zirconocene complex has also been reported, and a similar rationalisation offered for the stereochemical outcome.[69] The enantioselective reduction of alkenes with samarium cyclopentadienyl complexes has also been reported.[70]

The same titanium catalyst (**2.100**) has been used in the reduction of enamines with very good enantioselectivities (89–98% ee), including the reduction of enamine (**2.103**) to the amine (**2.106**).[71] One drawback associated with the use of catalyst (**2.100**) is that reaction times of several days are sometimes required for effective hydrogenation. This problem can be overcome by the use of the zirconium complex (**2.107**) activated with $[PhMe_2NH]^+[B(C_6F_5)_4]^-$.[72] The zirconium complex is converted into cationic metallocene hydride (**2.107**)* in the presence of the ammonium salt and hydrogen. It is believed that the enhanced electrophilic character of this catalyst enhances binding to olefins such that even tetrasubstituted derivatives such as 2,3-dimethylindene (**2.108**) undergo asymmetric hydrogenation in 17 hours with high ee.

2.4   Alkene Hydrogenation with Iridium Catalysts

The well-known cationic iridium(I)-cod complex (**2.110**) – Crabtree's catalyst – has been known for some time as a highly active hydrogenation catalyst.[73] Unlike rhodium and ruthenium complexes, (**2.110**) does not require the presence of chelating functionality and is effective in the hydrogenation of unfunctionalised tri- and tetrasubstituted alkenes. The Pfaltz group has pioneered the development of phosphinooxazoline (PHOX)-based chiral nonracemic mimics of this complex such as (**2.111**) that are highly effective catalysts for the asymmetric hydrogenation of aryl-substituted alkenes.[73] Highest conversions are achieved in weakly coordinating solvents such as dichloromethane and toluene using the bulky (tetrakis[3,5-bis(trifluoromethyl)phenyl])borate) (BARF) anion. This counterion improves the lifetime of the catalyst and allows preparation of these systems in the air without the need for dried solvents. The success obtained with PHOX catalysts of type (**2.111**) has led to the systematic variation of this structure in the search for improved ligands and a wide variety of *P,N*-ligands have been accessed where the core phenyl ring has been modified and the oxazoline replaced with other heterocycles. The phosphine has also been replaced with phosphinites and phosphites.[31] Other catalysts of this type with a wider scope than (**2.111**) include the readily available SimplePHOX ligands of general structure (**2.112**),[75] the pyridine-based phosphonites such as (**2.113**)[76] and phosphine-thiazoles such as (**2.114**) developed by Andersson.[77] Some work has also been directed towards the use of enantiopure

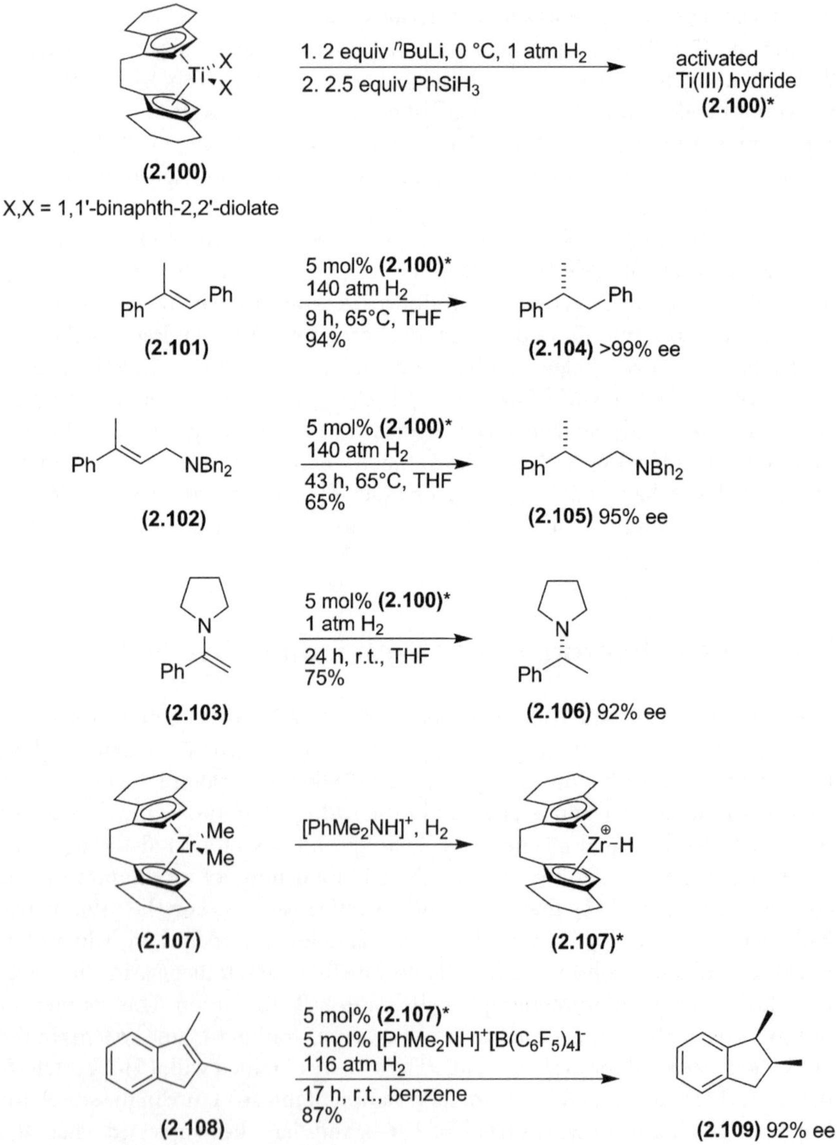

(2.100)

X,X = 1,1'-binaphth-2,2'-diolate

N-heterocyclic carbenes as ligands in this process. In particular complex **(2.115)** where the oxazoline moiety has been replaced with an imidazolylidine group has been shown to be as effective as some PHOX ligands in the hydrogenation of (*E*)-arylethenes.[78]

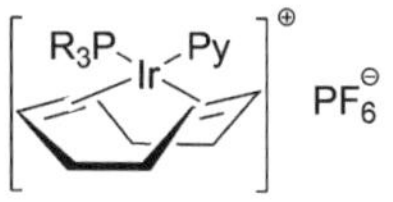

(2.110) Crabtree's catalyst

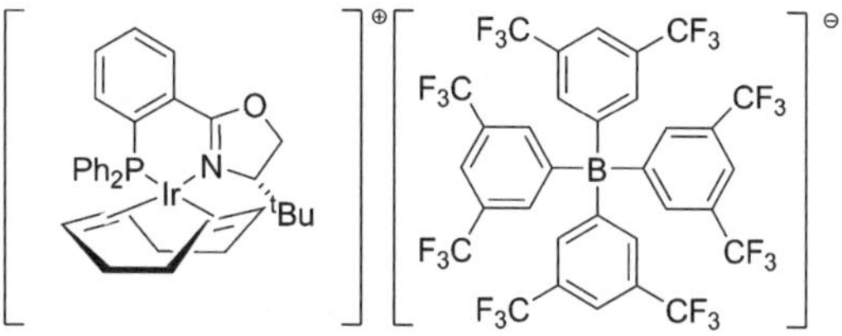

(2.111)

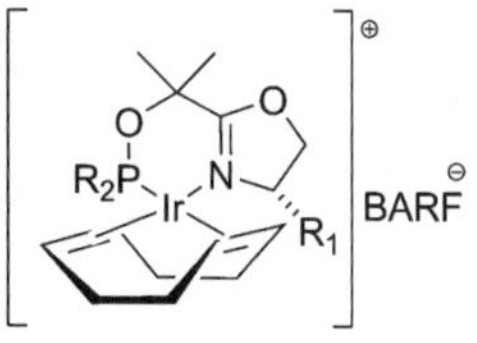

(2.112) SimplePHOX

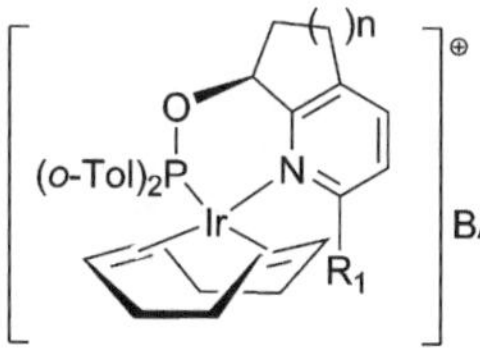

(2.113) n = 1,2

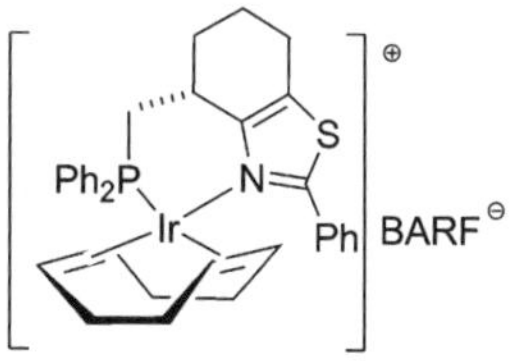

(2.114)

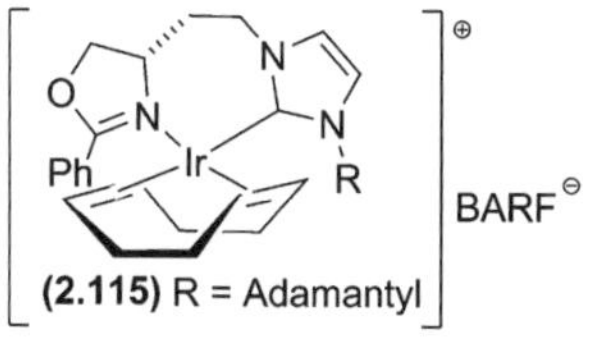

(2.115) R = Adamantyl

Catalysts such as (2.111) are highly effective at low catalyst loadings in the hydrogenation of aryl-containing trisubstituted alkenes, especially (E)-1,2-diphenylethenes such as (2.116) – probably the most common test substrates for catalyst systems of this type.[74a] (Z)-Aryl alkenes have proved to be more challenging substrates, but hydrogenation of trisubstituted arylalkene (2.118) can be achieved with good ee using 1 mol% of the Ir-SimplePHOX-type complex prepared from ligand (2.119).[74] This complex has also been applied to the highly enantioselective hydrogenation of the (E)-acrylic ester (2.121) and allylic alcohol (2.122). Both these transformations can be achieved with higher ee and lower catalyst loadings using the phosphine-thiazole complex (2.114). The scope of the majority of Ir-PHOX systems is largely limited to the hydrogenation of alkenes possessing adjacent aryl groups. Purely alkyl substituted alkenes and those with more remote aryl substituents do undergo asymmetric hydrogenation using bicyclic pyridine phosphinite-containing catalysts (2.113).[76a] For example, (E)-cyclohexylalkene (2.125) is converted into alkane (2.126) with high ee using bicyclic ligand (2.113). These pyridine-phosphite ligands display a broad scope and have been deployed with success in the asymmetric hydrogenation of cinnamate esters (2.121), allylic alcohols (2.122) and even furans to tetrahydrofurans.[76b]

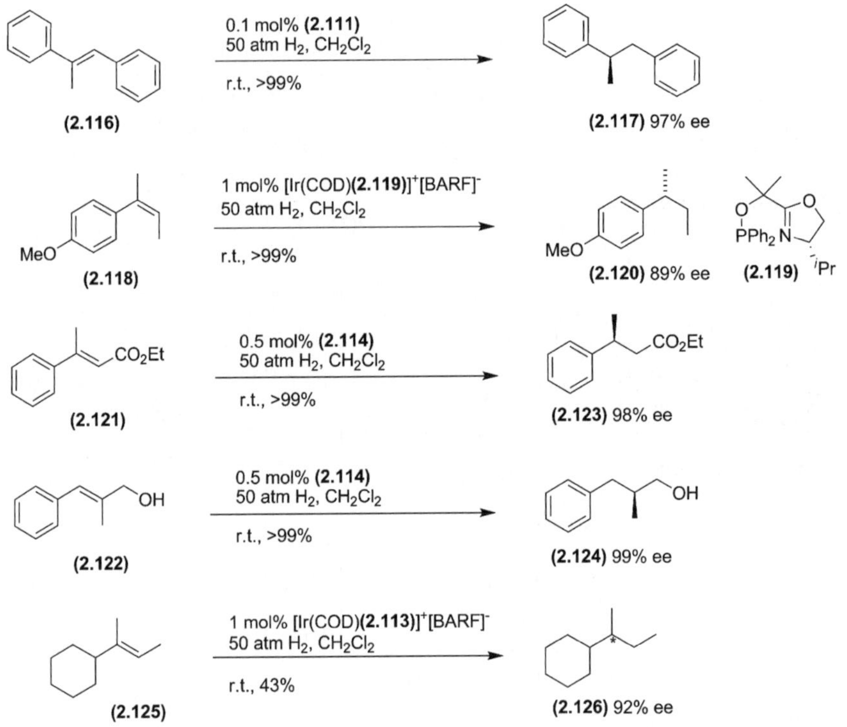

0.1 mol% (2.111)
50 atm H$_2$, CH$_2$Cl$_2$

r.t., >99%

(2.116)

(2.117) 97% ee

1 mol% [Ir(COD)(2.119)]$^+$[BARF]$^-$
50 atm H$_2$, CH$_2$Cl$_2$

r.t., >99%

(2.118)

(2.120) 89% ee   (2.119) ^{i}Pr

0.5 mol% (2.114)
50 atm H$_2$, CH$_2$Cl$_2$

r.t., >99%

(2.121)

(2.123) 98% ee

0.5 mol% (2.114)
50 atm H$_2$, CH$_2$Cl$_2$

r.t., >99%

(2.122)

(2.124) 99% ee

1 mol% [Ir(COD)(2.113)]$^+$[BARF]$^-$
50 atm H$_2$, CH$_2$Cl$_2$

r.t., 43%

(2.125)

(2.126) 92% ee

2.5   Alkene Hydrogenation with Organocatalysts

Nature achieves the hydrogenation of organic compounds with metalloenzymes in combination with cofactors such as NADH as hydride source. A biomimetic transfer hydrogenation of α,β-unsaturated aldehydes can be achieved using Hantzch esters as a hydride source in the presence of 5 mol% of activating ammonium salts such as dibenzylammonium trifluoroacetate as organocatalysts.[79] Enantioselective variants of this process have now been developed using enantiomerically pure imidazolidinone-based organocatalysts.[80] The group of List has achieved enantioselective ratios from 95:5 to 98:2 in the transfer hydrogenation of a variety of aromatic β,β—disubstituted aldehydes such as (2.127) using 10 mol% of the imidazolidinone salt (2.128) and a slight excess of Hantzch ester (2.129).[81] MacMillan has expanded the range of substrates to include β,β-dialkylsubstituted enals such as (2.131) using 20 mol% of imidazolidinone (2.132) and ester (2.133) as hydride source.[82] Both of these catalyst systems were inactive when applied to more demanding ketone substrates. However, MacMillan has shown that a variety of cyclohexenone and cyclopentenone substrates such as (2.134) do undergo highly enantioselective transfer hydrogenation using the furylimidazolinone (**2.135**) in combination with the *tert*-butyl Hantzch ester (**2.136**).[83]

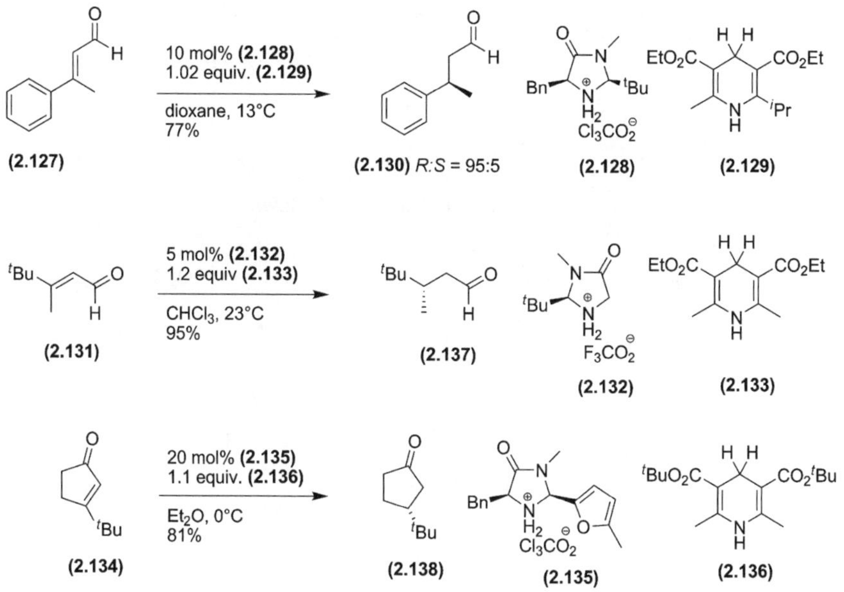

The reaction proceeds via formation of an iminium ion that is in rapid equilibrium with the geometrical isomer. The rate-determining step in this process is hydride transfer. As this occurs more rapidly onto the least hindered face of the (*E*)-isomer the absolute configuration of the major product is independent of the geometry of the enal/enone substrate. In contrast to metal-mediated hydrogenations, where olefin geometry dictates the mode of asymmetric induction, both (*E*)- and (*Z*)-alkene substrates converge to the same product enantiomer allowing the use of isomer mixtures.

2.6 Alkene Hydrosilylation

One of the most common methods for the preparation of enantiomerically enriched organosilanes is by palladium-catalysed asymmetric hydrosilylation of alkenes in the presence of trichlorosilane.[84] This area has been dominated by the use of monodentate phosphorus-based ligands and, in particular, Hayashi's MOP ligand/palladium catalyst combination offers a high level of enantioselectivity. The MOP ligands include MeO-MOP (**2.139**),[85] H-MOP ligands such as (**2.140**)[86] and MOP-Phen (**2.141**).[87] Other monodentate ligands capable of effecting highly enantioselective hydrosilylations include phosphoramidites such as (**2.142**)[88] and (**2.143**)[89] and the highly efficient arylmonophosphinoferrocene (MOPF) ligands, such as (**2.144**), capable of effecting the enantioselective hydrosilylation of styrene in minutes.[90]

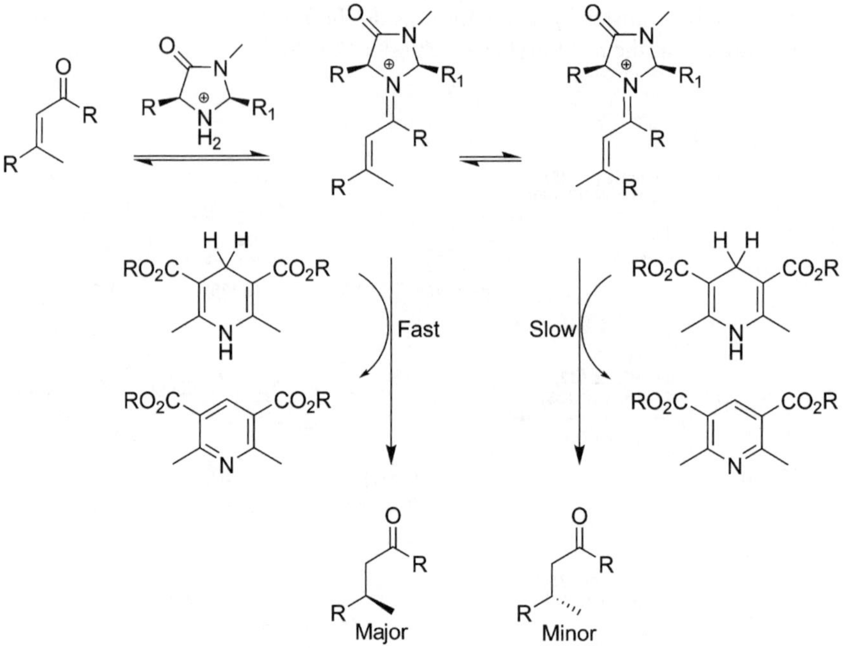

Representative alkene hydrosilylation reactions carried out with the MOP ligands are identified in the following schemes. The low catalyst loading in many cases is noteworthy. Norbornene (**2.145**), dihydrofuran (**2.146**), styrene (**2.147**) and cyclopentadiene (**2.148**) are all converted into the hydrosilylated derivatives with good to excellent enantioselectivity. In most cases, the trichlorosilanes were derivatised into the corresponding alcohols, although the trichloroallylsilane (**2.152**) was reacted with benzaldehyde to give the addition product (**2.156**).

The 1,4-disilylation of α,β-unsaturated ketones using a disilane with a palladium BINAP catalyst has also been achieved with good enantiomeric excess.[91] For example, with α,β-unsaturated ketone (**2.157**), the initially formed product (**2.158**) can be converted into the β-silyl ketone (**2.159**) by addition of methyllithium followed by hydrolysis. However, quenching the intermediate lithium enolate (formed on addition of MeLi) with an alkylating agent leads to an α-substituted product

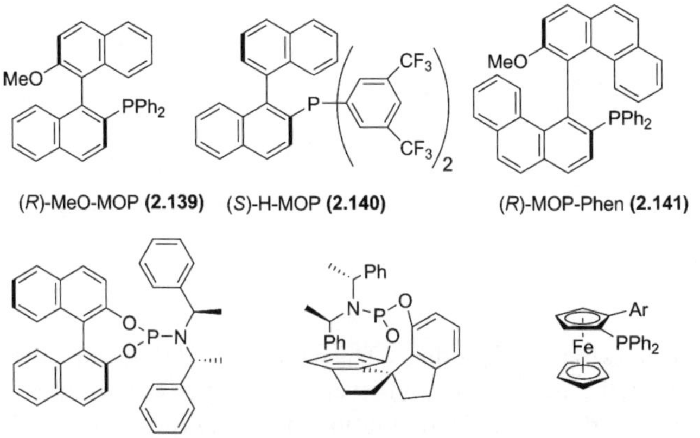

(*R*)-MeO-MOP **(2.139)**   (*S*)-H-MOP **(2.140)**   (*R*)-MOP-Phen **(2.141)**

(*S,R,R*)-**(2.142)**   (*R,R,R*)-**(2.143)**   (*S*)-**(2.144)**

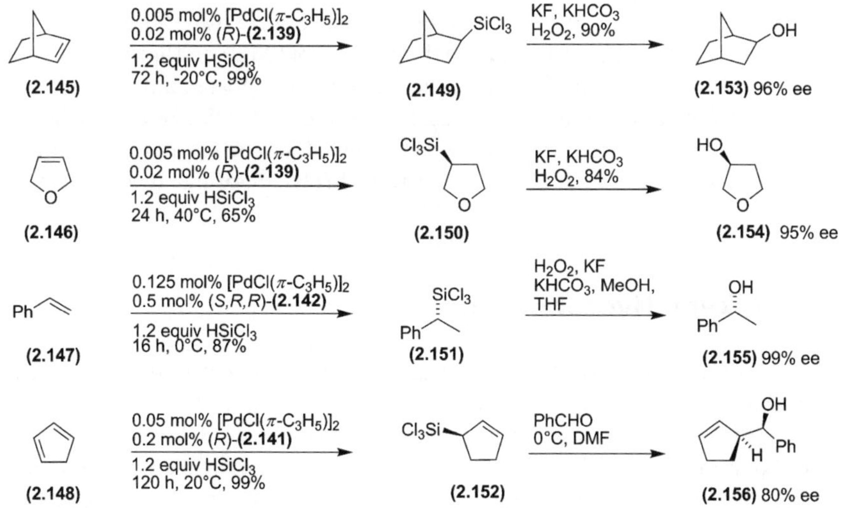

(**2.160**) with high *anti*-selectivity. The β-silyl ketones can be further converted into β-hydroxyketones by oxidation of the Si–C bond.

The other main method for the hydrosilylation of double bonds, that is restricted to the reduction of electrophilic alkenes, utilises catalytic quantities of ligated copper hydride in the presence of stoichiometric amounts of polymethylhydrosiloxane (PMHS).[92] Lipshutz and coworkers have shown that *in situ* generated copper hydride in combination with the chelating bisphosphines DTBM-SEGPHOS (**2.161**) or the analogue of JOSIPHOS PPF-P(*t*-Bu)$_2$ (**2.162**) is a highly effective catalyst for the asymmetric conjugate reduction of enones and enoates yielding silyl enol ethers

(2.157) → 0.5 mol% PdCl$_2$[(*R*)-BINAP], 1.8 -2.0 equiv PhCl$_2$SiSiMe$_3$, 80°C, benzene, 72% → **(2.158)** Cl$_2$PhSi / OSiMe$_3$

MeLi then H$_3$O$^+$, 90% → PhMe$_2$Si O Ph **(2.159)** 87% ee

MeLi MeI/THF 70% → PhMe$_2$Si O Ph **(2.160)** 85% ee >20:1 anti:syn

or ketene acetals as the initial products. The stability of [{DTBM-SEGPHOS}CuH] further enhances the utility of this catalyst and this reagent can be stored in solution for up to two months with little diminution in selectivity and activity.[93] Furthermore, this transformation can be performed under heterogenous conditions using copper impregnated in charcoal.[94] A variety of cyclic enones such as isopherone (2.163) is converted into the reduction product with high ee using Stryker's reagent (Ph$_3$P)CuH or *in situ* generated CuH in the presence of PMHS and ligand (2.161).[95] This ligand is generally ineffectual in the reduction of enoates and lactones, but the hydride/ligand combination arising from use of the JOSIPHOS-type ligand (2.162) effects highly enantioselective reduction of acyclic enoates and lactones such as (2.164)[96] and both *(E)*- and *(Z)*-β-silyl-α,β−unsaturated esters such as (2.165).[97]

2.7 Alkene Hydroboration

Whilst hydroboration of alkenes will occur without added catalyst, the reaction has been shown to be greatly accelerated by rhodium[98,99] and iridium complexes.[100] The use of a metal catalyst also has an affect on the regioselectivity of this process. The rhodium-catalysed hydroboration proceeds in a Markovnikov sense to give branched organoboranes while the use of iridium catalysts leads to formation of the linear isomer. This reversal in regioselectivity may indicate a difference in mechanism between the rhodium and iridium-catalysed procedures.[101] The use of an enantiomerically pure complex provides the opportunity for asymmetric induction in the hydroboration reaction. Rhodium complexes are employed in the majority of asymmetric hydroborations.[102] Some of the ligands examined for this reaction include BINAP,[103] QUINAP (2.169)[104,105] and Togni's pyrazole-containing ligand (2.170).[106,107] A standard procedure uses 1 mol% catalyst in THF at room temperature, although in order to achieve high selectivity with BINAP, −78°C was required. The suppression of the competing uncatalysed reaction requires the use of a relatively unreactive hydroborating reagent such as catecholborane (2.171). The intermediate boronate is converted, by hydrogen peroxide oxidation, into the

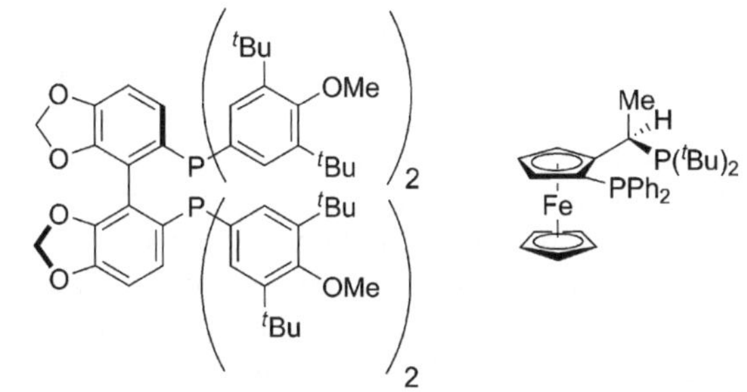

(2.161) (*R*)-DTBM-SEGPHOS   **(2.162)** (*R, S*)-PPF-P(*^t*Bu)₂

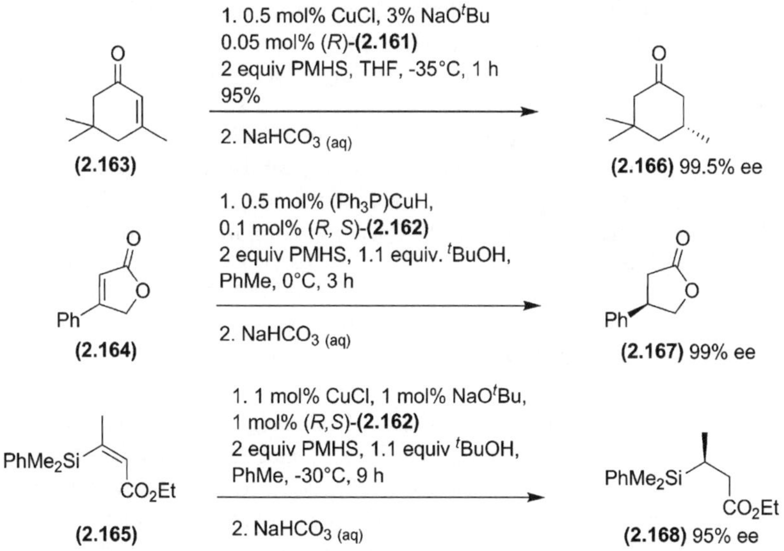

(2.163)

1. 0.5 mol% CuCl, 3% NaO*^t*Bu
0.05 mol% (*R*)-**(2.161)**
2 equiv PMHS, THF, -35°C, 1 h
95%

2. NaHCO₃ (aq)

(2.166) 99.5% ee

(2.164)

1. 0.5 mol% (Ph₃P)CuH,
0.1 mol% (*R, S*)-**(2.162)**
2 equiv PMHS, 1.1 equiv. *^t*BuOH,
PhMe, 0°C, 3 h

2. NaHCO₃ (aq)

(2.167) 99% ee

(2.165)

1. 1 mol% CuCl, 1 mol% NaO*^t*Bu,
1 mol% (*R,S*)-**(2.162)**
2 equiv PMHS, 1.1 equiv *^t*BuOH,
PhMe, -30°C, 9 h

2. NaHCO₃ (aq)

(2.168) 95% ee

corresponding alcohol with retention of configuration. Thus styrene (**2.147**) is converted into phenethyl alcohol (**2.155**) by this two-stage process.

An interesting development reported by John Brown is the conversion of the intermediate boronate into a primary amine. [108] This provides a route for the one-pot conversion of alkenes into enantiomerically enriched amines. Whilst the intermediate boronates cannot be directly converted into amines, treatment of the boronates with a Grignard reagent forms a borane, which is then converted into an amine using standard procedures. For example, the alkene (**2.172**) is converted into the amine (**2.173**) in a one-pot process. As catecholborane (**2.171**) is air-sensitive and somewhat unstable under these reaction conditions, some work has been directed towards the use of other more stable boranes, most notably pinacolborane (**2.174**). Styrene undergoes regioselective and enantioselective hydroboration with pinacolborane (**2.174**) in the presence of the catalyst prepared from [Rh(cod)$_2$]BF$_4$ and JOSIPHOS (**2.15**) to give the opposite enantiomer to that obtained using catecholborane and the same rhodium/ligand complex.[109] Use of iridium catalysts in this process led to a complete reversal in regioselectivity.

While styrene derivatives are the most common substrates in this process the enantioselective hydroboration/oxidation of other alkenes has been achieved. Meso bicyclic hydrazines such as (**2.175**) undergo highly enantioselective rhodium-catalysed hydroboration in the presence of the bisphosphine (*S,S*)-BDDP (**2.176**).[101] Use of the catalyst prepared from [Ir(cod)Cl]$_2$ and (*R,S*)-JOSIPHOS (**2.15**) leads to a reversal in the sense of asymmetric induction, but also a reduction in enantioselectivity. 3,3-Substituted cyclopropenes, such as (**2.177**), possessing an ester functionality also undergo rhodium-catalysed hydroboration with high ee. The ester moiety exerts a strong directing effect in this reaction giving the *cis*-product.

Use of bis(catecholato)diboron (**2.178**) in the rhodium-catalysed hydroboration results in enantioselective diboration and a number of simple *trans*-alkenes such as *trans*-β-methylstyrene (**2.179**) are converted into the 1,2-diol in high ee by diboration/oxidation using (nbd)Rh(acac) in combination with (*S*)-QUINAP (**2.169**) as catalyst.[110]

2.8   Hydroamination

Catalytic asymmetric hydroamination of alkenes can be achieved using early[111] and late transition metal catalysts[112] and lanthanide-based catalytic systems.[111a, 113] The most commonly used early transition metal catalysts are titanium and zirconium complexes comprising bis-aminophenolate or amidate ligands such as (**2.183**)[111b] while the late transition metal catalysts include the bridged dinuclear iridium complex (**2.184**).[112a] The most successful lanthanide catalysts include both metallocene (**2.185**)[112a] and nonmetallocene-based systems incorporating binaphthol or binaphthylamine ligands such as (**2.186**).[112c]

Both group 4-based and lanthanide systems have shown most promise as catalysts in the intramolecular hydroamination giving rise to pyrrolidines and piperidines. In general, the enantioselectivities observed in this process have been moderate, but cyclisation does occur with high ee in some cases. For example, pentenylamine (**2.187**) undergoes cyclohydroamination to give pyrrolidine (**2.188**) with 93% ee using the zirconocene catalyst (**2.183**).[111b] Similar cyclisations have been effected in up to 95% ee using the scandium-based hindered binaphtholate complex (**2.186**).[113c] The enantioselective formation of piperidine rings by hydroamination has proved challenging and the highest ees to date have been obtained during the cyclisation of octadienamine (**2.189**) in the presence of lanthanocine (**2.185**).

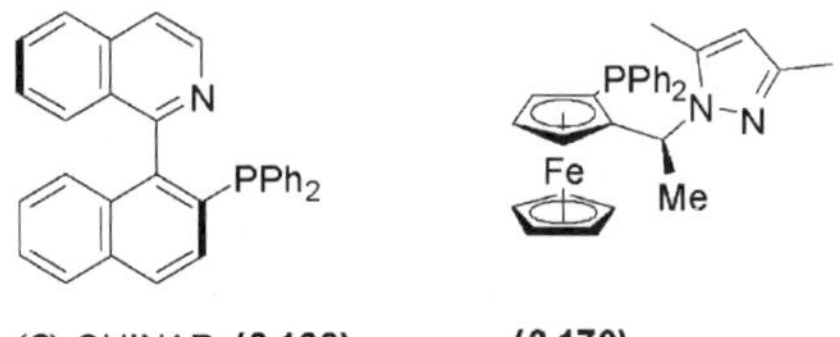

(*S*)-QUINAP **(2.169)**   **(2.170)**

(2.147)

1 mol% [Rh(cod)L]BF$_4$

O,O-BH **(2.171)**

20°C, THF
then, H$_2$O$_2$/NaOH

(2.155)

L	
BINAP	96% ee
QUINAP	88% ee
(2.170)	95% ee

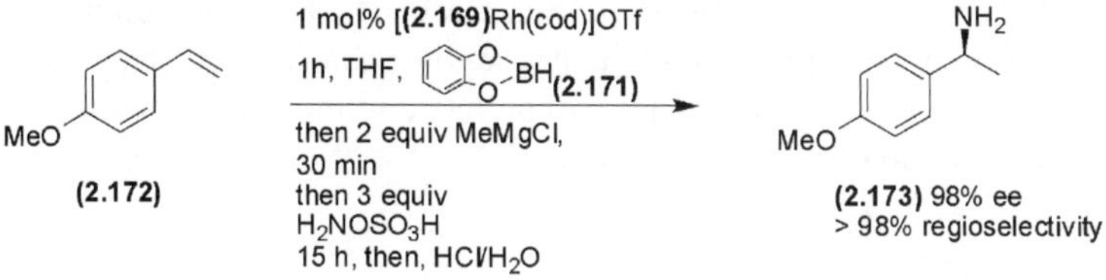

MeO

(2.172)

1 mol% [**(2.169)**Rh(cod)]OTf

1h, THF, O,O-BH **(2.171)**

then 2 equiv MeMgCl,
30 min
then 3 equiv
H$_2$NOSO$_3$H
15 h, then, HCl/H$_2$O

(2.173) 98% ee
> 98% regioselectivity

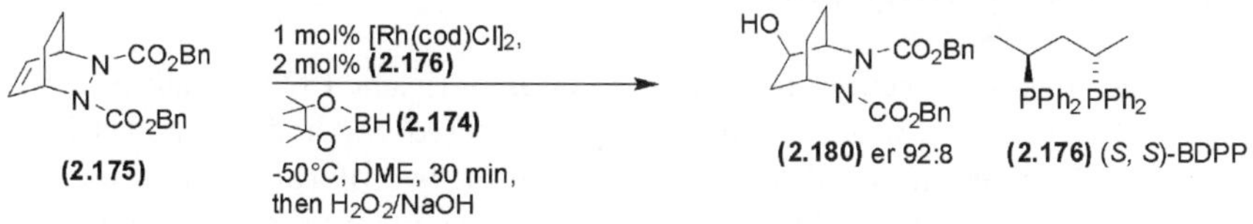

(2.175)

1 mol% [Rh(cod)Cl]$_2$,
2 mol% **(2.176)**

O,O-BH **(2.174)**

-50°C, DME, 30 min,
then H$_2$O$_2$/NaOH

(2.180) er 92:8   **(2.176)** (*S, S*)-BDPP

(2.177)

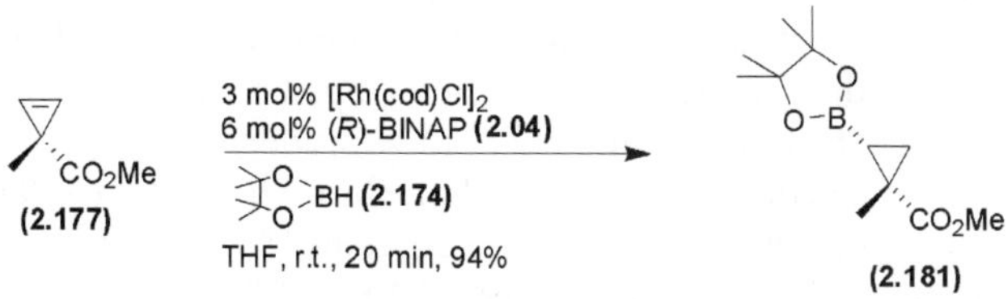

3 mol% [Rh(cod)Cl]$_2$
6 mol% (*R*)-BINAP **(2.04)**

O,O-BH **(2.174)**

THF, r.t., 20 min, 94%

(2.181)

Ph **(2.179)**

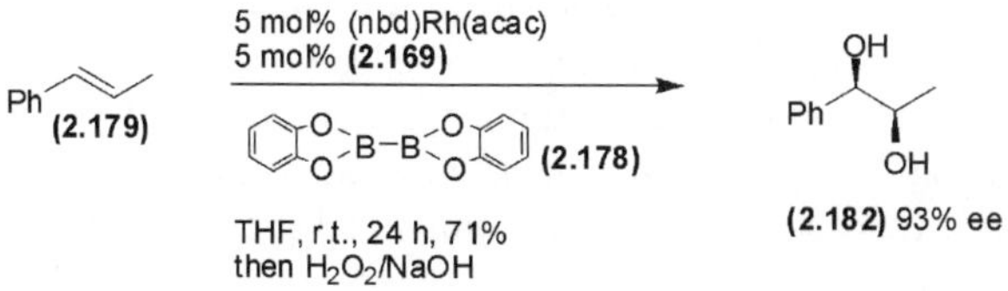

5 mol% (nbd)Rh(acac)
5 mol% **(2.169)**

O,O-B-B-O,O **(2.178)**

THF, r.t., 24 h, 71%
then H$_2$O$_2$/NaOH

(2.182) 93% ee

The intermolecular hydroamination has also received some attention and studies in this area have focussed on the use of iridium[111] and nickel catalysts.[114] Enantioselectivities for the intermolecular process are also generally moderate with ees observed between 60 and 70%. The highest enantioselectivities for this reaction have been obtained during the hydroamination of norbornene (**2.145**) with aniline in the presence of the iridium complex (**2.184**) and the fluoride ion source 1,1,1,3,3,3-hexakis(dimethylamino)diphosphazenium fluoride.[112a]

The group of Hartwig has shown that dienes undergo hydroamination with anilines in the presence of the palladium catalyst formed *in situ* from [{Pd(π-allyl)Cl}$_2$] and Trost's ligand (**2.190**).[115] Under optimum conditions cyclohexa-1,3-diene (**2.191**) undergoes conversion into the allylic amine (**2.192**) with 95% ee. $\alpha,\beta-$Unsaturated enones undergo highly enantioselective hydroamination in the presence of enantiomerically pure metal catalysts and organocatalysts. However, this process is considered as an enantioselective conjugate addition and is discussed in the appropriate section of this book (Section 11.4).

2.9 Hydroformylation

Hydroformylation reactions involve the addition of H_2 and CO to an alkene catalysed by a transition metal complex – typically based on rhodium, and it is an industrially important catalytic reaction. When the hydroformylation of unsymmetrical alkenes leads to a branched aldehyde as the major product, the potential for asymmetric induction in this process exists. Many phosphorus-based ligands have been examined for their ability to provide an asymmetric environment suitable for enantioselective hydroformylation.[116] Those ligands showing relatively good regioselectivity and enantioselectivities include the diphosphites (**2.195**),[117] furanoside diphosphite ligands such as (**2.196**), developed by Diéguez *et al.*,[118] and the bis(dialkylphospholano)ethane (Ph-bpe) (**2.197**).[119] The latter ligand has shown the best enantioselectivities to date for hydroformylation of allyl cyanide (**2.198**). Good results have been obtained using hybrid ligands bearing two different types of phosphorus. The phosphine phosphite ligand BINAPHOS (**2.199**) has been shown by Nozaki and coworkers to provide good enantioselectivity and regioselectivity across a range of simple and more complex alkenes.[120] These include cinnamyl alcohol (**2.200**) (which cyclises to give a hemi-acetal product)[121] and indene (**2.201**).[122] This ligand forms a well-organised active catalytic species with a phosphine in an equatorial position and a phosphite in an apical position around a rhodium atom. More recently, the mixed phosphorus phosphoramidite ligand (**2.202**) developed by Yan and Zhang has been used to effect the hydroformylation of styrene (**2.147**) and derivatives and vinyl acetate (**2.203**) with the highest ees observed to date.[123] BINAPHOS (**2.199**) has been incorporated into a cross-linked polymer matrix. Subsequent hydroformylation reactions were found to be only slightly less selective and active than their homogeneous counterparts.[124]

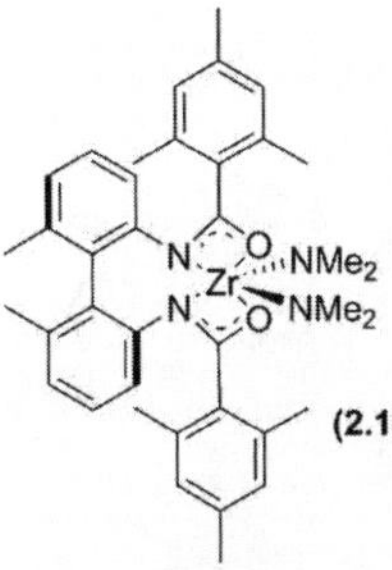

(2.183)

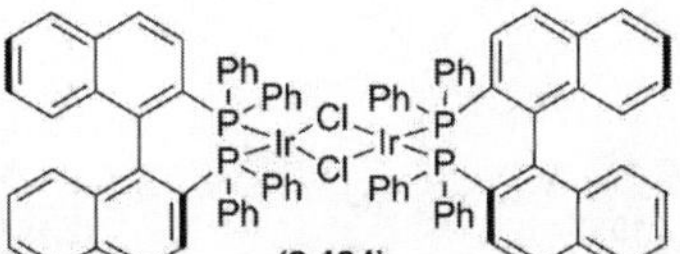

(2.184)

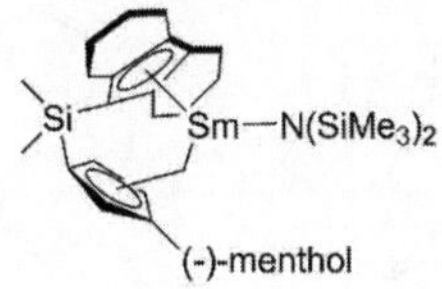
Si
Sm—N(SiMe3)2
(-)-menthol

(2.185)

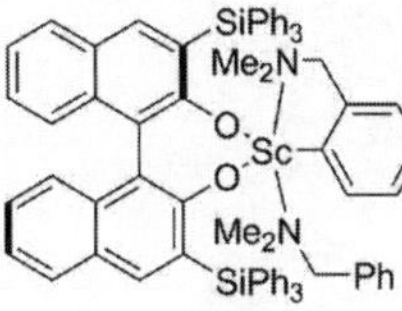
SiPh3
Me2N
O
Sc
O
Me2N
SiPh3
Ph

(2.186)

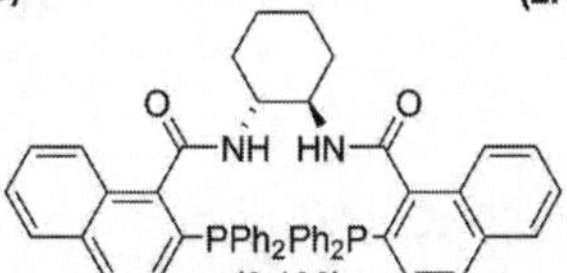
O
NH HN
O
PPh2Ph2P

(2.190)

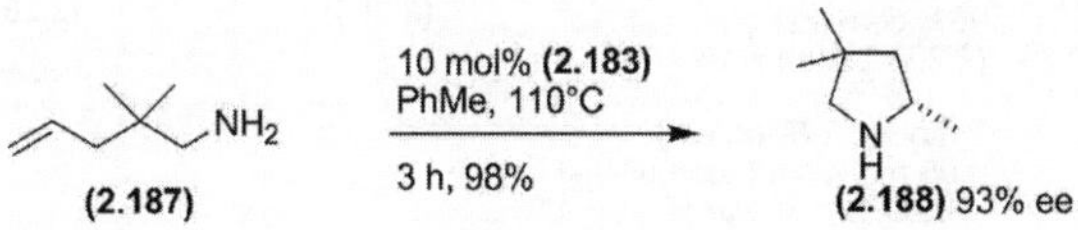
NH2
(2.187)
10 mol% (2.183)
PhMe, 110°C
3 h, 98%
N
H
(2.188) 93% ee

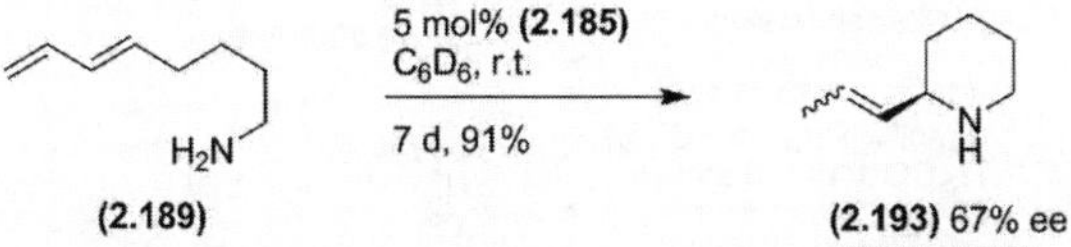
H2N
(2.189)
5 mol% (2.185)
C6D6, r.t.
7 d, 91%
N
H
(2.193) 67% ee
(E/Z = 97:3)

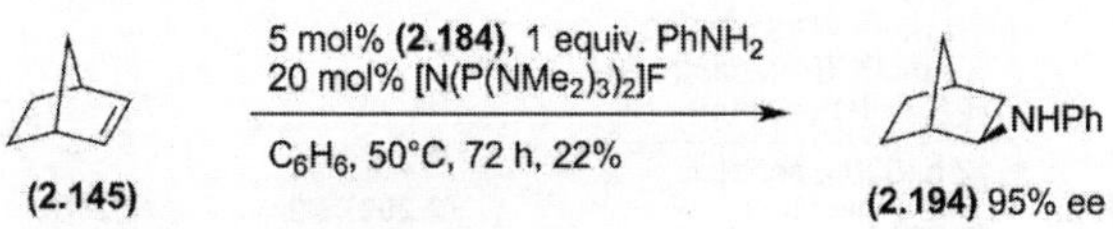
(2.145)
5 mol% (2.184), 1 equiv. PhNH2
20 mol% [N(P(NMe2)3)2]F
C6H6, 50°C, 72 h, 22%
NHPh
(2.194) 95% ee

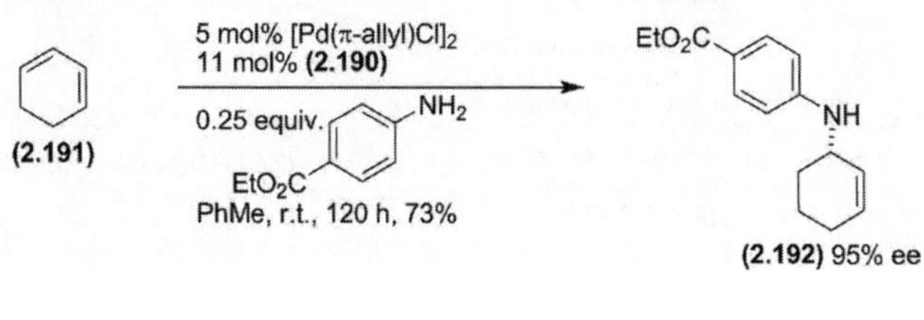
(2.191)
5 mol% [Pd(π-allyl)Cl]2
11 mol% (2.190)
0.25 equiv.
NH2
EtO2C
PhMe, r.t., 120 h, 73%
EtO2C
NH
(2.192) 95% ee

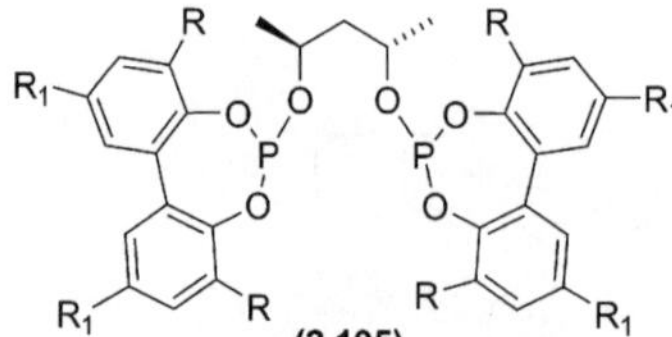

(2.195)

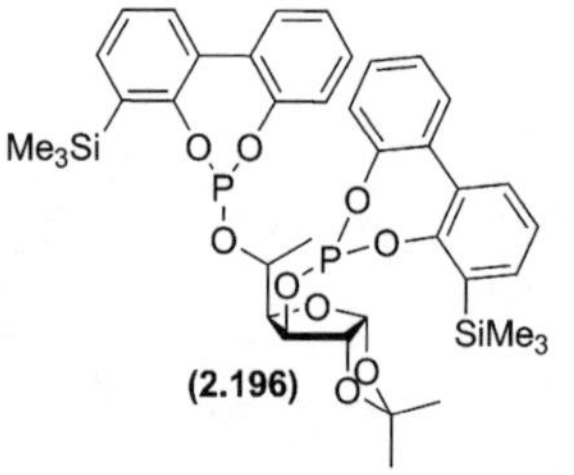

(2.196)

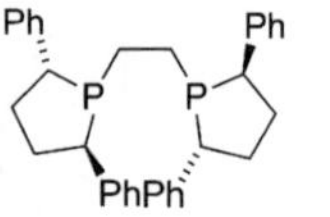

(*R, R*)-Ph-BPE **(2.197)**

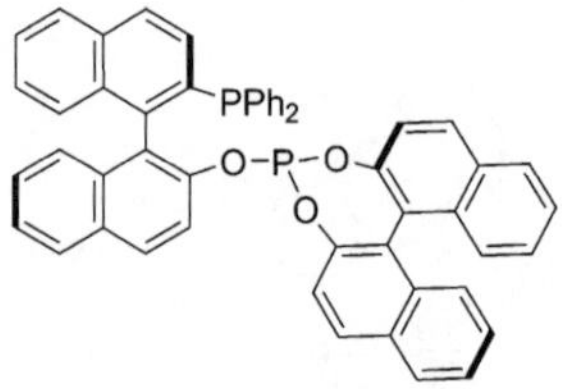

(*R,S*)-BINAPHOS **(2.199)**

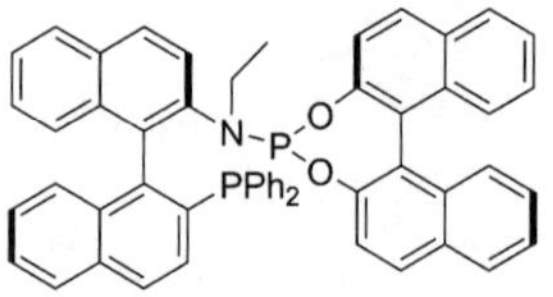

(2.202)

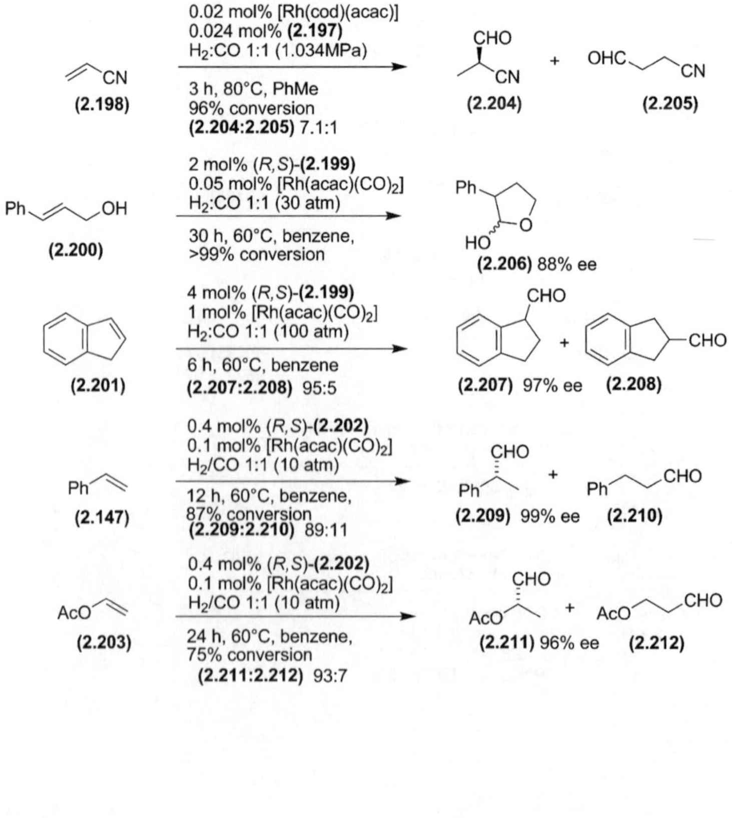

CN (2.198)	0.02 mol% [Rh(cod)(acac)] 0.024 mol% **(2.197)** H₂:CO 1:1 (1.034MPa) — 3 h, 80°C, PhMe 96% conversion **(2.204:2.205)** 7.1:1	CHO/CN **(2.204)** + OHC/CN **(2.205)**
Ph OH (2.200)	2 mol% (*R,S*)-**(2.199)** 0.05 mol% [Rh(acac)(CO)₂] H₂:CO 1:1 (30 atm) — 30 h, 60°C, benzene, >99% conversion	**(2.206)** 88% ee
(2.201)	4 mol% (*R,S*)-**(2.199)** 1 mol% [Rh(acac)(CO)₂] H₂:CO 1:1 (100 atm) — 6 h, 60°C, benzene **(2.207:2.208)** 95:5	**(2.207)** 97% ee **(2.208)**
Ph (2.147)	0.4 mol% (*R,S*)-**(2.202)** 0.1 mol% [Rh(acac)(CO)₂] H₂/CO 1:1 (10 atm) — 12 h, 60°C, benzene, 87% conversion **(2.209:2.210)** 89:11	CHO/Ph **(2.209)** 99% ee + Ph CHO **(2.210)**
AcO (2.203)	0.4 mol% (*R,S*)-**(2.202)** 0.1 mol% [Rh(acac)(CO)₂] H₂/CO 1:1 (10 atm) — 24 h, 60°C, benzene, 75% conversion **(2.211:2.212)** 93:7	CHO/AcO **(2.211)** 96% ee + AcO CHO **(2.212)**

As well as rhodium-catalysed hydroformylation, the use of platinum-catalysed hydroformylation has also been reported to give good enantioselectivity, although regioselectivity is typically poor. [125]

2.10   Hydroacylation of Alkenes

The addition of an aldehyde group across an alkene is a hydroacylation reaction. Whilst there is no hydrogen gas needed for these reactions, the process has some similarity to hydroformylation from a synthetic viewpoint, hence its mention in this chapter. In common with hydroformylations, catalytic asymmetric hydroacylations utilise enantiomerically pure rhodium complexes as catalysts. To date the catalytic asymmetric hydroacylation of alkenes has only been achieved in an intramolecular sense. [126, 127,128] 4-Substituted pentenal (2.213) and 3,4-disubstituted pentenal (2.214) undergo cyclisation to provide the cyclopentanones (2.215) and (2.216) using cationic rhodium catalysts. In a similar vein α−substituted 2-vinyl benzaldehydes such as (2.217) undergo asymmetric cyclisation to give enantiomerically enriched 3-substituted indanones with up to 99% ee. [129]

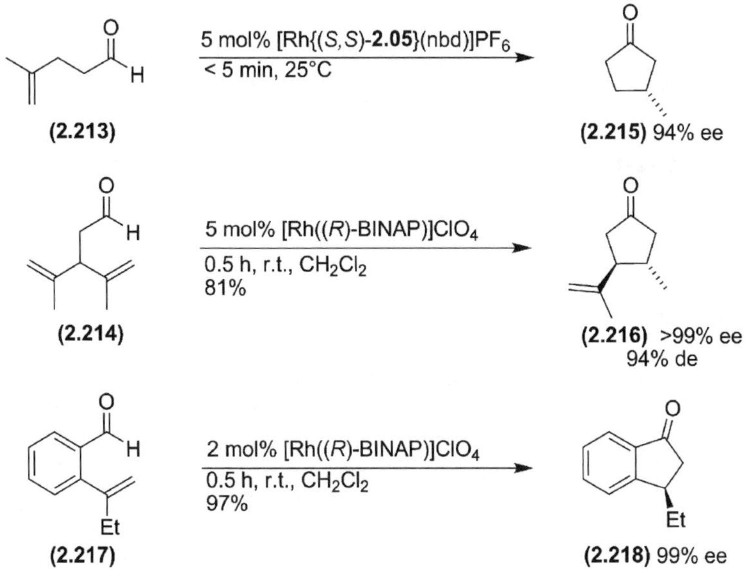

(2.213)

5 mol% [Rh{(S,S)-2.05}(nbd)]PF$_6$
< 5 min, 25°C

(2.215) 94% ee

(2.214)

5 mol% [Rh((R)-BINAP)]ClO$_4$
0.5 h, r.t., CH$_2$Cl$_2$
81%

(2.216) >99% ee
94% de

(2.217)

2 mol% [Rh((R)-BINAP)]ClO$_4$
0.5 h, r.t., CH$_2$Cl$_2$
97%

(2.218) 99% ee

2.11   Hydrocyanation of Alkenes

The hydrocyanation of vinylarenes[130] has been studied by a DuPont team using nickel catalysis.[131] The hydrocyanation of 6-methoxy-2-vinylnaphthalene (2.219) affords the product (2.220), where the enantiomeric excess is strongly dependent upon the electronic nature of the bisphosphinite ligand (2.12). Hydrolysis of the nitrile (2.220) affords the nonsteroidal anti-inflammatory drug Naproxen. This nickel-catalysed procedure has also been applied with some success to the regioselective, asymmetric hydrocyanation of 1,3-dienes such as 1-phenyl-1,3-butadiene (2.221) to give the 1,2-adduct (2.222) with ees between 50 and 83%.[132]

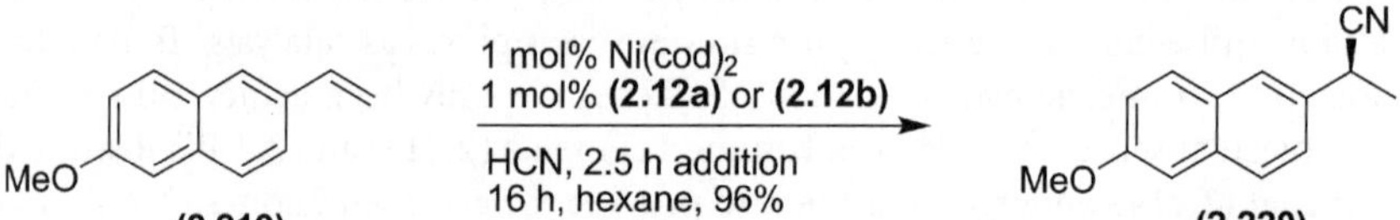

1 mol% Ni(cod)$_2$
1 mol% (2.12a) or (2.12b)

HCN, 2.5 h addition
16 h, hexane, 96%

(2.219)   (2.220)

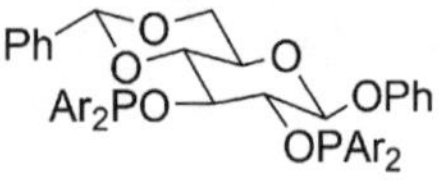

(2.12a) Ar = 3,5(CF$_3$)$_2$C$_6$H$_3$-   provides 85-91% ee
(2.12b) Ar = 3,5(CH$_3$)$_2$C$_6$H$_3$-   provides 16% ee

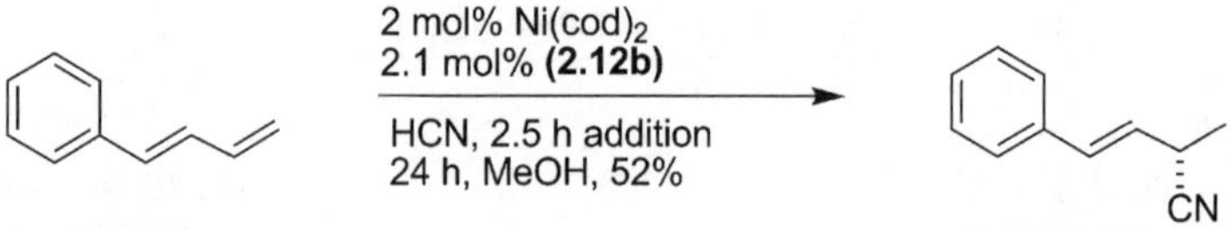

2 mol% Ni(cod)$_2$
2.1 mol% (2.12b)

HCN, 2.5 h addition
24 h, MeOH, 52%

(2.221)   (2.222) 83% ee

References

1. For some recent reviews on asymmetric hydrogenation see; (a) D. Arntz and A. Schafer in *Metal Promoted Selectivity in Organic Synthesis*, Ed. A.F. Noels, M. Graziani and A. J. Hubert, Kluwer Academic, Dordrecht, **1991**, 161. (b) H. Takaya, T. Ohta and R. Noyori in *Catalytic Asymmetric Synthesis*, Ed. I. Ojima, VCH, New York, **1993**, 1. (c) R. Noyori, *Asymmetric Catalysis in Organic Synthesis*, John Wiley & Sons, Inc., New York, **1994**, 16. (d) J. M. Brown, in *Comprehensive Asymmetric Catalysis*, Vol. 1, ed. E. N. Jacobsen, A. PFaltz and H. Yamamoto, Springer-Verlag, Berlin, **1999**, 121. (c) R. L. Halterman, in *Comprehensive Asymmetric Catalysis*, Vol. 1, ed. E. N. Jacobsen, A. Pfaltz and H. Yamamoto, Springer-Verlag, Berlin, **1999**, 183. (d) R. Noyori, *Angew. Chem., Int. Ed.*, **2002**, *41*, 2008. (e) W. S. Knowles, *Angew. Chem. Int. Ed.*, **2002**, *41*,

1998. (e) W. Tang and X. Zhang, *Chem. Rev.*, **2003**, *103*, 3029. (f) M. Kitamura and R. Noyori, in *Ruthenium in Organic Synthesis*, ed. S. Murahashi, Wiley-VCH, Weinheim, **2004**, 5.

2. L. Horner, H. Siegel and H. Büthe, *Angew. Chem., Int. Ed. Engl.*, **1968**, *7*, 942
3. W. S. Knowles and M. J. Sabacky, *J. Chem. Soc., Chem. Commun.*, **1968**, 1445
4. (a) T. P. Dang and H. B. Kagan, *J. Chem. Soc., Chem. Commun.*, **1971**, 481. (b) H. B. Kagan and T. P. Dang, *J. Am. Chem. Soc.*, **1972**, *94*, 6429.
5. (a) A. Miyashita, A. Yasuda, H. Takaya, K. Toriumi, T. Ito, T. Souchi and R. Noyori, *J. Am. Chem. Soc.*, **1980**, *102*, 7932. (b) M. J. Burk, *J. Am. Chem. Soc.*, **1991**, *113*, 8518. (c) M. D. Fryzuk and B. Bosnich, *J. Am. Chem. Soc.*, **1997**, *99*, 6262. (d) B.D. Vineyard, W.S. Knowles, M. J. Sabacky, G. L. Bachman and D. J. Weinkauff, *J. Am. Chem. Soc.*, **1977**, *99*, 5946. (e) T. Hayashi and M. Kumada, *Acc. Chem. Res.*, **1982**, *15*, 395. (f) A. Miyashita, H. Karino, J. -I. Shimamura, T. Chiba, K. Nagano, H. Nohira and H. Takaya, *Chem. Lett.*, **1989**, 1849. (g) P. A. McNeil, N. K. Roberts and B. Bosnich, *J. Am. Chem. Soc.*, **1981**, *103*, 2280. (h) U. Nagel, E. Kinzel, J. Andrade and G. Prescher, *Chem. Ber.*, **1986**, *119*, 3326. (i) T. Imamoto, J. Watanabe, Y. Wada, H. Masuda, H. Yamada, H. Tsuruta, S. Matsukawa and K. Yamaguchi, *J. Am. Chem. Soc.*, **1998**, *120*, 1635. (j) T.V. RajanBabu, T. A. Ayers, G.A. Halliday, K. K. You and J. C. Calabrese, *J. Org. Chem.*, **1997**, *62*, 6012. (k) F. Robin, F. Mercier, L. Ricard, F. Mathey and M. Spagnol, *Chem. Eur. J.*, **1997**, *3*, 1365. (l) J. Kang, J. H. Lee, S. H. Ahn and J. S. Choi, *Tetrahedron Lett.*, **1998**, *39*, 5523. (m) F.-Y. Zhang, C. -C. Pai and A. S. C. Chan, *J. Am. Chem. Soc.*, **1998**, *120*, 5808. (n) A. S. C. Chan, W. Hu, C. -C. Pai and C.-P. Lau, *J. Am. Chem. Soc.*, **1997**, *119*, 9570. (o) G. Zhu, P. Cao, Q. Jiang and X. Zhang, *J. Am. Chem. Soc.*, **1997**, *119*, 1799. (p) M. Sawamura, R. Kuwano and Y. Ito, *J. Am. Chem. Soc.*, **1995**, *117*, 9602. (q) A. Togni, C. Breutel, A. Schnyder, F. Spindler, H. Landert and A. Tijani, *J. Am. Chem. Soc.*, **1994**, *116*, 4062–4066.
6. T. Jerphagnon, J–L. Renaud and C. Bruneau, *Tetrahedron: Asymmetry*, **2004**, *15*, 2101.
7. (a) M. van den Berg, A. J. Minnaard, E. P. Schudde, J. Van Esch, A. H. M. De Vries, J. G. De Vries and B. L. Feringa, *J. Am. Chem. Soc.*, **2000**, *122*, 11539. (b) D. Peña, A. J. Minaard, J. G. de Vries and B. L. Feringa, *J. Am. Chem. Soc.*, **2002**, *124*, 14552. (c) D. Peña, A. J. Minarrd, A. H. M. de Vries, J. G. de Vires and B. L. Feringa, *Org. Lett.*, **2003**, *5*, 475. (d) D. Peña, A. J. Minaard, J. A. F. Boogers, A. H. M. De Vries, J. G. de Vries and B. L. Feringa, *Org. Biomol. Chem.*, **2003**, 1, 1087. (e) H. Bersnmann, M. van den Berg, R. Hoen, A. J. Minaard, G. Mehler, M. T. Reetz, J. G. De Vries and B. L. Ferfinga, *J. Org. Chem.*, **2005**, *70*, 943. (f) R. Hoen, J. A. F. Boogers, H. Bersnsmann, A. J. Minaard, A. Meetsma, T. D. Tiemersma-Wegmann, A. H. M. de Vries, J. G. de Vries and B. L. Feringa, *Angew. Chem. Int. Ed.*, **2005**, *44*, 4209.
8. M. T. Reetz and G. Mehler, *Angew. Chem. Int. Ed.*, **2000**, *39*, 3889. (b) M. T. Reetz, G. Mehler and A. Meiswinkel, *Tetrahedron Lett.*, **2002**, *43*, 7941. (c) M. T. Reetz, T. Sell, A. Meiswinkel and G. Mehler, *Angew. Chem. Int. Ed.*, **2003**, *115*, 790. (d) M. T. Reetz, L. J. Goossen, A. Meiswinkel, J. Patzold and J. F. Jensen, *Org. Lett.*, **2003**, *5*, 3099.
9. C. Claver, E. Fernandez, A. Gillon, K. Heslop, D. J. Hyett, A. Martorell, A. G. Orpen and P. G. Pringle, *Chem. Commun.*, **2000**, 961.
10. X. Jia, X. Li, L. Xu, Q. Shi, X. Yao and A. S. C. Chan, *J. Org. Chem.*, **2003**, *68*, 4539.

11. Q-H. Zeng, X. -P. Hu, Z. -C. Duan, X. -M. Liang and Z. Zheng, *J. Org. Chem.*, **2006**, *71*, 393.
12. M. T. Reetz and X. Li, *Angew. Chem. Int. Ed.*, **2005**, *44*, 2959.
13. K. A. Vallianatou, I. D. Kostas, J. Holz and A. Börner, *Tetrahedron Lett.*, **2006**, *47*, 7947.
14. (a) Z. Hua, V. C. Vassar and I. Ojima, *Org. Lett.*, **2003**, *5*, 3821. (b) C. Monti, C. Gennari and U. Piarulli, *Tetrahedron Lett.*, **2004**, *45*, 6859.
15. D. A. Evans, F. E. Michael, J. S. Tedrow and K. R. Campos, *J. Am. Chem. Soc.*, **2003**, *125*, 3534.
16. For reviews on combinatorial asymmetric catalysis see: (a) M. T. Reetz, *Angew. Chem. Int. Ed.*, **2001**, *40*, 284. (b) M. T. Reetz, *Angew. Chem.*, **2002**, *41*, 1335. (c) C. Gennari and U. Piarulli, *Chem. Rev.*, **2003**, *103*, 3071.
17. (a) M. T. Reetz and X. Li, *Tetrahedron*, **2004**, *60*, 9709. (b) R. Hoen, T. Tiemersma-Wegman, B. Procuranti, L. Lefort, J. G. de Vries, A. J. Minnaard and B. L. Feringa, *Org. Biomol. Chem.*, **2007**, *5*, 267.
18. M. J. Burk, S. Feng, M. F. Gross and W. Tumas, *J. Am. Chem. Soc.*, **1995**, *117*, 227.
19. W. S. Knowles, *Acc. Chem. Res.*, **1983**, *16*, 106.
20. (a) A. S. C. Chan, J. J. Pluth and J. Halpern, *J. Am. Chem. Soc.*, **1980**, *102*, 5952. (b) J. P. Halpern, *Science*, *217*, 401. (c) J. Halpern, *Pure Appl. Chem.*, **1983**, *55*, 99.
21. J. M. Brown and P. A. Chaloner, *J. Am. Chem. Soc.*, **1980**, *102*, 3040.
22. S. K. Armstrong, J. M. Brown and M. J. Burk, *Tetrahedron Lett.*, **1993**, *34*, 879.
23. (a) K. Rossen, *Angew. Chem. Int. Ed.*, **2001**, *40*, 4611. (b) P. J. Donoghue, P. Helquist and O. Wiest, *J. Org. Chem.*, **2007**, *72*, 839.
24. (a) I. D. Gridnev and T. Imamoto, *Acc. Chem. Res.*, **2004**, *37*, 633. (b) I. D. Gridnev, M. Yasutake, T. Imamoto and I. P. Beletskaya, *Proc. Nat. Acad. Sci.*, **2004**, *101*, 5385.
25. M. T. Reetz, A. Meiswinkel, G. Mehler, K. Angermund, M. Graf, W. Thiel, R. Mynott and D. G. Blackmond, *J. Am. Chem. Soc.*, **2005**, *127*, 10305.
26. A. Miyashita, H. Takaya, T. Souchi and R. Noyori, *Tetrahedron*, **1984**, *40*, 1245.
27. M. J. Burk, J. G. Allen and W. F. Kiesman, *J. Am. Chem. Soc.*, **1998**, *102*, 3040.
28. M. J. Burk, M. F. Gross and J. P. Martinez, *J. Am. Chem. Soc.*, **1995**, *117*, 9375.
29. R. Kuwano, S. Okuda and Y. Ito, *J. Org. Chem.*, **1998**, *63*, 3499–3503.
30. I. Dgridnev, M. Yasutake, N. Higashi and T. Imamoto, *J. Am. Chem.*, **2001**, *123*, 5268.
31. X. Cui and K. Burgess, *Chem. Rev.*, **2005**, *105*, 3272.
32. L. Panella, A. M. Aleixandre, G. J. Kruidhof, J. Robertus, B. L. Feringa, J. G. de Vries and A. J. Minaard, *J. Org. Chem.*, **2006**, *71*, 2026.
33. Q. Dai, W. Yang and X. Zhang, *Org. Lett.*, **2005**, *7*, 5343.
34. Y. Hsiao, N. R. Rivera, T. Rosner, S. W. Krska, E. Njolio, F. Wang, Y. Sun, J. D. Armstrong III, E. J. J. Grabowski, R. D. Tillyer, F. Spindler and C. Malan, *J. Am. Chem. Soc.*, **2004**, *126*, 9918.
35. G. –H. Hou, J. –H. Xie, L. –X. Wang and Q. –L. Zhou, *J. Am. Chem. Soc.*, **2006**, *128*, 11774.
36. M. J. Burk, C. S. Kalberg and A. Pizzano, *J. Am. Chem. Soc.*, **1998**, *120*, 4345.
37. (a) T. Chiba, A. Miyashita, H. Nohira and H. Takaya, *Tetrahedron Lett.*, **1991**, *32*, 4745. (b) M.J. Burk, F. Bienewald, M. Harris and A. Zanotti-Gerosa, *Angew. Chem., Int. Ed. Engl.*, **1998**, *37*, 1931.
38. T. Hayashi, N. Kawamura and Y. Ito, *Tetrahedron Lett.*, **1988**, *29*, 5969.

39. N.W. Boaz, *Tetrahedron Lett.*, **1998**, *39*, 5505.
40. W. Tang, D. Liu and X. Zhang, *Org. Lett.*, **2003**, *5*, 205.
41. L. Panella, B. L. Feringa, J. G. de Vries and A. J. Minnaard, *Org. Lett.*, **2005**, *7*, 4177.
42. M. T. Reetz, L. J. Goossen, A. Meiswinkel, J. Paetzold and J. F. Jensen, *Org. Lett.*, **2003**, *5*, 3099.
43. (a) J. B. Morgan and J. P. Morken, *J. Am. Chem. Soc.*, **2004**, *126*, 15338. (b) W. J. Moran and J. P. Morken, *Org. Lett.*, **2006**, *8*, 2413.
44. R. Noyori, *Chem. Soc. Rev.*, **1989**, *18*, 209.
45. T. Uemura, X. Zhang, K. Matsumura, N. Sayo, H. Kumobayashi, T. Ohta, K. Nozaki and H. Takaya, *J. Org. Chem.*, **1996**, *61*, 5510.
46. D.J. Bayston, J.L. Fraser, M.R. Ashton, A.D. Baxter, M. E. C. Polywka and E. Moses, *J. Org. Chem.*, **1998**, *63*, 3137.
47. J. P. Genêt, C. Pinel, V. Ratovelomanana-Vidal, S. Mallart, X. Pfister, L. Bischoff, M. C. Caño de Andrade, S. Darses, C. Galopin and J. A. Laffitte, *Tetrahedron: Asymmetry*, **1994**, *5*, 675.
48. J. P. Genêt, *Pure Appl. Chem.*, **2002**, *74*, 77.
49. J. Wu, X. Chen, R. Guo, C. Yeung and A. S. Chan, *J. Org. Chem.*, **2003**, *68*, 2490.
50. L. Qiu, F. Y. Kwong, J. Wu, W. H. Lam, S. Chan, W. –Y. Yu, Y. –M. Li, R. Guo, Z. Zhou and A. S. Chan, *J. Am. Chem. Soc.*, **2006**, *128*, 5955.
51. (a) H. Kawano, T. I. Kariya, Y. Ishii, M. Saburi, S. Yoshikawa, Y. Uchida and H. Kumobayashi, *J. Chem. Soc., Perkin Trans. 1*, **1989**, 1571. (b) R. Noyori, T. Ikeda, T. Ohkuma, M. Widhalm, M. Kitamura, H. Takaya, S. Akutagawa, N. Sayo, T. Saito, T. Taketomi and H. Kumobayashi, *J. Am. Chem. Soc.*, **1989**, *111*, 9134.
52. M. Kitamura, Y. Hsiao, M. Ohta, M. Tsukamoto, T. Ohta, H. Takaya and R. Noyori, *J. Org. Chem.* **1994**, *59*, 297.
53. T. Ohta, H. Takaya, M. Kitamura, K. Nagai and R. Noyori, *J. Org. Chem.*, **1987**, *52*, 3174.
54. M. Saburi, L. Shao, T. Sakurai and Y. Uchida, *Tetrahedron Letters*, **1992**, *33*, 7877.
55. H. Takaya, T. Ohta, N. Sayo, H. Kumobayashi, S. Akutagawa, S. Inoue, I. Kasahara and R. Noyori, *J. Am. Chem. Soc.*, **1987**, *109*, 1596.
56. M. Kitamura, K. Nagai, Y. Hsiao and R. Noyori, *Tetrahedron Lett.*, **1990**, *31*, 549.
57. M. Kitamura, I. Kasahara, K. Manabe, R. Noyori and H. Takaya, *J. Org. Chem.*, **1988**, *53*, 708.
58. J. W. Faller and M. Tokunaga, *Tetrahedron Lett.*, **1993**, *34*, 7359.
59. P. Le Gendre, F. Jérôme, C. Bruneau and P.H. Dixneuf, *J. Chem. Soc., Chem. Commu.*, **1998**, 533.
60. R. Kuwano and M. Kashiwabara, *Org. Lett.*, **2006**, *8*, 2653.
61. T. Ohta, T. Miyake and H. Takaya, *J. Chem. Soc., Chem. Commun.*, **1992**, 1725.
62. T. Ohta, T. Miyake, N. Seido, H. Kumobayashi and H. Takaya, *J. Org. Chem*, **1995**, *60*, 357.
63. M. J. Fehr, G. Consiglio, M. Scalone and R. Schmid, *J. Org. Chem.*, **1999**, *64*, 5768.
64. (a) M. Kitamura, M. Tsukamoto, Y. Bessho, M. Yoshimura, U. Kobs, M. Widhalm and R. Noyori, *J. Am. Chem. Soc.*, **2002**, *124*, 6649. (b) M. Tsukamoto, M. Yoshimura, K. Tsuda and M. Kitamura, **2006**, *62*, 5448.
65. E. Cesaroth, R. Ugo and H.B. Kagan, *Angew. Chem,. Int. Ed. Eng.*, **1979**, *18*, 779.

66. R. L. Halterman, K.P.C. Vollhardt, M.E. Welker, D. Bläser and R. Boese, *J. Am. Chem. Soc.*, **1987**, *109*, 8105

67. F. R. W. P. Wild, L. Zsolnai, G. Huttner and H. H. Brintzinger, *J. Organomet. Chem.*, **1982**, *232*, 233.

68. R. D. Broene and S. L. Buchwald, *J. Am. Chem. Soc.*, **1993**, *115*, 12569.

69. R. Waymouth and P. Pino, *J. Am. Chem. Soc.*, **1990**, *112*, 4911.

70. V. P. Conticello, L. Brard, M. A. Giardello, Y. Tsuji, M. Sabat, C. L. Stern and T. J. Marks, *J. Am. Chem. Soc.*, **1992**, *114*, 2761.

71. N. E. Lee and S. L. Buchwald, *J. Am. Chem. Soc.*, **1994**, *116*, 5985.

72. M. V. Troutman, D. H. Appella and S. L. Buchwald, *J. Am. Chem. Soc.*, **1999**, *121*, 4916.

73. R. H. Crabtree, *Acc. Chem. Res.*, **1979**, *12*, 331.

74. (a) A. Lightfoot, P. Schnider and A. Pfaltz, *Angew. Chem. Int. Ed.*, **1998**, *37*, 2897. (b) J. Blankenstein and A. Pfaltz, *Angew. Chem. Int. Ed.*, **2001**, *40*, 4445. (c) A. Pfaltz, J. Blankenstein, E. Hörmann, S. McIntyre, F. Menges, R. Hilgraf, M. Schönleber, S. P. Smidt, B. Wüstenberg and N. Zimmermann, *Adv. Synth. Catal.*, **2003**, *345*, 33.

75. S. P. Smidt, F. Menges and A. Pfaltz, *Org. Lett.*, **2004**, *6*, 2023.

76. (a) S. Bell, B. Wüstenberg, S. Kaiser, F. Menges, T. Netscher and A. Pfaltz, *Science*, **2006**, *311*, 642. (b) S. Kaiser, S. P. Smidt and A. Pfaltz, *Angew. Chem. Int. Ed.*, **2006**, *45*, 5194.

77. C. Hedberg, K. Källstrom, P. Brandt, L. K. Hansen and P. G. Andersson, *J. Am. Chem. Soc.*, **2006**, *128*, 2995.

78. (a) M. T. Powell, D. –R. Hou, M. C. Perry, X. Cui and K. Burgess, *J., Am. Chem. Soc.*, **2001**, *123*, 8878. (b) M. C. Perry, X. Cui, M. T. Powell, D.-R. Hou, J. H. Reibenspies and K. Burgess, *J. Am. Chem. Soc.*, **2003**, *125*, 113.

79. J. W. Yang, M. T. Hechavarria Fonseca and B. List, *Angew. Chem. Int. Ed.*, **2004**, *43*, 6660.

80. H. Adolfsson, *Angew. Chem. Int. Ed.*, **2005**, *44*, 3340.

81. J. W. Yang, M. T. Hechavarria Fonseca, N. Vignola and B. List, *Angew. Chem. Int. Ed.*, **2005**, *44*, 108.

82. S. G. Ouellet, J. B. Tuttle and D. W. C. MacMillan, *J.Am. Chem. Soc.*, **2005**, *127*, 32.

83. J. B. Tuttle, S. G. Oullet and D. W. C. MacMillan, *J. Am. Chem. Soc.*, **2006**, *128*, 12662.

84. T. Hayashi, in *Comprehensive Asymmetric Catalysis*, Vol. 1, ed. E. N. Jacobsen, A. Pfaltz and H. Yamamoto, Springer-Verlag, Berlin, **1999**, 319.

85. (a) Y. Uozumi, S.-Y. Lee and T. Hayashi, *Tetrahedron Lett.*, **1992**, *33*, 7185. (b) Y. Uozumi and T. Hayashi, *Tetrahedron Lett.*, **1993**, *34*, 2335.

86. T. Hayashi, S. Hiarte, K. Kitayama, H. Tsuji, A. Torii and Y. Uozumi, *J. Org. Chem.*, **2001**, *66*, 1441.

87. K. Kitayama, H. Tsuji, Y. Uozumi and T. Hayashi, *Tetrahedron Lett.*, **1996**, *37*, 4169.

88. J. F. Jensen, B. Y. Svendsen, T. V. La Cour, H. L. Pedersen and M. Johannsen, *J. Am. Chem. Soc.*, **2002**, *124*, 4558.

89. X. –X. Guo, J. –H. Xie, G. –H. Hou, W. –J. Shi, L. –X. Wang and Q. –L. Zhou, *Tetrahedron: Asymmetry*, **2004**, *15*, 2231.

90. H. L. Pedersen and M. Johannsen, *J. Org. Chem.*, **2002**, *67*, 7982.

91. Y. Matsumoto, T. Hayashi and Y. Ito, *Tetrahedron*, **1994**, *50*, 335.

92. O. Riant, N. Mostefaï and J. Courmacel, *Synthesis*, **2004**, 2943.

93. B. H. Lipshutz and B. A. Frieman, *Angew. Chem. Int . Ed.*, **2005**, *44*, 6345.
94. B. H. Lipshutz, B. A. Frieman and A. E. Tomaso Jr., *Angew. Chem. Int. Ed.*, **2006**, *45*, 1259.
95. B. H. Lipshutz, J. M. Servesko, T. B. Petersen, P. P. Papa and A. A. Lover, *Org. Lett.*, **2004**, *6*, 1273.
96. B. H. Lipshutz, J. M. Sevesko and B. R. Taft, *J. Am. Chem. Soc.*, **2004**, *126*, 8352.
97. B. H. Lipshutz, N. Tanaka, B. R. Taft and C. T. Lee, *Org. Lett.* **2006**, *8*, **1963**.
98. D. Männig and H. Nöth, *Angew. Chem., Int. Ed. Engl.*, **1985**, *24*, 878.
99. D. A. Evans, G. C. Fu and A. H. Hoveyda, *J. Am. Chem. Soc.*, **1992**, *114*, 6671.
100. D. A. Evans and G. C. Fu, *J. Am. Chem. Soc.*, **1991**, *113*, 4042.
101. A. P. Luna, M. Bonin, L. Micouin and H. P. Husson, *J. Am.Chem. Soc.*, **2002**, *124*, 12098.
102. (a) T. Hayashi, in *Comprehensive Asymmetric Catalysis*, Vol. *1*, ed. E. N. Jacobsen, A. Pfaltz and H. Yamamoto, Springer-Verlag, Berlin, **1999**, 351. (b) C. M. Crudden and D. Edwards, *Eur. J. Org. Chem.*, **2003**, 4695.
103. T. Hayashi, Y. Matsumoto and Y. Ito, *J. Am. Chem. Soc.*, **1989**, *111*, 3426.
104. J. M. Brown, D. I. Hulmes and T. P. Layzell, *J. Chem. Soc., Chem. Commun.*, **1993**, 1673.
105. J. M. Valk, G. A. Whitlock, T. P. Layzell and J. M. Brown, *Tetrahedron: Asymmetry*, **1995**, *6*, 2593.
106. A. Schnyder, L. Hintermann and A. Togni, *Angew. Chem. Int. Ed. Engl.*, **1995**, *34*, 931.
107. A. Schnyder, A. Togni and U. Wiesli, *Organometallics*, **1997**, *16*, 255.
108. E. Fernandez, M.W. Hooper, F.I. Knight and J.M. Brown, *J. Chem. Soc., Chem. Commun.*, **1997**, 173.
109. C. M. Crudden, Y. B. Hleba and A. C. Chen, *J. Am. Chem. Soc.*, **2004**, *126*, 9200.
110. J. B. Morgan, S. P. Miller and J. P. Morken, *J. Am. Chem. Soc.*, **2003**, *125*, 8702.
111. (a) K. C. Hultzch, *Org. Biomol. Chem.*, **2005**, *3*, 1819. (b) M. C. Wood, D. C. Leitch, C. S. Yeung, J. A. Kozak and L. L. Schafer, *Angew. Chem. Int. Ed.*, **2007**, *46*, 354.
112. (a) R. Dorta, P. Egli, F. Zürcher and A. Togni, *J. Am. Chem. Soc.*, **1997**, *119*, 10857. (b) P. W. Roesky and T. E. Müller, *Angew. Chem. Int. Ed.*, **2003**, *42*, 2708.
113. (a) S. Hong, A. M. Kawaoka and T. J. Marks, *J. Am. Chem. Soc.*, **2003**, *125*, 15878. (b) S. Hong and T. J. Marks, *Acc. Chem. Res.*, **2004**, *37*, 673. (b) D. Riegert, J. Collins, A. Meddour, E. Schulz and A. Trifonov, *J. Org. Chem.*, **2006**, *71*, 2514. (c) D. V. Gribkov, K. C. Hultzsch and F. Hampel, *J. Am. Chem. Soc.*, **2006**, *128*, 3748.
114. L. Fadini and A. Togni, *Chem. Commun.*, **2003**, 30.
115. O. Löber, M. Kawatsura and J. F. Hartwig, *J. Am. Chem. Soc.*, **2001**, *123*, 4366.
116. (a) S. Gladiali, J.C. Bayón and C. Claver, *Tetrahedron: Asymmetry*, **1995**, *6*, 1453. (b) K. Nozaki, in *Comprehensive Asymmetric Catalysis*, Vol. 1, ed. E. N. Jacobsen, A. Pfaltz and H. Yamamoto, Springer-Verlag, Berlin, **1999**, 381. (c) M. Diéguez, O. Pàmies and C. Claver, *Tetraehdron: Asymmetry*, **2004**, *15*, 2113.
117. G.J.H. Buisman, E.J. Vos, P.C.J. Kamer and P.W.N.M. van Leeuwen, *J. Chem. Soc., Dalton Trans.*, **1995**, 409.
118. (a) M. Diéguez, O. Pàmies, A. Ruiz, S. Castillón and C. Claver, *Chem. Commun.*, **2000**, 1607. (b) M. Diéguez, O. Pàmies, A. Ruiz, S. Castillón and C. Claver, *Chem. Eur. J.*,

2001, *7*, 3086. (c) M. Diéguez, O. Pàmies, A. Ruiz and C. Claver, *New. J. Chem.*, **2002**, *6*, 827.

119. A. T. Axtell, C. J. Cobley, J. Klosin, G. T. Whiteker, A. Zanotti-Gerosa and K. A. Abboud, *Angew. Chem. Int. Ed.*, **2005**, *44*, 5834.
120. K. Nozaki, N. Sakai, T. Nanno, T. Higashijima, S. Mano, T. Horiuchi and H. Takaya, *J. Am. Chem. Soc.*, **1998**, *120*, 4051.
121. K. Nozaki, W.-G. Li, T. Horiuchi and H. Takaya, *Tetrahedron Lett.*, **1997**, *38*, 4611.
122. N. Sakai, K. Nozaki and H. Takaya, *J. Chem. Soc., Chem. Commun.*, **1994**, 395.
123. Y. Yan and X. Zhang, *J. Am. Chem. Soc.*, **2006**, *128*, 7198.
124. K. Nozaki, Y. Itoi, F. Shibahara, E. Shirakawa, T. Ohta, H. Takaya and T. Hiyama, *J. Am. Chem. Soc.*, **1998**, *120*, 4051.
125. G. Consiglio, S.C.A Nefkens and A. Borer, *Organometallics*, **1991**, *10*, 2046.
126. X.-M. Wu, K. Funakoshi and K. Sakai, *Tetrahedron Lett.*, **1993**, *34*, 5927.
127. (a) R.W. Barnhart, X. Wang, P. Noheda, S. H. Bergens, J. Whelan and B. Bosnich, *J. Am. Chem. Soc.*, **1994**, *16*, 1821. (b) R. W. Barnhart, D.A. McMorran and B. Bosnich, *J. Chem. Soc., Chem. Commun.*, **1997**, 589–590.
128. M. Tanaka, K. Sakai and H. Suemune, *Curr. Org. Chem.*, **2003**, *7*, 353.
129. K. Kundu, J. V. McCullagh and A. T. Morehead Jr., *J. Am., Chem. Soc.*, **2005**, *127*, 16042.
130. T. V. RajanBabu and A. L. Casalnuvo, in *Comprehensive Asymmetric Catalysis*, Vol. 1, ed. E. N. Jacobsen, A. Pfaltz and H. Yamamoto, Springer-Verlag, Berlin, **1999**, 367.
131. A.L. Casalnuovo, T.V. RajanBabu, T.A. Ayers and T.H. Warren, *J. Am. Chem. Soc.*, **1994**, *116*, 9869.
132. B. Saha and T. V. RajanBabu, *Org. Lett.*, **2006**, *8*, 4657.

Chapter 3
Reduction of Ketones and Imines

The reduction of the carbonyl group (and related functionalities) by catalytic methods has been successfully achieved by a number of methods. Rhodium and ruthenium complexes are the most popular catalysts used in the hydrogenation of ketones. While most catalyst systems of this type require the presence of additional chelating functionality on the substrate the recent development of highly active ruthenium(diamine) complexes allows the reduction of simple unfunctionalised ketones. Ruthenium catalysts have also been applied, with much success, to the catalytic asymmetric transfer hydrogenation of ketones in the presence of alcohols or formate.

The use of alternative stoichiometric reducing agents including boranes and silanes in asymmetric reduction has been explored. The use of enantiopure oxazoborolidines in the presence of boranes is a powerful method for the reduction of ketones exhibiting high selectivity and wide scope, and high ees are also obtained using silanes in the presence of rhodium, iridium, titanium and copper catalysts.

This chapter also includes a discussion of the catalytic asymmetric reduction of the C=N bond of imines. This transformation has been achieved with high ee using both metal-based catalysts and also organic Brønsted acids.

3.1 Hydrogenation of Ketones

The reduction of ketones into enantiomerically enriched alcohols using hydrogen as the stoichiometric reductant is an appealing transformation. The reaction is atom economical, with no by-products, and has been achieved with very low catalyst loadings.

Ruthenium and rhodium complexes have maintained the best track record as ketone hydrogenation catalysts.[1] For high enantioselectivities, additional chelating functional groups are often needed. Typical substrates are represented by ketones (3.01) to (3.03), where esters, halides[2,3] and phosphonates[4,5] provide an additional donor group on the substrate. BINAP/ruthenium combinations, such as complex (3.07), have given consistently high enantioselectivities with such

Catalysis in Asymmetric Synthesis 2e   © 2009 Vittorio Caprio and Jonathan M.J. Williams

ketones. Variation of the counterions and method of catalyst construction can affect the enantioselectivity observed.

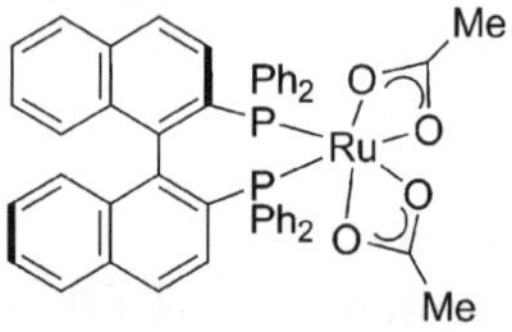

(3.07) [(*R*)-BINAP]Ru(OAc)$_2$

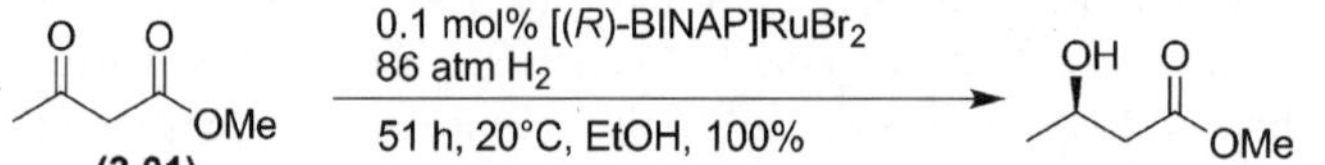

Not surprisingly, BINAP is not the only ligand to provide high selectivities in hydrogenation reactions. The alternative ligands/catalysts (**3.08**–**3.17**) also work well in related reactions.[6–15] The PhanePhos-based complex (**3.08**) and complex (**3.09**) are active in the hydrogenation of β-ketoesters such as (**3.01**)[6,7] while the BICHEP-Ru complex (**3.10**) effects the asymmetric hydrogenation of phenylgly-oxalates such as (**3.18**).[8] Other examples include the reduction of the β-keto sulfide (**3.19**) using SkewphosRuBr$_2$ (**3.11**),[9] and the trifluoroacetate (**3.20**) using the rhodium complex derived from the amidophosphine-phosphinite (**3.12**).[10] Amidophosphine-phosphinite ligands[11] are also effective for asymmetric hydro-genation of α–tertiary amino ketones, with particular success being achieved with the ligand (**3.13**).[12] Thus, ketone (**3.21**) is reduced with good enantioselectivity into the amino alcohol (**3.22**). Secondary amines also function as effective chelating groups in this reaction and monoprotected β–amino alcohols may be synthesised with high ee by reduction of β–secondary amino ketones such as (**3.23**) using the rhodium complex with (*S*,*R*)-Duanphos (**3.14**) as ligand.[13]

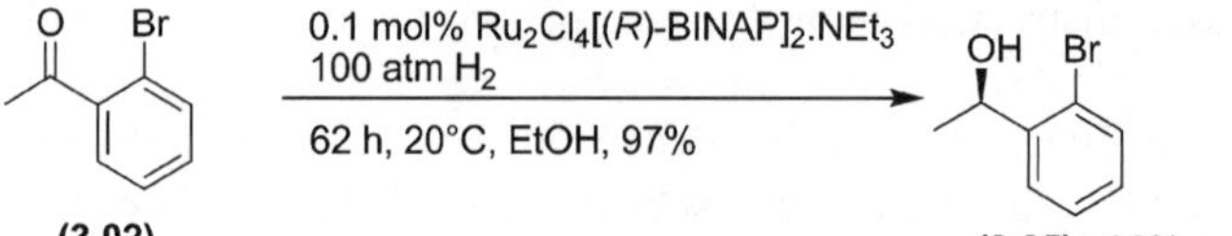

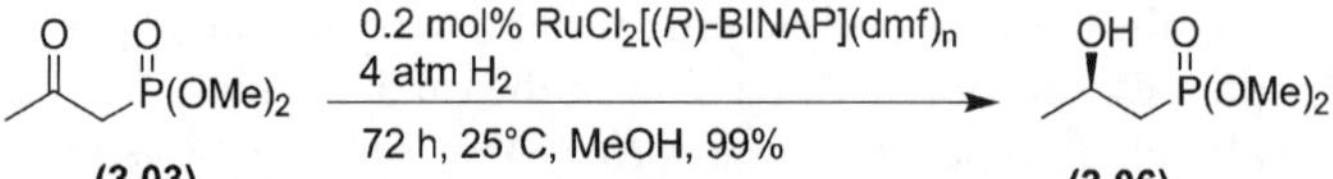

The hydrogenation of unfunctionalised ketones has, in the past, proved challeng-ing. However, Noyori and coworkers have discovered that the activity of ruthenium-based ketone hydrogenation catalysts can be greatly improved on incorporation of

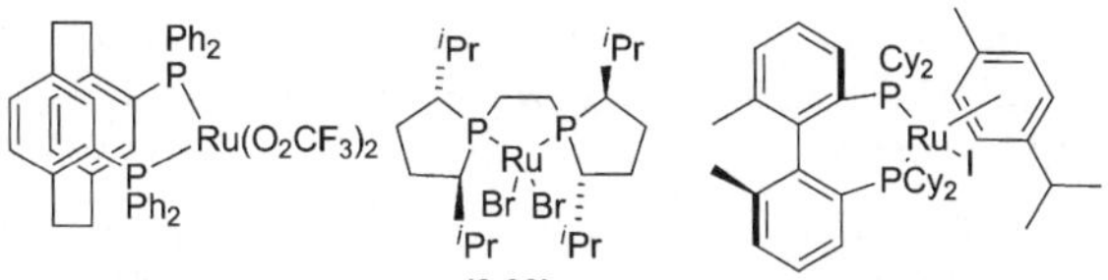

(3.08)　　(3.09)　　(3.10)

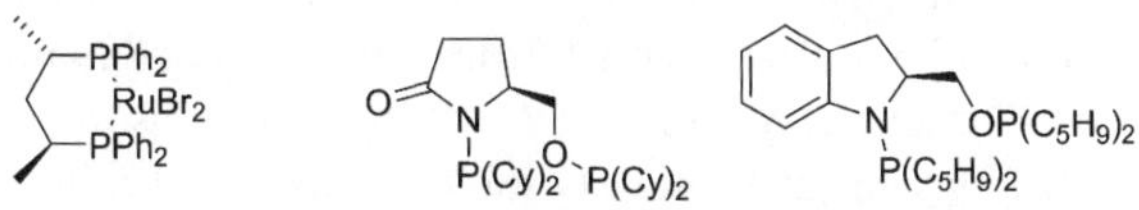

(3.11)　　(3.12)　　(3.13)

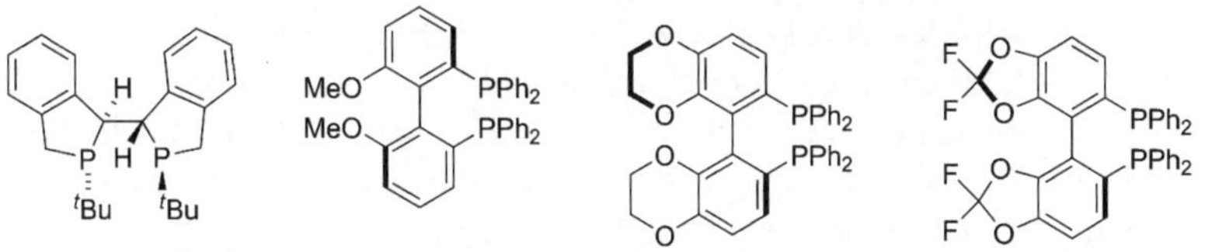

(S,R)-DUANPHOS (3.14)　(R)-MeO-BIPHEP (3.15)　(S)-SYNPHOS (3.16)　(S)-DIFLUOROPHOS (3.17)

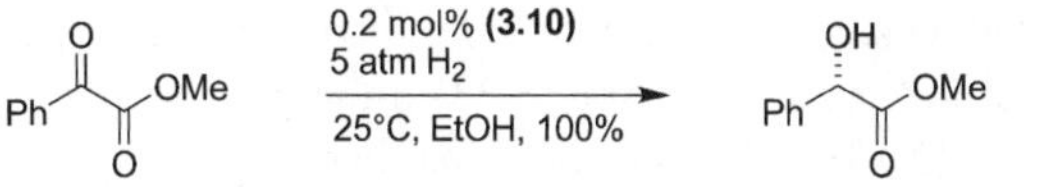

$$\begin{array}{c}\text{(3.18)}\end{array} \xrightarrow[\text{25°C, EtOH, 100\%}]{\substack{0.2 \text{ mol\% (3.10)}\\5 \text{ atm H}_2}} \text{(3.24) >99\% ee}$$

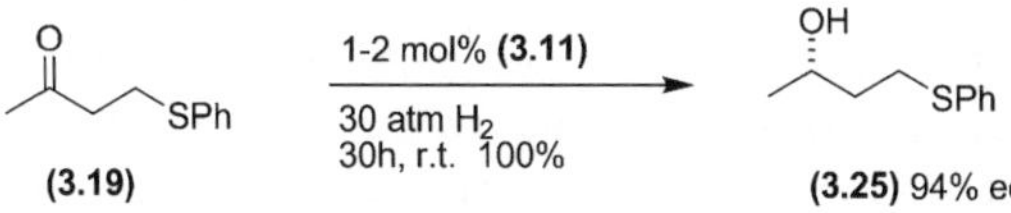

$$\text{(3.19)} \xrightarrow[\substack{30 \text{ atm H}_2\\30\text{h, r.t. }100\%}]{1\text{-}2 \text{ mol\% (3.11)}} \text{(3.25) 94\% ee}$$

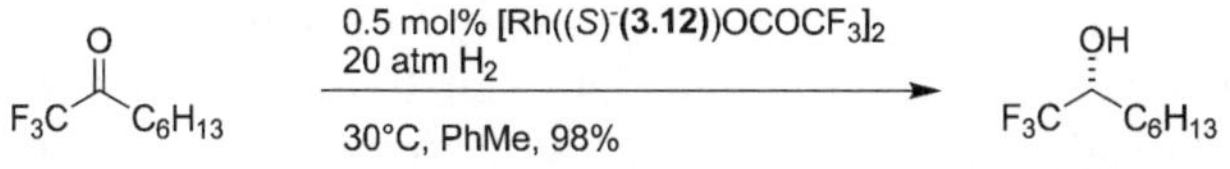

$$\text{(3.20)} \xrightarrow[\text{30°C, PhMe, 98\%}]{\substack{0.5 \text{ mol\% [Rh((S)(3.12))OCOCF}_3]_2\\20 \text{ atm H}_2}} \text{(3.26) 97\% ee}$$

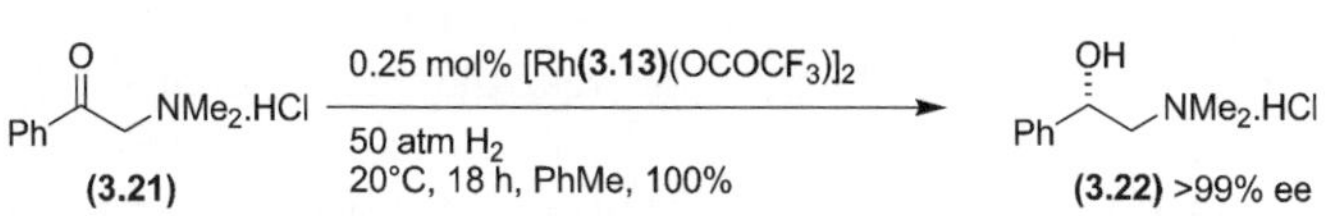

$$\text{(3.21)} \xrightarrow[\substack{50 \text{ atm H}_2\\20\text{°C, 18 h, PhMe, 100\%}}]{0.25 \text{ mol\% [Rh(3.13)(OCOCF}_3)]_2} \text{(3.22) >99\% ee}$$

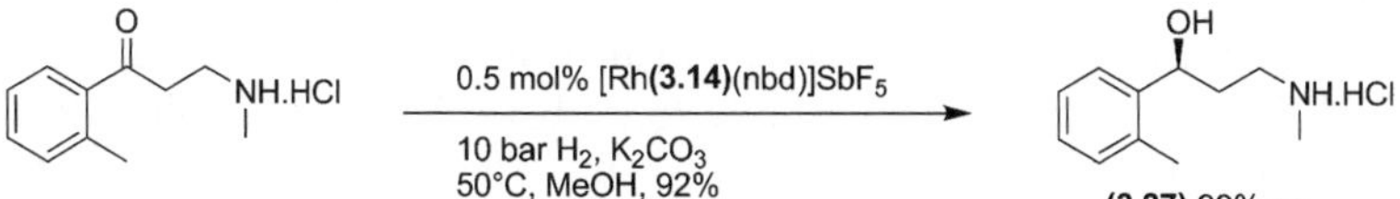

$$\text{(3.23)} \xrightarrow[\substack{10 \text{ bar H}_2, \text{ K}_2\text{CO}_3\\50\text{°C, MeOH, 92\%}}]{0.5 \text{ mol\% [Rh(3.14)(nbd)]SbF}_5} \text{(3.27) 99\% ee}$$

protic ligands. In particular, a variety of simple, alkyl, aryl and even olefinic ketones undergo rapid and enantioselective hydrogenation at the carbonyl group using ruthenium(xylBINAP)(1,2-diamine) catalysts[16] such as (3.28)[17] in isopropanol in the presence of catalytic amounts of base. The xylBINAP ligand and diamine unit have been replaced with a variety of chiral, nonracemic and achiral phosphines and diamines. The 1,1-dianisyl-2-isopropyl-1,2-ethylenediamine (DAIPEN) ligand in complex (3.28) can be replaced with 1,2-diphenylethylenediamine (DPEN),[17] 2-dimethylamino-1-phenylethylamine (DMAPEN)[18] and also bipyridines and α–picolylamine (PICA), to give catalysts such as (3.29).[19] While xylBINAP is the most commonly employed phosphine, this moiety has been exchanged with achiral bis(diphenylphosphino)benzophenone (DBPP) to give catalyst (3.30), which exhibits very high enantioselectivity in the hydrogenation of aryl ketones.[20] The scope of this reduction has been expanded even further by the development of catalyst (3.31) incorporating a η^1-BH_4 ligand that can be used under base-free conditions.[21] These complexes are highly active at very low catalyst loadings. For instance, 0.001 mol% of catalyst (3.28) effects reduction of acetophenone (3.32) in one hour to give (R)-1-phenylethanol with 99% ee.[17] These complex and related species are also highly active catalysts in the reduction of cyclopropyl ketones. Even sterically hindered alkyl ketones undergo asymmetric hydrogenation using ruthenium(diamine)catalysts and pinacolone (3.33) undergoes highly enantioselective hydrogenation using (3.29).[19] Furthermore, as these catalysts are highly reactive towards the carbonyl group they can be used to effect selective hydrogenation of olefinic ketones.[16,22] For instance (R)-carvone (3.34) is converted to the cis-alcohol (3.35) using a ruthenium((S)-BINAP)((S,S)-DPEN) system.[22]

It is thought that the high reactivity and chemoselectivity of these Ru(diamine) catalysts arises from activation of the ketone by ligand/substrate hydrogen bonding with a proton of the amine ligand, and hydrogenation by these complexes proceeds via an entirely different mechanism to reduction using traditional ruthenium complexes. Conventional hydrogenation occurs by formation of a metal–substrate complex, formed by interaction of ruthenium with the carbonyl oxygen and neighbouring functionality, followed by hydride transfer from ruthenium to the carbonyl group. In contrast, asymmetric reduction using Ru(diamine) complexes is thought to proceed by a pericyclic process involving simultaneous transfer of a hydride and a proton from the outer coordination sphere of the catalyst to the substrate and no metal–substrate complexation occurs.[16,23]

α-Substituted β-keto esters are particularly interesting substrates because the stereochemistry at the α-position is labile due to the low pKa of such substrates. However, after reduction of the ketone to an alcohol group, the stereochemistry is fixed. This provides the opportunity for a dynamic kinetic resolution, which has been achieved with remarkable efficiency.[24] The two enantiomers of starting material (3.38) and (3.39) are in equilibrium under the reaction conditions. The ruthenium catalyst reacts selectively with the (R)-enantiomer (3.38), and provides the β-keto ester (3.40) highly selectively, as shown in Figure 3.1. The major product

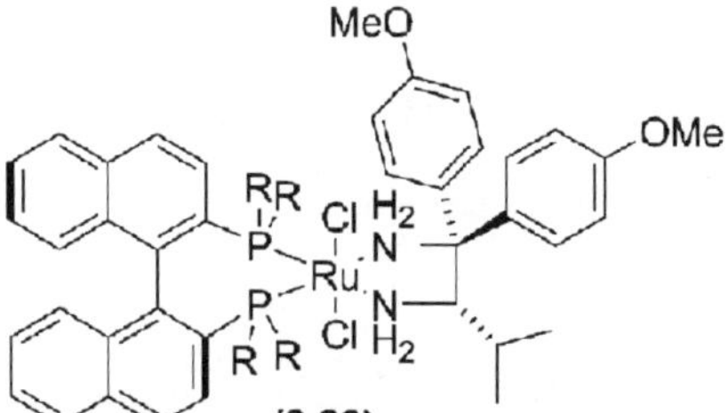

(3.28)

R = 3,5‑(CH$_3$)$_2$C$_6$H$_3$

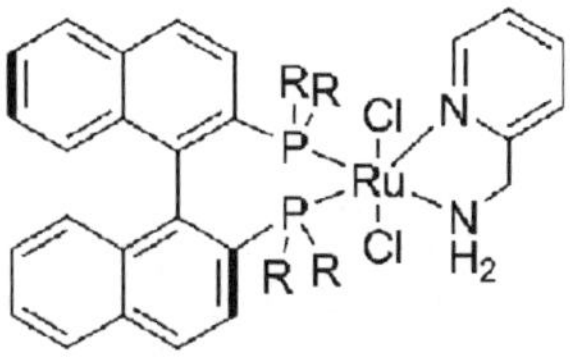

(3.29)

R = 4-CH$_3$C$_6$H$_4$

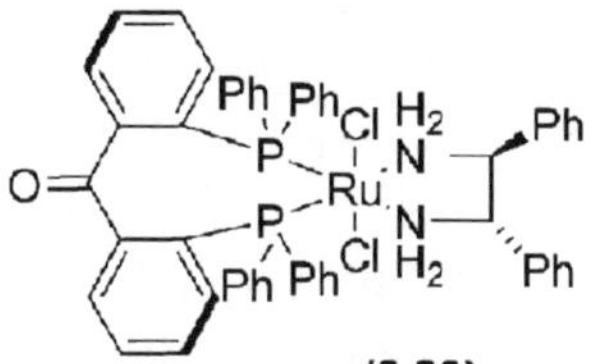

(3.30)

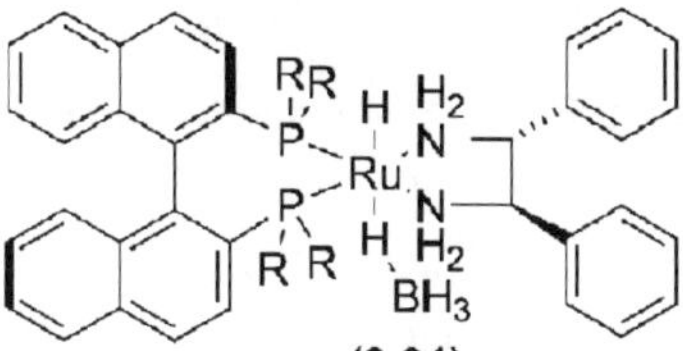

(3.31)

R = 4-CH$_3$C$_6$H$_4$

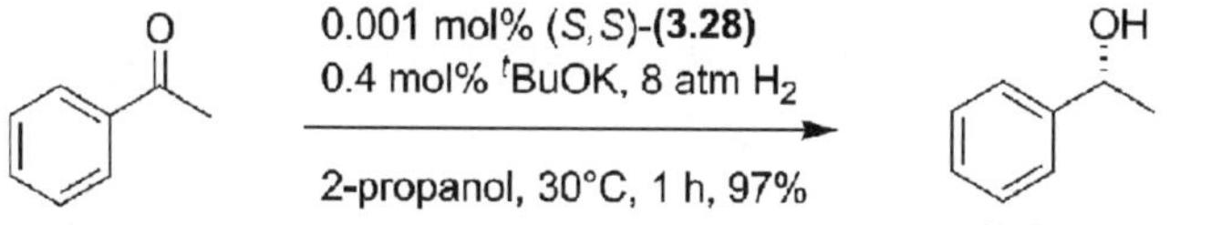

(3.32)

0.001 mol% (*S*,*S*)-**(3.28)**
0.4 mol% *t*BuOK, 8 atm H$_2$

2-propanol, 30°C, 1 h, 97%

(3.36) 99% ee

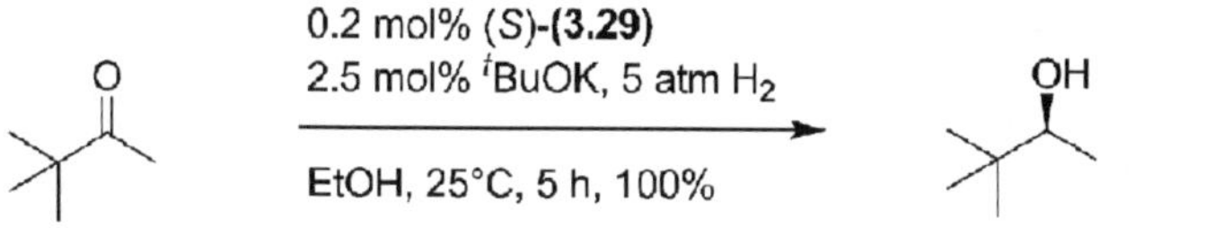

(3.33)

0.2 mol% (*S*)-**(3.29)**
2.5 mol% *t*BuOK, 5 atm H$_2$

EtOH, 25°C, 5 h, 100%

(3.37) 97% ee

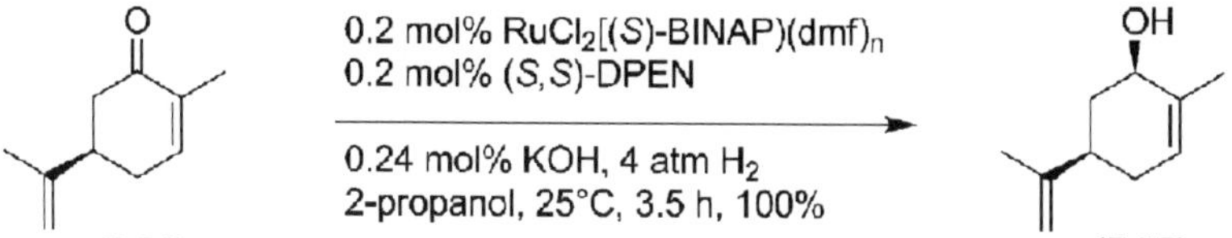

(3.34)

0.2 mol% RuCl$_2$[(*S*)-BINAP)(dmf)$_n$
0.2 mol% (*S*,*S*)-DPEN

0.24 mol% KOH, 4 atm H$_2$
2-propanol, 25°C, 3.5 h, 100%

(3.35)
(*cis*:*trans* - 100:0)

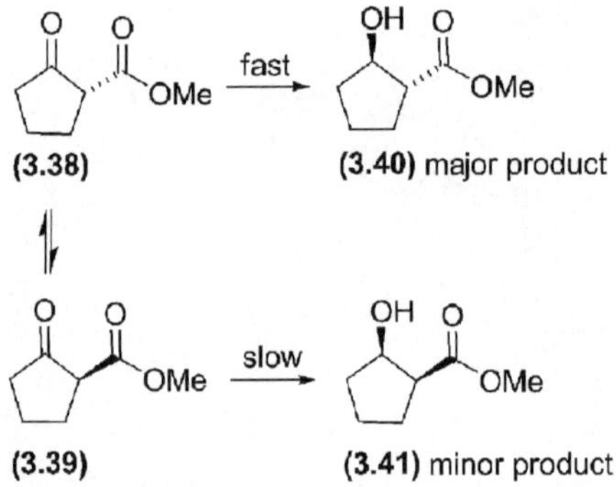

Figure 3.1 Dynamic kinetic resolution of β-keto esters

has been formed with 98% de and 92% ee using $\{RuCl[(R)\text{-BINAP}](C_6H_6)\}Cl$ in dichloromethane. In methanol the enantioselectivity is even higher (99% ee), but the diastereoselectivity is lower (92% de).[25]

The dynamic resolution strategy has been applied to related systems with success, for example the amide (3.42),[26] α-chloro-substituted compound (3.43),[27] and the α-amido β-ketophosphonate (3.44).[28] In each case a similar principle operates, where the starting material is rapidly racemising, and the catalyst selects only one enantiomer in the reduction process.

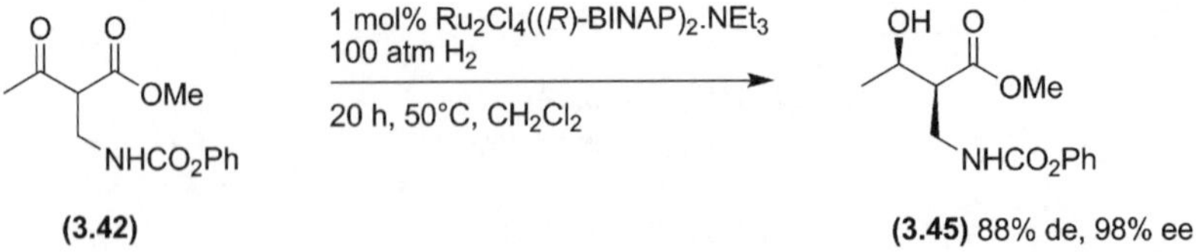

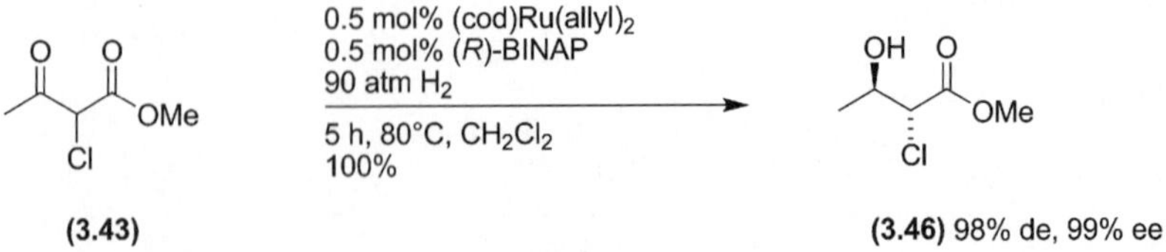

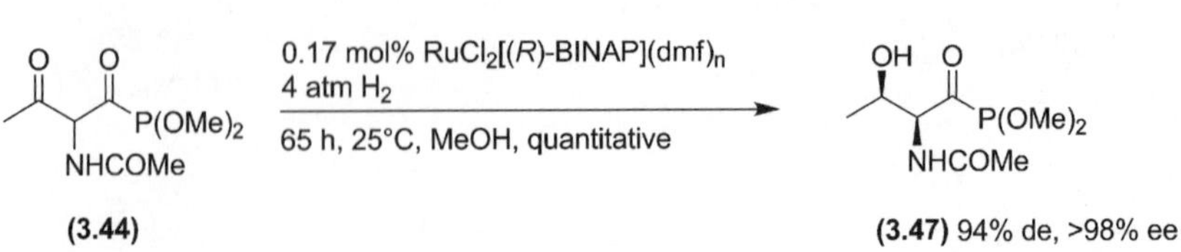

3.2 Hydrogenation and Transfer Hydrogenation of Imines and Related Compounds

The hydrogenation of imines and their derivatives provides a convenient route to enantiomerically enriched amines.[1c, 29] The reduction of the imino group has proved challenging and while a range of Rh and Ru catalysts has been used to great success in the hydrogenation of alkenes and ketones, only a relatively few of these systems has shown any promise in the asymmetric hydrogenation of C=N bonds. Best results have been obtained with substrates bearing additional chelating functionality on nitrogen.

Burk and Feaster used the rhodium complexes of the DUPHOS ligand (3.48) as catalysts to reduce hydrazones.[30] The levels of enantioselectivity were high for the reduction of *N*-acylhydrazones such as (3.49). The asymmetric transfer hydrogenation of imines has been achieved using rhodium-based catalysts. While no additional chelating functionality is required, high enantioselectivities have only been obtained using heterocyclic imines and iminium species as substrates. Rapid transfer hydrogenation of dihydroisoquinolines occurs in up to 99% ee using formic acid-triethylamine azeotrope as the hydrogen source in the presence of a rhodiumcyclopentadienyl complex bearing an enantiomerically pure diamine

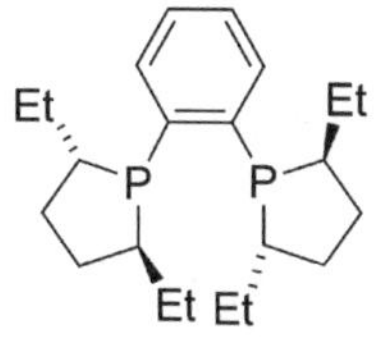

(3.48)

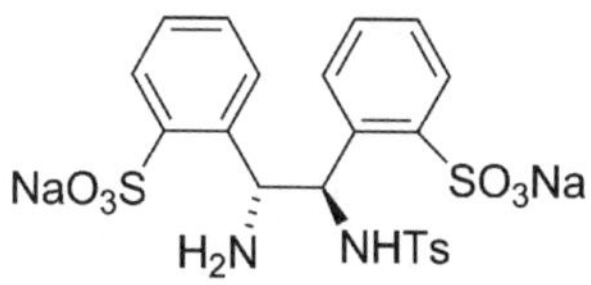

(3.51)

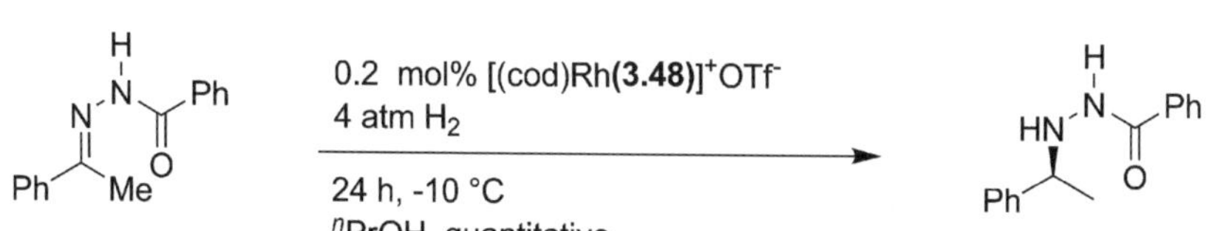

(3.49)

0.2 mol% [(cod)Rh(3.48)]⁺OTf⁻
4 atm H_2

24 h, -10 °C
^{n}PrOH, quantitative

(3.52) 95% ee

(3.50)

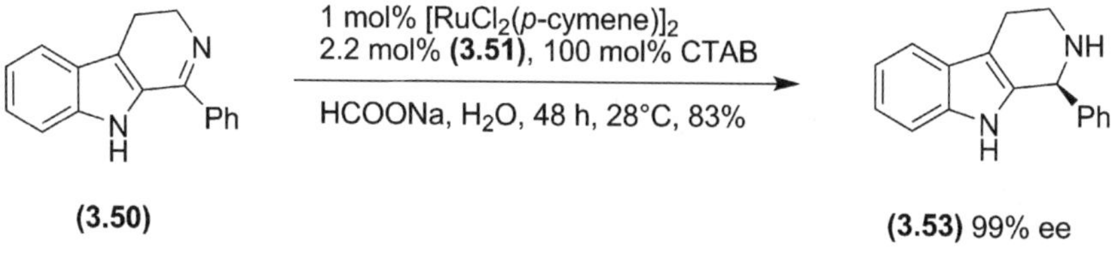

1 mol% [RuCl₂(p-cymene)]₂
2.2 mol% (3.51), 100 mol% CTAB

HCOONa, H_2O, 48 h, 28°C, 83%

(3.53) 99% ee

ligand in dichloromethane.[31] The transfer hydrogenation is of greater scope when performed in water in the presence of sodium formate, surfactants such as cetyltriethylammonium bromide (CTAB) and a water-soluble ruthenium complex.[32] In this procedure a range of dehydroisoquinolines, cyclic *N*-sulfonylimines and β–carbolines such as **(3.50)** are reduced to give products with high ee using the catalyst formed *in situ* from $[RuCl_2(p\text{-cymene})]_2$ and diamine **(3.51)**.

Group 4 metallocene complexes can also be used as catalysts in the reduction of C=N bonds. Willoughby and Buchwald employed the titanium-based Brintzinger catalyst **(3.54)** for the asymmetric reduction of imines.[33] The catalyst is activated by reduction to what is assumed to be the titanium(III) hydride species **(3.55)**. The best substrates for this catalyst are cyclic imines, which afford products with 95–98% ee. Various functional groups including alkenes, vinyl silanes, acetals and alcohols were not affected under the reaction conditions.[34] For example, the imine **(3.56)** was reduced with excellent enantioselectivity, without reduction of the alkene moiety.

Using the racemic substrate **(3.57)**, a highly selective kinetic resolution was achieved, affording the reduced product **(3.58)** and recovered starting material, both with very high enantioselectivity.[35] Mechanistic studies and further details of the scope of the reaction have also been published.[36,37]

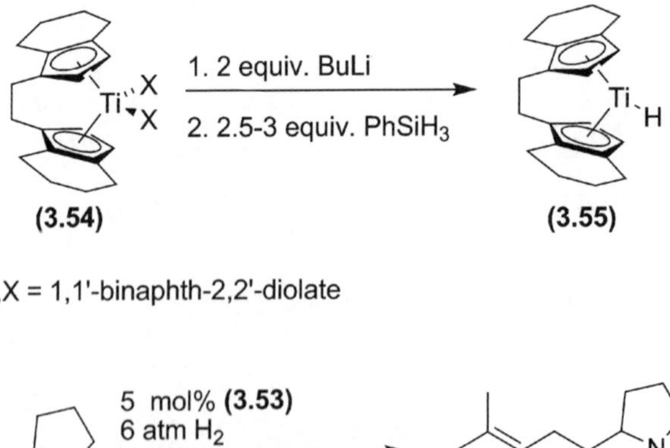
Ti
X
X
1. 2 equiv. BuLi
2. 2.5-3 equiv. PhSiH3
Ti
H

(3.54) **(3.55)**

X,X = 1,1'-binaphth-2,2'-diolate

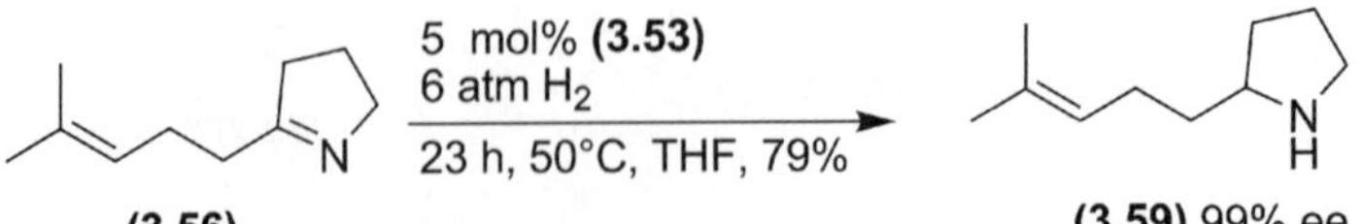
N
5 mol% (3.53)
6 atm H2
23 h, 50°C, THF, 79%
N
H

(3.56) **(3.59)** 99% ee

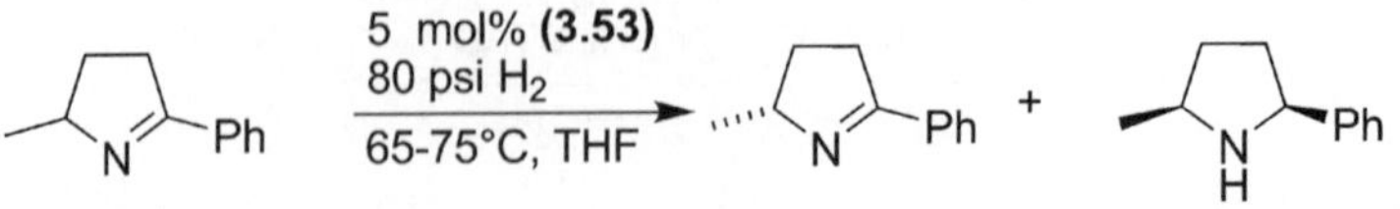
N
Ph
5 mol% (3.53)
80 psi H2
65-75°C, THF
N
Ph
+
N
H
Ph

(3.57) **(*R*)-(3.57)** 37% isolated 99% ee **(3.58)** 34% isolated 99% ee

The most successful imine hydrogenation catalysts are iridium-based and these have been used with great success in the hydrogenation of both cyclic and acyclic *N*-aryl imines. Morimoto has employed an iridium/BINAP complex in the hydrogenation of the dihydroisoquinoline **(3.60)**.[38] Research in this area has focussed on the use of catalysts bearing phosphineoxazoline (PHOX)-type ligands discussed in more detail in Chapter 2 (Section 2.4). The most successful ligands of this type include ferrocinyl-binaphane ligand **(3.62)**,[39] the diphenylphosphanylsulfoximine **(3.63)**[40] and spirobiindane (SIPHOX) ligands such as **(3.64)**.[41] All these ligands are effective in the iridium-catalysed hydrogenation of acyclic *N*-arylimines. Some of the highest ees to date have been obtained using binaphane ligand **(3.62)** and substrates such as **(3.65)**,[39] while the Ir complexes of SIPHOX such as **(3.64)** are capable of effecting hydrogenation of *N*-arylimines in up to 97% ee under only ambient pressure of hydrogen gas.[41]

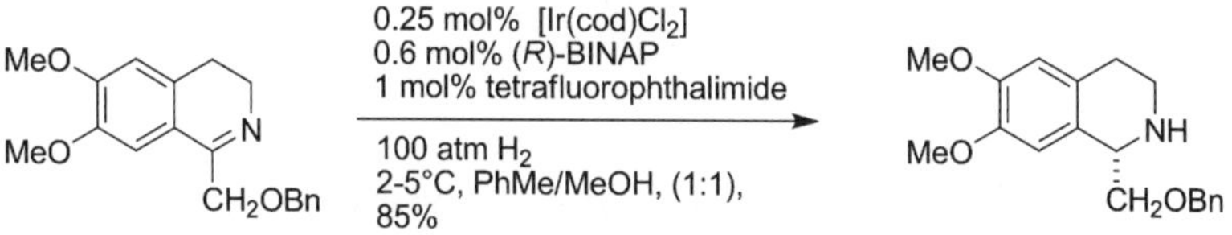
MeO
MeO
N
CH2OBn
0.25 mol% [Ir(cod)Cl2]
0.6 mol% (R)-BINAP
1 mol% tetrafluorophthalimide
100 atm H2
2-5°C, PhMe/MeOH, (1:1),
85%
MeO
MeO
NH
CH2OBn

(3.60) **(3.61)** 86% ee

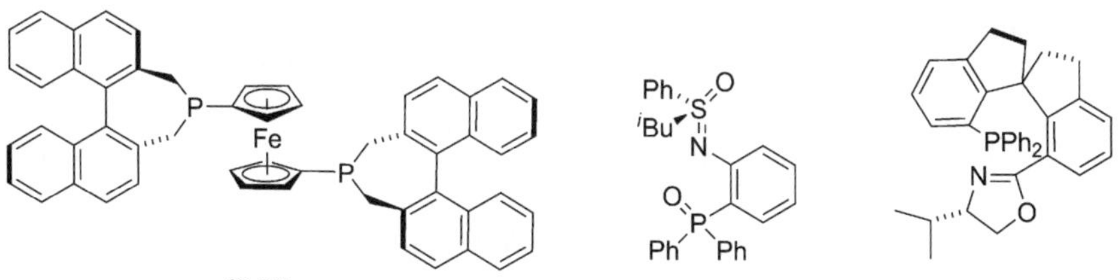
P
Fe
P
Ph
S
O
iBu
N
O
P
Ph
Ph
PPh2
N
O

(3.62) **(3.63)** **(3.64)**

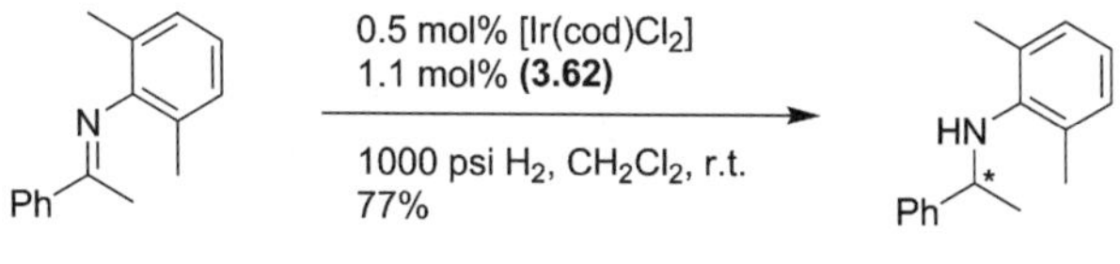
N
Ph
0.5 mol% [Ir(cod)Cl2]
1.1 mol% (3.62)
1000 psi H2, CH2Cl2, r.t.
77%
HN
*
Ph

(3.65)

(3.66)
absolute configuration
not determined

The asymmetric reduction of imines and iminium species can be achieved using organocatalysts. The transfer hydrogenation of imines is catalysed by acids and this has led to the development of biomimetic asymmetric reductions using enantioselective Brønsted acids[42] in combination with Hantzsch esters as a hydride source. In this process protonation of the imine forms an ion pair that undergoes face-selective reduction to give an enantiomerically enriched amine. Most success has been achieved using BINOL-derived phosphoric acid catalysts such as (3.67) and the highly enantioselective organocatalytic asymmetric transfer hydrogenation of aryl and alkyl ketimines such as (3.68) and (3.69) has been achieved by the groups of Rueping[43] and List[44] using this catalyst in combination with Hantschz ester (3.70). This methodology has been applied to the enantioselective reduction of the C=N bond of 2-aryl and alkylquinolines such as (3.73) giving access to enantioenriched tetrahydroquinoline alkaloids such as (+)-cuspareine (3.74).[45] MacMillan and coworkers have shown that this Brønsted-acid-catalysed transfer hydrogenation can also be applied to the reduction of iminium ions formed *in situ* during the reductive amination of alkyl and aryl ketones with aryl or heteroarylamines.[46] Thus, relatively complex enantioenriched N-heterocyclic amines such as (3.73) can be accessed in one pot by coupling of the requisite ketone and amine in the presence of triphenylsilyl-substituted phosphoric acid (3.79) and Hantzsch ester (3.70).

3.3 Transfer Hydrogenation of Ketones

The asymmetric transfer hydrogenation of ketones represents an alternative to conventional methods using H_2 gas.[1b,1c, 47] The hydride source is usually a donor alcohol such as isopropanol or, alternately, formate. The reaction can proceed with low conversion and ee when using alcohols as hydrogen donors owing to the reversibility of the process and the unfavourability of the ketone:alcohol equilibrium. Transfer hydrogenation utilising 2-propanol generally only proceeds in the presence of a relatively strong base and large excess of alcohol. Better selectivities and yields have been observed using sodium formate/triethylamine where the process is essentially irreversible. Iridium,[48] rhodium[49] and samarium[50] have all provided highly enantioselective catalysts for asymmetric transfer hydrogenation. However, ruthenium catalysts have enjoyed more attention.

Noyori and coworkers reported that monotosylated ruthenium(II) complex (3.80), gives very high enantioselectivity for many aryl/alkyl ketones, including various acetophenones (3.81) and cyclic ketones (3.82) and (3.83).[51] The same catalyst has been used in the reduction of α,β-acetylenic ketones such as (3.84).[52] Wills and coworkers have shown that much higher rates of transfer hydrogenation of aryl/alkyl ketones can be achieved using the tethered ruthenium(II) catalyst (3.85).[53] This catalyst effects the 100% conversion of acetophenone into the (S)-alcohol with 96% ee in three hours at a catalyst loading of 0.5 mol%. Catalyst loadings of down to 0.01% can be used with no attenuation of enantioselectivity, but reaction times are then increased to 80 hours. Tethered rhodium(III) catalysts such as (3.86) have also been developed by the group of Wills.[54] These catalysts are

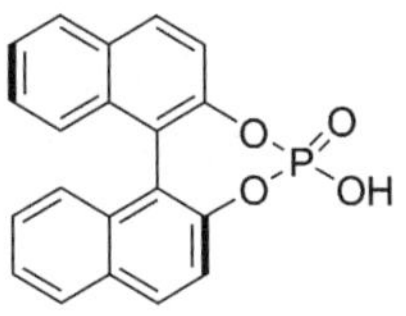

(3.67)

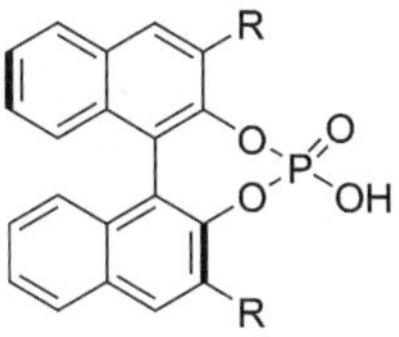

(3.75) R = 9-Phen

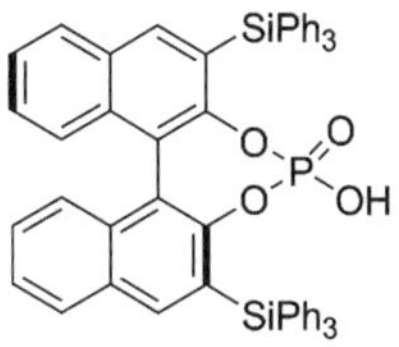

(3.79)

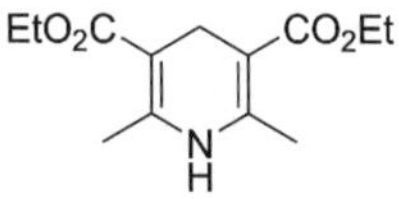

(3.70)

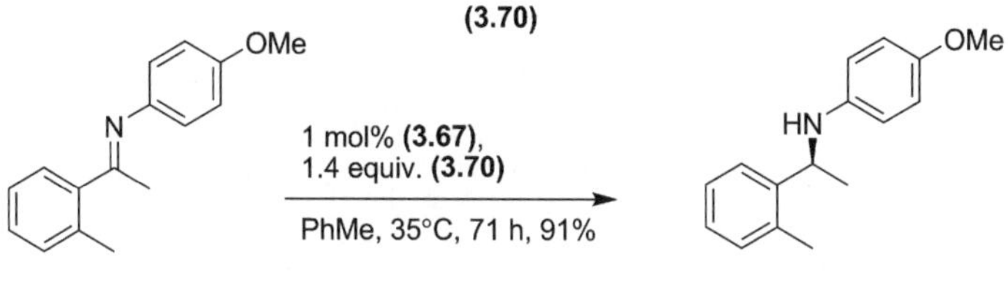

(3.68) → **(3.71)** 93% ee

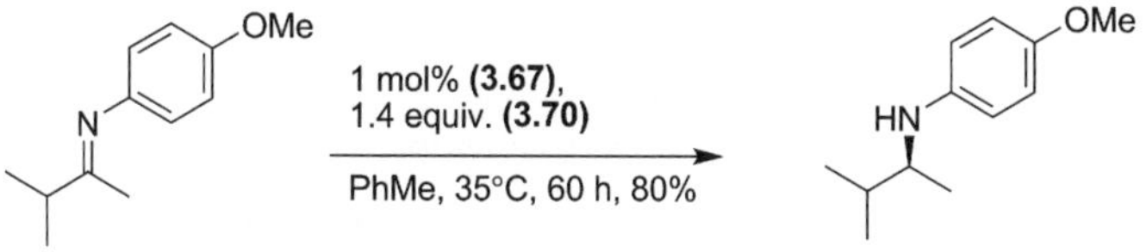

1 mol% **(3.67)**,
1.4 equiv. **(3.70)**

PhMe, 35°C, 71 h, 91%

(3.69) → **(3.72)** 90% ee

1 mol% **(3.67)**,
1.4 equiv. **(3.70)**

PhMe, 35°C, 60 h, 80%

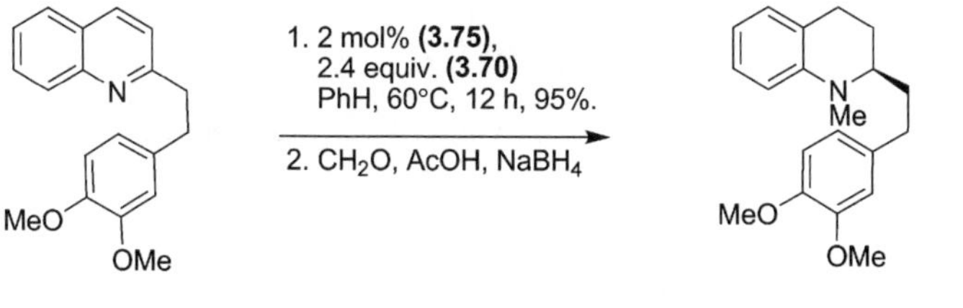

(3.73) → (+)-cuspareine **(3.74)** 90% ee

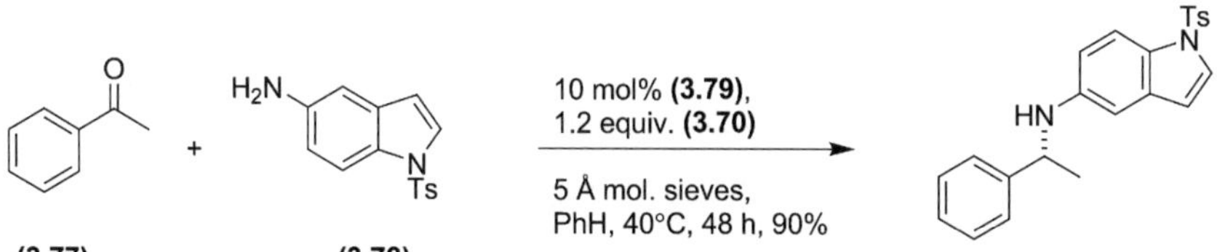

1. 2 mol% **(3.75)**,
   2.4 equiv. **(3.70)**
   PhH, 60°C, 12 h, 95%.
2. CH$_2$O, AcOH, NaBH$_4$

(3.77) + **(3.78)** → **(3.76)** 93% ee

10 mol% **(3.79)**,
1.2 equiv. **(3.70)**

5 Å mol. sieves,
PhH, 40°C, 48 h, 90%

effective for the reduction of a range of aryl/alkyl ketones and notably are capable of effecting the quantitative reduction of α-haloketones such as (3.87) that are reduced in poor yield with untethered diamine ligand (3.80). Furthermore catalyst (3.86) is highly effective in aqueous media in the presence of sodium formate.[54d]

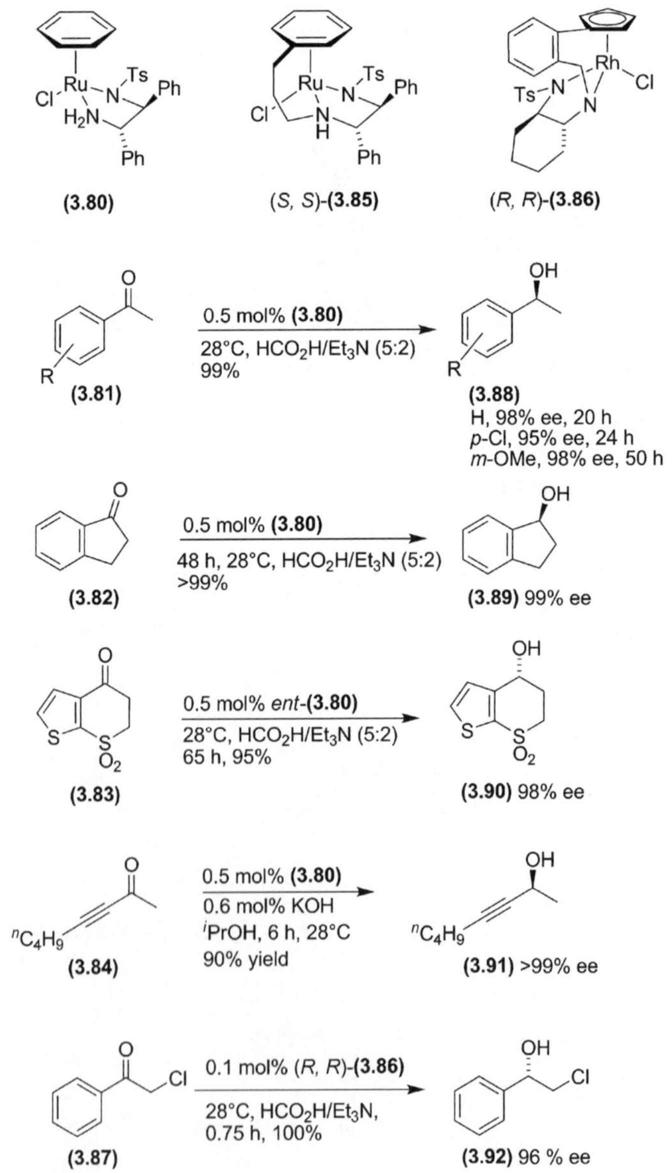

(3.80)   (S, S)-(3.85)   (R, R)-(3.86)

0.5 mol% (3.80)

28°C, HCO₂H/Et₃N (5:2)
99%

(3.81)

(3.88)
H, 98% ee, 20 h
p-Cl, 95% ee, 24 h
m-OMe, 98% ee, 50 h

0.5 mol% (3.80)

48 h, 28°C, HCO₂H/Et₃N (5:2)
>99%

(3.82)   (3.89) 99% ee

0.5 mol% ent-(3.80)

28°C, HCO₂H/Et₃N (5:2)
65 h, 95%

(3.83)   (3.90) 98% ee

0.5 mol% (3.80)
0.6 mol% KOH
ⁱPrOH, 6 h, 28°C
90% yield

(3.84)   (3.91) >99% ee

0.1 mol% (R, R)-(3.86)

28°C, HCO₂H/Et₃N,
0.75 h, 100%

(3.87)   (3.92) 96 % ee

The ruthenium catalysts (**3.80**),[55] (**3.85**) and (**3.86**), whilst particularly efficient for asymmetric transfer hydrogenations, are not the only ruthenium catalysts which display high enantioselectivity. The reduction of acetophenone into enantiomerically enriched phenethyl alcohol has also been achieved using the ruthenium complexes of the tetradentate ligand (**3.93**),[56] the alternative amino-sulfonamide ligand (**3.94**),[57] indanol (**3.95**),[58] oxazolines (**3.96**)[59] and (**3.97**),[60] amino alcohol (**3.98**)[61] and even simple, readily prepared dipeptide analogues such as (**3.99**).[62] Enantiomerically enriched PNNP ligand (**3.93**) has also been used in combination with the iridium hydride IrH(CO)(PPh$_3$)$_3$ to prepare an iridium hydride-based complex that is effective in the asymmetric transfer hydrogenation using isopropanol in the absence of base, and hence has the potential to effect reduction of base-sensitive ketones.[56c] Catalysts incorporating P,N- and N,N-ligands are generally restricted in scope to the transfer hydrogenation of aryl/alkyl ketones and often perform poorly using alkyl/alkyl ketones as substrates. Recently, Reetz and Li have shown that the ruthenium-based catalyst system utilising the all phosphorus BINOL-based phosphonite (**3.100**) displays much greater scope and can be used to effect highly enantioselective transfer hydrogenation of both aryl/alkyl ketones such as acetophenone and alkyl/alkyl ketones such as octanone (**3.101**) and even pinacolone (**3.102**) in the presence of isopropanol.[63]

3.4 Heterogeneous Hydrogenation

Problems associated with the high cost of large-scale catalytic asymmetric transformations and purification of the resulting products can, in principle, be overcome by the use of heterogeneous catalysts. The attachment of enantiomerically pure ligands to a solid support may aid in the recycling of expensive metal catalysts and also prevent leaching of metals.[64] The most common supports are polymers, metal surfaces and other inorganic solids. The majority of recent research into the development of polymer-based hydrogenation catalysts has focussed on the development of immobilised forms of Noyori's ruthenium(1,2-diamine) catalysts. An asymmetric heterogeneous transfer hydrogenation in water can be achieved using polymer-supported enantiomerically pure sulfonamides, prepared from quaternary ammonium salts of *p*-styrenesulfonyl chloride and catalytic amounts of *N*-tosyl-1,2-diphenylethylenediamine (TsDPEN), in the presence of [RuCl$_2$(*p*-cymene)]$_2$.[65] The supported catalyst shows similar activity to ruthenium complex (**3.80**) in the transfer hydrogenation of aryl/alkyl ketones, but vigorous stirring of this insoluble polymer is required to achieve high conversions. The catalyst is easily separated and may be recycled a number of times. Insoluble polymer–metal complexes, while easily recycled, often display poor activity owing to partial inaccessibility of the active sites. This problem has been overcome by the development

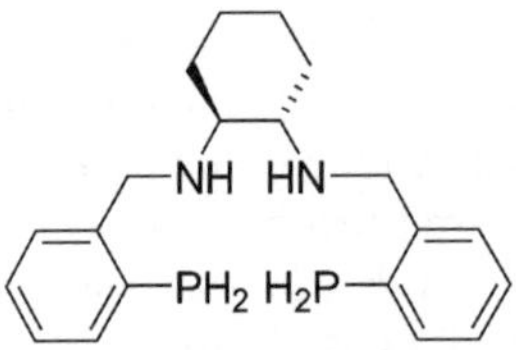

(3.93) up to 97% ee (*R*) 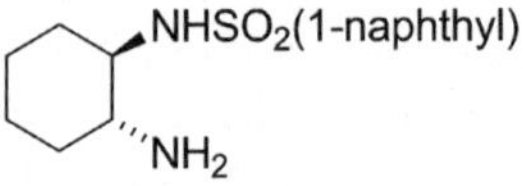   **(3.94)** up to 96% ee (*R*) 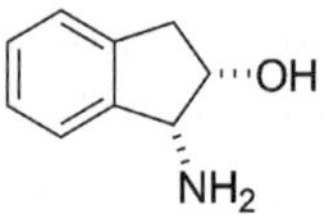   **(3.95)** up to 91% ee (*S*)

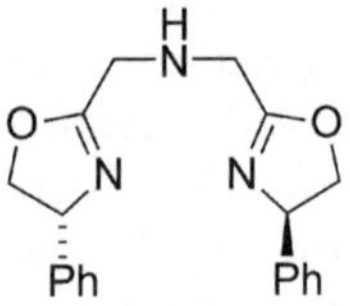

(3.96) up to 94% ee (*R*) 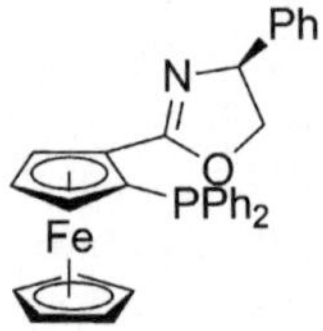   **(3.97)** up to 98% ee (*S*) 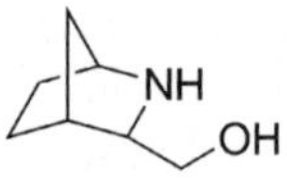   **(3.98)** up to 95% ee (*S*)

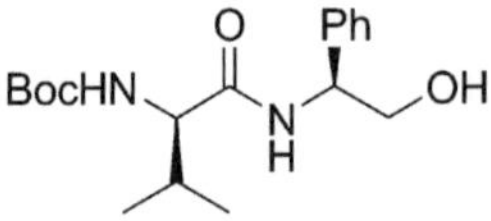

(3.99) up to 95% ee (*R*)

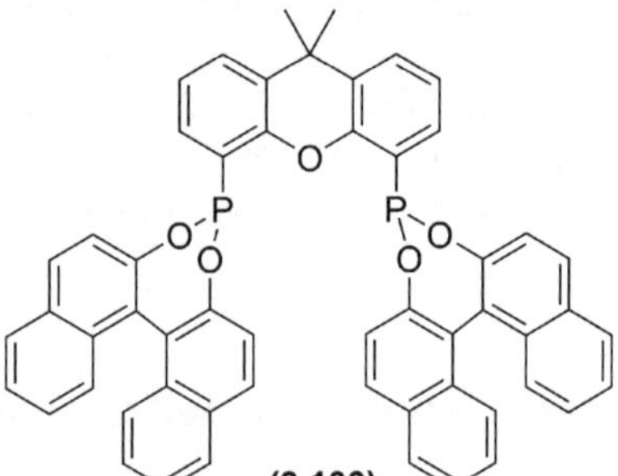

(3.100)

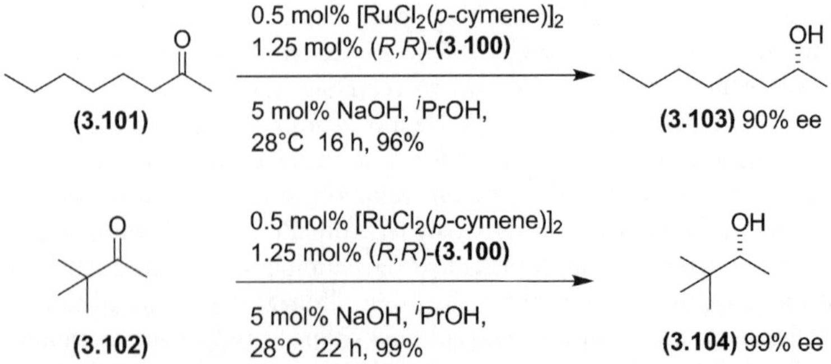

(3.101)

0.5 mol% [RuCl$_2$(*p*-cymene)]$_2$
1.25 mol% (*R,R*)-**(3.100)**

5 mol% NaOH, ^{i}PrOH,
28°C 16 h, 96%

(3.103) 90% ee

(3.102)

0.5 mol% [RuCl$_2$(*p*-cymene)]$_2$
1.25 mol% (*R,R*)-**(3.100)**

5 mol% NaOH, ^{i}PrOH,
28°C 22 h, 99%

(3.104) 99% ee

of soluble polymer supports that are insoluble in specific solvents. Polyethylene glycol-supported 1,2-diphenylethylenediamine (DPEN) is soluble in polar organic solvents and the immobilised catalyst prepared by reaction with chiral nonracemic ruthenium complexes such as that prepared from PhanePhos (**3.105**) effects quantitative and highly enantioselective hydrogenation of aryl/alkyl ketones such as acetonaphthone (**3.106**). [66] The catalyst is reisolated by precipitation on addition of diethyl ether. The activity of BINAP-based catalysts is retained on incorporation into a polymer and the supported ruthenium complex prepared from poly(BINAP), DPEN and $[RuCl_2(C_6H_6)]_2$ catalyses the highly enantioselective hydrogenation of acetophenone and acetonaphthone.[67] While this polymeric complex is soluble in a variety of organic solvents, it can be precipitated in methanol for reuse. Dendrimers can also be functionalised with enantiomerically pure ligands and metal complexes to provide recyclable asymmetric catalysts. Enantioselective transfer hydrogenation of aryl/alkyl ketones in water has been achieved using hydrophobic dendritic catalysts formed by linking Fréchet-type dendrons to chiral nonracemic 1,2-diaminocyclohexane and rhodium(III) complexes.[68] The hydrophobic catalyst is found to be dissolved in the liquid substrate and is reisolated on addition of hexane. A similar recoverable dendritic system, generated with TsDPEN as the enantiomerically pure diamine, catalyses the transfer hydrogenation of acetophenone in up to 97% ee using formic acid-triethylamine as hydride source in dichloromethane.[69]

Alternatively, the surface of a metal can be modified with an enantiomerically pure additive. For example, in 1956 palladium metal/silk fibroin was used for the hydrogenation of alkenes with moderate enantioselectivity.[70] However, most success with modified metal surfaces has been achieved in the hydrogenation of ketones. The reduction of β-keto esters with a Raney nickel/tartaric acid/sodium bromide catalyst provides good enantioselectivities.[71,72] For example, β-keto ester (3.108) affords the β-hydroxy ester (3.109). Platinum metal modified with cinchona alkaloids has been successfully used with α-keto esters. The reaction yields product with up to 90% ee in the reduction of α-keto ester (3.110).[73,74]

The use of polymer-stabilised colloidal platinum clusters (containing cinchonidine) in acetic acid provides up to 97.6% ee in the reduction of methyl pyruvate ($MeCOCO_2Me$).[75] Polyvinylpyrrolidine is used as the polymer, and the particle size is small.[76] The use of nanoparticles as a support for catalysts is an attractive concept owing to the large surface areas of these materials. However, such supported catalysts of this type are often difficult to recover. Hu and coworkers have achieved the immobilisation of ruthenium(phosphine)(1,2-diamine) catalysts onto magnetite nanoparticles and shown that these supported complexes (**3.112**) catalyse the hydrogenation of aromatic ketones such as acetonaphthone (**3.106**) with high ee and conversion.[77] The catalyst is easily removed by decantation using an external magnet and may be reused up to 14 times.

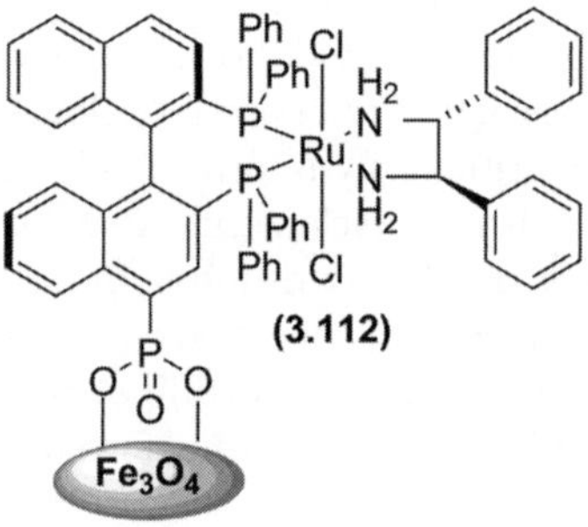
Ph
Cl
Ph
H2
P
N
Ru
P
N
Ph
H2
Ph
Cl
(3.112)
P
O
O
O
Fe3O4

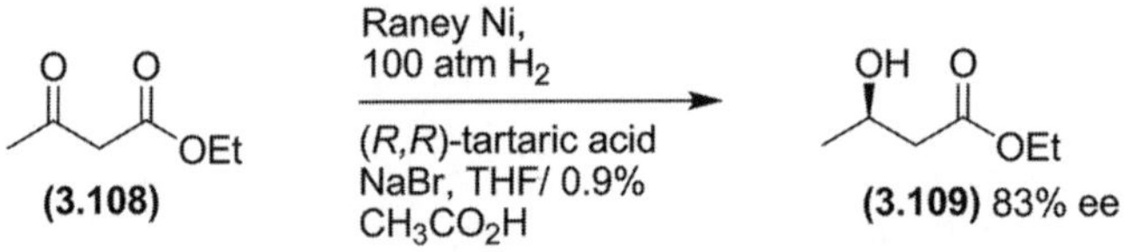
O
O
OEt
(3.108)
Raney Ni,
100 atm H2
(R,R)-tartaric acid
NaBr, THF/ 0.9%
CH3CO2H
OH
O
OEt
(3.109) 83% ee

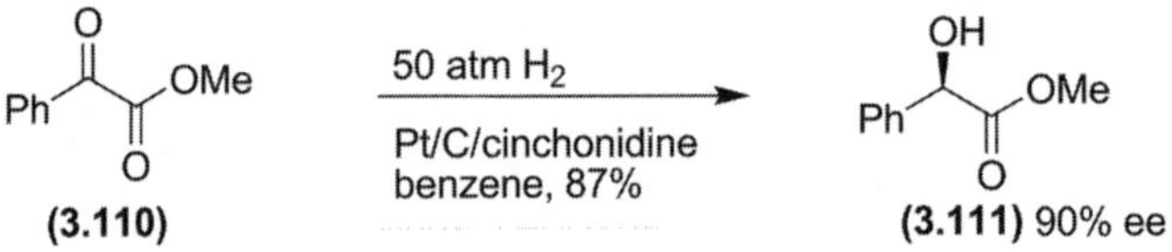
O
Ph
OMe
O
(3.110)
50 atm H2
Pt/C/cinchonidine
benzene, 87%
OH
Ph
OMe
O
(3.111) 90% ee

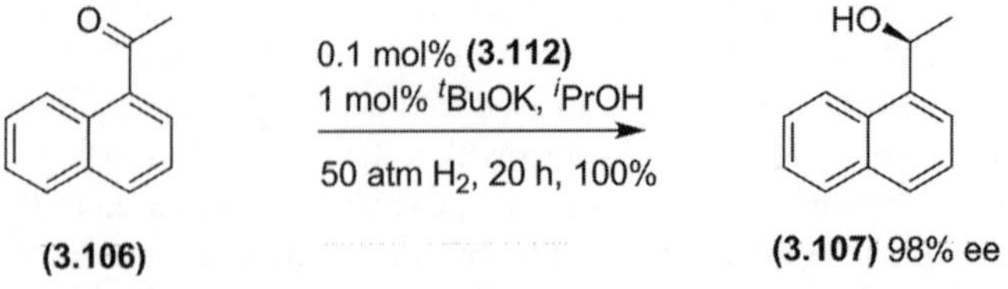
O
(3.106)
0.1 mol% (3.112)
1 mol% tBuOK, iPrOH
50 atm H2, 20 h, 100%
HO
(3.107) 98% ee

Other inorganic supports have also been used to prepare highly active heterogeneous hydrogenation catalysts. For, example TsDPEN can be immobilised on amorphous silica gel and the complex formed in the presence of $[RuCl_2(p\text{-}cymene)]_2$ effects the transfer hydrogenation of a variety of aryl/alkyl ketones with ees up to 97% in water using formic acid/triethylamine as hydride source.[78] Furthermore, the BINAP-based phosphonic acids used to synthesise immobilised catalyst (**3.112**) can also be used to prepare an inorganic–organic hybrid material by reaction with $Zr(O^tBu)_4$. The resulting porous solid catalyses the conversion of a variety of aryl/alkyl ketones into the corresponding alcohols with ees between 91 and 99%.[79]

3.5 Reduction of Ketones Using Enantioselective Borohydride Reagents

Enantiomerically pure boranes have a long history in the reduction of prochiral ketones.[80] Amongst the early results using stoichiometric oxazaborolidines, the work of Itsuno is of particular interest. For example, acetophenone (**3.32**) could be reduced with the oxazaborolidines (**3.113**)[81] or (**3.114**)[82] where the ratio of amino alcohol to borane was 1:2, implying that one equivalent of oxazaborolidine and one equivalent of borane were present in the transition state. Itsuno also reported that the oxazaborolidine reagent (**3.114**) could be used catalytically in the reduction of prochiral ketoxime ethers.[83]

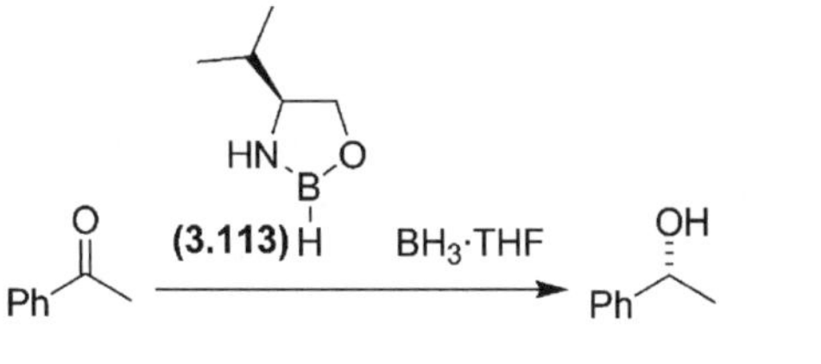

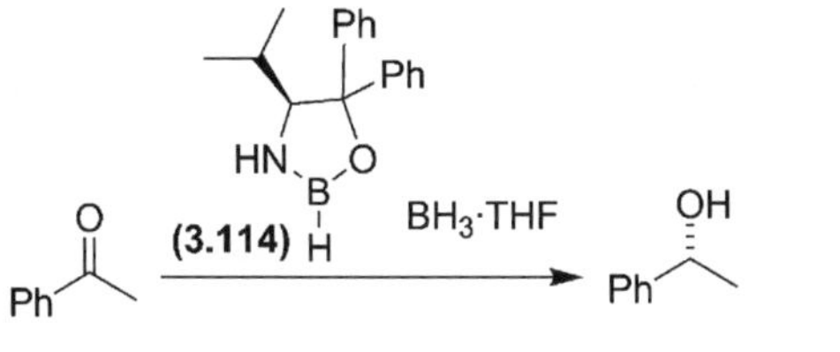

In 1987, Corey, Bakshi and Shibata demonstrated that the enantioselective reduction of ketones could be catalysed by oxazaborolidines. They showed that acetophenone was reduced only slowly by BH_3·THF alone, and that oxazaborolidine (**3.114**) alone did not cause reduction. However, in combination they reduced acetophenone in one minute at room temperature. Using just 2.5 mol% of oxazaborolidine, and stoichiometric BH_3·THF still provided excellent enantioselectivity.[84] In the same initial communication, proline-derived oxazaborolidine (**3.115**) was identified as a catalyst that was suitable for the reduction of a range of ketones, including acetophenone (**3.32**) and ketones (**3.116**) and (**3.117**).

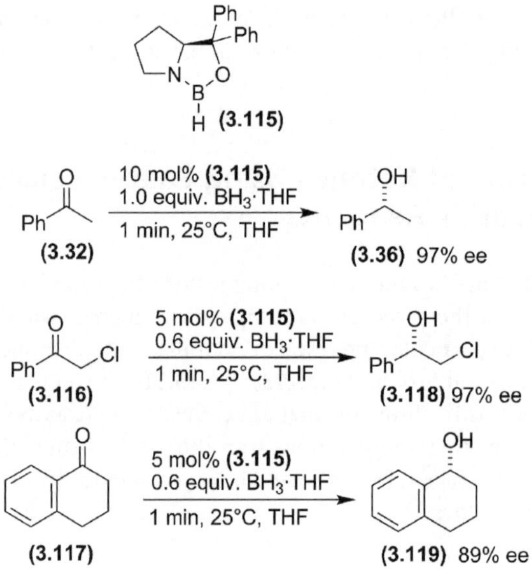

Catalyst (**3.115**) is often referred to as the CBS catalyst after the names of the original authors (Corey, Bakshi and Shibata). A catalytic cycle was proposed which explains the experimental observations. The oxazaborolidine interacts reversibly with borane, which then allows complexation of the ketone to give the key intermediate (**3.122**), as depicted in Figure 3.2. In this process the catalyst acts as both a Lewis acid and Lewis base activating the borane towards hydride delivery and the ketone towards reduction by interaction with the boron in the oxazaborolidine. This dual activation and enhanced steric bulk of the pyrrolidine moiety leads to

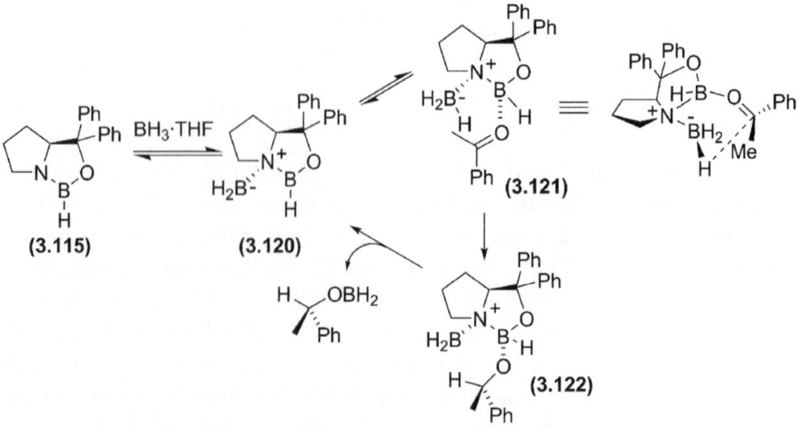

Figure 3.2   Mechanism of oxazaborolidine-catalysed reduction of ketones

an efficient and highly enantioselective carbonyl reduction in which the catalyst recycles. Furthermore, the availability of both enantiomers of the oxazaborolidine (generally prepared from an amino alcohol and a borane or boronic acid) and the predictability of the sense of stereoinduction make this a powerful method for the synthesis of enantioenriched active alcohols.[85] These catalysts also exhibit exceptionally wide scope and have been used effectively in the asymmetric reduction of a range of simple and complex functionalised ketones.[85f]

β-Alkylated oxazaborolidines such as (**3.123**)[86] and (**3.124**)[87] are more stable to air and moisture than (**3.115**) and exhibit better selectivities in some cases. BH_3·THF is not the only borane carrier used and both BH_3·DMS and catecholborane can be used as the stoichiometric reductant. As the latter reagent is effective at low temperature and is less reactive than BH_3 it can be used to effect reduction of the ketone group of α,β–enones such as (**3.125**) with no hydroboration of the alkene moiety.[88] Catecholborane can also be used as a stoichiometric reductant in the asymmetric reduction of halogenated ketones (**3.126**) and (**3.127**).

(3.123) **(3.124)**

(3.125) **(3.133)** 92% ee

(3.127) **(3.131)** 94% ee **(3.132)** fluoxetine (Prozac)

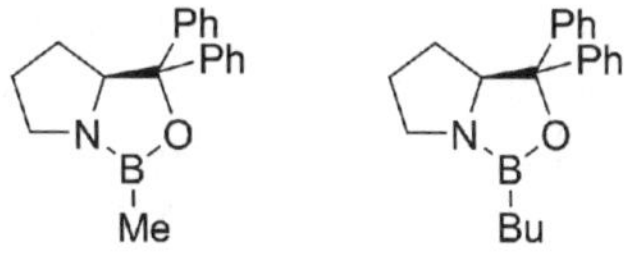

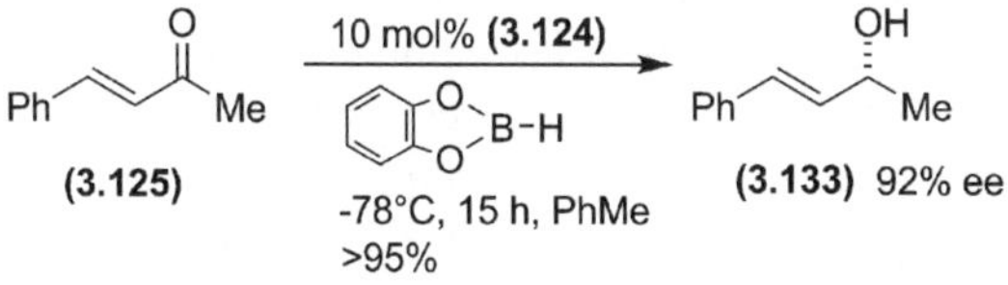

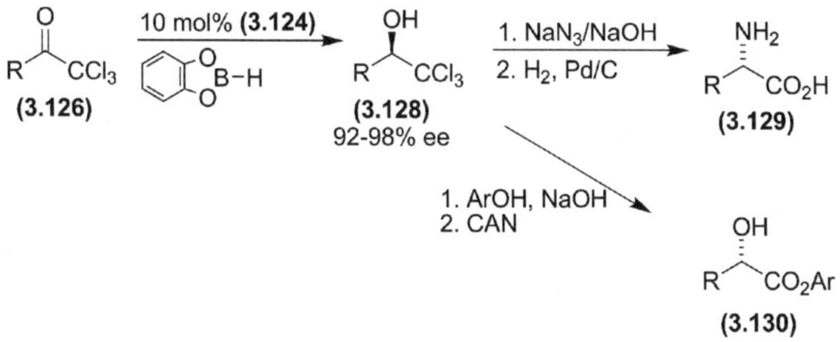

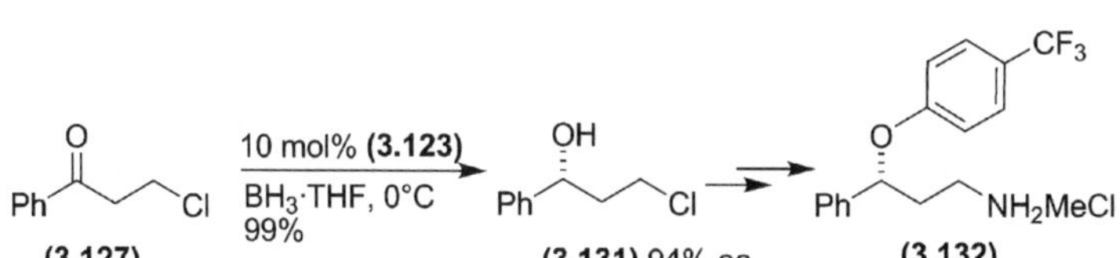

The reduction product (**3.128**) serves as a precursor for α-amino acids (**3.129**) and α–hydroxy esters (**3.130**) while the reduction product (**3.131**) is a precursor to the anti-depressive drug fluoxetine (**3.132**).

A great many other amino alcohols have been examined for their ability to provide good oxazaborolidine catalysts. It is difficult to know which amino alcohol provides the very best enantioselectivity over a range of ketone reductions, since not all amino alcohols have been tried with all ketones.[85] As an example, the four-membered ring analogue (**3.134**) gives high enantioselectivity in the catalytic reduction of acetophenone (**3.32**).[89]

The catalytic asymmetric reduction of ketones can also be achieved using alternate boron-based reagents. For example, the reduction of ketones by catecholboranes has been achieved enantioselectively using enantiomerically pure titanium complexes as the catalyst.[90] The most successful of these approaches centres on the use of borohydrides in the presence of enantiopure catalysts. Activated borohydride reagent (**3.135**) prepared *in situ* from tetrahydrofurfuryl alcohol (THFA) (**3.136**), ethanol and sodium borohydride has been used as an alternative stoichiometric reducing agent in conjunction with β-ketoiminatocobalt(II) complexes of general structure (**3.137**) in $CHCl_3$.[91] The active species is thought to be the cobalt hydride (**3.138**) formed by reaction of (**3.137**) with borohydride (**3.135**) and chloroform. While this reaction is generally performed in chloroform, this solvent functions as an activator and the reaction proceeds with high ee when chloroform is only used in catalytic amounts in THF.[92] This cobalt-catalysed procedure has been applied to the highly enantioselective reduction of aryl/alkyl and aryl/aryl ketones and symmetrical and unsymmetrical 1,3-diketones. For example, ketone (**3.139**) undergoes enantioselective reduction using 1.5 equivalents of sodium borohydride, ethanol, THFA and cobalt complex (**3.140**).[93] Dibenzoylmethane (**3.141**) is reduced to the C_2-symmetric diol (**3.142**) under similar conditions using catalyst (**3.143**).[94] 2-Substituted symmetrical 1,3-diketones such as (**3.144**) undergo reductive desymmetrisation using complex (**3.143**) to give the *anti*-aldol product (**3.145**) predominantly.[95] A kinetic resolution of unsymmetrical 2-alkyl-1,3-diketones can be achieved using catalyst (**3.143**) to give the *anti*-aldol products with up to 98% ee and 47% conversion[96] and a dynamic kinetic resolution of 2-substituted 3-phenyl-3-ketoesters such as (**3.146**) occurs in the presence of sodium methoxide to give the 3-hydroxyester (**3.147**) in excellent yield.[97] As this process proceeds via the formation of the enolate, in a similar fashion to the dynamic kinetic resolution of β-ketoesters discussed in Section 3.1, near quantitative yields of product can be obtained.

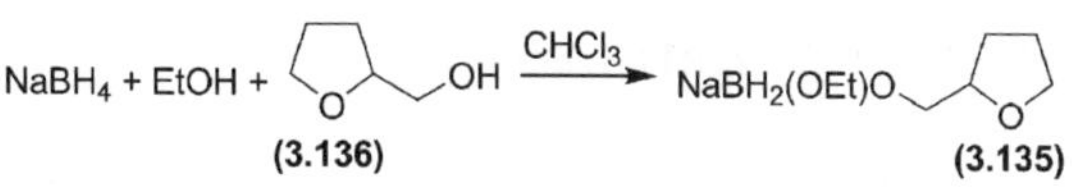

$$NaBH_4 + EtOH + \text{(3.136)} \xrightarrow{CHCl_3} NaBH_2(OEt)O\text{(3.135)}$$

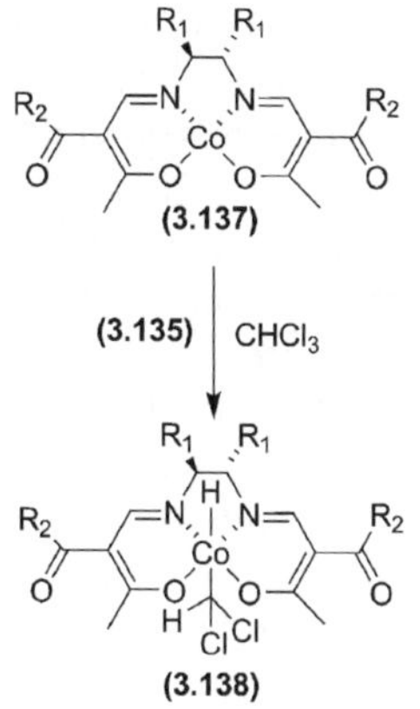

(3.137)

(3.135) | CHCl$_3$

(3.138)

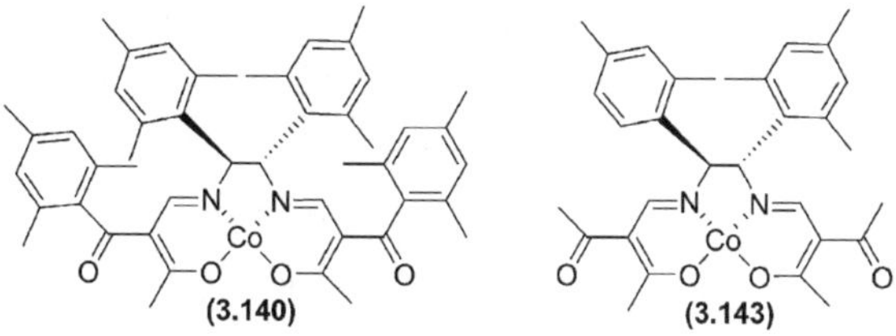

(3.140)   (3.143)

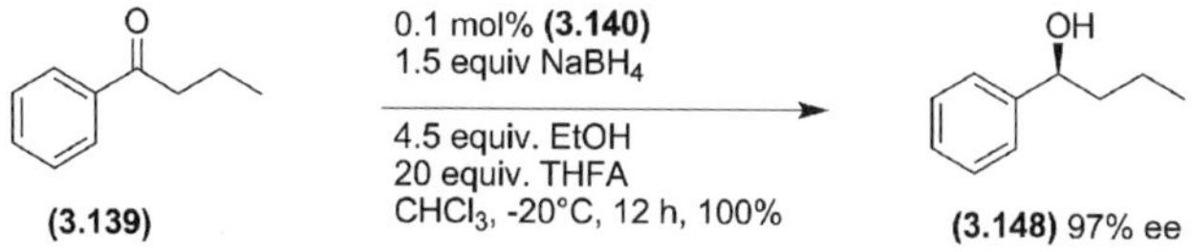

(3.139)

0.1 mol% **(3.140)**
1.5 equiv NaBH$_4$

4.5 equiv. EtOH
20 equiv. THFA
CHCl$_3$, -20°C, 12 h, 100%

(3.148) 97% ee

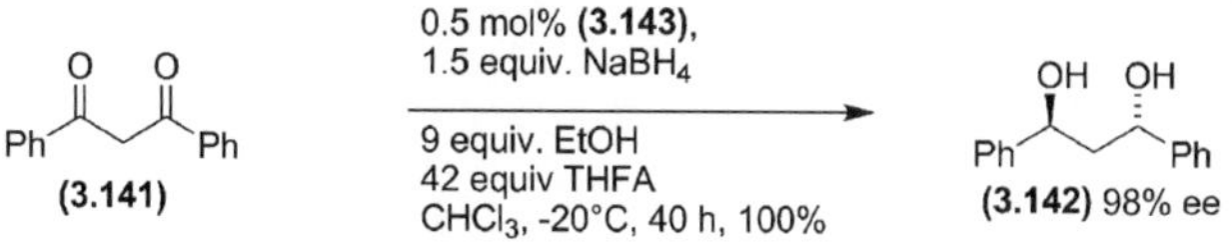

Ph $\quad$ Ph
(3.141)

0.5 mol% **(3.143)**,
1.5 equiv. NaBH$_4$

9 equiv. EtOH
42 equiv THFA
CHCl$_3$, -20°C, 40 h, 100%

Ph $\quad$ Ph
(3.142) 98% ee

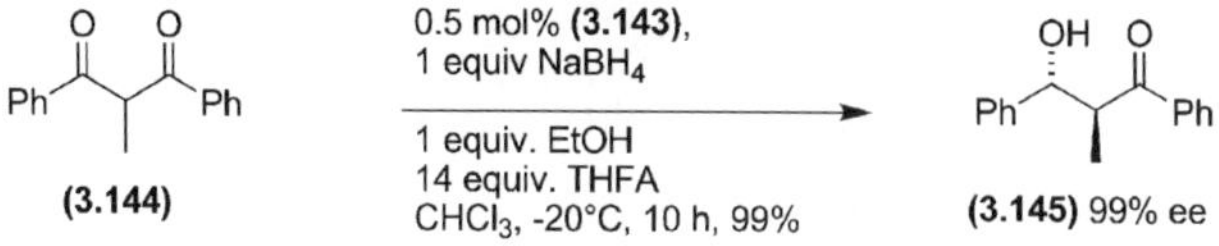

Ph $\quad$ Ph
(3.144)

0.5 mol% **(3.143)**,
1 equiv NaBH$_4$

1 equiv. EtOH
14 equiv. THFA
CHCl$_3$, -20°C, 10 h, 99%

Ph $\quad$ Ph
(3.145) 99% ee

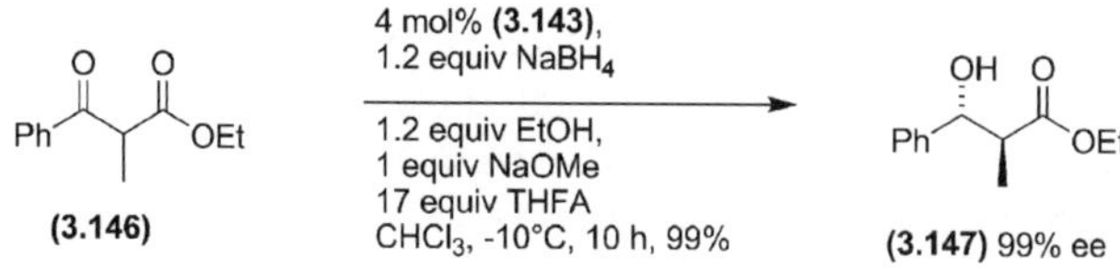

Ph $\quad$ OEt
(3.146)

4 mol% **(3.143)**,
1.2 equiv NaBH$_4$

1.2 equiv EtOH,
1 equiv NaOMe
17 equiv THFA
CHCl$_3$, -10°C, 10 h, 99%

Ph $\quad$ OEt
(3.147) 99% ee

3.6 Hydrosilylation of Ketones

The hydrosilylation of ketones provides an alternative to direct reduction with H_2. In terms of synthesis, the initially formed silyl ethers are almost invariably hydrolysed into the corresponding alcohols.[98] The rhodium-catalysed hydrosilylation of ketones has received more attention than other metals. One of the earliest ligands to give good enantioselectivity in this reaction was Brunner's Pythia ligand, (3.149), which provides the asymmetric environment for the hydrosilylation of acetophenone (3.32).[99] This paved the way for the examination of other nitrogen-based ligands in this reaction. In 1989, three groups reported the use of oxazoline ligands (3.150),[100] (3.151)[101] and (3.152)[102] for rhodium-catalysed hydrosilylation. These were amongst the first reports of the now ubiquitous oxazoline moiety in asymmetric catalysis.[103,104] The PYBOX ligands in particular have been extended to a wide range of ketone substrates, with most products having high ee (generally 63–99% ee).[105] The sense of asymmetric induction provided by these oxazoline ligands is very sensitive to the details of the ligand structure. For instance, introduction of a methyl group into the 6-position of the pyridine ring in the PYBOX ligand also inverts the stereochemistry of the subsequently formed phenethyl alcohol.[106]

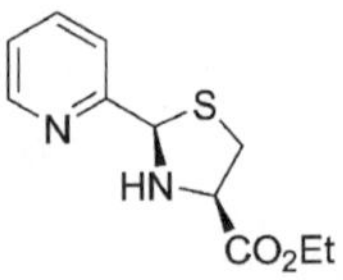

(3.149) up to 97.6% ee (*S*)

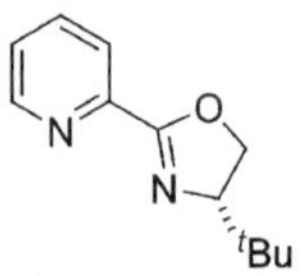

(3.150) up to 91% ee (*R*)

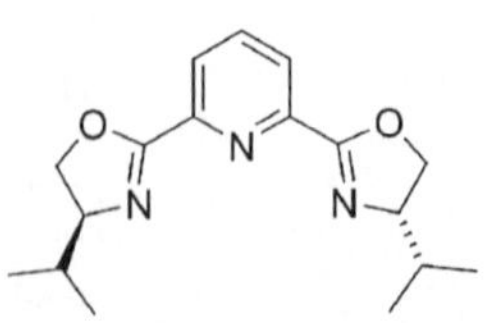

(3.151) up to 95% ee (*S*)

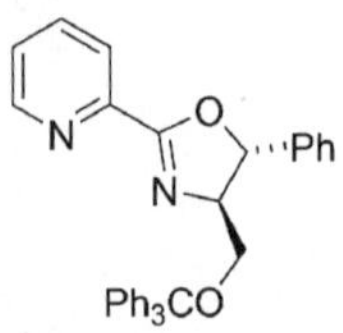

(3.152) up to 80% ee (*R*)

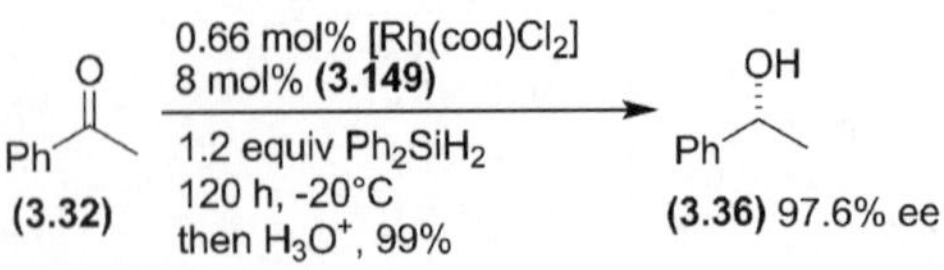

0.66 mol% [Rh(cod)Cl₂]
8 mol% (3.149)

1.2 equiv Ph₂SiH₂
120 h, -20°C
then H₃O⁺, 99%

Ph (3.32) → Ph (3.36) 97.6% ee

Alternative bis-oxazolines,[107,108] as well as phosphino-oxazolines (see Section 10.2)[109,110] have also been used, although these ligands were generally less selective. However, Uemura and coworkers reported interesting results with the oxazolinylferrocene-phosphine ligand (3.153). Whilst the rhodium-catalysed reaction of acetophenone (**3.32**) provides mainly the (*R*)-alcohol (**3.36**) as product, the corresponding reaction with an iridium catalyst provides the (*S*)-alcohol.[111] The iridium-catalysed system works well on many other ketones, including aryl/alkyl ketone (**3.154**) and α,β-unsaturated ketone (**3.155**), but very low selectivity is observed with 2-octanone as the substrate (19% ee in the product).

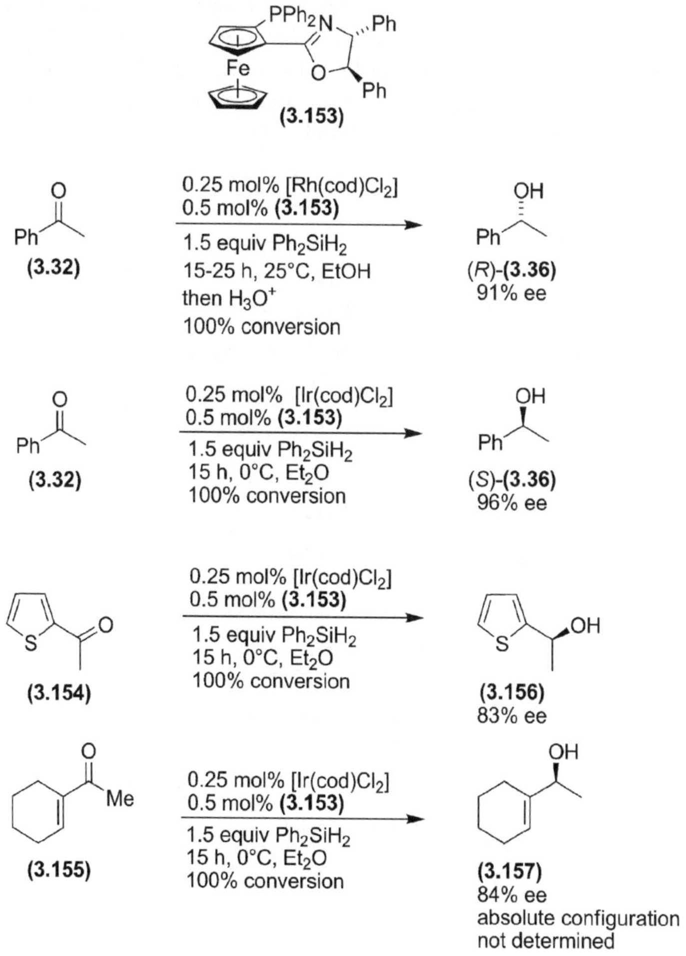

(3.153)

O Ph⤴ **(3.32)**	0.25 mol% [Rh(cod)Cl$_2$] 0.5 mol% **(3.153)** ———————————→ 1.5 equiv Ph$_2$SiH$_2$ 15-25 h, 25°C, EtOH then H$_3$O$^+$ 100% conversion	OH Ph⤴ (*R*)-**(3.36)** 91% ee
O Ph⤴ **(3.32)**	0.25 mol% [Ir(cod)Cl$_2$] 0.5 mol% **(3.153)** ———————————→ 1.5 equiv Ph$_2$SiH$_2$ 15 h, 0°C, Et$_2$O 100% conversion	OH Ph⤴ (*S*)-**(3.36)** 96% ee
(3.154)	0.25 mol% [Ir(cod)Cl$_2$] 0.5 mol% **(3.153)** ———————————→ 1.5 equiv Ph$_2$SiH$_2$ 15 h, 0°C, Et$_2$O 100% conversion	OH **(3.156)** 83% ee
O ⤴Me **(3.155)**	0.25 mol% [Ir(cod)Cl$_2$] 0.5 mol% **(3.153)** ———————————→ 1.5 equiv Ph$_2$SiH$_2$ 15 h, 0°C, Et$_2$O 100% conversion	OH **(3.157)** 84% ee absolute configuration not determined

Alkyl/alkyl ketones are challenging substrates for rhodium-catalysed asymmetric hydrosilylation and are generally reduced with low enantioselectivities using oxazoline-based ligands. However a number of alternative ligand systems have been used successfully in the rhodium-catalysed hydrosilylation of alkyl/alkyl ketones. For example, the δ–keto ester (**3.158**) undergoes enantioselective hydrosilylation followed by cyclisation to give the lactone (**3.159**) using the *trans*-chelating diphosphine EtTRAP (**3.160**).[112,113] The rhodium complex formed from the planar enantiomerically pure P,N-ligand (**3.161**) provides high ees in the asymmetric hydrosilylation of both aryl/alkyl ketones and a small number of *n*-alkyl methyl ketones, cyclohexyl methyl and even adamantyl methyl ketone (**3.162**).[114] The mixed phosphorus/sulfur ligands developed by Evans for use in alkene hydrogenations (see Section 2.1) are also highly effective in the asymmetric hydrosilylation of ketones.[115] In particular the cationic rhodium complex derived from (**3.163**) displays a wide scope and has been used to effect highly enantioselective reduction of aryl/alkyl ketones, cyclic aromatic ketones, dialkyl ketones such as (**3.102**) and also β-keto esters such as (**3.164**).

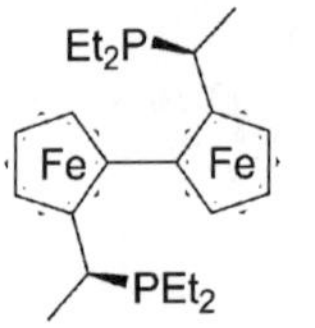

(3.160) 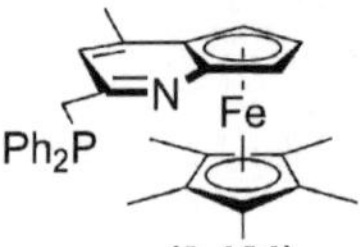   **(3.161)** 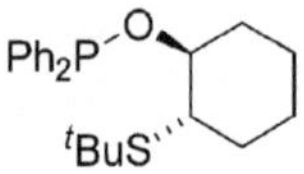   **(3.163)**

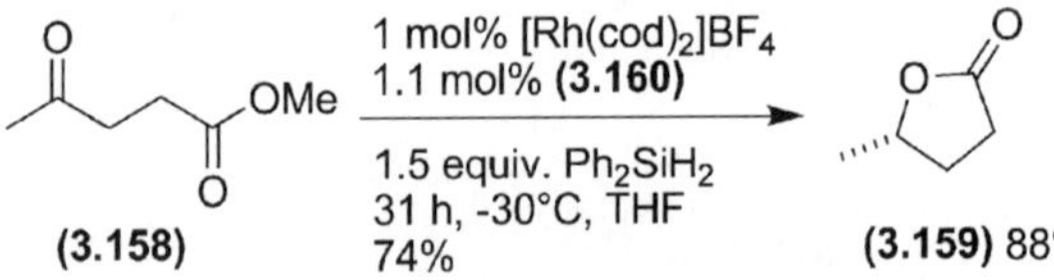

1 mol% [Rh(cod)₂]BF₄
1.1 mol% (**3.160**)

1.5 equiv. Ph₂SiH₂
31 h, -30°C, THF
74%

(3.158)   **(3.159)** 88% ee

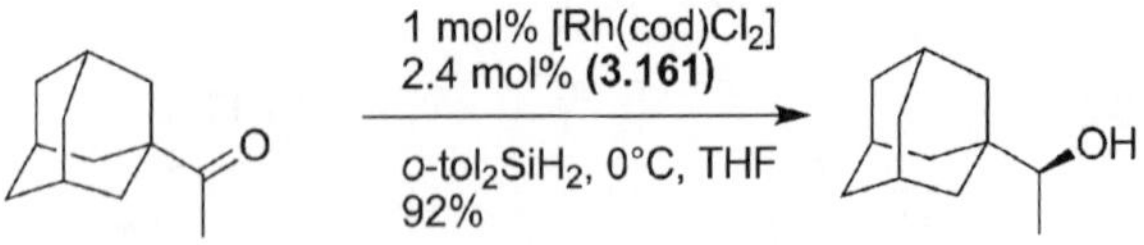

1 mol% [Rh(cod)Cl₂]
2.4 mol% (**3.161**)

o-tol₂SiH₂, 0°C, THF
92%

(3.162)   **(3.165)** 96% ee

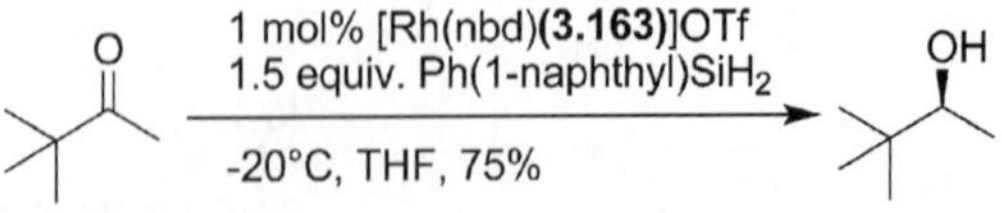

1 mol% [Rh(nbd)(**3.163**)]OTf
1.5 equiv. Ph(1-naphthyl)SiH₂

-20°C, THF, 75%

(3.102)   **(3.104)** 98% ee

1 mol% [Rh(nbd)(**3.163**)]OTf
1.5 equiv. Ph(1-naphthyl)SiH₂

-20°C, THF, 94%

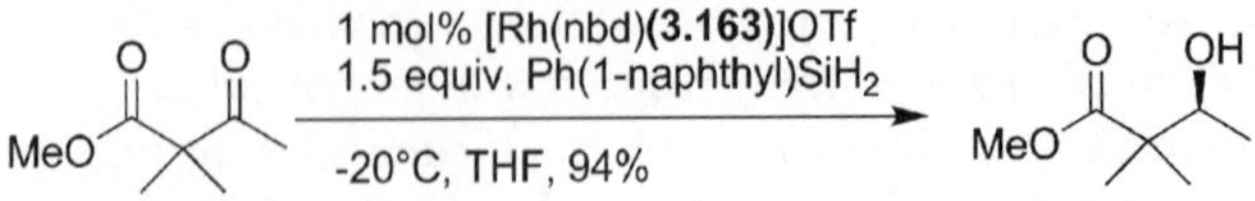

(3.164)   **(3.166)** 94%ee

N-heterocyclic carbenes are another class of ligands showing good scope in this process. Most success has been achieved using bidentate ligands including one or two carbenes. The rhodium catalysts prepared from the axially chiral bis(diamino)carbene (**3.167**)[116] with a BINAP-based backbone and mixed oxazoline-carbene ligands such as (**3.167**) have been used to hydrosilylate aryl/alkyl ketones and a small number of alkyl/alkyl ketones with good to high ee.[117] The selectivity of the rhodium complex derived from (**3.168**) improves with increasing steric bulk of the alkyl/alkyl ketone substrate. Thus, while 2-butanone is hydrosilylated with an ee of only 65%, *tert*-butylmethyl ketone (**3.102**) undergoes reduction with 95% ee.

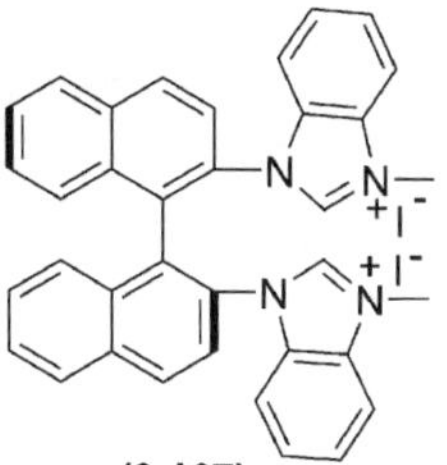
N
N
+ I⁻
N
N
+ I⁻

(3.167)

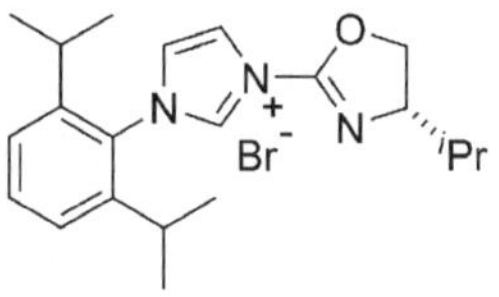
N
N
+
Br⁻
O
N
ⁱPr

(3.168)

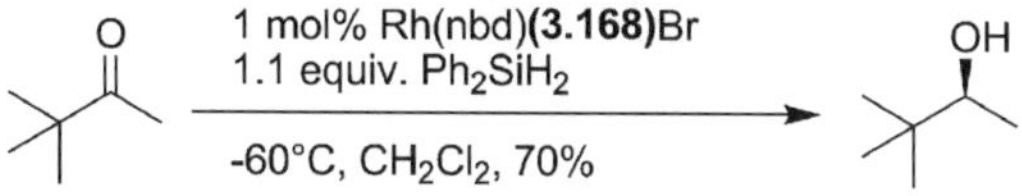
O
1 mol% Rh(nbd)(3.168)Br
1.1 equiv. Ph2SiH2
-60°C, CH2Cl2, 70%
OH

(3.102) **(3.104)** 95% ee

The hydrosilylation of a symmetrical ketone with a prochiral silane leads to the possibility of asymmetric induction in the newly formed silicon stereocentre. In their best example, Takaya and coworkers reported essentially complete control of asymmetric induction using the CyBINAP ligand (**3.169**) and a rhodium catalyst.[118] Pentan-3-one (**3.170**) and prochiral silane (**3.171**) are converted into the alkoxysilane (**3.172**), where the asymmetry is associated with the silicon atom.

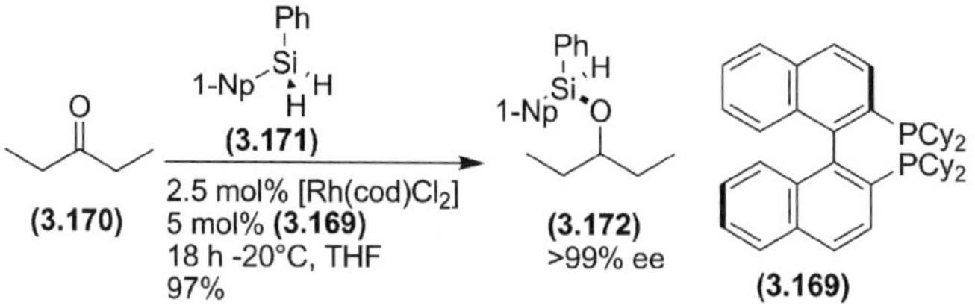
Ph
Si
1-Np
H
H
(3.171)
O
2.5 mol% [Rh(cod)Cl2]
5 mol% (3.169)
18 h -20°C, THF
97%
(3.170)
Ph
H
Si
1-Np
O
(3.172)
>99% ee
PCy2
PCy2
(3.169)

As well as rhodium-catalysed hydrosilylation, asymmetric ruthenium[119] and titanium-catalysed[120,121] hydrosilylation have also been reported. Amongst these, Buchwald's report of the hydrosilylation of ketones using titanocene catalysts and inexpensive polymethylhydrosiloxane (PMHS) appear to be the most general.[122]

Reactions proceed via formation of the active titanium hydride catalyst (**3.173**) formed either by treatment of the precatalyst (**3.174**) with two equivalents of butyllithium, followed by addition of polymethylhydrosiloxane,[122a] or by reaction of difluoride (**3.175**) with $PhSiH_3$ in the presence of methanol.[122b] The greatest selectivities are obtained using conjugated carbonyls as substrates. Ketones (**3.32**) and (**3.117**) are reduced with good enantioselectivity using active catalyst prepared from precatalyst (**3.174**). Similar ees are obtained using (**3.173**) prepared from difluoride (**3.175**) as catalyst precursor, but at lower catalyst loadings and shorter reaction times. For example α,β-enone (**3.155**) is reduced in the presence of only 2 mol% catalyst in five hours.

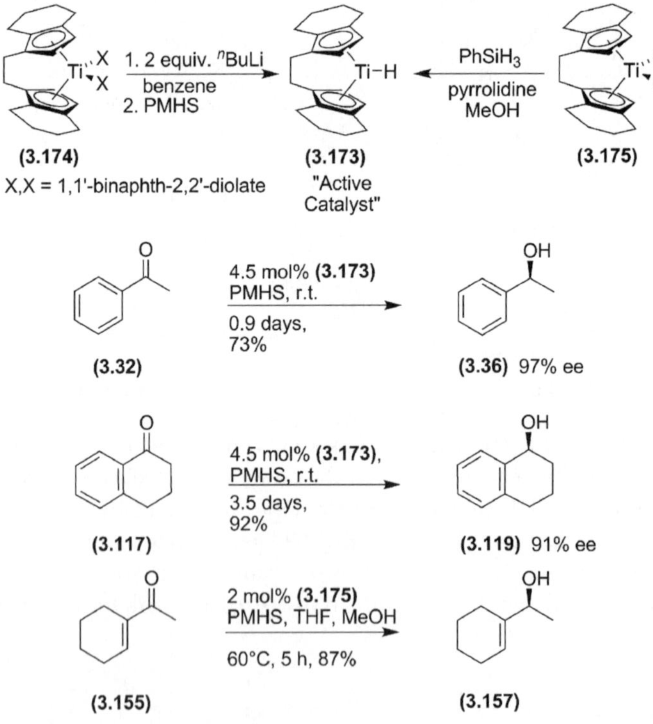

1. 2 equiv. ⁿBuLi
benzene
2. PMHS
PhSiH3
pyrrolidine
MeOH
(3.174)
X,X = 1,1'-binaphth-2,2'-diolate
(3.173)
"Active Catalyst"
(3.175)
4.5 mol% (3.173)
PMHS, r.t.
0.9 days,
73%
(3.32)
(3.36) 97% ee
4.5 mol% (3.173),
PMHS, r.t.
3.5 days,
92%
(3.117)
(3.119) 91% ee
2 mol% (3.175)
PMHS, THF, MeOH
60°C, 5 h, 87%
(3.155)
(3.157)

The ligated copper hydride-based asymmetric hydrosilylation of electrophilic double bonds developed by Lipshutz (see Section 2.6) can also be applied to the reduction of aryl/alkyl and heteroaryl ketones and high ees in this process have been achieved using very low catalyst loadings.[123] While heteroaryl ketones are reduced with low enantioselectivity using rhodium-based catalysts, acetylpyridines, furans such as (**3.176**), thiazoles such as (**3.177**) and isoxazoles are reduced with high ee using catalytic quantities of CuH generated *in situ* by reaction of CuCl and PMHS in the presence of SEGPHOS (**3.178**).

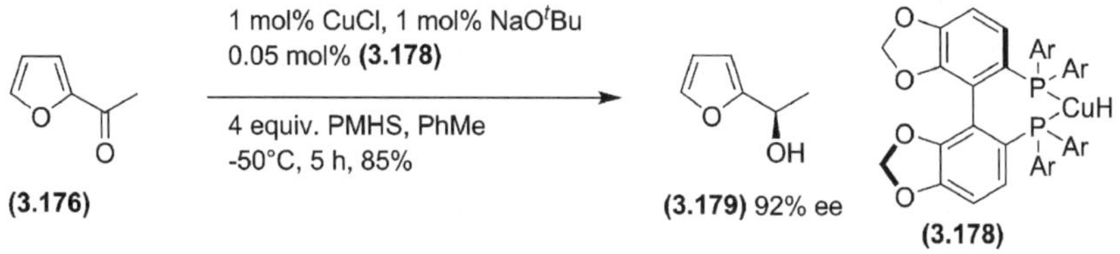

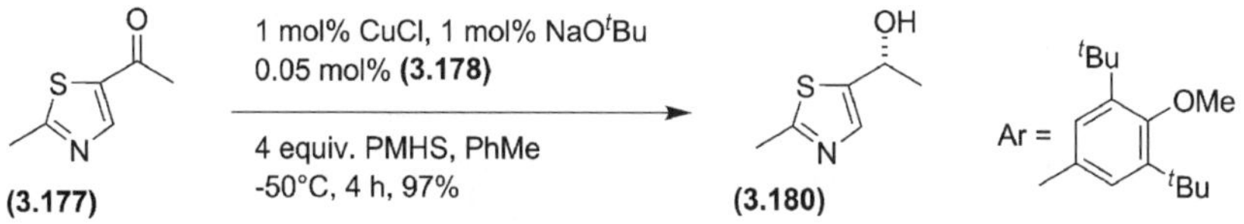

(3.176)   1 mol% CuCl, 1 mol% NaO^{t}Bu
0.05 mol% **(3.178)**

4 equiv. PMHS, PhMe
-50°C, 5 h, 85%

(3.179) 92% ee

(3.178)

(3.177)   1 mol% CuCl, 1 mol% NaO^{t}Bu
0.05 mol% **(3.178)**

4 equiv. PMHS, PhMe
-50°C, 4 h, 97%

(3.180)

Ar =

Copper-based asymmetric hydrosilylation of aryl/alkyl ketones can also be achieved in up to 92% ee using 1 mol% CuF in the presence of BINAP and PhSiH$_3$.[124] In contrast to the method developed by Lipshutz, this process does not require the use of anaerobic reaction conditions and is actually enhanced when performed under oxygen.

3.7   Hydrosilylation of Imines and Nitrones

Although there has been little research into the hydrosilylation of functionality other than carbonyls (and alkenes), what work has been published provides some exceptionally good examples of enantioselective catalysis.[98]

The Buchwald group has published a superb example of an asymmetric catalytic reaction using the active titanocene catalyst **(3.173)** derived from precatalyst **(3.175)**;[125,126] with catalyst loadings as low as 0.02 mol% (substrate:catalyst ratio of 5,000:1), very high ees are obtained for the hydrosilylation of a wide range of imine substrates, including the cyclic imine **(3.181)**, and the acyclic imines **(3.182)** and **(3.183)**. The asymmetric reduction of *N*-arylketimines can be achieved using this catalyst system at slightly higher catalyst loadings with isobutylamine as an additive.[127] Cyclohexyl-substituted *N*-aryl imines such as **(3.184)** and even straight chain derivatives such as **(3.185)** undergo enantioselective hydrosilylation using 2 mol% catalyst, however *N*-arylimines derived from aromatic ketones such as acetophenone are reduced with poor ee.

Some recent work has been directed towards the use of organocatalysts, in particular Lewis basic pipecolinic formamides, in the asymmetric hydrosilylation of *N*-arylimines.[128] These catalysts function by activating the silane and exhibit broad substrate scope. For example formamide **(3.186)** effects enantioselective hydrosilylation of aryl-derived ketimines along with isopropyl-substituted imine **(3.187)** and α,β-unsaturated imine **(3.188)**.

Imines derived from aryl/alkyl ketones can be reduced enantioselectively using the CuH-based catalytic system developed by Lipshutz. In general, high ees can be obtained (94–99% ee) using this procedure, but it is limited to the hydrosilylation of *N*-dixylylphosphinylimines such as (**3.189**). [129]

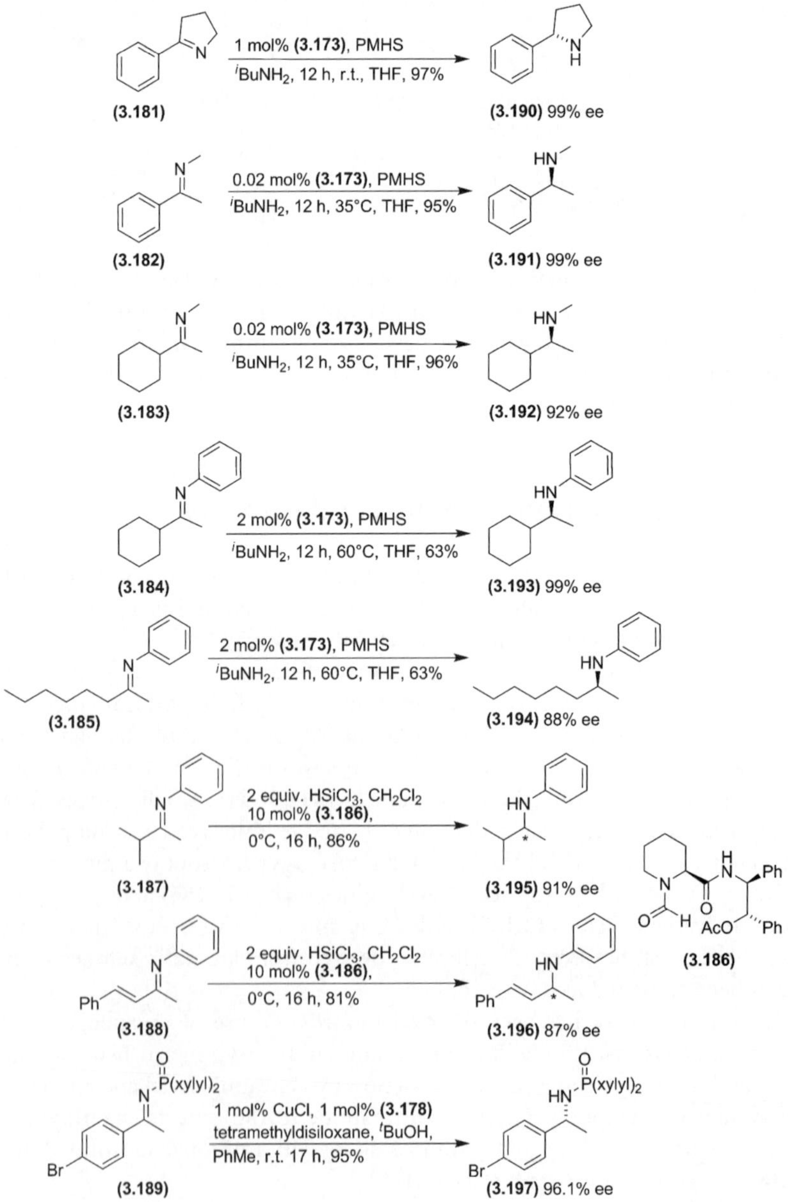

(3.181)

1 mol% (**3.173**), PMHS
*i*BuNH₂, 12 h, r.t., THF, 97%

(3.190) 99% ee

(3.182)

0.02 mol% (**3.173**), PMHS
*i*BuNH₂, 12 h, 35°C, THF, 95%

(3.191) 99% ee

(3.183)

0.02 mol% (**3.173**), PMHS
*i*BuNH₂, 12 h, 35°C, THF, 96%

(3.192) 92% ee

(3.184)

2 mol% (**3.173**), PMHS
*i*BuNH₂, 12 h, 60°C, THF, 63%

(3.193) 99% ee

(3.185)

2 mol% (**3.173**), PMHS
*i*BuNH₂, 12 h, 60°C, THF, 63%

(3.194) 88% ee

(3.187)

2 equiv. HSiCl₃, CH₂Cl₂
10 mol% (**3.186**),
0°C, 16 h, 86%

(3.195) 91% ee

(3.186)

(3.188)

2 equiv. HSiCl₃, CH₂Cl₂
10 mol% (**3.186**),
0°C, 16 h, 81%

(3.196) 87% ee

(3.189)

1 mol% CuCl, 1 mol% (**3.178**)
tetramethyldisiloxane, *t*BuOH,
PhMe, r.t. 17 h, 95%

(3.197) 96.1% ee

The hydrosilylation of nitrones provides a route to *N,N*-disubstituted hydroxylamines. The use of an Ru/TolBINAP complex was reported to give the best results in the asymmetric hydrosilylation of nitrones.[130] Nitrone (**3.198**) was converted into hydroxylamine (**3.199**) in reasonable yield and good enantioselectivity.

References

1. For reviews see: (a) R. Noyori, *Asymmetric Catalysis in Organic Synthesis*, John Wiley & Sons, Inc., New York, **1994**, 56. (b) T. Ohkuma and R. Noyori, in *Comprehensive Asymmetric Catalysis*, Vol. 1, ed. E. N. Jacobsen, A. Pfaltz and H. Yamamoto, Springer-Verlag, Berlin, **1999**, 199.

2. M. Kitamura, T. Okuma, S. Inoue, N. Sayo, H. Kumobayashi, S. Akutagawa, T. Ohta, H. Takaya and R. Noyori, *J. Am. Chem. Soc.*, **1988**, *110*, 629.

3. R. Noyori, T. Okuma, M. Kitamura, H. Takaya, N. Sayo, H. Kumobayashi and S. Akutagawa, *J. Am. Chem. Soc.*, **1987**, *109*, 5856.

4. M. Kitamura, M. Tokunaga and R. Noyori, *J. Am. Chem. Soc.*, **1995**, *117*, 2931.

5. I. Gauter, V. Ratovelomanana-Vidal, P. Savignac, J.-P. Genêt, *Tetrahedron Lett.*, **1996**, *37*, 7721.

6. P. J. Pye, K. Rossen, R. A. Reamer, R. P. Volante and P. J. Reider, *Tetrahedron Lett.*, **1998**, *39*, 4441.

7. M. J. Burk, T. G. P. Harper and C. S. Kalberg, *J. Am.Chem. Soc.*, **1995**, *117*, 4423.

8. T. Chiba, A. Miyashita, H. Nohira and H. Takaya, *Tetrahedron Lett.*, **1993**, *34*, 2351.

9. D. Blanc, J.-C. Henry, V. Ratovelomanana-Vidal and J.-P. Genet, *Tetrahedron Lett.*, **1997**, *38*, 6603.

10. Y. Kuroki, Y. Sakamaki and K. Iseki, *Org. Lett.*, **2001**, *3*, 457.

11. A. Roucoux, F. Agbossou, A. Mortreux and F. Petit, *Tetrahedron: Asymmetry*, **1993**, *4*, 2279.

12. C. Pasquier, S. Naili, L. Pelinski, J. Brocard, A. Mortreux and F. Agbossou, *Tetrahedron: Asymmetry*, **1998**, *9*, 193.

13. D. Liu, W. Gao, C. Wang and X. Zhang, *Angew. Chem. Int. Ed.*, **2005**, *44*, 1687.

14. J. P. Genêt, *Pure Appl. Chem.*, **2002**, *74*, 77.

15. S. Jeulin, N. Champion, P. Dellis, V. Ratovelomanana-Vidal and J. P. Genêt, *Synthesis*, **2005**, 3666.

16. R. Noyori and T. Ohkuma, *Angew. Chem. Int. Ed.*, **2001**, *40*, 40.

17. T. Ohkuma, M. Koizumi, H. Doucet, T. Pham, M. Kozawa, K. Murata, E. Katayama, T. Yokozawa, T. Ikariya and R. Noyori, *J. Am. Chem. Soc.*, **1998**, *120*, 13529.

18. N. Arai, H. Ooka, K. Azuma, T. Yabuuchi, N. Kurono, T. Inoue and T. Ohkuma, *Org. Lett.*, **2007**, *9*, 939.

19. T. Ohkuma, C. A. Sandoval, R. Srinivasan, Q. Lin, Y. Wei, K. Muñiz and R. Noyori, *J. Am. Chem. Soc.*, **2005**, *127*, 8288.
20. K. Mikami, K. Wakabayashi and K. Aikawa, *Org. Lett.*, **2006**, *8*, 1517.
21. T. Ohkuma, M. Koizumi, K. Muñiz, G. Hilt, C. Kaboto and R. Noyori, *J. Am. Chem. Soc.*, **2002**, *124*, 6508.
22. T. Ohkuma, H. Ikehira, T. Ikariya and R. Noyori, *Synlett*, **1997**, 467.
23. C. A. Sandoval, T. Ohkuma, K. Muñiz and R. Noyori, *J. Am. Chem. Soc.*, **2003**, *125*, 13490.
24. R. Noyori and H. Takaya, *Acc. Chem. Res.*, **1990**, *23*, 245.
25. For a mathematical treatment of dynamic resolution reactions, see: M. Kitamura, M. Tokunga and R. Noyori, *J. Am. Chem. Soc.*, **1993**, *115*, 144.
26. K. Mashima, K.-H. Kusano, N. Sato, Y.-I. Matsumura, K. Nozaki, H. Kumobayashi, N. Sayo, Y. Hori, T. Ishizaki, S. Akutagawa and H. Takaya, *J. Org. Chem.*, **1994**, *59*, 3064.
27. J.-P. Genêt, M.C. Caño de Andrade and V. Ratovelomanana-Vidal, *Tetrahedron Lett.*, **1995**, *36*, 2063.
28. M. Kitamura, M. Tokunaga, T. Pham, W.D. Lubell and R. Noyori, *Tetrahedron Lett.*, **1995**, *36*, 5769.
29. (a) C. Bolm, *Angew. Chem., Int. Ed. Engl.*, **1993**, *32*, 232. (b) H.-U. Blaser and S. Spindler, in *Comprehensive Asymmetric Catalysis*, Vol. 1, ed. E. N. Jacobsen, A. Pfaltz and H. Yamamoto, Springer-Verlag, Berlin, **1999**, 247. (b) V. I. Tararov and A. Börner, *Synlett*, **2005**, 203.
30. M. J. Burk and J. E. Feaster, *J. Am. Chem. Soc.*, **1992**, *114*, 6266.
31. J. Mao and D. C. Baker, *Org. Lett.*, **1999**, *1*, 841.
32. J. Wu, F. Wang, Y. .Ma, X. Cui, L. Cun, J. Zhu, J. Deng and B. Yu, *Chem. Commun.*, **2006**, 1766.
33. C. A. Willoughby and S. L. Buchwald, *J. Am. Chem. Soc.*, **1992**, *114*, 7562.
34. C. A. Willoughby and S. L. Buchwald, *J. Am. Chem. Soc.*, **1993**, *58*, 7627.
35. A. Viso, N.E. Lee and S. L. Buchwald, *J. Am. Chem. Soc.*, **1994**, *116*, 9373.
36. C. A. Willoughby and S. L. Buchwald, *J. Am. Chem. Soc.*, **1994**, *116*, 11703.
37. C. A. Willoughby and S. L. Buchwald, *J. Am. Chem. Soc.*, **1994**, *116*, 8952.
38. T. Morimoto, N. Suzuki and K. Achiwa, *Tetrahedron: Asymmetry*, **1998**, *9*, 183.
39. D. Xaio and X. Zhang, *Angew. Chem. Int. Ed.*, **2001**, *40*, 3425.
40. C. Moessner and C. Bolm, *Angew. Chem. Int. Ed.*, **2005**, *44*, 7564.
41. S.-F. Zhu, J.-B. Xie, Y.-Z. Zhang, S. Li and Q.-L. Zhou, *J. Am. Chem. Soc.*, **2006**, *128*, 12886.
42. For general reviews on the use of enantiomerically enriched Brønsted acids see: (a) P. R. Schreiner, *Chem. Soc. Rev.*, **2003**, *32*, 289. (b) P. M. Pihko, *Angew. Chem. Int. Ed.*, **2004**, *43*, 2062. (c) C. Bolm, T. Rantanen, I. Schiffers and L. Zani, *Angew. Chem. Int. Ed.*, **2005**, *44*, 1758. (d) J. Seayad and B. List, *Org. Biomol. Chem.*, **2005**, *3*, 719.
43. M. Rueping, E. Sugiono, C. Azap, T. Theissmann and M. Bolte, *Org. Lett.*, **2005**, *7*, 3781.
44. S. Hoffmann, A. M. Seayad and B. List, *Angew. Chem. Int. Ed.*, **2005**, *44*, 7424.
45. M. Rueping, A. P. Antonchick and T. Theissman, *Angew. Chem. Int. Ed.*, **2006**, *45*, 3683.
46. R. I. Storer, D. E. Carrera, Y. Ni and D. C. MacMillan, *J. Am. Chem. Soc.*, **2006**, *128*, 84.

47. (a) M. J. Palmer and M. Wills, *Tetrahedron: Asymmetry*, **1999**, *10*, 2045. (b) S. Gladiali, and E. Alberico, *Chem. Soc. Rev.*, **2006**, *35*, 226.
48. D. Müller, G. Umbricht, B. Weber and A. Pfaltz, *Helv. Chim. Acta*, **1991**, *74*, 232.
49. P. Gamez, F. Fache, P. Mangeney and M. Lemaire, *Tetrahedron Lett.*, **1993**, *34*, 6897.
50. D. A. Evans, S. G. Nelson, M. R. Gagné, A. R. Muci, *J. Am. Chem. Soc.*, **1993**, *115*, 9800.
51. A. Fujii, S. Hashiguchi, N. Uematsu, T. Ikariya and R. Noyori, *J. Am. Chem. Soc.*, **1996**, *118*, 2521.
52. K. Matsumura, S. Hashiguchi, T. Ikariya and R. Noyori, *J. Am. Chem. Soc.*, **1997**, *119*, 8738.
53. (a) A. M. Hayes, D. J. Morris, G. J. Clarkson and M. Wills, *J. Am. Chem. Soc.*, **2005**, *127*, 7318. (b) D. J. Morris, A. M. Hayes and M. Wills, *J. Org. Chem.*, **2006**, *71*, 7035.
54. (a) D. S. Matharu, D. J. Morris, A. M. Kawamoto, G. J. Clarkson and M. Wills, *Org. Lett.*, **2005**, *7*, 5489. (b) D. S. Matharu, D. J. Morris, G. J. Clarkson and M. Wills, *Chem. Commun.*, **2006**, 3232.
55. (a) R. Noyori and S. Hashiguchi, *Acc. Chem. Res.*, **1997**, *30*, 97. (b) K. J. Haack, S. Hashiguchi, A. Fujii, T. Ikariya and R. Noyori, *Angew. Chem., Int. Ed., Engl.*, **1997**, *36*, 285.
56. (a) J.-X. Gao, T. Ikariya and R. Noyori, *Organometallics*, **1996**, *15*, 1087. (b) H. Zhang, C.-B. Yang, Y. Y. Li, Z.-R. Dong, J.-X. Gao, H. Nakamura, K. Murata, and T. Ikariya, *Chem. Commun.*, **2003**, 142. (c) Z.-R. Dong, Y.-Y. Li, J.-S. Chen, B.-Z. Li, Y. Xing and J.-X. Gao, *Org. Lett.*, **2005**, *7*, 1043.
57. K. Püntener, L. Schwink and P. Knochel, *Tetrahedron Lett.*, **1996**, *37*, 8165.
58. (a) M. Palmer, T. Walsgrove and M. Wills, *J. Org. Chem.*, **1997**, *62*, 5226. (b) M. J. Palmer, J. A. Kenny, T. Walsgrove, A. M. Kawamoto and M. Wills, *J. Chem. Soc. Perkin. Trans. 1*, **2002**, 416.
59. T. Sammakia and E. L. Stangeland, *J. Org. Chem.*, **1997**, *62*, 6104.
60. Y. Jiang, Q. Jiang and X. Zhang, *J. Am. Chem. Soc.*, **1998**, *120*, 3817.
61. D. A. Alonso, D. Guijarro, P. Pinho, O. Temme and P. G. Andersson, *J. Org. Chem.*, **1998**, *63*, 2749.
62. I. M. Pastor, P. Västilä and H. Adolfsson, *Chem. Commun.*, **2003**, 2046.
63. M. T. Reetz and X. Li, *J. Am. Chem. Soc.*, **2006**, *128*, 1044.
64. For recent general reviews on the development and use of supported enantiomerically enriched catalysts see: (a) C. E. Song and S. Lee, *Chem. Rev.*, **2002**, *102*, 3495. (b) N. E. Leadbetter and M. Marco, *Chem. Rev.*, **2002**, *102*, 3217. (c) L.-X. Dai, *Angew. Chem. Int. Ed.*, **2004**, *43*, 5726. (d) K. Ding, Z. Wang, X. Wang, Y. Liang and X. Wang, *Chem. Eur. J.*, **2006**, *12*, 5188. (e) M. Tada and Y. Iwasawa, *Chem. Commun.*, **2006**, 2833. (f) M. Heitbaum, F. Glorius and I. Escher, *Angew. Chem. Int. Ed.*, **2006**, *45*, 4732.
65. Y. Arakawa, N. Hraguchi and S. Itsuno, *Tetrahedron Lett.*, **2006**, *47*, 3239.
66. (a) X. Li, W. Chen, W. Hems, F. King and J. Xiao, *Org. Lett.*, **2003**, *5*, 4559. (b) X. Li, W. Chen, W. Hems, F. King and J. Xiao, *Tetrahedron Lett.*, **2004**, *45*, 951.
67. (a) H.-B. Yu, Q.-S. Hu and L. Pu, *Tetrahedron Lett.*, **2000**, *41*, 1681. (b) H.-B. Yu, Q.-S. Hu and L. Pu, *J. Am. Chem. Soc.*, **2000**, *122*, 6500.
68. L. Jiang, T.-F. Wu, Y.-C. Chen, J. Zhu and J.-G. Deng, *Org. Biomol. Chem.*, **2006**, *4*, 3319.

69. Y.-C. Chen, T.-F. Wu, J. G. Deng, H. Liu, Y.-Z. Jiang, M. C. K. Choi and A. S. C. Chan, *Chem. Commun.*, **2001**, 1488.
70. S. Akabori, S. Sakurai, Y. Izumi and Y. Fujii, *Nature*, **1956**, *178*, 323.
71. Y. Izumi, *Adv. Catal.*, **1983**, *32*, 215.
72. M. Nakahata, M. Imaida, H. Ozaki, T. Harada and A. Tai, *Bull. Chem. Soc. Jpn.*, **1982**, *55*, 2186.
73. Y. Orito, S. Imai and S. Niwa, *J. Chem. Soc. Jpn.*, **1980**, 670.
74. M. Garland and H.-U. Blaser, *J. Am. Chem. Soc.*, **1990**, *112*, 7048.
75. X. Zuo, H. Liu and M. Liu, *Tetrahedron Lett.*, **1998**, *39*, 1941.
76. H.-U. Bläser, *Tetrahedron: Asymmetry*, **1991**, *2*, 843.
77. A. Hu, G. T. Yee and W. Lu, *J. Am. Chem. Soc.*, **2005**, *127*, 12486.
78. (a) P. N. Liu, J. G. Deng, Y. Q. Tu and S. H. Wang, *Chem. Commun.*, **2004**, 2070. (b) P. N. Liu, P.-M. Gu, J.-G. Deng, Y.-Q. Tu and Y.-P. Ma, *Eur. J. org. Chem.*, **2005**, 3221.
79. A. Hu, H. L. Ngo and W. Lin, *J. Am. Chem. Soc.*, **2003**, *125*, 11490.
80. For an overview, see; G. Procter, *Asymmetric Synthesis*, Oxford University Press, Oxford, **1996**, 161.
81. A. Hirao, S. Itsuno, S. Nakahama and N. Yamazaki, *J. Chem. Soc., Chem. Commun.*, **1981**, 315.
82. S. Itsuno, M. Nakano, K. Miyazaki, H. Masuda and K. Ito, *J. Chem. Soc., Perkin Trans. 1*, **1983**, 1673.
83. S. Itsuno, Y. Sakurai, K. Ito, A. Hirao and S. Nakahama, *Bull. Chem. Soc. Jpn.*, **1987**, 395.
84. E. J. Corey, R. K. Bakshi and S. Shibata, *J. Am. Chem. Soc.*, **1987**, *109*, 5551.
85. For reviews see: (a) S. Wallbaum and J. Martens, *Tetrahedron: Asymmetry*, **1992**, *3*, 1475. (b) B. B. Lohray and V. Bhushan, *Angew. Chem. Int. Ed. Engl.*, **1992**, *31*, 729. (c) L. Deloux and M. Srebnik, *Chem. Rev.*, **1993**, *93*, 763. (d) E. J. Corey and C. J. Helal, *Angew. Chem. Int. Ed.*, **1998**, *37*, 1986. (e) S. Itsuno, in *Comprehensive Asymmetric Catalysis*, Vol. *1*, ed. E. N. Jacobsen, A. Pfaltz and H. Yamamoto, Springer-Verlag, Berlin, **1999**, 289. (f) B. T. Cho, *Tetrahedron: Asymmetry*, **2006**, *62*, 7621.
86. E. J. Corey, R. K. Bakshi, S. Shibata, C.-P. Chen, V. K. Singh, *J. Am. Chem. Soc.*, **1987**, *109*, 7925.
87. E. J. Corey and J. O. Link, *Tetrahedron Lett.*, **1990**, *31*, 601.
88. E. J. Corey and R. K. Bakshi, *Tetrahedron Lett.*, **1990**, *31*, 611.
89. W. Behnen, C. Dauelsberg, S. Wallbaum and J. Martens, *Synth. Commun.*, **1992**, *22*, 2143.
90. G. Giffels, C. Dreisbach, U. Kragl, M. Weigerding, H. Waldmann and C. Wandrey, *Angew. Chem. Int. Ed. Engl.*, **1995**, *34*, 2005.
91. T. Yamada, T. Nagata, K. D. Sugi, K. Yorozu, T. Ikeno, Y. Ohtsuka, D. Miyazaki and T. Mukaiyama, *Chem. Eur. J.*, **2003**, *9*, 4485.
92. A. Kokura, S. Tanaka, H. Teraoka, A. Shibahara, T. Ikeno, T. Nagata and T. Yamada, *Chem. Lett.*, **2007**, 26.
93. K. D. Sugi, T. Nagata, T. Yamada and T. Mukaiyama, *Chem. Lett.*, **1996**, 737.
94. Y. Ohtsuka, T. Kubota, T. Ikeno, T. Nagata and T. Yamada, *Synlett*, **2000**, 535.
95. Y. Ohtsuka, K. Koyasu and T. Yamada, *Org. Lett.*, **2001**, *3*, 2543.
96. Y. Ohtsuka, K. Koyasu, D. Miyazaki, T. Ikeno and T. Yamada, *Org. Lett.*, **2001**, *3*, 3421.

97. Y. Ohtsuka, D. Miyazaki, T. Ikeno and T. Yamada, *Chem. Lett.*, **2002**, 24.

98. (a) H. Brunner, H. Nishiyama and K. Itoh in *Catalytic Asymmetric Synthesis*, VCH, New York, **1993**, 303. (b) H. Ishiyama, in *Comprehensive Asymmetric Catalysis*, Vol. *1*, ed. E. N. Jacobsen, A. Pfaltz and H. Yamamoto, Springer-Verlag, Berlin, **1999**, 267. (c) J.-F. Carpentier and V. Bette, *Curr. Org. Chem.*, **2002**, *6*, 913. (d) O. Riant, N. Mostefaï and J. Courmarcel, *Synthesis*, **2004**, 2943.

99. H. Brunner, R. Becker and G. Riepl, *Organometallics*, **1984**, *3*, 1354.

100. H. Brunner and U. Obermann, *Chem. Ber.*, **1989**, *122*, 499.

101. H. Nishiyama, H. Sakaguchi, T. Nakamura, M. Horihata, M. Kondo and K. Itoh, *Organometallics*, **1989**, *8*, 847.

102. G. Balavoine, J. C. Clinet and I. Lellouche, *Tetrahedron Lett.*, **1989**, *38*, 5141.

103. The first report of an oxazoline ligand in asymmetric catalysis was from Brunner's group: H. Brunner, U. Obermann and P. Wimmer, *J. Organomet. Chem.*, **1986**, *316*, C1.

104. For a review on the use of bis-oxazolines in asymmetric catalysis, see; A. K. Ghosh, P. Mathivanan and J. Cappiello, *Tetrahedron: Asymmetry*, **1998**, *9*, 1.

105. H. Nishiyama, M. Kondo, T. Nakamura and K. Itoh, *Organometallics*, **1991**, *10*, 500.

106. H. Brunner and P. Brandl, *J. Organomet. Chem.*, **1990**, *390*, C81.

107. G. Helmchen, A. Krotz, K.-T. Ganz and D. Hansen, *Synlett*, **1991**, 257.

108. Y. Imai, W. Zhang, T. Kida, Y. Nakatsuji and I. Ikedan, *Tetrahedron: Asymmetry*, **1996**, *7*, 2453.

109. L. M. Newman, J. M. J. Williams, R. McCague and G. A. Potter, *Tetrahedron: Asymmetry*, **1996**, *7*, 1597.

110. T. Langer, J. Janssen and G. Helmchen, *Tetrahedron: Asymmetry*, **1996**, *7*, 1599.

111. Y. Nishibayashi, K. Segawa, H. Takada, K. Ohe and S. Uemura, *J. Chem. Soc., Chem. Commun.*, **1996**, 847.

112. M. Sawamura, R. Kuwano, J. Shirai and Y. Ito, *Synlett*, **1995**, 347.

113. M. Sawamura, R. Kuwano and Y. Ito, *Angew. Chem., Int. Ed. Engl.*, **1994**, *33*, 111.

114. B. Tao and G. C. Fu, *Angew. Chem., Int. Ed.*, **2002**, *41*, 3892.

115. D. A. Evans, F. E. Michael, J. S. Tedrow and K. R. Campos, *J. Am. Chem. Soc.*, **2003**, *125*, 3534.

116. W. L. Duan, M. Shi and G.-B. Rong, *Chem. Commun.*, **2003**, 2916.

117. (a) L. H. Gade, V. César and S. Bellemin-Laponnaz, *Angew. Chem. Int, Ed.*, **2004**, *43*, 1014. (b) V. César, S. Bellemin-Laponnaz, H. Wadepohl and L. H. Gade, *Chem. Eur. J.*, **2005**, *11*, 2862.

118. T. Ohta, M. Ito, A. Tsuneto and H. Takaya, *J. Chem. Soc., Chem. Commun.*, **1994**, 2525.

119. (a) G. Zhu, M. Terry, X. Zhang, *J. Organomet. Chem.*, **1997**, *547*, 97. (b) Y. Nishibayashi, I. Takei, S. Uemura and H. Masanobu, *Organometallics*, **1998**, *17*, 3420. (c) C. Moreau, C. G. Frost and B. Murrer, *Tetrahedron Lett.*, **1999**, *40*, 5617. (d) C. Song, C. Ma, Y. Ma, W. Feng, S. Ma, Q. Chai and M. B. Andruss, *Tetrahedron Lett.*, **2005**, *46*, 3241.

120. J.-I. Sakaki, W. B. Schweizer and D. Seebach, *Helv. Chim. Acta*, **1993**, *76*, 2654.

121. H. Imma, M. Mori and T. Nakai, *Synlett*, **1996**, 1229.

122. (a) M. B. Carter, B. Schiøtt, A. Gutiérrez and S. L. Buchwald, *J. Am. Chem. Soc.*, **1994**, *116*, 11667. (b) J. Yun and S. L. Buchwald, *J. Am. Chem. Soc.*, **1999**, *121*, 5640.

123. (a) B. H. Lipshutz, K. Noson and W. Chrisman, *J. Am. Chem. Soc.*, **2001**, *123*, 12917. (b) B. H. Lipshutz, A. Lower and K. Noson, *Org. Lett.*, **2002**, 4045. (c) B. H. Lipshutz, A. Lower, R. J. Kucejko and K. Noson, *Org. Lett.*, **2006**, 2969.

124. S. Sirol, J. Courmacel, N. Mostefai and O. Riant, *Org. Lett.*, **2001**, 4111.

125. X. Verdageur, U. E. W. Lange, M. T. Reding and S. L. Buchwald, *J. Am. Chem. Soc.*, **1994**, *116*, 11667.

126. X. Verdaguer, U. E. W. Lange and S. L. Buchwald, *Angew. Chem., Int. Ed. Engl.*, **1998**, *37*, 1103.

127. M. C. Hansen and S. L. Buchwald, *Org. Lett.*, **2000**, *2*, 713.

128. (a) Z. Wang, X. Ye, S. Wei, P. Wu, A. Zhang and J. Sun, *Org. Lett.*, **2006**, *8*, 999. (b) Z. Wang, M. Cheng, P. Wu, W. Wei and J. Sun, *Org. Lett.*, **2006**, *8*, 3045.

129. B. H. Lipshutz and H. Shimizu, *Angew. Chem. Int. Ed.*, **2004**, *43*, 2228.

130. S.-I. Murahashi, S. Watanabe and T. Shiota, *J. Chem. Soc., Chem. Commun.*, **1994**, 725.

Chapter 4
Epoxidation

This chapter considers the catalytic enantioselective formation of epoxides from alkenes and aldehydes.[1] Most of the individual sections are ordered by catalyst class rather than substrate type. However, a number of generalisations regarding substrate scope can be made. Despite the very high selectivities obtained with many substrates, there is still no universal catalyst that is effective for all classes of alkene structure. The titanium/tartrate-based catalysts developed by Sharpless are highly effective for the epoxidation of primary, *trans*-allylic alcohols, while enantiomerically pure vanadium-based catalysts have been used to good effect in the epoxidation of *cis*-allylic alcohols and homoallylic substrates. Both these catalyst systems proceed via metal coordination to the alcohol moiety and thus cannot be used to effect epoxidation of unfunctionalised alkenes. Chiral nonracemic metal(salen) complexes are effective catalysts in the epoxidation of unfunctionalised alkenes, but scope is largely limited to *cis*-alkenes and trisubstituted substrates and unfunctionalised terminal olefins are generally poor substrates. However, enantiopure metal porphyrins are beginning to show promise as catalysts for the asymmetric epoxidation of the latter class of alkene. A separate subchapter is devoted to the enantioselective epoxidation of electron deficient olefins – especially α,β-enones and some progress has been made in this area using oxidants in the presence of enantiomerically pure Lewis acids or organocatalysts for instance. The most successful organocatalyst-based asymmetric epoxidation utilises dioxiranes derived from ketones and the scope of this process continues to expand. The catalytic asymmetric epoxidation of aldehydes using enantiopure sulfonium ylides is also included and this chapter concludes with a discussion of asymmetric aziridinations of both alkenes and imines.

4.1 Epoxidation of Allylic Alcohols

The asymmetric epoxidation of allylic alcohols using titanium tetraisopropoxide/tartrate ester/*tert*-butylhydroperoxide was developed by Sharpless during the 1980s to become one of the most important methods of asymmetric catalysis. Whilst there

Catalysis in Asymmetric Synthesis 2e   © 2009 Vittorio Caprio and Jonathan M.J. Williams

has been little further significant development of this reaction in recent years, this is simply a testament to the fact that the reaction already works well. There are several reviews available detailing the Sharpless asymmetric epoxidation reaction.[1,2]

The first reports from the Sharpless group described the stoichiometric use of the 'catalyst',[3] however, the truly catalytic variant of the reaction was found to be more general in the presence of activated molecular sieves.[4] The benefits of using catalytic amounts of the titanium/tartrate combination include not only reduced cost, but also an easier work-up procedure. This is particularly true for water soluble epoxide products such as glycidol.

The mechanism of the Sharpless asymmetric epoxidation reaction has been a topic of some debate.[5] It is clear that the isopropoxy ligands can be displaced by the diol (tartrate), the hydroperoxide and the allylic alcohol substrate. In fact, at the point of epoxidation, a catalytic ensemble containing two titanium atoms seems likely and the asymmetric environment provided by the tartrate influences the stereoselectivity of the epoxidation. Figure 4.1 represents a likely assembly in the Sharpless epoxidation reaction and this dimeric structure has been supported by X-ray crystallographic studies.[6] Examination of this dimer readily reveals the scope of the asymmetric epoxidation. The R^1 substituent is oriented away from the catalyst and thus *trans*-allylic alcohols undergo highly enantioselective epoxidation. *cis*-Isomers also undergo asymmetric epoxidation, but with generally lower selectivity especially when R^2 is a bulky substituent. Primary allylic alcohols are the most common substrates for this reaction. Substituents other than hydrogen at R^5 undergo steric interactions with the ligand and tertiary allylic alcohols are very poor substrates for this reaction. Secondary allylic alcohols do undergo epoxidation when $R^5 = H$ and this facial discrimination between enantiomers provides the basis for a kinetic resolution procedure.

Despite the complexity of the active catalyst, the sense of asymmetric induction in Sharpless asymmetric epoxidation reactions can be reliably predicted using the model shown in Figure 4.2. In order for the model to predict the stereochemical outcome correctly, only two points need to be remembered. The allylic hydroxy group resides in the bottom right corner and D-(-)-diethyl tartrate (which has the (S,S)-configuration) attacks from above the plane.

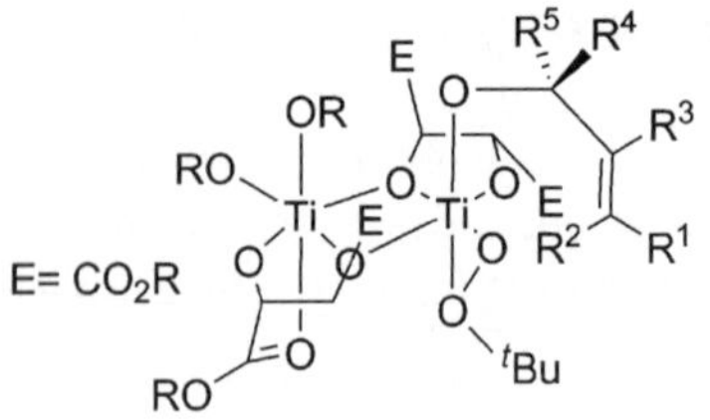

Figure 4.1 Possible assembly in the Sharpless Asymmetric Epoxidation

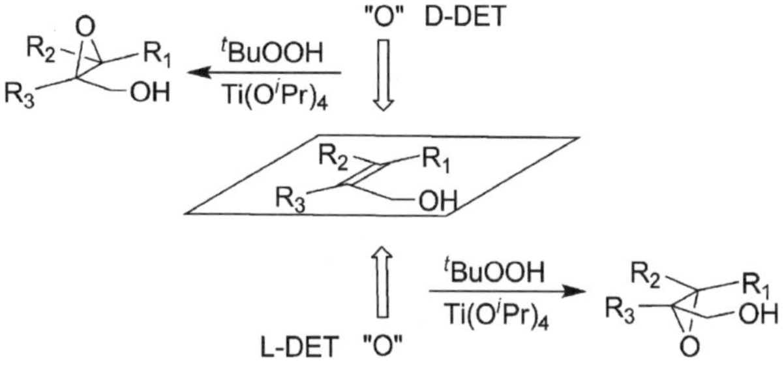

Figure 4.2 Schematic representation of the Sharpless Asymmetric Epoxidation. Abbreviation: DET = diethyl tartrate

Throughout this section, allylic alcohols will be drawn in this way, such that it is clear how the stereochemistry can be related to the model. Thus, allyl alcohol (**4.01**) is converted into (*R*)-glycidol (**4.02**) using the D-(-)-diethyl tartrate/titanium tetraisopropoxide combination. However, the use of L-(+)-diethyltartrate as the ligand affords the opposite enantiomer, (*S*)-glycidol.

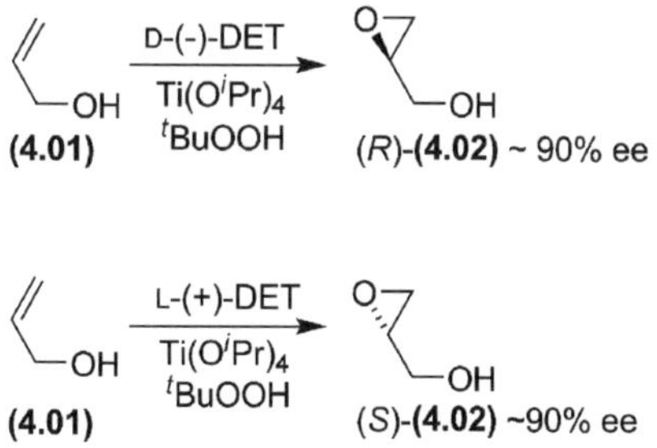

Variously substituted primary allylic alcohols all obey the model. This method can be used to form epoxides with tertiary and quaternary stereocentres and is also compatible with a variety of functional groups including, alcohols, ethers, amine groups and other alkenyl and alkynyl moieties. [2f] A range of examples is provided by the conversions of the allylic alcohols (**4.03–4.08**) into the corresponding epoxides (**4.09–4.14**) with high ee. In the case of the diene (**4.07**), it is noteworthy that only the alkene that is part of the allylic alcohol unit undergoes epoxidation. Conjugated dienes such as (**4.08**) also undergo selective epoxidation at the double bond allylic to the directing hydroxyl group. [7] The use of dimethyl tartrate (DMT), diethyltartrate (DET) and diisopropyl esters (DIPT) are all fairly common, and generally give similar levels of selectivity, although optimal selectivities can be obtained by screening the various esters.

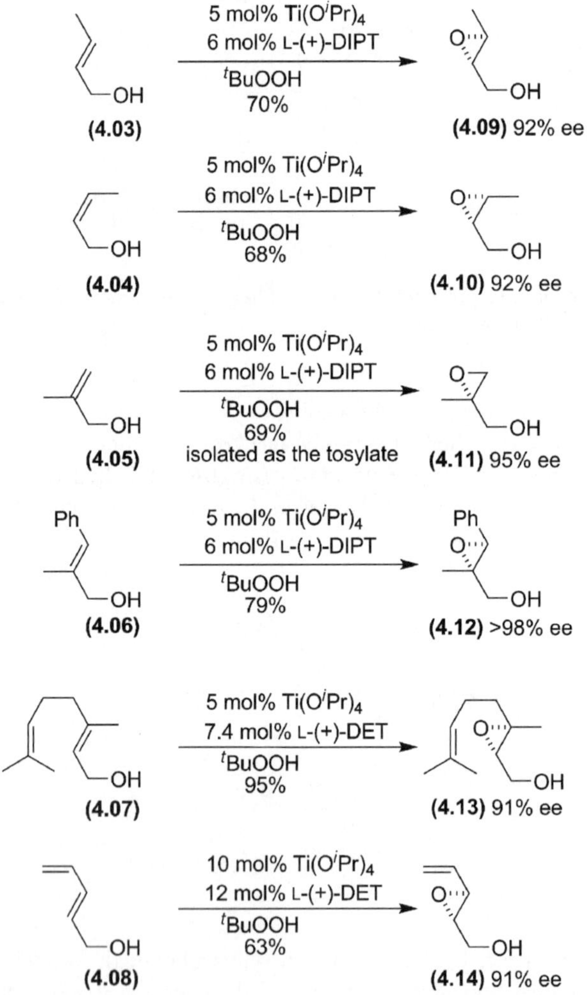

5 mol% Ti(O^{i}Pr)$_4$
6 mol% L-(+)-DIPT
^{t}BuOOH
70%

(4.03)

(4.09) 92% ee

5 mol% Ti(O^{i}Pr)$_4$
6 mol% L-(+)-DIPT
^{t}BuOOH
68%

(4.04)

(4.10) 92% ee

5 mol% Ti(O^{i}Pr)$_4$
6 mol% L-(+)-DIPT
^{t}BuOOH
69%
isolated as the tosylate

(4.05)

(4.11) 95% ee

Ph

5 mol% Ti(O^{i}Pr)$_4$
6 mol% L-(+)-DIPT
^{t}BuOOH
79%

(4.06)

Ph

(4.12) >98% ee

5 mol% Ti(O^{i}Pr)$_4$
7.4 mol% L-(+)-DET
^{t}BuOOH
95%

(4.07)

(4.13) 91% ee

10 mol% Ti(O^{i}Pr)$_4$
12 mol% L-(+)-DET
^{t}BuOOH
63%

(4.08)

(4.14) 91% ee

Heterogenous versions of the Sharpless epoxidation, commonly using polymer-linked tartrates, have also been successful.[8] The polymer scaffold chain length has been shown to influence the mode of enantioselection in some cases. For instance, while *trans*-2-hex-2-en-1-ol is converted to the expected (2S,3S)-epoxide with 93% ee using the tartrate ester prepared from L-(+)-tartrate and poly(ethylene)glycol monomethyl ether with a molecular weight of 75 (MPEG 75), surprisingly, use of ligand prepared from MPEG 2000 results in formation of the (2R,3R)-epoxide with 75% ee.[9] This reversal of enantioselectivity is postulated to arise from formation of

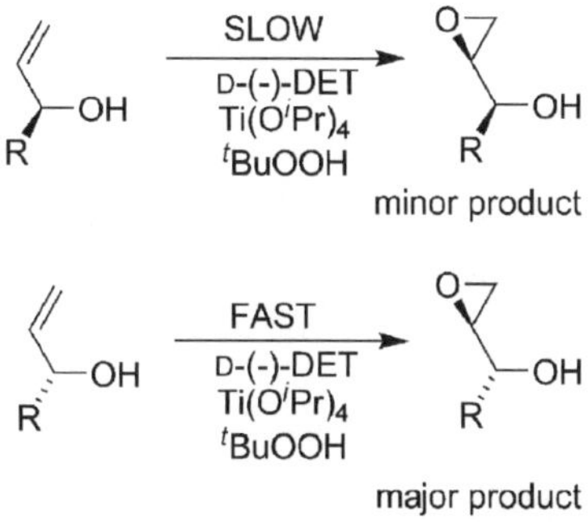

Figure 4.3 Sharpless kinetic resolution of racemic substrates. Abbreviation: DET = diethyl tartrate

a monomeric 2:1 Ti–ligand complex in the presence of long chain MPEG that has an alternate asymmetric environment around the metal than the normal 2:2 dimeric complexes observed in traditional Sharpless epoxidations (see Figure 4.1).[10]

There can be no doubt that the reliability of the Sharpless reaction amongst many different classes of allyl alcohol contributes to its success as a synthetic tool in asymmetric synthesis. Another remarkable attribute of the Sharpless asymmetric epoxidation is the very high level of discrimination between enantiomers of secondary allylic alcohols leading to the wide use of this system for the kinetic resolution of these substrates.[2] The kinetic resolution (Figure 4.5) was first reported using stoichiometric amounts of titanium/tartrate,[11] but catalytic amounts of titanium may also be employed.[12,13]

Examples of the kinetic resolution include the reaction of the secondary alcohols (**4.15**) and (**4.16**) which show large differences in the rates of the faster and slower reacting enantiomers. In the case of substrate (**4.16**), the $k$rel (= k_{fast}/k_{slow}) is as high as 700! k_{rel} is also known as the selectivity factor, S (see Section 12.8). Whilst these results are obtained under stoichiometric conditions, kinetic resolutions are successful under catalytic conditions, especially when the alkene possesses a bulky group remote from the hydroxy-containing substituent.[4b] In general, as with other kinetic resolutions, best results are obtained by stopping the reaction before 50% conversion, if it is the product that is needed. If it is the recovered starting material which is needed, then the reaction should be allowed to proceed beyond 50% conversion.

The substrate (**4.19**) is achiral, but epoxidation of the enantiotopic alkenes occurs selectively, to give the expected product (**4.20**), where a new chiral centre is formed at the secondary alcohol, as well as the one associated with the epoxide.[14] The substrate (**4.21**) represents a real challenge for a selective epoxidation reaction. There are two enantiomers of the substrate, each of which has two different alkenes which could react. Each alkene group has two diastereotopic faces. In all,

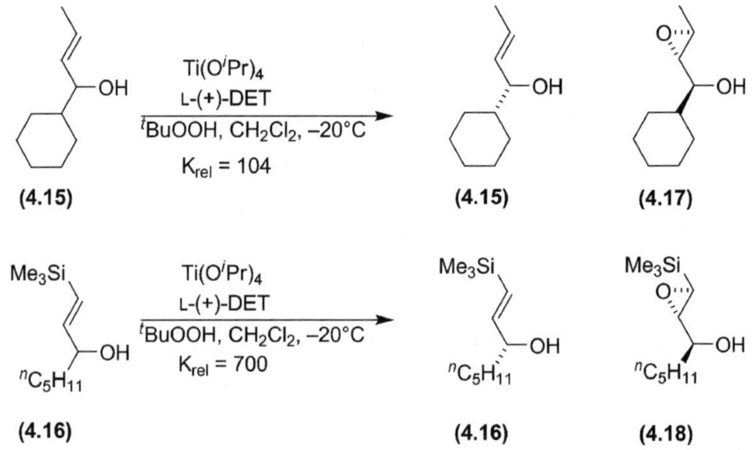

(4.15)

Ti(O^{i}Pr)$_4$
L-(+)-DET
^{t}BuOOH, CH$_2$Cl$_2$, –20°C
K$_{rel}$ = 104

(4.15)   **(4.17)**

(4.16)

Ti(O^{i}Pr)$_4$
L-(+)-DET
^{t}BuOOH, CH$_2$Cl$_2$, –20°C
K$_{rel}$ = 700

Me$_3$Si ^{n}C$_5$H$_{11}$

(4.16)   **(4.18)**

there are eight possible outcomes, of which only one product, epoxide (4.22), is observed in practice.[15] This result was also obtained using stoichiometric titanium reagent, but the underlying principles would be the same for the catalytic variant.

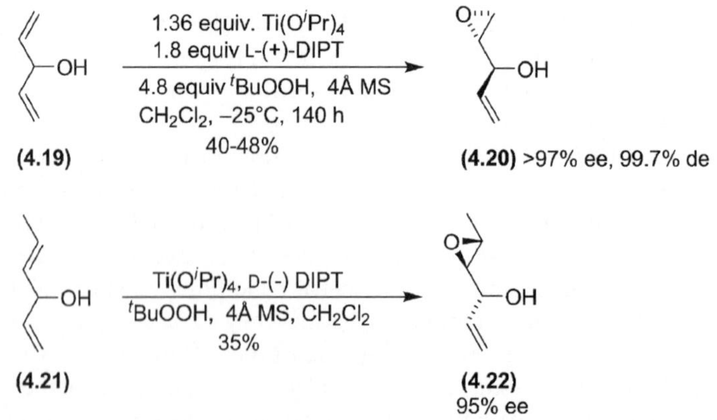

(4.19)

1.36 equiv. Ti(O^{i}Pr)$_4$
1.8 equiv L-(+)-DIPT
4.8 equiv ^{t}BuOOH, 4Å MS
CH$_2$Cl$_2$, –25°C, 140 h
40–48%

(4.20) >97% ee, 99.7% de

(4.21)

Ti(O^{i}Pr)$_4$, D-(-) DIPT
^{t}BuOOH, 4Å MS, CH$_2$Cl$_2$
35%

(4.22)
95% ee

Other metal-based epoxidation catalysts have been explored to overcome some of the limitations of the Sharpless procedure. One drawback with the Sharpless asymmetric epoxidation is the slightly lower ees often obtained when using *cis*-olefin substrates. The group of Yamamoto have achieved highly enantioselective epoxidations of *cis*-alkenes using vanadium(V) oxytriisopropoxide in the presence of C$_2$-symmetric bishydroxamic acid ligands such as (4.23).[16] In contrast to the Sharpless procedure this process is not hampered by the presence of air or moisture and can be performed using aqueous hydroperoxides as cooxidants. Even *cis*-alkenes with relatively bulky substituents such as (**4.24**) and (**4.25**) undergo oxidation with high ee using this procedure. The greater scope of this vanadium-catalysed epoxidation is attributed to the flexibility of the active vanadium catalyst formed *in situ.*

The Sharpless epoxidation proceeds poorly with homoallylic alcohols and some work has been directed towards the development of zirconium and vanadium-based catalytic systems capable of oxidising these substrates in an enantioselective

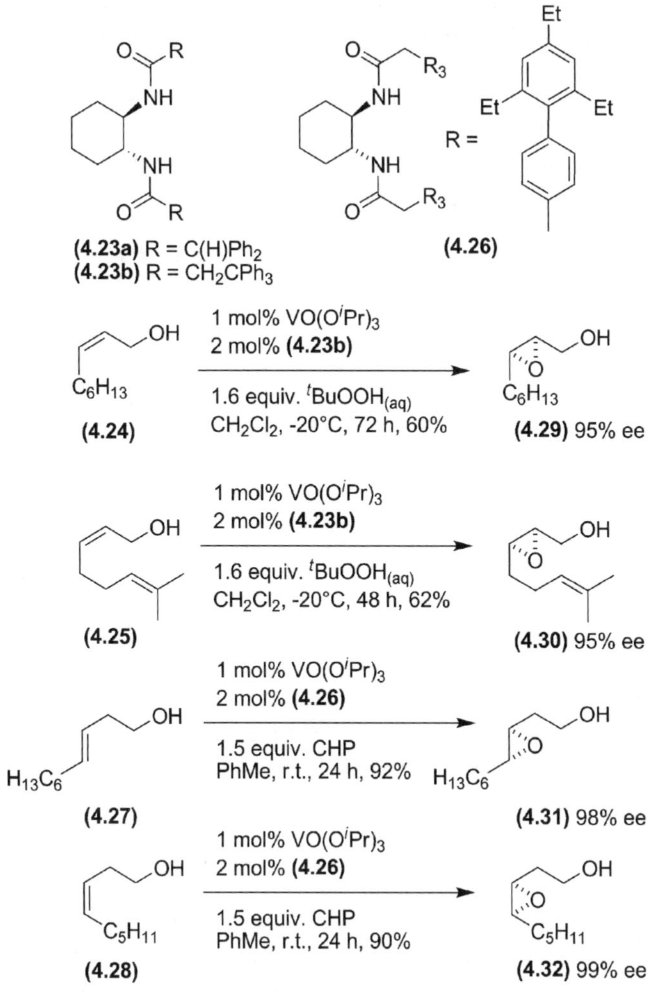

(**4.23a**) R = C(H)Ph$_2$
(**4.23b**) R = CH$_2$CPh$_3$

(**4.26**)

(**4.24**) → (**4.29**) 95% ee

1 mol% VO(O^{i}Pr)$_3$
2 mol% (**4.23b**)

1.6 equiv. ^{t}BuOOH$_{(aq)}$
CH$_2$Cl$_2$, -20°C, 72 h, 60%

(**4.25**) → (**4.30**) 95% ee

1 mol% VO(O^{i}Pr)$_3$
2 mol% (**4.23b**)

1.6 equiv. ^{t}BuOOH$_{(aq)}$
CH$_2$Cl$_2$, -20°C, 48 h, 62%

(**4.27**) → (**4.31**) 98% ee

1 mol% VO(O^{i}Pr)$_3$
2 mol% (**4.26**)

1.5 equiv. CHP
PhMe, r.t., 24 h, 92%

(**4.28**) → (**4.32**) 99% ee

1 mol% VO(O^{i}Pr)$_3$
2 mol% (**4.26**)

1.5 equiv. CHP
PhMe, r.t., 24 h, 90%

CHP = cumene hydroperoxide

manner.[17] The best results to date have been obtained by the group of Yamamoto using the hindered hydroxamic acid ligand (**4.26**) and VO(O^{i}Pr)$_3$.[18] This catalyst system is capable of oxidising both *cis-* and *trans-*homoallylic alcohols such as (**4.27**) and (**4.28**).

4.2 Epoxidation with Metal(salen) Complexes

Metal complexes of enantiomerically pure N,N'-ethylenebis(salicylideneaminato) (salen) complexes in combination with stoichiometric oxidants currently provide the most selective method for the catalytic asymmetric epoxidation of unfunctionalised alkenes.[1,19] The use of C_2-symmetric salen complexes of manganese(III) were reported independently in 1990 by Jacobsen and coworkers[20] and Katsuki and coworkers.[21] The first generation catalysts are represented by the general structure (**4.33**). The complex with R = ^{t}Bu is known as Jacobsen's catalyst.[22] All of the first generation catalysts are composed of a enantiopure diamine core and possess large substituents at the 3/3′ and 5/5′ positions. Subsequently Katsuki and coworkers developed second generation catalysts such as (**4.34**) with axially chiral groups at the 3/3′ positions.[23]

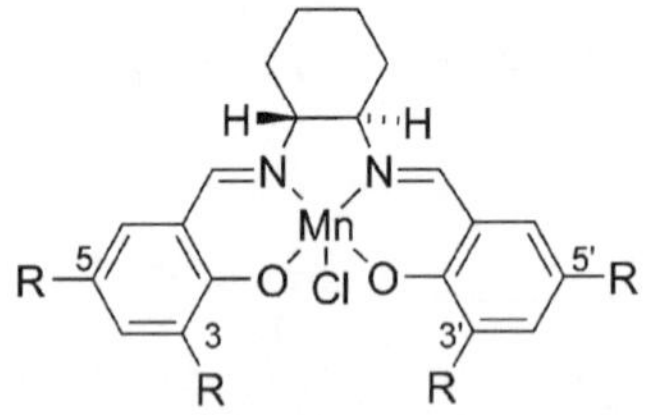

(4.33) 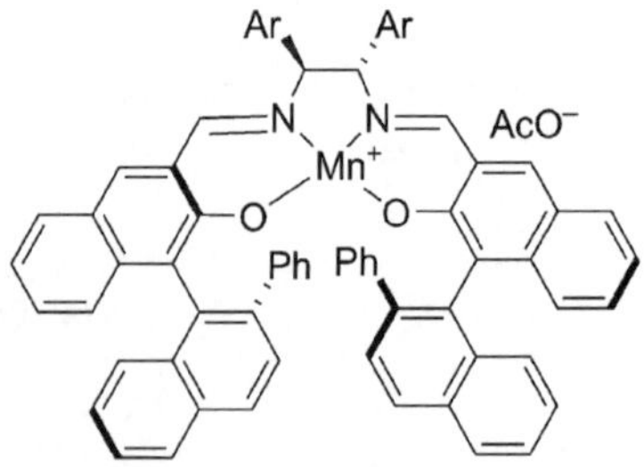   (4.34)

A variety of catalysts with differing aromatic substituents and ethylenediamine cores have been prepared and tested, but few perform better than Jacobsen's catalyst. These catalysts are not general for all alkene substrates and the best results have been achieved using *cis-*alkenes, trisubstituted alkenes and some tetrasubstituted derivatives. Especially high ees have been obtained with conjugated alkenes, in particular chromenes and β-substituted styrenes while *trans-*olefins and terminal olefins are poor substrates for this process.

The details of the manganese(salen) catalysed epoxidation remain a topic of some debate.[19i] It has been established that the epoxidation proceeds via formation of an oxo Mn(V) complex by reaction of the Mn(III) complex with the oxidant.[24] However, a number of other oxidising species have also been postulated to take part.[25] The reaction is not always diastereoselective, especially using conjugated alkene substrates where mixtures of *cis* and *trans*-epoxides often result and, while a concerted mechanism has been proposed for the oxidation of nonconjugated alkenes, a radical-based process has been proposed for the epoxidation of conjugated alkenes. The involvement of a radical intermediate in a step-wise process neatly explains the isomerisation, and also explains why it is the better radical-stabilising group that becomes stereochemically scrambled.[26] Thus, oxidation of an Mn(III) complex (**4.35**) to an Mn(V) complex (**4.36**) provides a catalytic species capable of oxidising an alkene (e.g. alkene (**4.37**)). The reaction then can proceed either via a concerted transition state (**4.38**) or a benzylic radical intermediate (**4.39**) that can then undergo direct collapse to the *cis*-epoxide (**4.40**), or rotate before collapse, affording the *trans*-epoxide (**4.41**), as shown in Figure 4.4. The involvement of manganaoxetane complexes, formed by [2 + 2] addition, has also been proposed[27] and disputed.[28]

The catalyst adopts a nonplanar, bent structure and asymmetric induction in this process is postulated to arise from side on approach of the alkene from the least hindered side, with the largest substituent pointing away from the bulky substituents of the aromatic ring (Figure 4.5).[27a, 29] The approach vector depicted

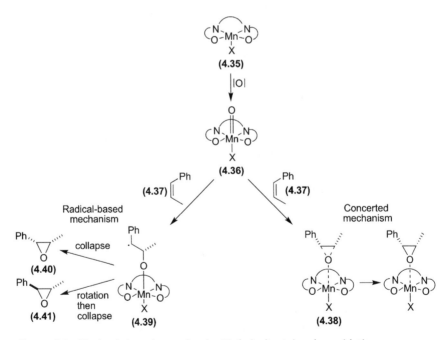

Figure 4.4 Mechanistic pathways for the Mn(salen)-catalysed epoxidations

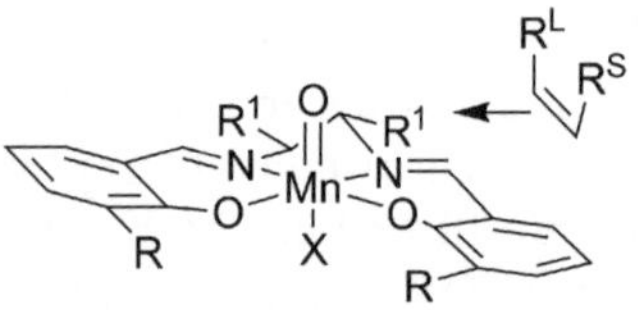

Figure 4.5　Trajectory of approach of alkene to oxo Mn(salen)

explains the somewhat poor selectivities obtained with *trans*-alkenes, where one substituent must be oriented towards the ligand.

Most frequently, iodosylbenzene (PhIO) or sodium hypochlorite (NaOCl) are used as the stoichiometric oxidant, although alternative reagents have been used, including hydrogen peroxide,[30] periodate,[31] dimethyldioxirane,[32] and an *m*CPBA/*N*-methylmorpholine-*N*-oxide combination which allows the use of a lower temperature and provides higher enantioselectivities.[33]

Catalyst (*t*Bu-**4.33**) is effective in the epoxidation of many *cis*-alkenes, such as the acyclic alkenes (**4.42**) and also cyclic alkenes, especially 2,2-dimethylchromenes such as (**4.43**).[34] Enhanced enantioselectivities are often observed using Katsuki's second generation catalysts (**4.34**) where the phenyl substituents of the BINAP moiety undergo greater steric interactions with the alkene substrate. Dihydronaphthalene (**4.44**) has been epoxidised with record ees using this catalyst system.[35]

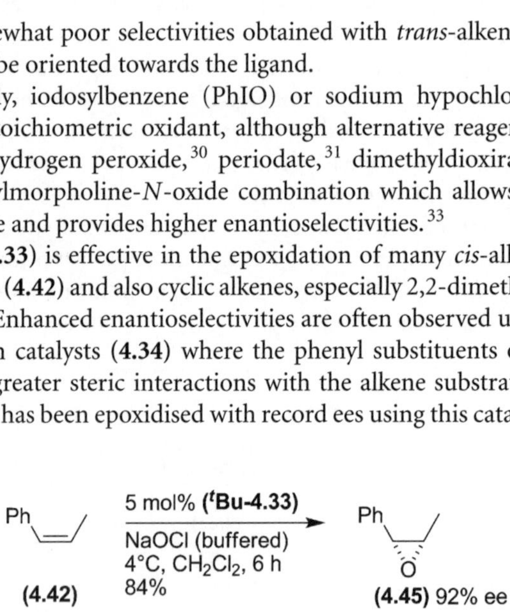

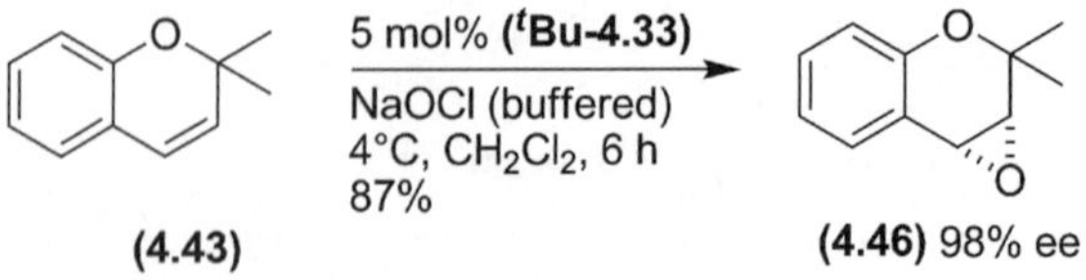

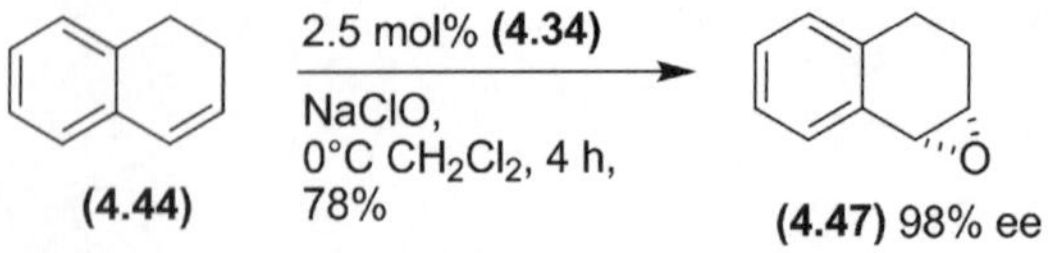

Tri-substituted alkenes are also good substrates for the metal(salen)-catalysed epoxidation.[36] The addition of 4-phenyl pyridine *N*-oxide was found to be beneficial in using lower catalytic loadings in industrial applications.[37,38] This is exemplified by the epoxidation of alkene (**4.48**). Cyclic tetrasubstituted alkenes have also been epoxidised with high enantiomeric excess. The best results seem to be with chromene substrates such as compound (**4.49**). [39] Di- and tri-substituted alkenes on a chromene scaffold undergo epoxidation with high enantioselectivity (94 to >98% ee). [40] Enynes generally give a mixture of *cis* and *trans*-epoxides. Katsuki and coworkers have converted the mixture of epoxides (**4.51**) and (**4.52**) into an alcohol (**4.53**) containing one stereocentre by reaction with lithium aluminium hydride, and used the products in synthetic procedures to give enantiomerically enriched insect pheromones. [41]

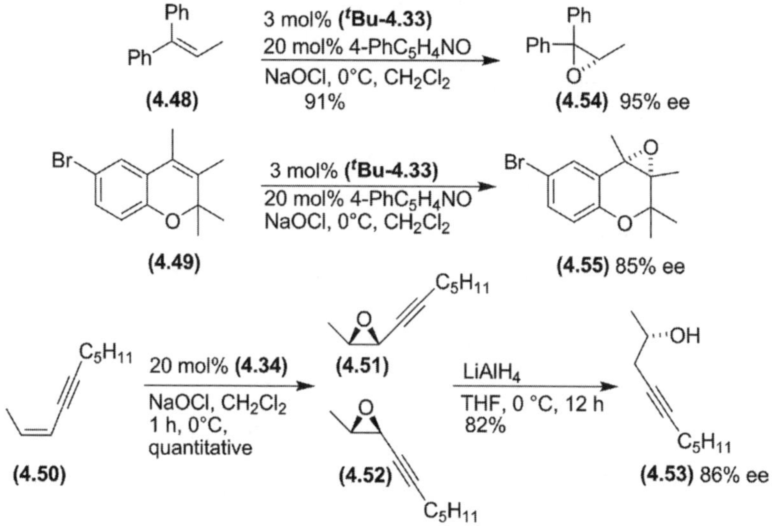

Trans-alkenes are poor substrates for the manganese(salen)-catalysed asymmetric epoxidation. One strategy developed to overcome this is the use of additives that promote the formation of *trans*-epoxides during the epoxidation of *cis*-alkenes. Jacobsen and coworkers have discovered that enantiomerically pure quaternary ammonium salts are particularly effective in this regard. [42] It is not clear why these salts reverse the normal diastereoselectivity of the reaction, but the epoxidation of *cis*-stilbene to *trans*-stilbene oxide can be achieved with >96:4 *trans*-selectivity and 90% ee using catalyst (**'Bu-4.33**). Phenolate additives can also be used to effect enantio- and diastereoselective formation of *trans*-epoxides from *cis*-alkenes. [43] In this case it is proposed that the *trans*-selectivity arises from a radical-based epoxidation that predominates in the presence of the phenolate anion. An alternate strategy for the synthesis of *trans*-epoxides is to use a catalyst better able to accommodate the approach of a *trans*-alkene. Katsuki's second generation catalysts such as (**4.56**), where one phenyl ring of the BINAP moiety is omitted, adopt a more deeply stepped conformation than Jacobsen's catalyst and have been used to effect asymmetric epoxidation of *trans*-β-alkyl substituted styrene derivatives such as (**4.57**).[44] The development and use of stable oxo chromium(salen) complexes has been reported by Bousquet and Gilheaney and, in contrast to manganese(salen) complexes, the chromium derivatives are more effective in the epoxidation of *trans*-alkenes.[19i, 45] For instance, *trans*-β-methylstyrene is epoxidised with up to 86% ee using chromium(salen) complexes such as (**4.58**). This selectivity is proposed to arise from the enhanced nonplanarity of these catalysts.

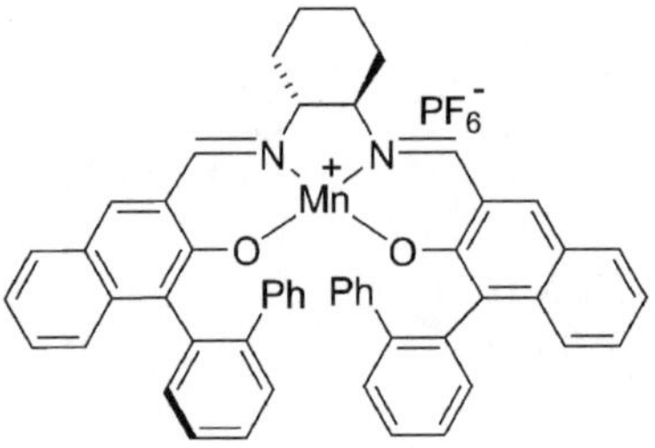

(**4.56**) 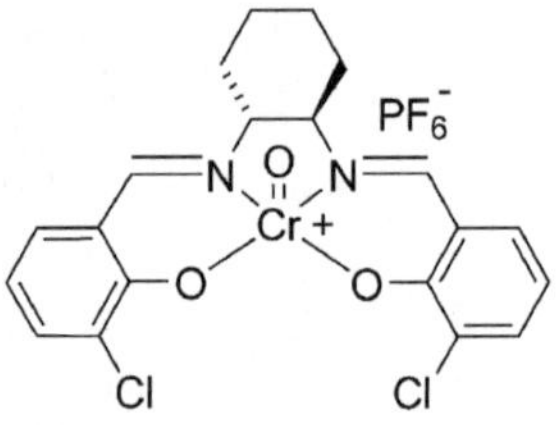   (**4.58**)

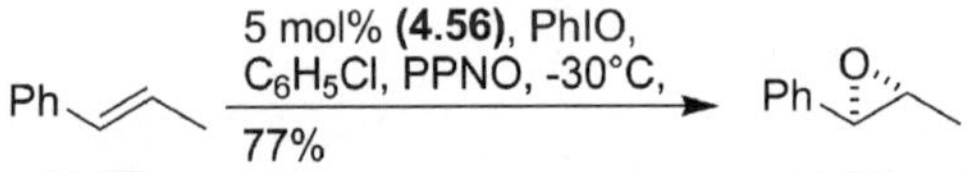

PPNO = 4-phenylpyridine-*N*-oxide

Dienes[46] have been used as substrates to give high enantioselectivity in the product mono-epoxides. Hentemann and Fuchs have shown that dienylsulfones are especially good substrates, at least in cyclic cases.[47] Thus, dienes (**4.60**) and (**4.61**) are converted into the mono-epoxides (**4.62**) and (**4.63**) as single regioisomers with excellent enantioselectivity.

Terminal olefins have also proved to be difficult substrates for the metal(salen)-catalysed epoxidation. The use of *m*CPBA as the stoichiometric oxidant allows the use of lower temperatures and has been used to give good enantioselectivities in the epoxidation of mono-substituted alkenes, such as styrenes (**4.64**) using catalyst (**4.65**), which give fairly good enantioselectivities in the formation of mono-substituted epoxides (**4.66**), although the enantioselectivity is less impressive than it is for many other classes of alkene epoxidation. The *m*CPBA must be added once the solution has been cooled, since premixing NMO and *m*CPBA does not give any epoxidation.

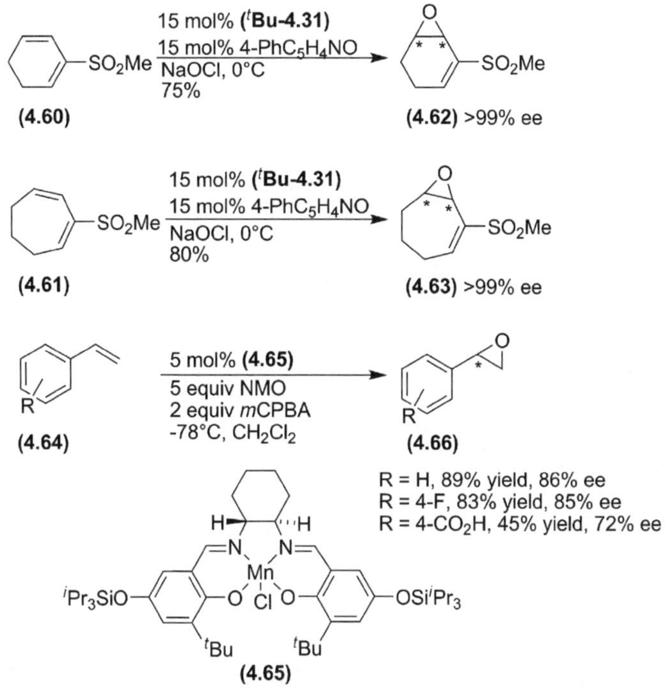

Racemic substrates have been employed in Mn(salen)-catalysed epoxidation reactions. Reasonable to good kinetic resolution of racemic chromanes (selectivity factor, $S = 2.7$–9.3)[48] and racemic allenes[49] have been achieved using enantiomerically pure manganese(salen) complexes to catalyse epoxidation.

Enol ethers are interesting substrates for 'epoxidations' since α-hydroxy ketones or the corresponding acetals are isolated, depending on the choice of solvent. Katsuki has used enol ethers as substrates, including the cyclic enol ether (4.67), which affords the hydroxy acetal product (4.68).[50] Adam has used silyl enol ethers and silyl ketene acetals as substrates. A typical example is provided by the asymmetric oxidation of silyl enol ether (4.69), generating the α-hydroxy ketone (4.70) after a suitable work up.[51]

A number of other transition metal(salen) complexes have been investigated. A ruthenium(salen)-catalysed asymmetric epoxidation of wide scope has been developed by Katsuki that proceeds under irradiation with visible light.[52] *Trans-* and terminal olefins are epoxidised with good ee but the selectivities obtained with *cis*-substrates are not as good as those achieved using Mn(salen) complexes. The use of palladium(salen) complexes as epoxidation catalysts has also been explored, but relatively poor selectivities have been obtained to date.[53]

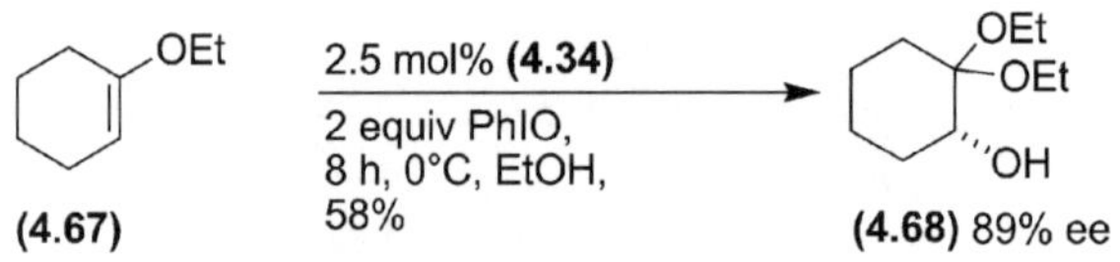

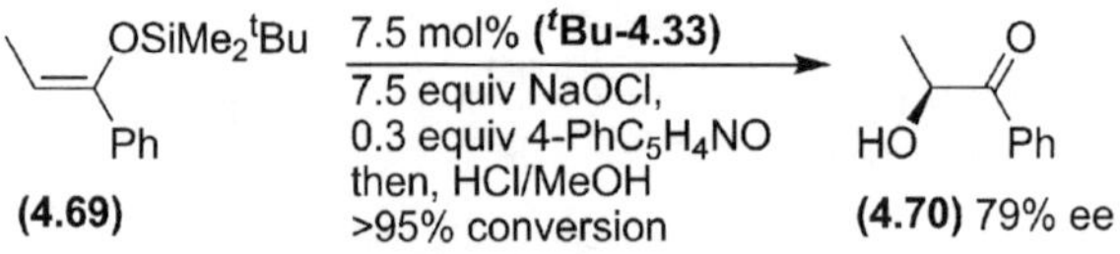

Reaction scheme: (4.67) with OEt, 2.5 mol% (4.34), 2 equiv PhIO, 8 h, 0°C, EtOH, 58% gives (4.68) 89% ee

Reaction scheme: (4.69) with OSiMe$_2$^{t}Bu and Ph, 7.5 mol% (^{t}Bu-4.33), 7.5 equiv NaOCl, 0.3 equiv 4-PhC$_5$H$_4$NO then, HCl/MeOH, >95% conversion gives (4.70) 79% ee

Recoverable salen complexes immobilised on organic polymer[54] and inorganic supports[54] have been shown to exhibit selectivities comparable or better than the homogenous versions. Manganese(salen) ligands possessing perfluoroalkyl groups (R = C$_8$F$_{17}$ in Jacobsen's ligand (4.33)) have been used successfully to effect asymmetric epoxidations under fluorous biphasic conditions.[56] Indene and triphenethene can be epoxidised with good ee using this process. At the end of the reaction the fluorous layer retains the catalyst and can be reused.

4.3 Epoxidation Using Metal–Porphyrin-Based Catalysts

Some work has been directed towards mimicking the action of heme-containing cytochrome p-450 enzymes using synthetic metalloporphyrins as epoxidation catalysts.[1b, 19a, 57] These catalysts show useful substrate scope, being most effective in the asymmetric epoxidation of terminal, unfunctionalised alkenes which are poor substrates for both the Sharpless and metal(salen)-catalysed epoxidation. The best results have been obtained using enantiomerically pure manganese and iron porphyrins, in particular those containing 'strapped' loops such as (4.71) developed by the group of Collman.[58] This catalyst, in combination with the 1,5-dicyclohexylimidazole as a donor ligand to protect the open face of the porphyrin, effects epoxidation of styrene to give (R)-(+)-styrene oxide with 69% ee. Greater success has been achieved using double faced 'BINAP-strapped' iron porphyrins such as (4.72). Catalyst (4.72a) has been used to effect asymmetric epoxidation of unfunctionalised terminal olefins such as *tert*-butylethylene (4.73) and trimethylsilylethylene (4.74),[59] while the most enantioselective porphyrin-mediated epoxidation of styrene to date has been effected using catalyst (4.72b), resulting in formation of (R)-styrene oxide (4.76) with 96% ee.[60] Interestingly, despite the structural similiarites of iron porphyrins (4.72a) and (4.72b) these catalysts display opposite modes of enantioselection, and epoxidation of styrene using (4.72a) gives (S)-styrene oxide with 83% ee.[59]

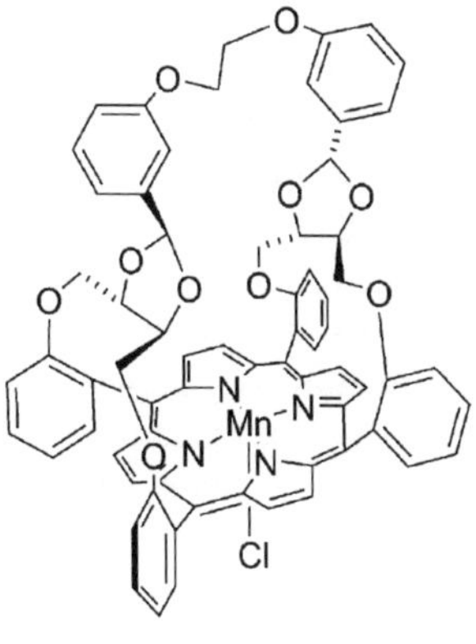

(4.71)

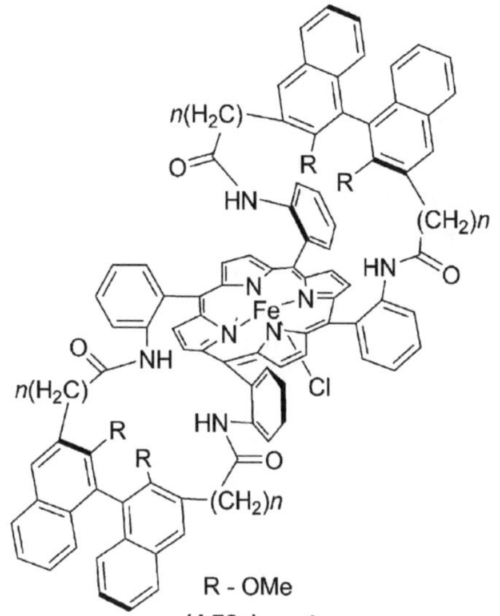

R - OMe
(4.72a) $n = 0$
(4.72b) $n = 1$

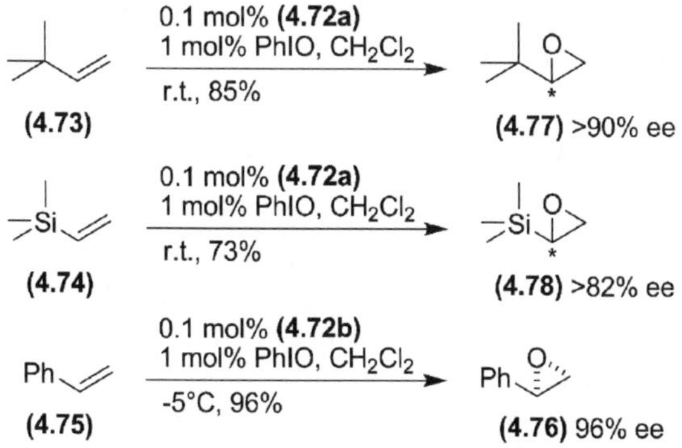

(4.73)

0.1 mol% **(4.72a)**
1 mol% PhIO, CH$_2$Cl$_2$
r.t., 85%

(4.77) >90% ee

(4.74)

0.1 mol% **(4.72a)**
1 mol% PhIO, CH$_2$Cl$_2$
r.t., 73%

(4.78) >82% ee

(4.75)

0.1 mol% **(4.72b)**
1 mol% PhIO, CH$_2$Cl$_2$
-5°C, 96%

(4.76) 96% ee

4.4 Other Metal-Catalysed Epoxidations of Unfunctionalised Olefins

Whilst the Sharpless epoxidation with titanium catalysts and the Jacobsen–Katsuki epoxidation with manganese(salen) complexes are at the forefront of enantioselective epoxidation with metal catalysts, there are alternative systems available.[1b,19c] Ruthenium pyridinebisoxazoline (PYBOX) complexes have been independently reported, using either phenyliodinium diacetate[61] or sodium periodate[62] as oxidant, to give moderately good enantioselectivity in the epoxidation of stilbene and a wide variety of aromatic olefins are epoxidised with moderate to good ee using Ru(PYBOX) and Ru(pyridinebisoxazine) catalysts in combination with the cheaper and more environmentally benign hydrogen peroxide as oxidant.[63] Strukul and coworkers have reported the use of enantiomerically pure platinum complexes incorporating bis-aryl diphosphine ligands for the epoxidation of a variety of unfunctionalised terminal olefins that also uses hydrogen peroxide as oxidant.[64] Best results are obtained with diene substrates that undergo selective epoxidation at the terminal position with up to 98% ee.

Enantiomerically pure manganese complexes using ligands other than the salen structure have been reported,[65,66] but so far with lower enantioselectivities. Better results have been achieved using molybdenum complexes bearing hydroxamic acid ligands and TBHP or cumylhydroperoxide as oxidant. This system has been used to effect the epoxidation of a range of olefins with up to 96% ee.

4.5   Epoxidation of Electron-Deficient Alkenes

α,β-Unsaturated ketones can be epoxidised by the Weitz–Scheffer process using peroxides in the presence of base and a number of asymmetric variants of this reaction have been investigated.[67]

4.5.1   Epoxidation Using Lanthanum-BINOL Catalysts

Aryl and alkyl ketones are epoxidised with high ee using bimetallic lanthanum and ytterbium-BINOL catalysts, prepared from equimolar quantities of Ln(O^{i}Pr)$_3$ and BINOL in the presence of stoichiometric oxidants such as TBHP.[1b, 67,68] The rate of this reaction is somewhat low, but can be improved by addition of stoichiometric quantities of water,[69] triphenylphosphine oxide[70] or triphenylarsene oxide[71] and a range of aryl *trans*-α,β-enones such as chalcone (**4.79**), and *trans*-alkyl ketones such as (**4.80**) undergo highly enantioselective epoxidation using the latter additive in reasonable reaction times. *Cis*-enones have proved to be more difficult substrates under these conditions, but can be epoxidised using the ytterbium catalyst prepared from BINOL (**4.81**).[68b] For example, enone (**4.82**) is converted into epoxide (**4.83**) in 94% ee using this catalyst. While amides are epoxidised smoothly and in high ee using these catalysts,[72] epoxidation of α,β-unsaturated esters has proved problematic owing to competing transesterification with the oxidant to give acylhydroperoxides. This problem has been overcome using α,β-unsaturated imidazoles as substrates that undergo highly enantioselective epoxidation and are converted into the α,β-epoxy ester on treatment with methanol.[72b, 73] N-Acylpyrroles also act as effective ester surrogates in this procedure.[74]

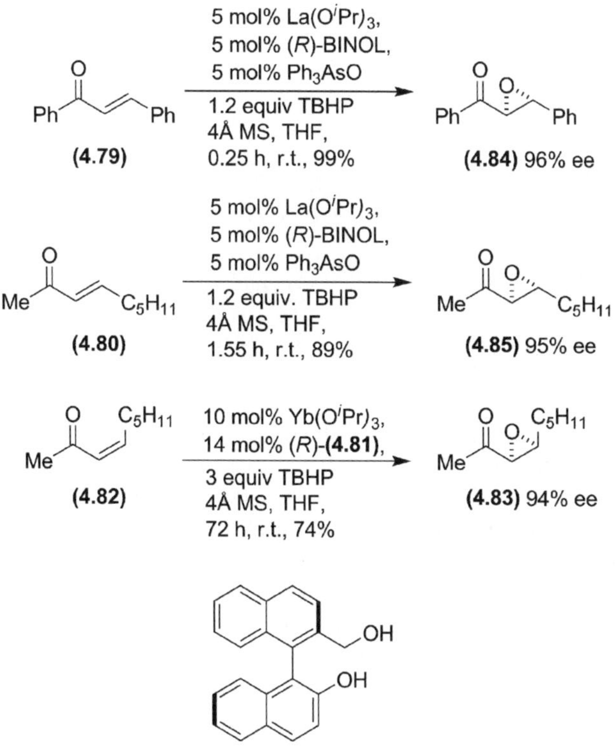

This type of epoxidation is proposed to proceed via formation of a lanthanide peroxide (**4.86**) that coordinates to the carbonyl group and undergoes 1,4-addition to give an enolate (**4.87**) (Figure 4.6).[71,74] Cyclisation of the enolate forms the epoxide product and regenerates the catalyst.

4.5.2 Julia–Colonna Epoxidation

The epoxidation of α,β-unsaturated ketones catalysed by polyamino acids is known as the Julia–Colonna epoxidation.[67,75] This three-phase procedure utilises aqueous hydrogen peroxide as oxidant along with a water immiscible solvent and solid poly-L-leucine as a catalyst and is mainly effective in the epoxidation of chalcone (**4.79**) and derivatives.[76]

There have been several modifications of the Julia–Colonna procedure, most particularly by Roberts and coworkers, who have developed a two-phase system by performing the reaction in the absence of water.[77] Replacement of aqueous

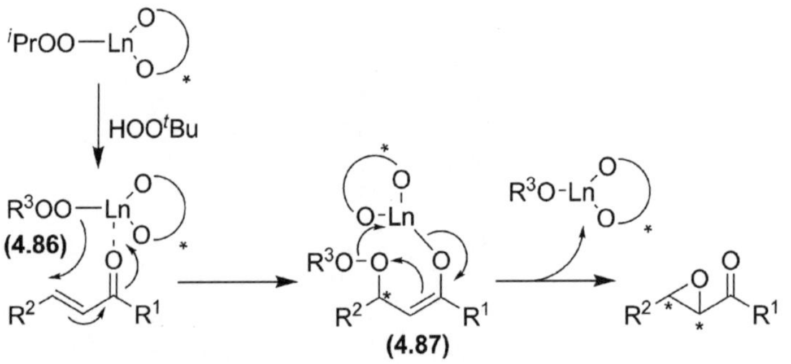

Figure 4.6 Mechanism of Ln-BINOL catalysed epoxidation

sodium hydroxide/hydrogen peroxide by urea-hydrogen peroxide (UHP) and 1,8-diazabicyclo[5.4.0]undec-7-ene (DBU) provides a more reactive system that tolerates alkyl vinyl ketones as well as aryl vinyl ketones. Immobilisation of the insoluble polyleucine on silica gel also leads to improvements in selectivity and reactivity and, in addition, allows more facile recycling of the catalyst.[78] A soluble catalyst, that is also readily recycled, can be obtained by binding polyleucine to polyethylene glycol.[79] Polyleucine has been discovered to sequester hydrogen peroxide effectively to give a reactive polyamino acid/peroxide gel that can be used to effect epoxidation of enones and also vinyl sulfones such as (**4.88**) with high ee.[80] Roberts and coworkers have shown how the chalcone epoxides can be used in natural product synthesis.[81]

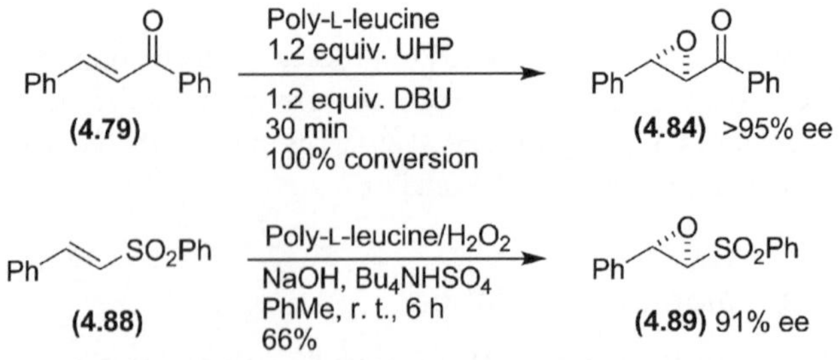

This epoxidation proceeds by conjugate addition of hydroperoxide anion followed by cyclisation of the resulting enolate in similar fashion to the Wietz–Scheffer process. It is proposed that the polyamino acid, present as an α–helix, forms a simple active site near the N-terminus within which the epoxidation occurs, and that catalysis arises from stabilisation of the intermediate enolate by interaction with an oxy-anion hole formed within this active site.[82]

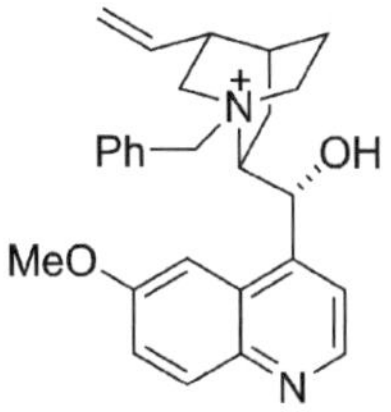

(4.90) 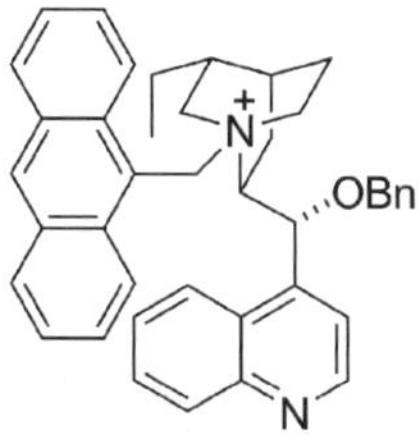   **(4.91)** 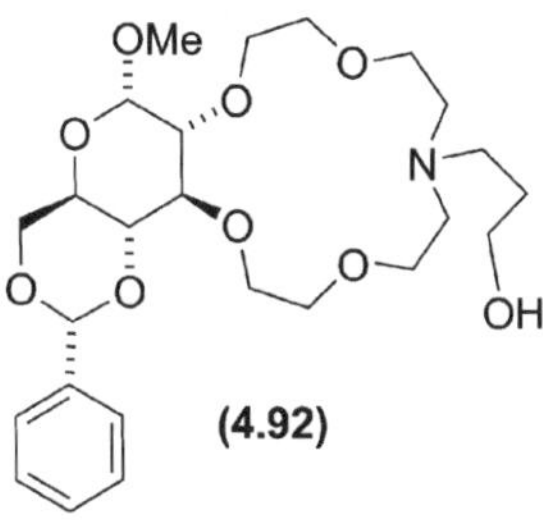   **(4.92)**

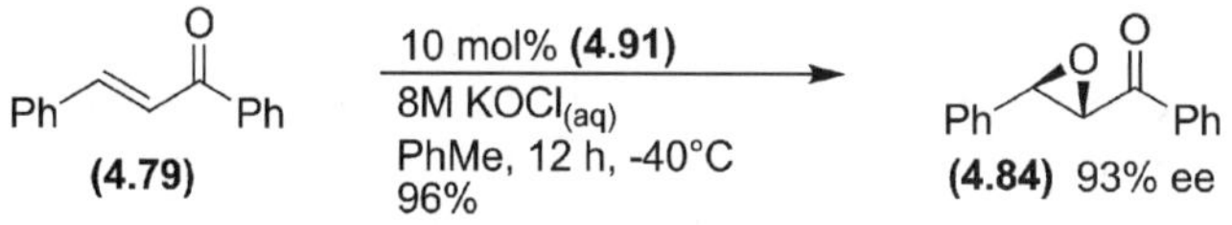

4.5.3   Phase-Transfer-Catalysed Epoxidation

Asymmetric phase-transfer-catalysed reactions in epoxidation and in other processes are of growing interest, and have recently been showing their ability to impart high levels of enantioselectivity.[1b, 67d, 83] Quinine and cinchonine-derived quaternary ammonium salts such as **(4.90)** have been widely used since the pioneering work of Wynberg and Greijdanus.[84,85] Replacement of the benzyl group with a 9-anthracenylmethyl moiety yields catalysts that display good selectivity and a relatively wide scope.[86] Corey and Zhang have reported high ees in the epoxidation of chalcone **(4.79)** and a variety of other aryl vinyl ketones using catalyst **(4.91)** in combination with aqueous potassium hypochlorite.[87] Carbohydrate-derived aza-crown ethers have been investigated as alternate phase-transfer catalysts in the epoxidation of chalcone and derivatives, and ees up to 92% have been obtained in the epoxidation of chalcone using catalyst **(4.92)** in combination with TBHP.[88]

4.5.4   Epoxidation with Organocatalysts

Lattanzi and coworkers have discovered that commercially available α,α-diphenylprolinol **(4.93)** catalyses the epoxidation of chalcone and derivatives with

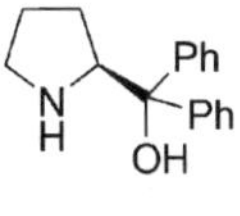

(4.93) 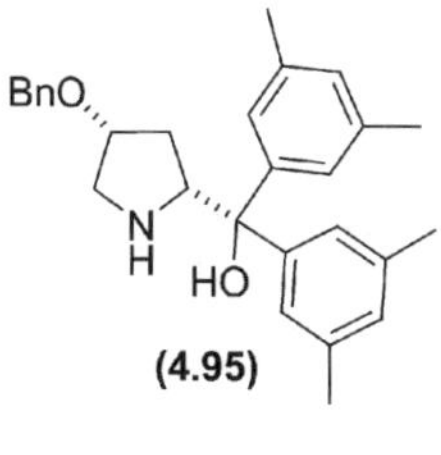   **(4.95)** 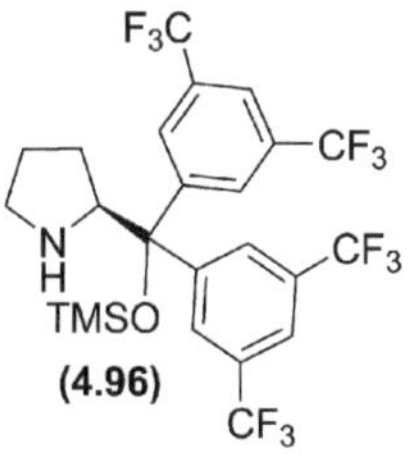   **(4.96)**

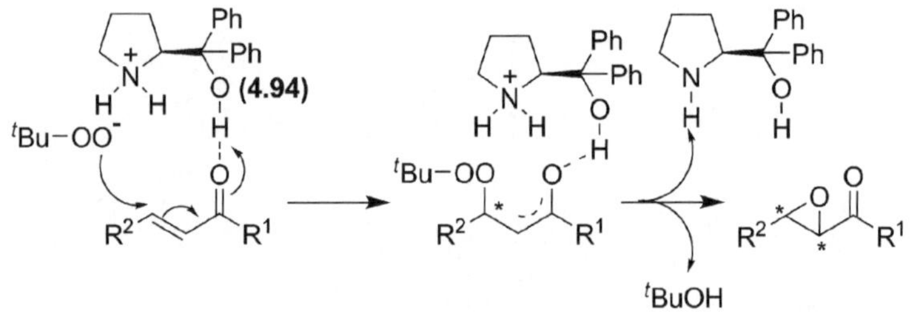

Figure 4.7 Mechanism of prolinol-catalysed epoxidation

up to 80% ee in the presence of TBHP.[89] It is postulated that prolinol (**4.93**) functions as a bifunctional catalyst acting to activate the hydroperoxide towards nucleophilic attack by formation of an ammonium hydroperoxide (**4.94**) and also the carbonyl group of the enone via H-bonding of the free hydroxyl group to the carbonyl oxygen (Figure 4.7).

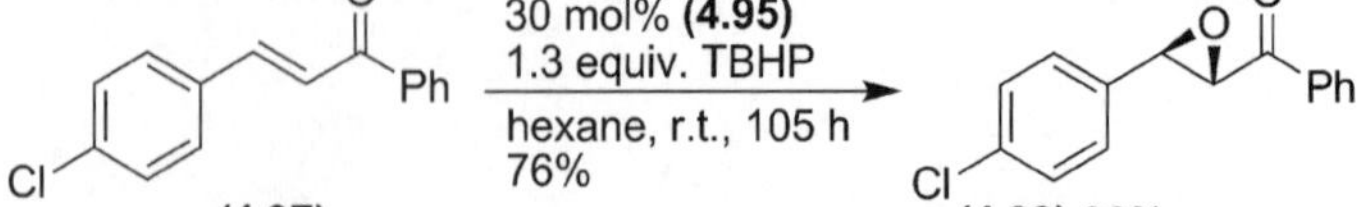

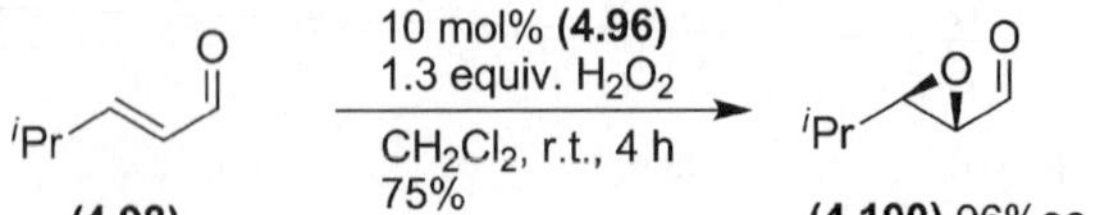

Prolinols bearing electron-rich aromatic moieties have been found to be more active catalysts than (**4.93**). For instance, while 20–30 mol% of (**4.93**) is usually employed, only 10 mol% of the catalyst with 3,4,5-trimethoxyphenyl substituents is required for formation of epoxides with similar ees.[90] Better selectivities have been obtained using C-4 substituted prolinols such as (**4.95**) and a variety of chalcone derivatives such as (**4.97**) can be converted into epoxides with ees between 90 and 96% using this catalyst in combination with TBHP. While the above organocatalytic methods are limited to the oxidation of α,β-unsaturated ketone and carboxylic acid derivatives, proline (**4.96**) effects highly enantioselective epoxidation of α,β-unsaturated aldehydes in the presence of hydrogen peroxide as oxidant at relatively low catalyst loadings.[91] In this case, the mechanism is postulated to centre on the formation of an α,β-unsaturated iminium ion (see Section 2.5) followed by conjugate addition of hydrogen peroxide. Both aryl- and alkyl-substituted enals such as (**4.98**) are tolerated as substrates to give epoxides with 94–98% ee. Similarly, a variety of α,β-enals with both aryl and aliphatic substituents at the β-position are epoxidised with ees ranging from 85 to 97% using the imidazolidinones discussed in Section 2.5 as organocatalysts and [(nosylimino)iodo]benzene (NsNIPh) as a slow release oxidant.[92]

4.6 Epoxidation with Iminium Salts

Iminium salts such as (**4.101**) are converted into oxaziridinium salts (**4.102**) on oxidation using oxone ($KHSO_5$) in aqueous media, and these species are able to effect the epoxidation of alkenes (**4.103**) under basic conditions (Figure 4.8).[67d, 93,94]

Several groups have investigated the use of chiral, nonracemic iminium salts as asymmetric epoxidation catalysts.[1b] In general, the highest ees are obtained with trisubstituted, cyclic alkenes and terminal alkenes are epoxidised with relatively low selectivities. One of the earliest asymmetric iminium salts developed was (**4.105**), which was used to effect epoxidation of stilbene with 33% ee.[95] Since then, a wide range of enantiomerically pure iminium salts have been investigated, but, in general, only moderate enantioselectivities have been obtained. Bulman Page and coworkers have developed iminium salts such as (**4.106**) and (**4.107**) where the asymmetric control element is close to the point of oxygen transfer and these are the most selective catalysts to date.[96] Catalyst (**4.106**) effects epoxidation of 1-phenylcyclohexene (**4.108**) and 1-phenyl-3,4-dihydronaphthalene (**4.109**) with high ee. Tetraphenylphosphonium monoperoxybisulfate (TPPP) also functions as an effective oxidant in this process, allowing the reaction to be carried out in organic solvents at lower temperature and in the absence of base. The performance of a number of iminium salt catalysts is enhanced under these conditions.[97] For example, iminium salt (**4.107**) is an effective catalyst for the epoxidation of *cis*-alkenes, in particular benzopyrans such as (**4.110**) using TPPP in chloroform at $-40°$C.[98]

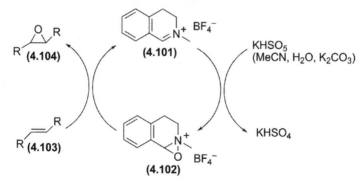

Figure 4.8 Mechanism of iminium-salt catalysed epoxidation

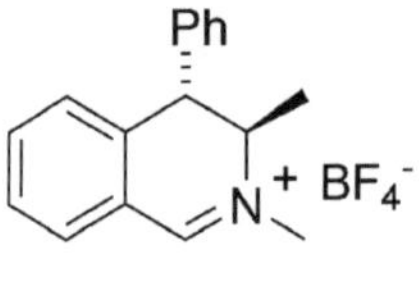

(4.105)

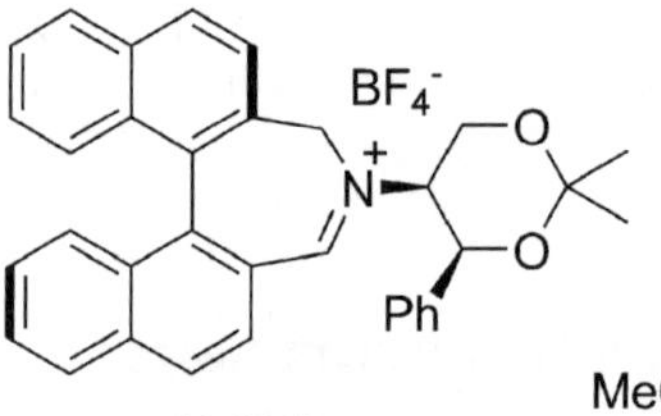

(4.106)

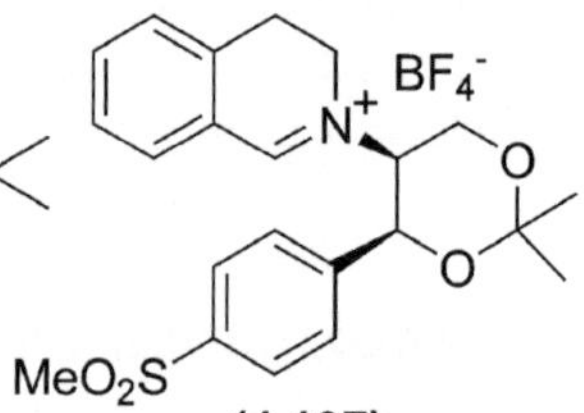

(4.107)

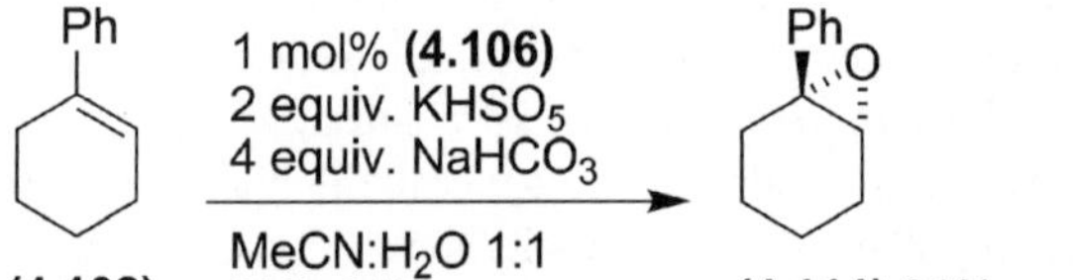

(4.108)

1 mol% **(4.106)**
2 equiv. KHSO$_5$
4 equiv. NaHCO$_3$
_____
MeCN:H$_2$O 1:1
0°C, 64%

(4.111) 91% ee

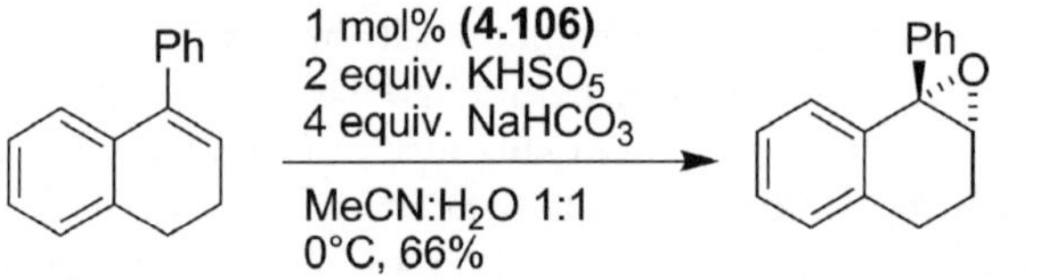

(4.109)

1 mol% **(4.106)**
2 equiv. KHSO$_5$
4 equiv. NaHCO$_3$
_____
MeCN:H$_2$O 1:1
0°C, 66%

(4.112) 95% ee

(4.110)

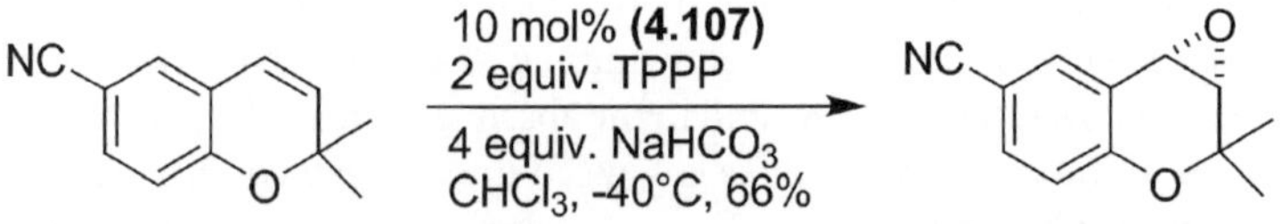

10 mol% **(4.107)**
2 equiv. TPPP
_____
4 equiv. NaHCO$_3$
CHCl$_3$, -40°C, 66%

(4.113) 97% ee

4.7   Epoxidation with Ketone Catalysts

Alkenes can be epoxidised by dioxiranes, and this process can be achieved enantioselectively if an enantiomerically pure dioxirane is employed.[1b, 19a, 67d, 99] Since ketones can be oxidised to dioxiranes with oxone under conditions where alkenes are not directly epoxidised, this allows the following catalytic cycle to be established, where a ketone **(4.114)** is converted into a dioxirane **(4.115)**. This oxidises the alkene **(4.103)** into an epoxide **(4.104)** (Figure 4.9). There are clear parallels between this mechanism and the iminium-catalysed epoxidation described in Section 4.6. This catalytic cycle was first demonstrated by Curci and coworkers in 1984 using an enantiomerically pure ketone.[100] However, it was not until 1996 that high levels of enantioselectivity were reported.

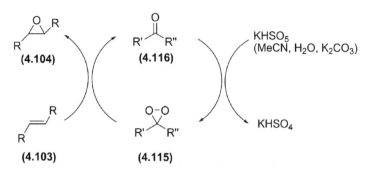

Figure 4.9 Catalytic cycle of the ketone-catalysed epoxidation

The choice of ketone is governed by its ability to form a dioxirane quickly (an electron withdrawing group in the α-position is helpful) and not to be prone to either racemisation or Baeyer–Villiger oxidation. The intermediate dioxirane must also be willing to donate an oxygen to the alkene substrate.

Nonfunctionalised *trans*- and trisubstiuted olefins are the best substrates for this process, which is also tolerant of a variety of functional groups. Terminal olefins and *cis*-olefins are generally epoxidised with low ee. Yang and coworkers have used the catalyst (**4.116**) with great effect in the epoxidation of stilbenes (**4.117**).[99c, 101,102] Armstrong and Hayter have used the α-fluorinated tropinone-derived ketone (**4.118**) which provides good selectivity for the epoxidation of phenylstilbene (**4.119**).[103] The highest selectivities have been achieved using the fructose-derived ketone catalyst (**4.120**), although this is typically used in 20–30 mol% catalyst loading. For example, stilbene (**4.121**) is epoxidised with high selectivity.[104] The reaction has also been successfully applied to the mono-epoxidation of many conjugated dienes, all with excellent enantioselectivity.[105] A typical example is provided by the epoxidation of diene (**4.122**) into the mono-epoxide (**4.123**). Some modification to this class of ketones has been made in an effort to improve the scope of the asymmetric epoxidation. Highly enantioselective epoxidation of both *cis*-alkenes with conjugating aromatic groups such as *cis*-β-methylstyrene (**4.57**)[106] occurs using catalyst (**4.124**) incorporating an oxazolidinone moiety and this ketone has also been used to effect the epoxidation of terminal alkenes such as styrene (**4.75**) and derivatives.[107] Dioxiranes are electrophilic reagents and therefore epoxidise electron-deficient alkenes slowly – often undergoing competing decomposition under the reaction conditions. However, more reactive catalysts are formed from ketones (**4.125**) and (**4.126**) and these have been used to enantioselectively epoxidise α,β-enones in 82–96 % ee[108] and *trans*- and trisubstituted cinnamates in up to 97% ee, respectively.[109]

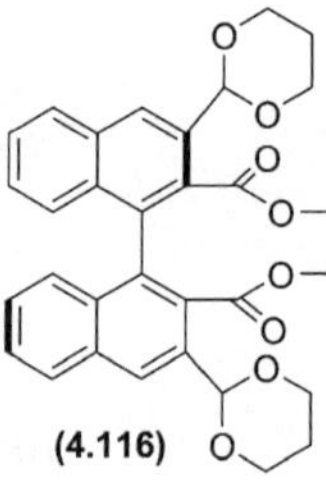

(4.116)

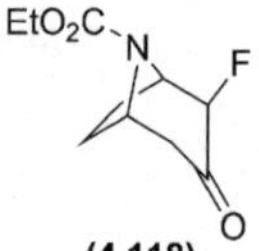

EtO$_2$C
N
F
O

(4.118)

(4.120)

NBoc

(4.124)

HO

(4.125)

AcO
OAc

(4.126)

R
R
(4.117)
10 mol% (4.116)
10 equiv. Oxone
31 equiv. NaHCO$_3$
Na$_2$EDTA
20 h, 0°C, DME/H$_2$O
>90%
R
R
(4.127)
R = H, 84% ee
R = ^{t}Bu, 95% ee

Ph
Ph
Ph
(4.119)
10 mol% (4.118)
10 equiv. Oxone
5-15 equiv. NaHCO$_3$
MeCN/Na$_2$EDTA$_{(aq)}$
100%
Ph
Ph
Ph
(4.128) 83% ee

Ph
Ph
(4.121)
30 mol% (4.120)
1.38 equiv. KHSO$_4$
5.8 equiv. K$_2$CO$_3$
Na$_2$B$_4$O$_7$.10H$_2$O in MeCN
0°C, EDTA$_{(aq)}$
75%
Ph
Ph
(4.129) 97% ee

Et
OTBS
(4.122)
25 mol% (4.120)
1.38 equiv. KHSO$_4$
5.8 equiv. K$_2$CO$_3$
Na$_2$B$_4$O$_7$.10H$_2$O in MeCN
0°C, EDTA$_{(aq)}$
81%
Et
OTBS
(4.123) 96% ee
(+ trace bis-epoxide)

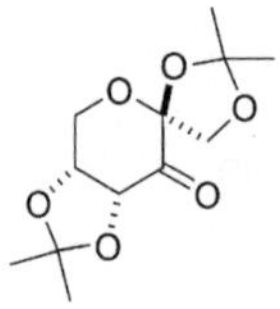

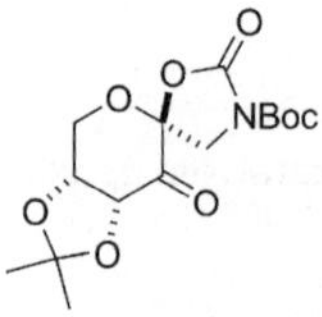

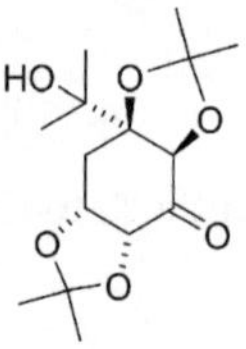

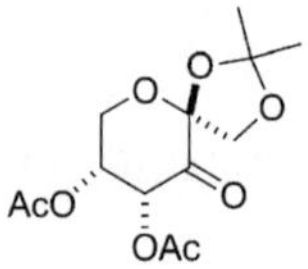

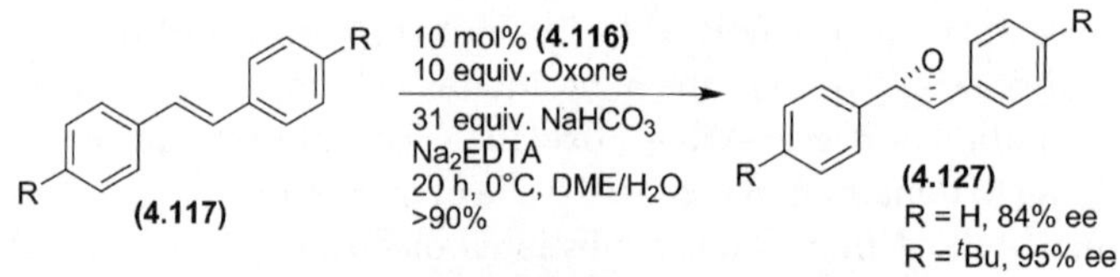

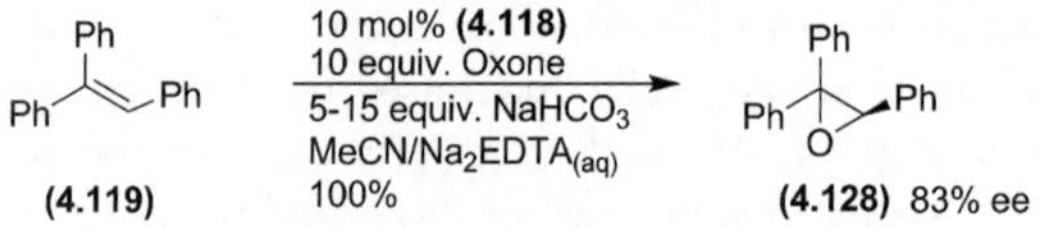

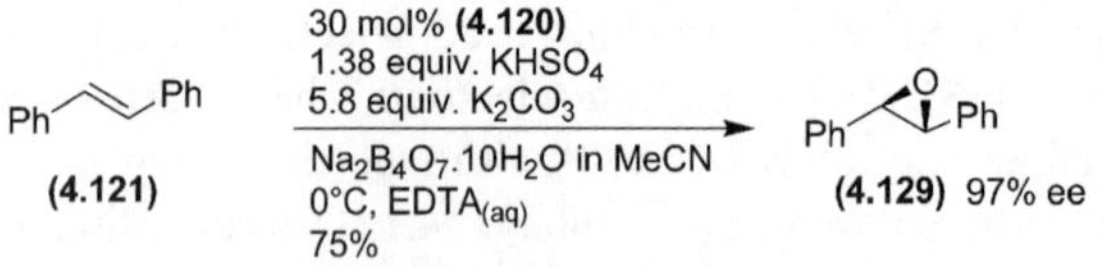

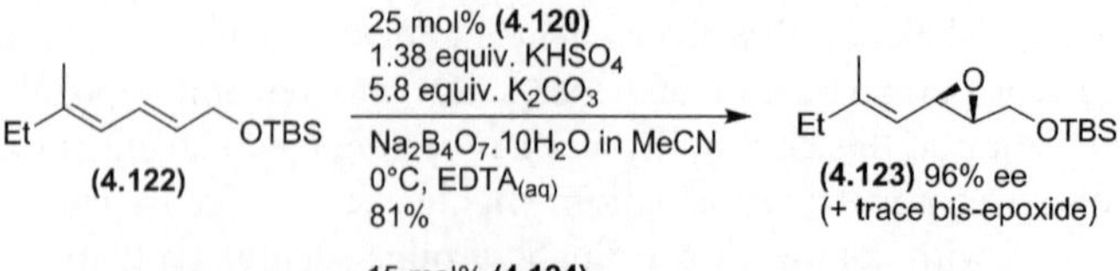

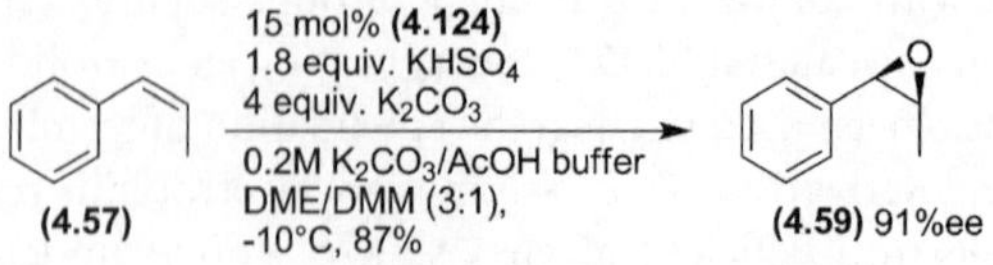

(4.57)
15 mol% (4.124)
1.8 equiv. KHSO$_4$
4 equiv. K$_2$CO$_3$
0.2M K$_2$CO$_3$/AcOH buffer
DME/DMM (3:1),
-10°C, 87%
(4.59) 91%ee

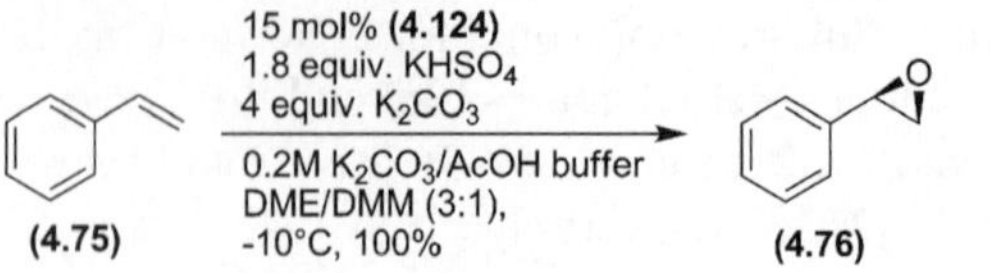

(4.75)
15 mol% (4.124)
1.8 equiv. KHSO$_4$
4 equiv. K$_2$CO$_3$
0.2M K$_2$CO$_3$/AcOH buffer
DME/DMM (3:1),
-10°C, 100%
(4.76)

4.8   Epoxidation of Aldehydes

An alternate strategy for the synthesis of epoxides is by the reaction of sulfonium ylides with aldehydes. Aggarwal and coworkers have developed a catalytic asymmetric variant of this process using camphorsulphonic acid-derived sulfonium ylides such as **(4.130)** and **(4.131)**.[1b, 67a, 110]

(4.130)   **(4.131)**

The ylide, of general structure **(4.132)**, is formed by reaction of the corresponding sulphide **(4.133)** with a metallocarbenoid resulting from transition metal salt-mediated decomposition of a diazo compound **(4.134)** or *N*-tosyl hydrazone salt **(4.135)**. Reaction of the sulfonium ylide **(4.132)** with aldehyde **(4.136)** yields the *trans*-epoxide **(4.137)** as the major product and sulfide **(4.133)** which is then returned to the catalytic cycle (Figure 4.10).

This reaction proceeds with high ee using aromatic aldehydes such as benzaldehyde **(4.138)** and derivatives using tosyl hydrazone salt **(4.139)** and thus has utility in the synthesis of enantiopure stilbene oxide derivatives.[111] High ees are also obtained with some heteroaromatic aldehydes such as 2-furaldehyde **(4.140)**. Basic heteroaromatic aldehydes are poor substrates and the reaction proceeds with low yield and/or diastereoselectivity using aliphatic aldehydes and most α,β-enones as substrates. These limitations have been overcome using a stoichiometric variant of this process.[112]

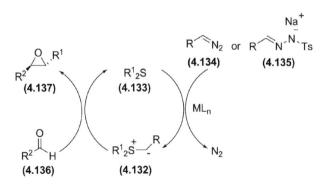

Figure 4.10   Catalytic cycle of the sulfonium ylide catalysed epoxidation of aldehydes

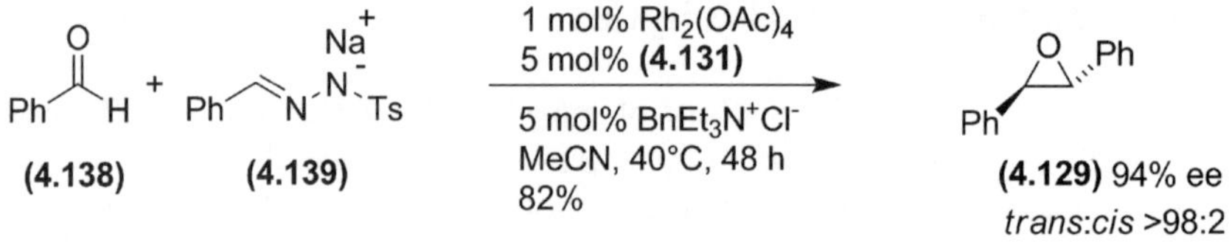

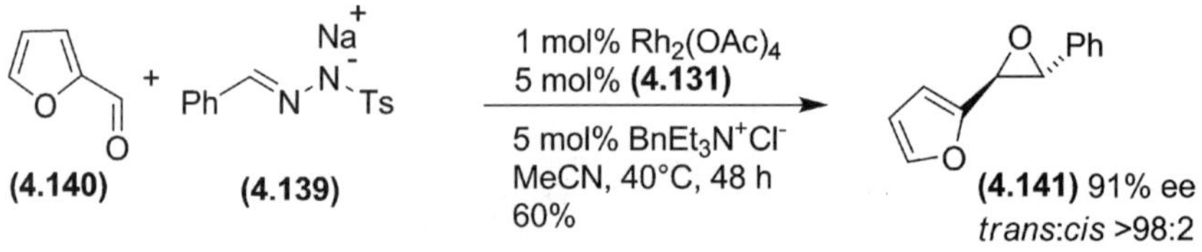

Ph
H
O
+
Na+
Ph
N
N-
Ts
(4.138)
(4.139)
1 mol% Rh2(OAc)4
5 mol% (4.131)
5 mol% BnEt3N+Cl-
MeCN, 40°C, 48 h
82%
O
Ph
Ph
(4.129) 94% ee
trans:cis >98:2

O
O
+
Na+
Ph
N
N-
Ts
(4.140)
(4.139)
1 mol% Rh2(OAc)4
5 mol% (4.131)
5 mol% BnEt3N+Cl-
MeCN, 40°C, 48 h
60%
O
Ph
O
(4.141) 91% ee
trans:cis >98:2

4.9 Aziridination of Alkenes

So far aziridination reactions have, in some ways, had more in common with cyclopropanation reactions (see Section 9.1) than with epoxidation reactions. Nevertheless, the aziridination reaction is more synthetically akin to epoxidation, and on that basis, is included in the present chapter. Aziridines may be prepared by nitrene transfer to alkenes or by carbene transfer to imines and both approaches have been performed in an enantioselective sense using enantiomerically pure metal-based catalysts.[113]

Copper complexes of bis-oxazoline (BOX) ligands (**4.142**), which are very efficient cyclopropanation catalysts, are also competent in aziridination reactions. Evans and coworkers have shown that high selectivities are obtained using *trans*-alkenes, and the 'nitrene' source PhI=NTs.[114] The substrates (**4.75**) and (**4.143**) both undergo enantioselective aziridination, but it is cinnamate esters such as (**4.143**) that undergo more selective reaction. Rigidification of the bis-oxazoline by the incorporation of rings has led to the development of ligands such as (**4.144**) that can be used to effect highly enantioselective aziridination of chalcones such as (**4.145**).[115] Jacobsen and coworkers employed diimine ligands (**4.146**), which worked especially well with *cis*-alkene substrates especially chromenes such as (**4.147**), although enantioselectivities are much lower using simple alkenes such as styrene (30–87%) were obtained.[116,117] Bisimines incorporating axially chiral BINAP[118] and biaryl moieties such as (**4.148**)[119] have also been investigated as ligands in the copper-catalysed aziridination and give high ees with both chromenes and *trans*-cinnamate esters.

Most manganese(III)(salen) complexes have generally been found to be less effective for aziridination than for epoxidation, both in terms of enantioselectivity and yield.[120] However, the salen complex (**4.149**) has been shown to give high enantioselectivity for some simple alkenes such as styrene (**4.75**) that undergo

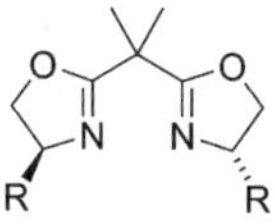

(4.142)

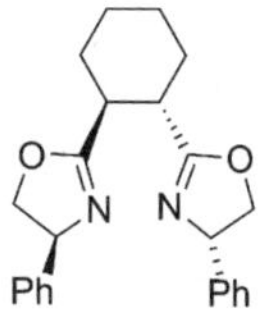

(4.144)

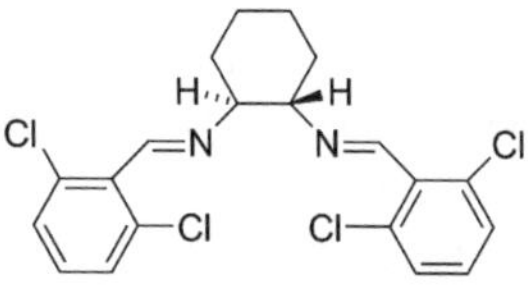

(4.146)

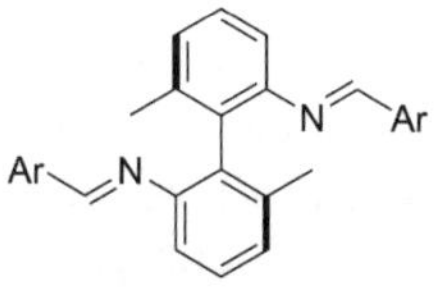

(4.148)

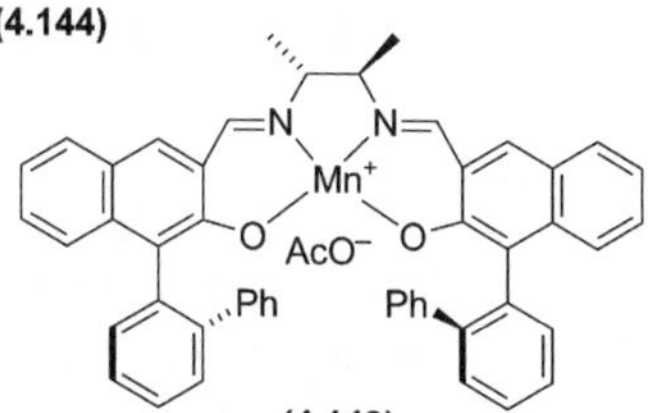

(4.149)

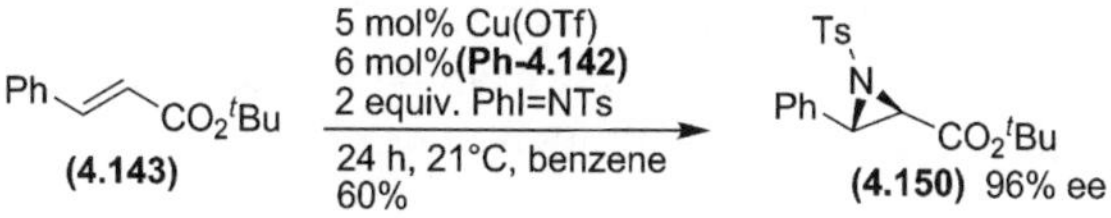

$Ph\diagdown\diagup CO_2^tBu$ (4.143)

5 mol% Cu(OTf)
6 mol% **(Ph-4.142)**
2 equiv. PhI=NTs
24 h, 21°C, benzene
60%

→

(4.150) 96% ee

$Ph\diagdown$ (4.75) as a solvent

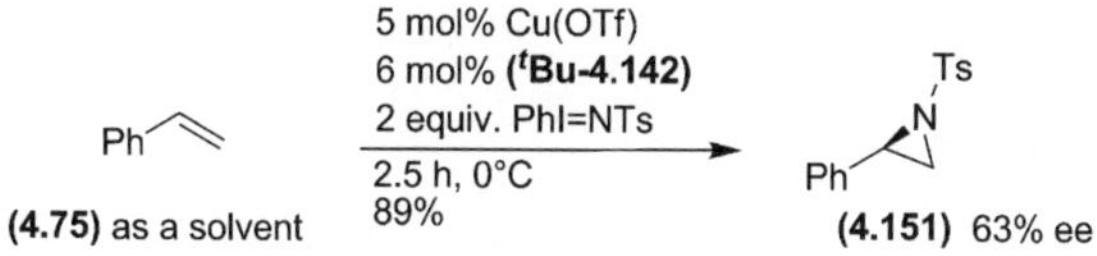

5 mol% Cu(OTf)
6 mol% (**^{t}Bu-4.142**)
2 equiv. PhI=NTs
2.5 h, 0°C
89%

→

(4.151) 63% ee

(4.145)

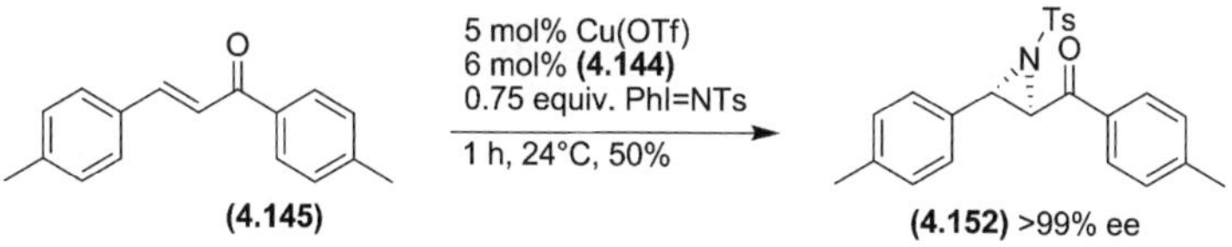

5 mol% Cu(OTf)
6 mol% (4.144)
0.75 equiv. PhI=NTs
1 h, 24°C, 50%

→

(4.152) >99% ee

(4.147)

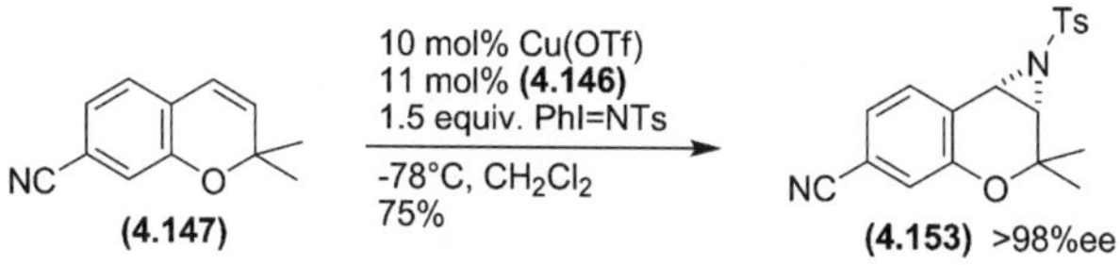

10 mol% Cu(OTf)
11 mol% (4.146)
1.5 equiv. PhI=NTs
-78°C, CH_2Cl_2
75%

→

(4.153) >98%ee

$Ph\diagdown$ (4.75)

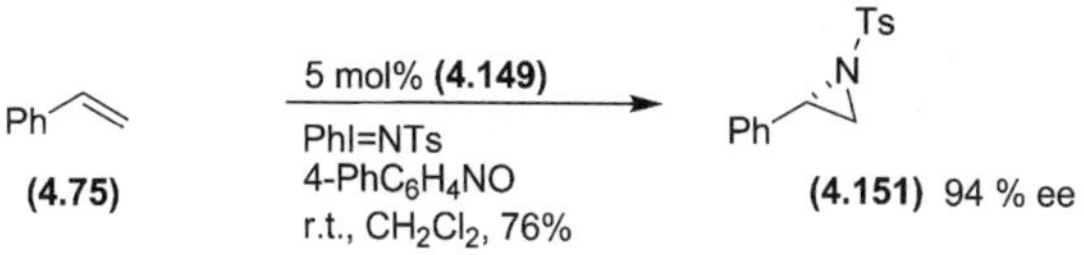

5 mol% (4.149)

PhI=NTs
4-PhC_6H_4NO
r.t., CH_2Cl_2, 76%

→

(4.151) 94 % ee

aziridination with only moderate ee using other catalyst systems. [121] An asymmetric rhodium-catalysed aziridination reaction has also shown encouraging levels of enantioselectivity (up to 73% ee). [122]

4.10   Aziridination of Imines

An alternative approach to aziridine synthesis involves transfer of a carbenoid species to imines. Jacobsen achieved the first asymmetric aziridination of imines by transfer of copper carbenoids derived from copper bis-oxazoline catalysts and ethyl diazoacetate onto imines, but this process only proceeds with moderate yield and selectivity. [123] Better results have been achieved by addition of ethyl diazoacetate to imines in the presence of enantiopure Lewis acids such as the boron-based catalysts prepared from vaulted biaryls such as VAPOL (**4.154**) and B(OPh)$_3$. [124] A range of aryl and alkyl *N*-benzylaldimines, for example (**4.155**) and (**4.156**), undergo aziridination to give *cis*-aziridines with high ee using this procedure.

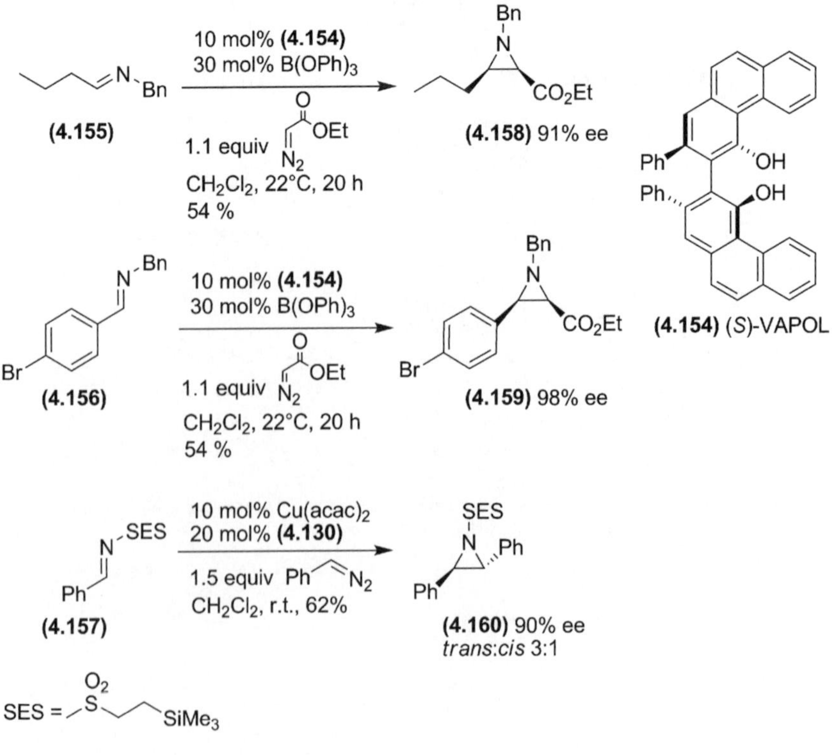

SES = $\overset{O_2}{\underset{}{S}}$∕∕∕SiMe$_3$

The sulfonium ylide-mediated epoxidation procedure developed by Aggarwal (see Section 4.8) has been adapted to the enantioselective synthesis of aziridines from imines bearing electron withdrawing groups on nitrogen. [109b, 109e, 125] The reaction proceeds in moderate to good yield and high ee using aromatic aldimines bearing electron withdrawing groups on nitrogen such as *N*-SES imine (**4.157**), but diastereoselectivities are rather low. [126]

References

1. For recent general reviews on the asymmetric epoxidation of alkenes see: (a) B. S Lane and K. Burgess, *Chem. Rev.*, **2003**, *103*, 2457. (b) Q.-H. Xia, H.-Q. Ge, C.-P. Ye, Z.-M. Liu and K.-X. Su, *Chem. Rev.*, **2005**, *105*, 1603.

2. (a) R. A. Johnson and K. B. Sharpless, *Comprehensive Organic Synthesis*, Vol. *7*, ed. B.M. Trost, Pergamon, Oxford, **1991**, 389. (b) R. A. Johnson and K.B. Sharpless in *Catalytic Asymmetric Synthesis*, ed. I. Ojima, VCH, New York, **1993**, 103. (c) T. Katsuki and V. S. Martin, *Organic Reactions*, **1996**, *48*, 1. (d) T. Katsuki, in *Comprehensive Asymmetric Catalysis*, Vol. *2*, ed. E. N. Jacobsen, A. Pfaltz and H. Yamamoto, Springer-Verlag, Berlin, **1999**, 621. (e) K. B. Sharpless, *Angew. Chem., Int. Ed.*, **2002**, *41*, 2024. (f) D. J. Ramón and M. Yus, *Chem. Rev.*, **2006**, *106*, 2126.

3. T. Katsuki and K. B. Sharpless, *J. Am. Chem. Soc.*, **1980**, *102*, 5974.

4. (a) R. M. Hanson and K. B. Sharpless, *J. Org. Chem.*, **1986**, *51*, 1922. (b) Y. Gao, R.M. Hanson, J.M. Klunder, S.Y. Ko, H. Masamune and K.B. Sharpless, *J. Am. Chem. Soc.*, **1987**, *109*, 5765.

5. (a) S. S. Woodard, M. G. Finn and K. B. Sharpless, *J. Am. Chem. Soc.*, **1991**, *113*, 106.

6. I. D. Williams, S. F. Pedersen, K. B. Sharpless and S. J. Lippard, *J. Am. Chem. Soc.*, **1984**, *106*, 6430.

7. R. Martín, A. Moyano, M. Pericàs and A. Riera, *Org. Lett.*, **2000**, *2*, 93.

8. (a) J. K. Karjalainen, O. E. O. Hormi and D.C. Sherrington, *Tetrahedron:Asymmetry*, **1998**, *9*, 1563. (b) T. J. Dickerson, N. N. Reed and K. D. Janda, *Chem. Rev.*, **2002**, *102*, 3325. (c) C. E. Song and S. E. Lee, *Chem. Rev.*, **2002**, *102*, 3495. (d) H. Guo, X. Shi, Z. Qiao, S. Hou and M. Wang, *Chem. Commun.*, **2002**, 118. (e) H.-C. Guo, X.-Y. Shi, X. Wang, S.-Z. Liu and M. Wang, *J. Org. Chem.*, **2004**, *69*, 2042.

9. N. N. Reed, T. J. Dickerson, G. E. Boldt and K. D. Janda, *J. Org. Chem.*, **2005**, *70*, 1728.

10. L. D.-L. Lu, R. A. Johnson, M. G. Finn and K. B. Sharpless, *J. Org. Chem.*, **1984**, *49*, 728.

11. V. S. Martin, S. S. Woodard, T. Katsuki, Y. Yamada, M. Ikeda and K. B. Sharpless, *J. Am. Chem. Soc.*, **1981**, *103*, 6237.

12. P. R. Carlier, W. S. Mungall, G. Schröder and K. B. Sharpless, *J. Am. Chem. Soc.*, **1988**, *110*, 2978.

13. Y. Kitano, T. Matsumoto and F. Sato, *Tetrahedron*, **1988**, *44*, 4072.

14. S. L. Schreiber, T. S. Schreiber and D. B. Smith, *J. Am. Chem. Soc.*, **1987**, *109*, 1525.

15. K. B. Sharpless, C. H. Behrens, T. Katsuki, A. W. M. Lee, V. S. Martin, M. Takatani, S. M. Viri, F .J. Walker and S. S. Woodard, *Pure Appl. Chem.*, **1983**, *55*, 589. (b) M. G. Finn and K. B. Sharpless, *J. Am. Chem. Soc.*, **1991**, *113*, 113.

16. W. Zhang, A. Basak, Y. Kosugi, Y. Hoshino and H. Yamamoto, *Angew. Chem. Int. Ed.*, **2005**, *44*, 4389.

17. (a) T. Okachi, N. Murai and M. Onaka, *Org. Lett.*, **2003**, *5*, 85. (b) N. Makita, Y. Hoshino and H. Yamamoto, *Angew. Chem. Int. Ed.*, **2003**, *42*, 941.

18. W. Zhang and H. Yamamoto, *J. Am. Chem. Soc.*, **2007**, *129*, 286.

19. For reviews, see; (a) E.N. Jacobsen in *Catalytic Asymmetric Synthesis*, ed. I. Ojima, VCH, New York, **1993**, 159. (b) T. Katsuki, *Coord. Chem. Rev.* **1995**, *140*, 189. (c) E. N. Jacobsen in *Comprehensive Organometallic Chemistry II*, Vol. *12*, ed. G. Wilkinson, F.G.A. Stone, E.W. Abel and L.S. Hegedus, Pergamon, New York, **1995**, Chapter 11.1. (d) P. J. Pospisil, D.H. Carsten and E. N. Jacobsen, *Chem. Eur. J.* **1996**, *2*, 974. (e) E. N. Jacobsen and M. H. Wu, in *Comprehensice Asymmetric Catalysis*, Vol. *2*, ed. E. N. Jacobsen, A. Pfaltz and H. Yamamoto, Springer-Verlag, Berlin, **1999**, 649. (f) M. Bandini, P. G. Cozzi and A. Umani-Ronchi, *Chem. Commun.*, **2002**, 919. (g) M. Corsi, *Synlett*, **2002**, 2127. (h) T. Katsuki, *Synlett*, **2003**, 281. (i) E. McGarrigle and D. G. Gilheany, *Chem. Rev.*, **2005**, *105*, 1563.

20. W. Zhang, J. L. Loebach, S.R. Wilson and E. N. Jacobsen, *J. Am. Chem. Soc.*, **1990**, *112*, 2801.

21. R. Irie, K. Noda, Y. Ito, N. Matsumoto and T. Katsuki, *Tetrahedron Lett.*, **1990**, *31*, 7345.

22. J. F. Larrow, E. N. Jacobsen, Y. Gao, Y. Hong, X. Nie and C. M. Zepp, *J. Org. Chem.*, **1994**, *59*, 1939.

23. H. Sasaki, R. Irie and T. Katsuki, *Synlett*, **1994**, 356.

24. (a) D. Feichtinger and D. A. Plattner, *Angew. Chem., Int. Ed. Engl.*, **1997**, *36*, 1718. (b) D. Feichtinger and D. A. Plattner, *Chem. Eur. J.*, **2001**, *7*, 591.

25. (a) W. Adam, C. Mock-Knoblauch, C. R. Saha-Möller and M. Herderich, *J. Am. Chem. Soc.*, **2000**, *122*, 9685. (b) W. Adam, K. J. Roschmann, C. R. Saha-Möller and D. Seebach, *J. Am. Chem. Soc.*, **2002**, *124*, 5068.

26. (a) N. Hosoya, A. Hatayama, K. Yanai, H. Fujii, R. Irie and T. Katsuki, *Synlett*, **1993**, 641. (b) M. Palucki, N.S. Finney, P. J. Pospisil, M. L. Güler, T. Ishida and E. N. Jacobsen, *J. Am. Chem. Soc.*, **1998**, *120*, 948, and references therein.

27. (a) T. Hamada, T. Fukuda, H. Imanishi and T. Katsuki, *Tetrahedron*, **1996**, *52*, 515. (b) C. Linde, M. Arnold, P.-O. Norrby and B. Åkermark, *Angew. Chem., Int. Ed. Engl.*, **1997**, *36*, 1723. (c) W. Adam, R. T. Fell, V. R. Stegmann, C. R. Saha-Möller, *J. Am. Chem. Soc.*, **1998**, *120*, 708.

28. (a) N. S. Finney, P. J. Pospisil, S. Chang, M. Palucki, R.G. Konsler, K.B. Hansen and E. N. Jacobsen, *Angew. Chem., Int. Ed. Engl.*, **1997**, *36*, 1720. (b) M. Palucki, N. S. Finney, P. J. Pospisil, M. L. Güler, T. Ishida and E. N. Jacobsen, *J. Am. Chem. Soc.*, **1998**, *120*, 948.

29. (a) T. Hamada, R. Irie and T. Katsuki, *Synlett*, **1994**, 479. (b) Y. Noguchi, R. Irie, T. Fukuda and T. Katsuki, *Tetrahedron Lett.*, **1996**, *37*, 4533.

30. P. Pietikäinen, *Tetrahedron Lett.*, **1994**, *35*, 941.

31. P. Pietikäinen, *Tetrahedron Lett.*, **1995**, *36*, 319.

32. W. Adam, J. Jekö, A. Lévai, C. Nemes, T. Patonay and P. Sebök, *Tetrahedron Lett.*, **1995**, *36*, 3669.

33. M. Palucki, G. J. McCormick and E. N. Jacobsen, *Tetrahedron Lett.*, **1995**, *36*, 5457.
34. E. N. Jacobsen, W. Zhang, A. R. Muci, J.R. Ecker and L. Deng, *J. Am. Chem. Soc.*, **1991**, *113*, 7063.
35. H. Sasaki, R. Irie, T. Hamada, K. Suzuki and T. Katsuki, *Tetrahedron*, **1994**, *50*, 11827.
36. B. D. Brandes and E.N. Jacobsen, *J. Org. Chem.*, **1994**, *59*, 4378.
37. D. Bell, M. R. Davies, F. J. L. Finney, G. R. Green, P. M. Kincey and I. S. Mann, *Tetrahedron Lett.*, **1996**, *37*, 3895.
38. C. H. Senanayake, G. B. Smith, K. M. Ryan, L. E. Fredenburgh, J. Liu, F. E. Roberts, D.L. Hughes, R. D. Larsen, T.R. Verhoeven and P. J. Reider, *Tetrahedron Lett.*, **1996**, *37*, 3271.
39. B. D. Brandes and E. N. Jacobsen, *Tetrahedron Lett.*, **1995**, *36*, 5123.
40. N. H. Lee, A. R. Muci and E. N. Jacobsen, *Tetrahedron Lett.*, **1991**, *32*, 5055.
41. (a) T. Hamada, K. Daikai, R. Irie and T. Katsuki, *Synlett*, **1995**, 407. (b) T. Hamada, K. Daikai, R. Irie and T. Katsuki, *Tetrahedron: Asymmetry*, **1995**, *6*, 2441.
42. S. Chang, J. M. Galvin and E. N. Jacobsen, *J. Am. Chem. Soc.*, **1994**, *116*, 6937.
43. C. Linde, M. F. Anderlund and B. Åkermark, *Tetrahedron Lett.*, **2005**, *46*, 5597.
44. H. Nishikori, C. Ohta and T. Katsuki, *Synlett*, **2000**, 1557.
45. (a) C. Bousquet and D.G. Gilheany, *Tetrahedron Lett.*, **1995**, *36*, 7739. (b) K. M. Ryan, C. Bousquet and D. G. Gilheany, *Tetrahedron Lett.*, **1999**, *40*, 3613. (c) C. P. O'Mahony, E. M. McGArrigle, M. F. Renehan, K. M. Ryan, N. J. Kerrigan, C. Bousquet and D. C. Gilheany, *Org. Lett.*, **2001**, *3*, 3435.
46. (a) S. Chang, N.H. Lee and E. N. Jacobsen, *J. Org. Chem.*, **1993**, *58*, 6939. (b) D. Mikame, T. Hamda, R. Irie and T. Katsuki, *Synlett*, **1995**, 827.
47. M. F. Hentemann and P. L. Fuchs, *Tetrahedron Lett.*, **1997**, *38*, 5615.
48. S. L. Vander Velde and E. N. Jacobsen, *J. Org. Chem.*, **1995**, *60*, 5380.
49. Y. Noguchi, H. Takiyama and T. Katsuki, *Synlett*, **1998**, 543.
50. T. Fukuda and T. Katsuki, *Tetrahedron Lett.*, **1996**, *37*, 4389–492
51. (a) W. Adam, R. T. Fell, C. Mock-Knoblauch, C. R. Saha-Möller, *Tetrahedron Lett.*, **1996**, *37*, 6531. (b) W. Adam, R. T. Fell, V. R. Stegmann and C. R. Saha-Möller, *J. Am. Chem. Soc.*, **1998**, *120*, 708.
52. (a) T. Takeda, R. Irie, Y. Shinoda and T. Katsuki, *Synlett*, **1999**, 1157. (b) K. Nakata, T. Takeda, J. Mihara, T. Hamada, R. Irie and T. Katsuki, *Chem. Eur. J.*, **2001**, *7*, 3776.
53. X.-G. Zhou, J.-S. Huang, X.-Q. Yu, Z.-Y. Zhou and C. M. Che, *J. Chem. Soc., Dalton Trans*, **2000**, 1075.
54. (a) T. S. Reger and K. D. Janda, *J. Am. Chem. Soc.*, **2000**, *122*, 6929. (b) H. Zhang, Y. Zhang and C. Li, *Tetrahedron:Asymmetry*, **2005**, *16*, 2417. (c) M. Holbach and M. Weck, *J. Org. Chem.*, **2006**, *71*, 1825. (d) K. Smith, C.-H. Liu and G. A. El-Hiti, *Org. Biomol. Chem.*, **2006**, *4*, 917.
55. (a) S. Xiang, Y. Zhang, Q. Xin and C. Li, *J. Chem. Soc., Chem. Commun.*, **2002**, *22*, 2696. (b) S. Battacharjee and J. A. Anderson, *J. Chem. Soc., Chem Commun.*, **2004**, *5*, 554.
56. (a) G. Pozzi, F. Cinato, F. Montanari and S. Quici, *J. Chem. Soc., Chem. Commun.*, **1998**, 877. (b) M. Cavazzini, A. Manfredi, F. Montanari, S. Quici and G. Pozzi, *J. Chem. Soc., Chem. Commun.*, **2000**, 2171.

57. J. P. Collman, X. Zhang, V. J. Lee, E. S. Uffelman and J. I. Brauman, *Science*, **1993**, *261*, 1401. (b) E. Rose, B. Andrioletti, S. Zrig and M. Quelquejeu, *Chem. Soc. Rev.*, **2005**, *34*, 573.

58. J. P. Collman, V. J. Lee, C. J. Kellen-Yuen, X. Zhang, J. A. Ibers and J. I. Braun, *J. Am. Chem. Soc.*, **1995**, *117*, 692.

59. J. P. Collman, Z. Wang, A. Straumanis, M. Quelquejeu and E. Rose, *J. Am. Chem. Soc.*, **1999**, *121*, 460.

60. E. Rose, Q.-Z. Ren and B. Andrioletti, *Chem. Eur. J.* **2004**, *10*, 224.

61. (a) H. Nishiyama, T. Shimada, H. Itoh, H. Sugiyama and Y. Motoyama, *J. Chem. Soc., Chem. Commun.*, **1997**, 1863. (b) M. K. Tse, S. Bhor, M. Klawonn, C. Döbler and M. Beller, *Tetrahedron Lett.*, **2003**, *44*, 7479.

62. N. End and A. Pfaltz, *J. Chem. Soc., Chem. Commun.*, **1998**, 589.

63. (a) M. K. Tse, C. Döbler, S. Bhor, M. Klawonn, W. Mägerlein, H. Hugl and M. Beller, *Angew. Chem. Int. Ed.*, **2004**, *43*, 5255. (b) M. K. Tse, S. Bhor, M. Klawonn, G. Anilkumar, H. Jiao, C. Döbler, A. Spannenberg, W. Mägerlein, H. Hugl and M. Beller, *Chem. Eur. J.*, **2006**, *12*, 1855. (c) M. K. Tse, S. Bhor, M. Klawonn, G. Anilkumar, H. Jiao, A. Spannenberg, C. Döbler, W. Mägerlein, H. Hugl and M. Beller, *Chem. Eur. J.*, **2006**, *12*, 1875.

64. M. Colladon, A. Scarso, P. Sgarbossa, R. A. Michelin and G. Strukul, *J. Am. Chem. Soc,*. **2006**, *128*, 14006.

65. T. Mukaiyama, T. Yamada, T. Nagata and K. Imagawa, *Chem.Lett.*, **1993**, 327.

66. C. Bolm, D. Kadereit and M. Valacchi, *Synlett*, **1997**, 687.

67. (a). V. K. Aggarwal, in *Comprehensive Asymmetric Catalysis*, Vol. *2*, ed. E. N. Jacobsen, A. Pfaltz and H. Yamamoto, Springer-Verlag, Berlin, **1999**, 679. (b) M. J. Porter and J. Skidmore, *J. Chem. Soc., Chem. Commun.*, **2000**, 1215. (c) S. Ebrahim and M. Wills, *Tetrahedron: Asymmetry*, **1997**, *8*, 3163. (d) A. Berkessel and H. Gröger, *Asymmetric Organocatalysis*, Wiley-VCH, Weinheim, **2005**, 277.

68. (a) M. Bougauchi, S. Watanabe, T. Arai, H. Sasai and M. Shibasaki, *J. Am. Chem. Soc.*, **1997**, *119*, 2329. (b) S. Watanabe, T. Arai, H. Sasai, M. Bougauchi and M. Shibasaki, *J. Org. Chem.*, **1998**, *63*, 8090.

69. S. Watanabe, Y. Kobayashi, T. Arai, H. Sasai, M. Bougauchi and M. Shibasaki, *Tetrahedron Lett.*, **1998**, *39*, 7321.

70. K. Daikai, M. Kamaura, and J. Inanaga, *Tetrahedron Lett.*, **1998**, *39*, 7321.

71. T. Nemoto, T. Ohshima, K. Yamaguchi and M. Shibasaki, *J. Am. Chem. Soc.*, **2001**, *123*, 2725.

72. (a) T. Nemoto, H. Kakei, V. Gnanadesikan, S. Tosaki, T. Ohshima and M. Shibasaki, *J. Am. Chem. Soc.*, **2002**, *124*, 14544. (b) T. Ohshima, T. Nemoto, S. Tosaki, H. Kakei, V. Gnanadesikan and M. Shibasaki, *Tetrahedron*, **2003**, *59*, 10485.

73. T. Nemoto, T. Ohshima and M. Shibasaki, *J. Am. Chem. Soc.*, **2001**, *123*, 9474.

74. S. Matsunaga, T. Kinoshita, S. Okada, S. Harada and M. Shibasaki, *J. Am. Chem. Soc.*, **2004**, *126*, 7559.

75. C. Lauret and S. M. Roberts, *Aldrichimica Acta*, **2002**, *35*, 47.

76. S. Banfi, S. Colonna, H. Molinari, S. Julia and J. Quixer, *Tetrahedron*, **1984**, *40*, 5207.

77. (a) P.A. Bentley, S. Bergeron, M. W. Cappi, D. E. Hibbs, M.B. Hursthouse, T.C. Nugent, R. Pulido, S. M. Roberts and L. E. Wu, *J. Chem. Soc., Chem. Commun.*, **1997**, 739.

(b) J. V. Allen, M. W. Cappi, P. D. Cary, S. M. Roberts, N. M. Williamson and L. E. Wu, *J. Chem. Soc., Perkin Trans. 1.*, **1997**, 3297. (c) J. V. Allen, S. Bergeron, M. J. Griffiths, S. Mukherjee, S. M. Roberts, N. M. Williamson and L. E. Wu, *J. Chem. Soc. Perkin Trans. 1*, **1998**, 3171. (d) J. V. Allen, K.-H. Drauz, R. W. Flood, S. M. Roberts and J. Skidmore, *Tetrahedron Lett.*, **1999**, *40*, 5417.

78. (a) T. P. Geller and S. M. Roberts, *J. Chem. Soc., Perkin Trans. 1*, **1999**, 1397. (b) L. Carde, H. Davies, T. P. Geller and S. M. Roberts, *Tetrahedron Lett.*, **1999**, *40*, 5421.

79. R. W. Flood, T. P. Geller, S. A. Petty, S. M. Roberts, J. Skidmore and M. Volk, *Org. Lett.*, **2001**, *3*, 683.

80. J.-M. Lopez-Pedrosa, M. R. Pitts, S. M. Roberts, S. Saminathan and J. Whittall, *Tetrahedron Lett.*, **2004**, 5073.

81. For example, see; M.W. Cappi, W.-P. Chen, R.W. Flood, Y.-W. Liao, S.M. Roberts, J. Skidmore, J.A. Smith and N.M. Williamson, *J. Chem. Soc., Chem. Commun.*, **1998**, 1159.

82. (a) D. R. Kelly and S. M. Roberts, *J. Chem. Soc., Chem. Commun.*, **2004**, 2018. (b) D. R. Kelly, A. Meek and S. M. Roberts, *J. Chem. Soc., Chem..Commun.*, **2004**, 2021.

83. (a) M. J. O'Donnell in *Catalytic Asymmetric Synthesis*, ed. I. Ojima, VCH, New York, **1993**, 89. (b) A. Nelson, *Angew. Chem. Int. Ed.*, **1999**, *38*, 1583.

84. For examples, see (a) H. Wynberg and B. Greijdanus, *J. Chem. Soc., Chem. Commun.*, **1978**, 427. (b) G. Macdonald, L. Alcaraz, N. J. Lewis and R. J. K. Taylor, *Tetrahedron Lett.*, **1998**, *39*, 5433. (c) L. Alcaraz, G. Macdonald, J. P. Ragot, N. Lewis and R. J. K. Taylor, *J. Org. Chem.*, **1998**, *63*, 3526. (d) S. Arai, H. Tsuge, M. Oku, M. Miura and T. Shioiri, *Tetrahedron*, **2002**, *58*, 1623.

85. For a review, see; H. Wynberg, *Top. Stereochem.*, **1986**, *16*, 87.

86. B. Lygo and P. G. Wainwright, *Tetrahedron Lett.*, **1998**, *39*, 1599.

87. E. J. Corey and F.-Y. Zhang, *Org. Lett.*, **1999**, *1*, 1287.

88. (a) T. Bakó, P. Bakó, G. Keglevich, P. Bombicz, M. Kubinyi, K. Pál, S. Bodor, A. Makó and L. Tőke, *Tetrahedron: Assymetry*, **2004**, *15*, 1589. (b) P. Bakó, A. Makó, G. Keglevich, M. Kubinyi and K. Pál, *Tetrahedron: Asymmetry*, **2005**, *16*, 1861.

89. A. Lattanzi, *Org. Lett.*, **2005**, *7*, 2579.

90. A. Lattanzi and A. Russo, *Tetrahedron*, **2006**, *62*, 12264.

91. M. Marigo, J. Franzen, T. B. Pulsen, W. Zhuang and A. Jørgensen, *J. Am. Chem. Soc.*, **2005**, *127*, 6964.

92. S. Lee and D. W. C. MacMillan, *Tetrahedron*, **2006**, *62*, 11413.

93. (a) X. Lusinchi and G. Hanquet, *Tetrahedron*, **1997**, *53*, 13727, and references therein.

94. P. C. Bulman Page, D. Barros, B. R. Buckley and B. A. Marples, *Tetrahedron: Asymmetry*, **2005**, *16*, 3488.

95. L. Bohé, G. Hanquet, M. Lusinchi and X. Lusinchi, *Tetrahedron Lett.*, **1993**, *34*, 7271.

96. P. C. Bulman Page, B. R. Buckley and A. J. Blacker, *Org. Lett.*, **2004**, *6*, 1543.

97. P. C. Bulman Page, B. R. Buckley, D. Barros, A. J. Blacker, B. A. Marples and M. R. J. Elsegood, *Tetrahedron*, **2007**, *63*, 5386.

98. P. C. Bulman Page, B. R. Buckley, H. Heaney and A. J. Blacker, *Org. Lett.*, **2005**, *7*, 375.

99. (a) M. Frohn and Y. Shi, *Synthesis*, **2000**, 1979. (b) Y. Shi, *Acc. Chem. Res*, **2004**, *37*, 488. (c) D. Yang, *Acc. Chem. Res*, **2004**, *37*, 497.

100. R. Curci, M. Fiorentino and M. R. Serio, *J. Chem. Soc., Chem. Commun.*, **1984**, 155.
101. D. Yang, X.-C. Wang, M.-K. Wong, Y.-C. Yip and M.-W. Tang, *J. Am. Chem. Soc.*, **1996**, *118*, 11311.
102. D. Yang, Y.-C. Yip, M.-W. Tang, M.-K. Wong, J.-H. Zheng and K.-K. Cheung, *J. Am. Chem. Soc.*, **1996**, *118*, 491.
103. A. Armstrong and B.R. Hayter, *J. Chem. Soc., Chem. Commun.*, **1998**, 621.
104. Z.-X. Wang, Y. Tu, M. Frohn and Y. Shi, *J. Org. Chem.*, **1997**, *62*, 2328.
105. M. Frohn, M. Dalkiewicz, Y. Tu, Z.-X. Wang and Y. Shi, *J. Org. Chem.*, **1998**, *62*, 2948.
106. H. Tian, X. She, L. Shu, H. Yu and Y. Shi, *J. Am. Chem. Soc.*, **2000**, *122*, 11551.
107. H. Tian, X. She, J. Xu and Y. Shi, *Org. Lett.*, **2001**, *3*, 1929.
108. Z.-X. Wang, S. M. Miller, O. P. Anderson and Y. Shi, *J. Org. Chem.* **1999**, *64*, 6443.
109. X.-Y. Wu, X. She and Y. Shi, *J. Am. Chem. Soc.*, **2002**, *124*, 8792.
110. (a) V. K. Aggarwal, J. G. Ford, A. Thompson, R. V. H. Jones and M. C. H. Standen, *J. Am. Chem. Soc.*, **1996**, *118*, 7004. (b) V. K. Aggarwal, *Synlett*, **1998**, 329. (c) V. K. Aggarwal, J. N. Harvey and J. Richardson, *J. Am. Chem. Soc.*, **2002**, *124*, 5747. (d) V. K. Aggarwal and J. S. Richardson, *J. Chem. Soc., Chem. Commun.*, **2003**, 2644. (e) V. K. Aggarwal and C. L. Winn, *Acc. Chem. Res.*, **2004**, *37*, 611.
111. V. K. Aggarwal, E. Alonso, I. Bae, G. Hynd, K. M. Lydon, M. J. Palmer, M. Patel, M. Porcelloni, J. Richardson, R. A. Stenson, J. R. Studley, J.-L. Vasse and C. L. Winn, *J. Am. Chem. Soc*, **2003**, *125*, 10926.
112. V. K. Aggarwal, I. Bae, H.-Y. Lee., J. Richardson and D. T. Williams, *Angew. Chem. Int. Ed.*, **2003**, *42*, 3274.
113. (a) H. M. I. Osborn and J. Sweeney, *Tetrahedron: Asymmetry*, **1997**, *8*, 1693. (b) E. N. Jacobsen, in *Comprehensive Asymmetric Catalysis*, Vol. 2, ed. E. N. Jacobsen, A. Pfaltz and H. Yamamoto, Springer-Verlag, Berlin, **1999**, 607. (c) P. Müller and C. Fruit, *Chem. Rev.*, **2003**, *103*, 2905.
114. (a) D. A. Evans, M.M. Faul, M. T. Bilodeau, B.A. Anderson and D.M. Barnes, *J. Am. Chem. Soc.*, **1993**, *115*, 5328. (b) G. Desimoni, G. Faita and K. A. Jørgensen, *Chem Rev.*, **2006**, 3561.
115. (a) J. Xu, L. Ma and P. Jiao, *J. Chem. Soc., Chem. Commun.*, **2004**, **1616**. (b) L. Ma, D.-M. Du and J. Xu, *J. Org. Chem.*, **2005**, *70*, 10155. (c) L. Ma, P. Jiao, Q. Zhang, D.-M. Du and J. Xu, *Tetrahedron: Asymmetry*, **2007**, *18*, 878.
116. Z. Li, K. R. Conser and E. N. Jacobsen, *J. Am. Chem. Soc.*, **1993**, *115*, 5326.
117. Z. Li, R. W. Quan and E. N. Jacobsen, *J. Am. Chem. Soc.*, **1995**, *117*, 5889.
118. M. Shi, C. J. Wang and A. S. C. Chan, *Tetrahedron: Asymmetry*, **2001**, *12*, 3105.
119. K. M. Gillespie, C. J. Sanders, P. O'Shaughnessy, I. Westmoreland, C. P. Thickitt and P. Scott, *J. Org. Chem.*, **2002**, *67*, 3450.
120. K. Noda, N. Hosoya, R. Irie, Y. Ito, T. Katsuki, *Synlett*, **1993**, 469.
121. H. Nishikori and T. Katsuki, *Tetrahedron Lett.*, **1996**, *37*, 9245.
122. P. Müller, C. Baud and Y. Jacquier, *Tetrahedron*, **1996**, *52*, 1543.
123. K. B. Hansen, N. S. Finney and E. N. Jacobsen, *Angew. Chem., Int. Ed. Engl.*, **1995**, *34*, 676.

124. (a) J. C. Antilla and W. D. Wulff, *J. Am. Chem. Soc.*, **1999**, *121*, 5099. (b) J. C. Antilla and W. D. Wulff, *Angew. Chem. Int. Ed.*, **2000**, *39*, 4518. (c) Z. Lu, Y. Zhang and W. D. Wulff, *J. Am. Chem. Soc.*, **2007**, *129*, 7185.
125. V. K. Aggarwal, M. Ferrara, C. J. O'Brien, A. Thompson, R. V. H. Jones and R. Fieldhouse, *J. Chem. Soc., Perkin. Trans. 1*, **2001**, 1635.
126. V. K. Aggarwal, A. Thompson, R. V. H. Jones and M. C. H. Standen, *J. Org. Chem.* **1996**, *61*, 8368.

Chapter 5
Further Oxidation Reactions

There are many other oxidation reactions apart from epoxidation. Alkenes can be oxidised to diols directly using the osmium-catalysed asymmetric dihydroxylation procedure developed by Sharpless, which uses cinchona alkaloids as the framework for enantiomerically pure ligands and stoichiometric quantities of an oxidant. The highest levels of selectivity are observed with *trans-* and trisubstituted and some terminal alkenes while the dihydroxylation of *cis*-alkenes generally only proceeds with moderate ee. A catalytic asymmetric amino-hydroxylation can result when using a nitrogen source as oxidant in this procedure and a range of alkenes undergo this transformation with high ee.

The third subsection of this chapter discusses the α-funtionalisation of aldehydes and ketones. α-Oxidation, amination and halogenation have recently been achieved with high levels of enantioselectivity using enantiopure Lewis acids, or by generation of chiral nonracemic metal enolates, in the presence of a suitable electrophilic heteroatom source. Similar levels of selectivity in this transformation are obtained via the intermediacy of chiral enamines generated using organocatalysts.

The final sections in this chapter consider the oxidation of C–H bonds, the Baeyer–Villiger oxidation of ketones and the oxidation of sulfides to sulfoxides, all of which have been achieved in asymmetric fashion using metal-based catalysts.

5.1 Dihydroxylation

The asymmetric dihydroxylation of alkenes (the AD reaction) using osmium catalysts was discovered and developed by Sharpless, and now represents one of the most impressive achievements of asymmetric catalysis. The majority of early results did not use catalytic systems: however, a breakthrough in the catalytic asymmetric dihydroxylation reaction was reported by Sharpless and coworkers in 1988.[1,2]

The ligands used by the Sharpless group are based on dihydroquinidine (DHQD) (**5.01**) and dihydroquinine (DHQ) (**5.02**). Dihydroquinidine and dihydroquinine are diastereomers, although their derivatives behave as 'pseudo-enantiomers' in

Catalysis in Asymmetric Synthesis 2e   © 2009 Vittorio Caprio and Jonathan M.J. Williams

osmium-catalysed dihydroxylation reactions, providing opposite and approximately equal selectivity.

Early ligands were simple esters of dihydroquinidine and dihydroquinine, such as acetates and p-chlorobenzoates, and gave very good enantioselectivity.[3] Nevertheless, the C_2-symmetric ligands that use a phthalazine spacer unit have proved to be the most generally applicable ligands, working well with several classes of alkene.[4] Dihydroquinidine and dihydroquinine can both be attached to the phthalazine spacer, for example, illustrated by (DHQD)$_2$PHAL **(5.03)**. The ligands accelerate the rate of the osmium-catalysed dihydroxylation, which is particularly helpful in asymmetric catalysis. This ligand, and its pseudo-enantiomeric partner (DHQ)$_2$PHAL **(5.04)** provide excellent enantioselectivity in the dihydroxylation reaction. The ligands used in the asymmetric dihydroxylation reactions have been immobilised onto silica,[5] polymers[6] and also in ionic liquids[7] for recovery and reuse.[8] The usual stoichiometric oxidant used in these reactions is now potassium ferricyanide ($K_3Fe(CN)_6$) and this is used in the commerically available AD-mix. AD-mix-α contains $K_3Fe(CN)_6$, K_2CO_3 and (DHQ)$_2$-PHAL along with involatile potassium-osmate(VI) dihydrate. The alternative AD-mix-β contains the (DHQD)$_2$-PHAL ligand.

Trans-disubstituted alkenes often afford especially good selectivities, using the AD-mix reagents, as represented by the conversion of alkene **(5.05)** into either enantiomer of the diol **(5.06)**.

Not surprisingly, for such an important process as the AD process, there has been considerable debate over the mechanism of the reaction. The two main pathways are either concerted [3+2] cycloaddition of the osmium tetroxide **(5.07)** with alkene, or [2+2] cycloaddition followed by subsequent conversion into the osmium(VI) glycolate **(5.08)**, as identified in Figure 5.1. It seems reasonable that the ligand adds before the cycloaddition, and Lohray has indicated that this is the case.[9] While Sharpless initially presented evidence that the [2+2] mechanism is more likely[10] more recent theoretical studies indicate that the [2+2] pathway proceeds through a much higher energy barrier than [3+2] cycloaddition[11] and the matching of kinetic isotope effects calculated for the cycloaddition pathway with those observed experimentally further support the postulate that a concerted [3+2] pathway operates.[12]

Once the osmium(VI) glycolate **(5.08)** is formed, the catalytic cycle is completed by reoxidation with the stoichiometric oxidant followed by hydrolysis (Figure 5.2). Water is required to liberate the diol from the osmium and methanesulfonamide is often used as an additive, especially when using internal olefins, to increase the rate of this hydrolysis.[13] It has been found that the rate of osmium glycolate hydrolysis is also enhanced under basic conditions. As the basicity of the AD reaction mixture decreases during the reaction, the maintenance of a constant pH of ca. 12, using an automatic titrator, leads to improvement in the reaction rate allowing the methanesulfonamide to be omitted.[14]

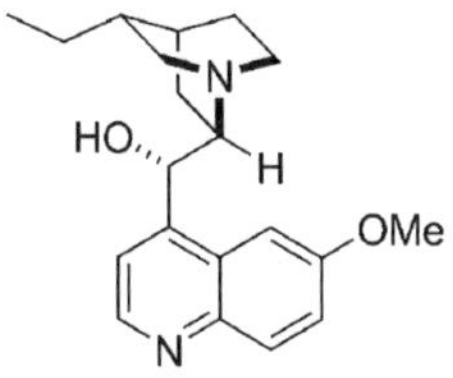

(5.01) DHQD

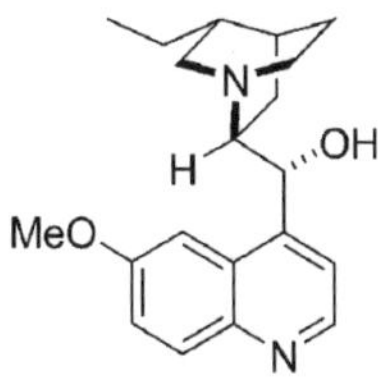

(5.02) DHQ

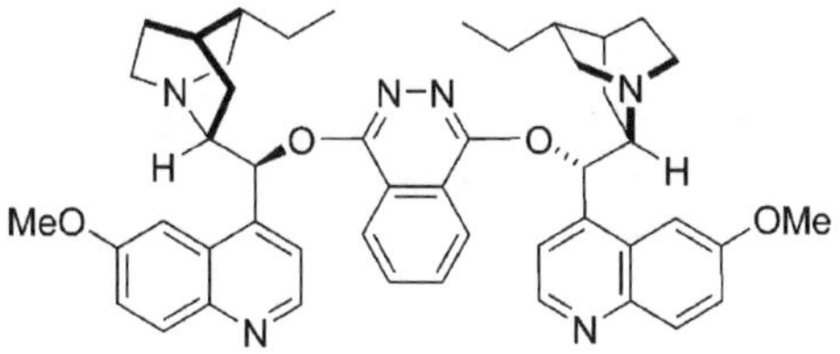

(5.03) (DHQD)₂PHAL

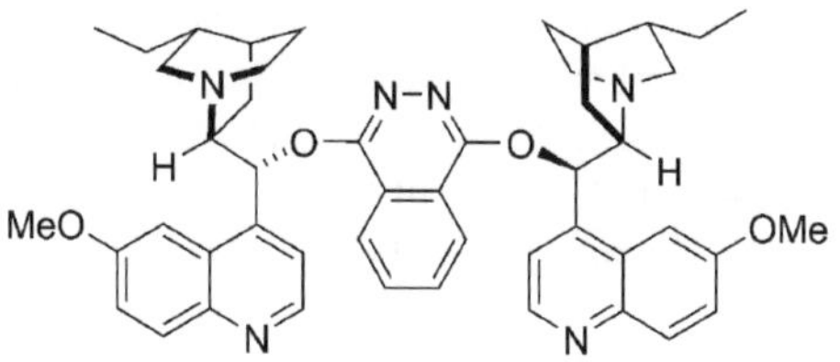

(5.04) (DHQ)₂PHAL

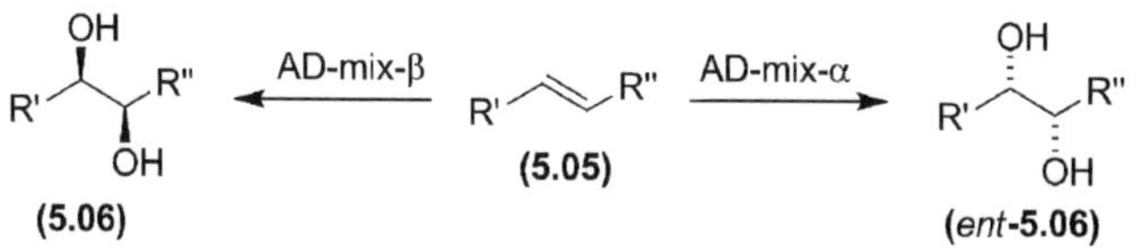

(5.06)

(5.05)

(ent-5.06)

Enantioselectivity in this process arises from reaction of the alkene and osmium tetroxide within an asymmetric environment created by the ligand. However, the precise shape of this binding pocket and the nature of the interactions between the substrate and ligand has been the subject of some debate. Corey has advanced a model based on a U-shaped ligand conformation initially derived from an inspection of molecular models [15] while Sharpless has proposed an L-shaped active site based on the results of molecular mechanics studies. [16]

Fortunately a mnemonic device, consisting of a plane divided into four quadrants, can be used to reliably predict the sense of asymmetric induction in the

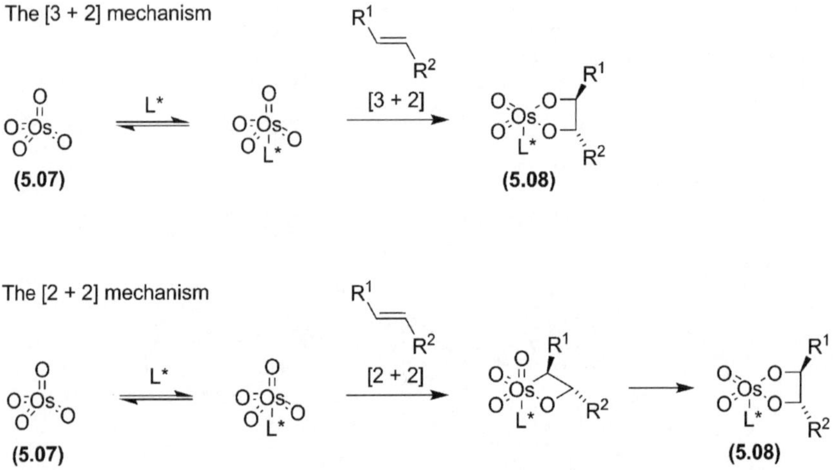

The [3 + 2] mechanism

The [2 + 2] mechanism

Figure 5.1   Possible mechanisms of osmium glycolate formation

product, as shown in Figure 5.3.[17] The south east (SE) quadrant presents a steric barrier, preferring the smallest substituent (usually a hydrogen atom). The north west (NW) quadrant also presents a steric barrier, but a smaller one than in the SE. It is clear that (E)-disubstituted alkenes fit this mnemonic device very well, since the substituents can fit interchangably in the available south west (SW) and north east (NE) quadrants.

The NE quadrant is available for moderately sized substituents whilst the SW quadrant is considered as an attractive area, with a preference for flat

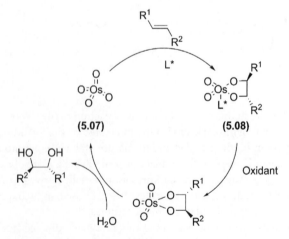

Figure 5.2   Catalytic cycle of dihydroxylation

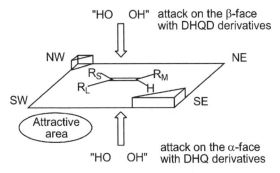

"HO   OH" attack on the β-face
with DHQD derivatives

NW   NE

R_S R_M
R_L H

SW   SE

Attractive
area

attack on the α-face
"HO   OH" with DHQ derivatives

Figure 5.3 Predictive model for the asymmetric dihydroxylation reaction. Abbreviations: DHQ = dihydroquinine, DHQD = dihydroquinidine

aromatic substituents. Aliphatic substituents are also accepted in the SW quadrant, but $ROCH_2$ and methyl groups disfavour this region.[18] Substrates fitted into the mnemonic according to these rules are then oxidised from above (the β-face) using DHQD ligands and from below (the α-face) using DHQ ligands.

Specific examples are given by the dihydroxylation of the *trans*-disubstituted alkenes **(5.09)**–**(5.11)** using AD-mix-β, which afford diols **(5.12)**–**(5.14)** in very high enantiomeric excess. Dihydroxylation of stilbene **(5.09)** has been developed such that 1 kg of substrate can undergo dihydroxylation in a 5 litre vessel.[19] The corresponding AD-mix-α containing $(DHQ)_2PHAL$ often gives slightly lower selectivities. In the last three cases, 99.5% ee, 93% ee and 95% ee, respectively are obtained.

Ph   Ph

(5.09)

1 equiv. $MeSO_2NH_2$
————————
AD-mix-β, 0 °C,
$^tBuOH/H_2O$ (1:1)

OH
Ph   Ph
OH
(5.12) 99.8% ee

C_4H_9 C_4H_9

(5.10)

1 equiv $MeSO_2NH_2$
————————
AD-mix-β, 0°C,
$^tBuOH/H_2O$ (1:1)

OH
C_4H_9 C_4H_9
OH
(5.13) 97% ee

Ph CO_2Et

(5.11)

1 equiv $MeSO_2NH_2$
————————
AD-mix-β, r.t.,
$^tBuOH/H_2O$ (1:1)

OH
Ph CO_2Et
OH
(5.14) 97% ee

Asymmetric dihydroxylation reactions can be carried out on substrates containing heteroatoms. [20,21] In the case of allylic halides, including cinnamyl chloride (5.15), NaHCO$_3$ is required in addition to K$_2$CO$_3$ in the AD-mix. This suppresses hydrolysis of the starting material and ring closure of the product. In the case of allylic sulfides, such as compound (5.16), the alkene is selectively oxidised in preference to the sulfur. Heteroaromatic moieties are also tolerated although it has been observed that blocking of the nitrogen in pyrrolyl and pyridyl-substituted substrates by *N*-protection or *N*-oxide formation is required for AD using PHAL ligands. [22]

Allylic alcohols can also be dihydroxylated with good enantioselectivity, although the corresponding ethers or esters generally give higher enantioselectivity. [23] Thus, the allylic alcohol (5.17) undergoes dihydroxylation with good enantioselectivity, but the corresponding benzoate ester affords a product with 99% ee. Tertiary allylic alcohols can also be used as substrates for asymmetric dihydroxylation. [24] Corey and coworkers used a ligand closely related to the (DHQD)$_2$-PHAL ligand, where the linker is a pyridazine unit (the benzene ring is not included). [25] In their examples, they found that *p*-methoxybenzoate derivatives of allylic alcohols and *p*-methoxyphenyl ether derivatives of homoallylic alcohols were the most suitable.

Dienes are interesting substrates for the dihydroxylation reaction. In cases where the diene is symmetrical the enediols are isolated in good yield and enantioselectivity. [26] For example, diene (5.21) affords the enediol (5.24).

Similarly, enynes, including substrate (5.22), are selectively oxidised in the alkene moiety to give products which have been shown to be synthetically useful. [27]

However, for nonsymmetrical dienes, there is an added issue of regioselectivity. The reactions tend to proceed to retain any conjugation (as in product (**5.26**)), although other electronic effects are also operative in some cases.[28]

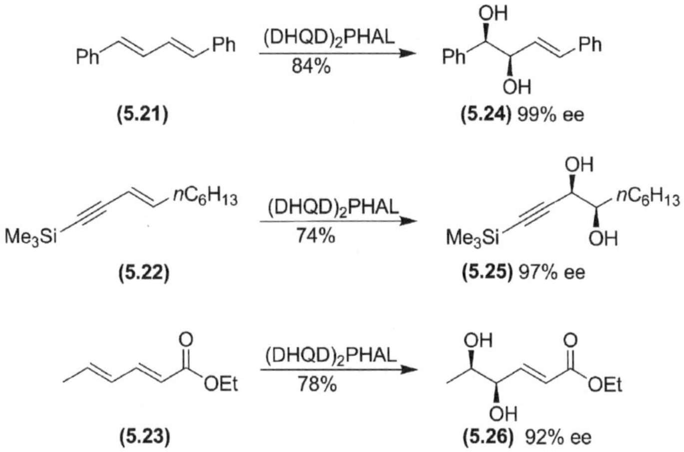
Ph
Ph
(DHQD)$_2$PHAL
84%
OH
Ph
Ph
OH
(5.21)
(5.24) 99% ee
nC_6H_{13}
Me$_3$Si
(DHQD)$_2$PHAL
74%
OH
nC_6H_{13}
Me$_3$Si
OH
(5.22)
(5.25) 97% ee
O
OEt
(DHQD)$_2$PHAL
78%
OH
O
OEt
OH
(5.23)
(5.26) 92% ee

Hydroxy lactones are formed from β,γ- and γ,δ-unsaturated esters under the Sharpless AD reaction conditions, where the diol closes selectively to provide γ-lactones.[29] The enantiomeric excess can be excellent, as with the unsaturated ester (**5.27**), which produces the γ-lactone (**5.28**).

A similar approach has also been used with vinyl silanes possessing a pendant ester group.[30]

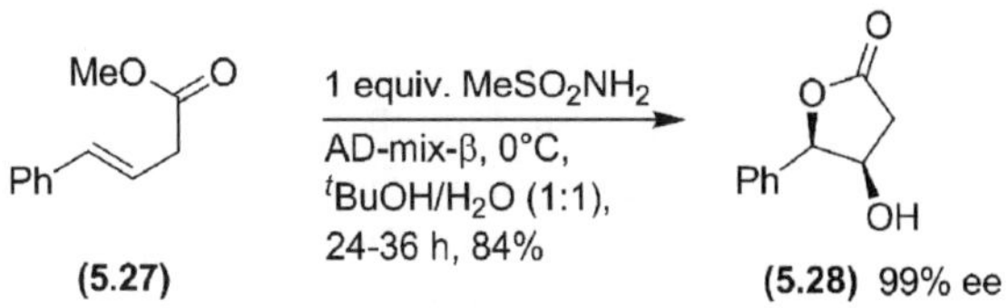
MeO
O
Ph
1 equiv. MeSO$_2$NH$_2$
AD-mix-β, 0°C,
^{t}BuOH/H$_2$O (1:1),
24-36 h, 84%
O
O
Ph
OH
(5.27)
(5.28) 99% ee

Trisubstituted alkenes are also good substrates for the AD reaction, and the (DHQD)$_2$PHAL and (DHQ)$_2$PHAL ligands generally give the best selectivities.[31] The acyclic trisubstituted alkene (**5.29**) and the cyclic trisubstituted alkene (**5.30**) both afford the corresponding diols with very high enantioselectivity.

Enol ethers represent an interesting class of alkene, since the isolated products are α-hydroxy ketones.[32] (*E*)-enol ethers tend not to give such high selectivities as (*Z*)-enol ethers, such as compound (**5.31**). However, both geometries of a given enol ether substrate afford the same enantiomer of α-hydroxy ketone preferentially.

Cyclic enol ethers are also good substrates for the Sharpless AD reaction with ee being dependant on the chain length of the enol ether side chain.[33] Thus, while the pentyl enol ether (5.32) is converted into α-hydroxyketone (5.36) with 94% ee, the corresponding methyl enol ether is oxidised with only 83% ee.

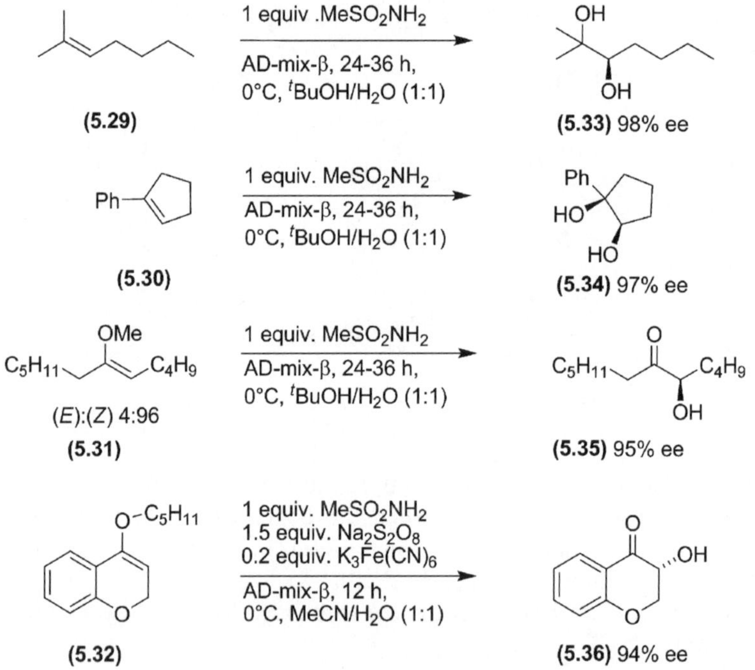

For terminal alkenes, enantioselectivities are somewhat more variable. In some, but not all cases, the alternative pyrimidine ligands (DHQD)$_2$PYR (5.37) and (DHQ)$_2$PYR (5.38)[34] and the anthraquinone ligands (DHQD)$_2$AQN (5.39) and (DHQ)$_2$AQN (5.40)[35] provide superior results. The AQN-based ligands have shown utility in the dihydroxylation of simple allylic-substituted terminal olefins such as (5.41), while the pyrimidine ligands give better results with more sterically congested alkenes such as (5.42).

PHAL ligands have been observed to perform poorly on substrates possessing pyridine rings (*vide supra*), however both PYR and AQN ligands have been used to effect highly selective AD reaction of 1-aryl-1′-pyridyl-substituted terminal alkenes such as (5.43).[36]

Tetrasubstituted enol ethers such as (5.44) undergo dihydroxylation using PYR ligands with good ee, but for other tetrasubstituted olefins, enantioselectivities tend to be reduced and isolated yields also suffer.[37]

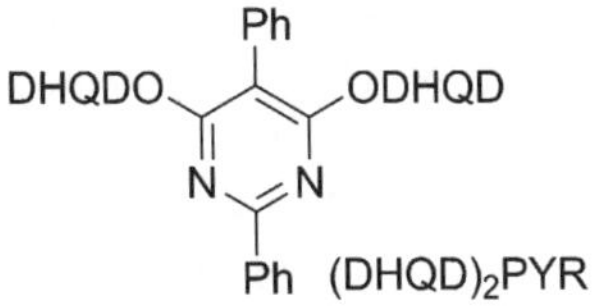

DHQDO—[Ph]—ODHQD

N N

Ph (DHQD)₂PYR

(5.37)

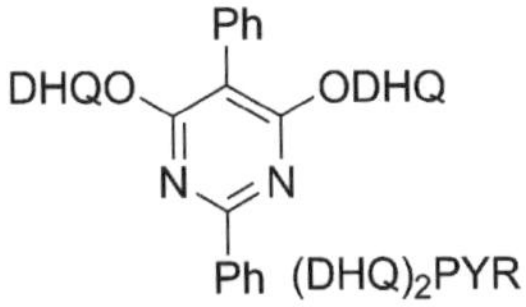

$$\text{(DHQD)}_2\text{PYR}$$

DHQO—[Ph]—ODHQ

N N

Ph (DHQ)₂PYR

(5.38)

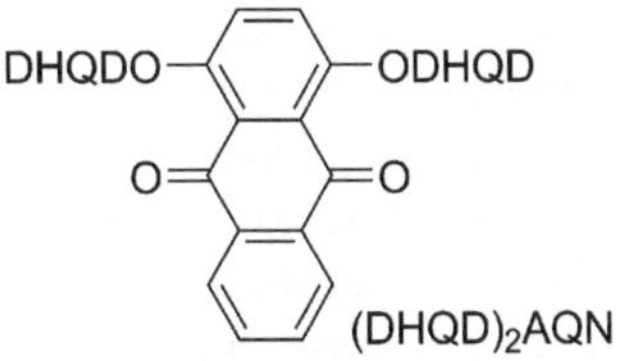

DHQDO——ODHQD

O==O

(DHQD)₂AQN

(5.39)

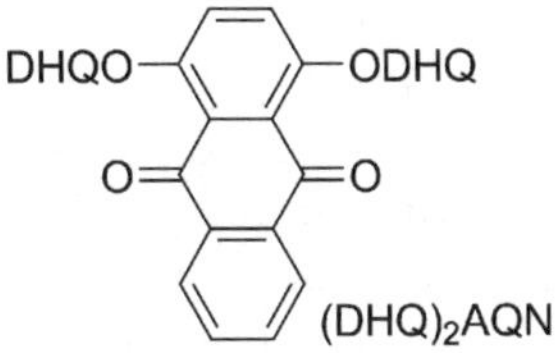

DHQO——ODHQ

O==O

(DHQ)₂AQN

(5.40)

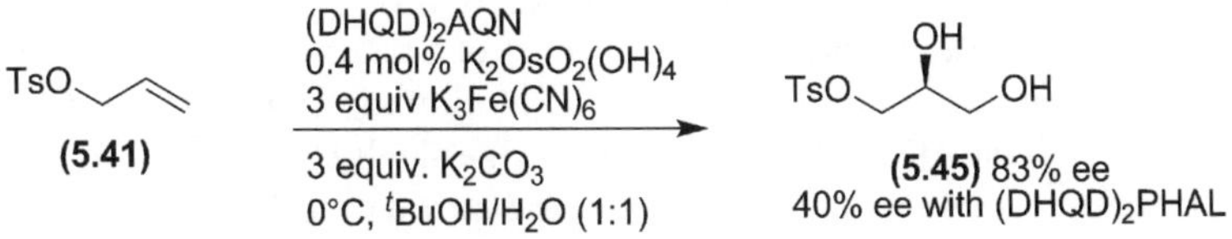

TsO~~

(5.41)

(DHQD)₂AQN
0.4 mol% K₂OsO₂(OH)₄
3 equiv K₃Fe(CN)₆

3 equiv. K₂CO₃
0°C, ᵗBuOH/H₂O (1:1)

→

TsO——OH with OH

(5.45) 83% ee
40% ee with (DHQD)₂PHAL

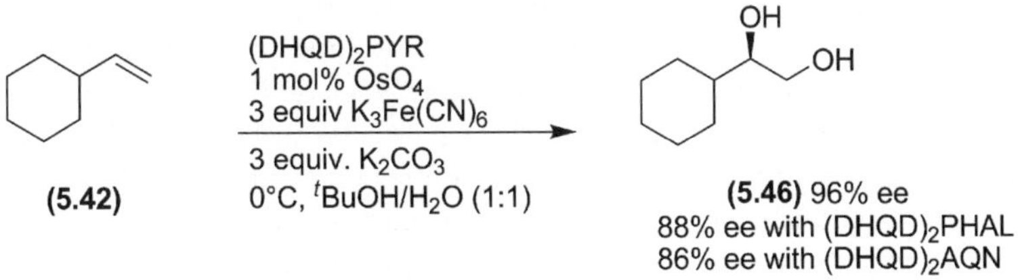

(5.42)

(DHQD)₂PYR
1 mol% OsO₄
3 equiv K₃Fe(CN)₆

3 equiv. K₂CO₃
0°C, ᵗBuOH/H₂O (1:1)

→

(5.46) 96% ee
88% ee with (DHQD)₂PHAL
86% ee with (DHQD)₂AQN

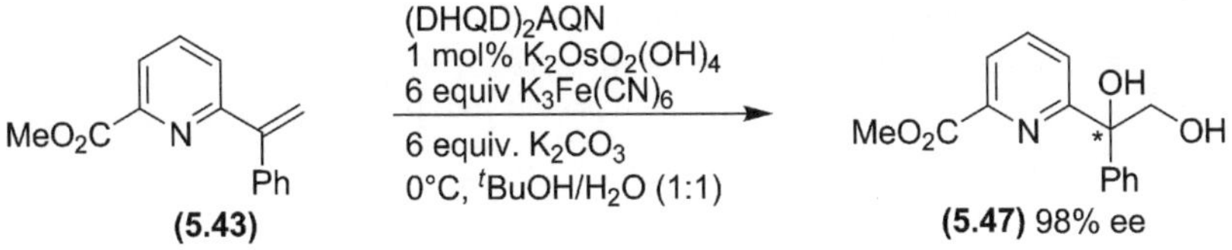

MeO₂C—N— —Ph

(5.43)

(DHQD)₂AQN
1 mol% K₂OsO₂(OH)₄
6 equiv K₃Fe(CN)₆

6 equiv. K₂CO₃
0°C, ᵗBuOH/H₂O (1:1)

→

MeO₂C—N—*—OH
Ph

(5.47) 98% ee

OTBDMS
Ph

(5.44)

(DHQD)₂PYR
1 mol% OsO₄
3 equiv K₃Fe(CN)₆

3 equiv K₂CO₃
0°C, ᵗBuOH/H₂O (1:1)

→

O
Ph
OH

(5.48) 95% ee
93% ee with (DHQD)₂PHAL

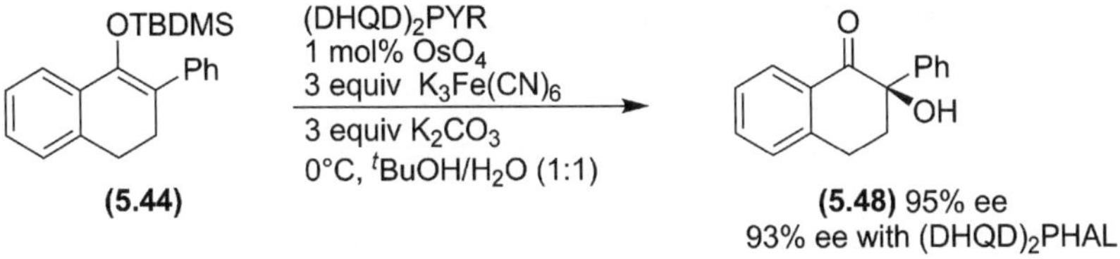

(Z)-disubstituted-alkenes are not usually good substrates for the Sharpless AD reaction, and although alternative ligands have been used,[38,39] enantioselectivities are moderate, with only a few exceptions.

In the Sharpless asymmetric dihydroxylation of alkenes that already contain a stereocentre, the possibility of matched and mismatched selectivity arises. There have been several reports of this phenomenon. For example, Cha and coworkers showed that, in the absence of a ligand, the alkene **(5.49)** has a slight preference for formation of diastereomer **(5.50)** in the dihydroxylation reaction.[40] The use of (DHQD)$_2$PHAL **(5.03)** reinforces this preference (matched), whereas the (DHQ)$_2$PHAL **(5.04)** ligand represents a mismatched case and overturns the underlying selectivity, affording diastereomer **(5.51)** as the major product. There are also cases where the intrinsic facial bias of the substrate overrides that of the catalysts. For instance, during a synthesis of (-)-swainsonine Pyne and coworkers observed that olefin **(5.52)** undergoes dihydroxylation to diastereomer **(5.53)** in similar ratios using either AD-mix-α or AD-mix-β.[41]

In fact, the rate at which one enantiomer of a substrate reacts may be different from the other one, and hence a kinetic resolution of racemic substrates is also possible with the AD reaction. In contrast to the highly efficient kinetic resolution of secondary alcohols by Sharpless asymmetric epoxidation, kinetic resolution via AD have given good, but not exceptional, discrimination. Some of the more selective examples of kinetic resolution include the axially chiral alkene **(5.55)**.[42] Some of the highest K_{rel} values have been obtained by Corey and coworkers during the kinetic resolution of the protected allylic alcohols, such as **(5.56)**, using ligand **(5.57)**. Tuck and coworkers have performed a detailed study of this kinetic resolution and concluded that the transition state for the AD reaction is reactant-like.[43] Thus, the success of the resolution is highly dependant on the structure of the substrate. Specifically, the kinetic resolution of two enantiomers is more successful when one face of the alkene is sterically hindered and strong enantiodirecting groups such as arenes are present. As an example, phenyl-substituted bicyclic alkene **(5.58)** undergoes effective kinetic resolution using (DHQD)$_2$PHAL, but resolution of the methyl alkene **(5.59)** is ineffective.

5.2 Aminohydroxylation

In 1996, Sharpless reported that modification of the AD reaction, by the inclusion of a nitrogen source, which also functions as an oxidant, gives an asymmetric aminohydroxylation (AA).[44,45]

The most common nitrogen sources used are all N-halogenated species. The first modification reported by Sharpless and coworkers employed chloramine-T trihydrate (TsNClNa.3H$_2$O) as the nitrogen source. Subsequently, alternative reagents

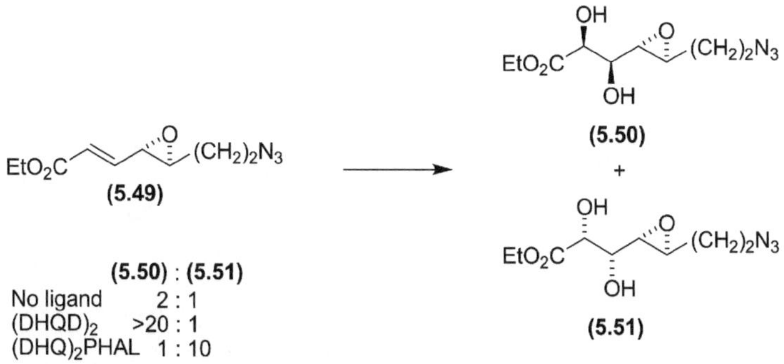

(5.49)

(5.50) : **(5.51)**
No ligand   2 : 1
(DHQD)₂   >20 : 1
(DHQ)₂PHAL   1 : 10

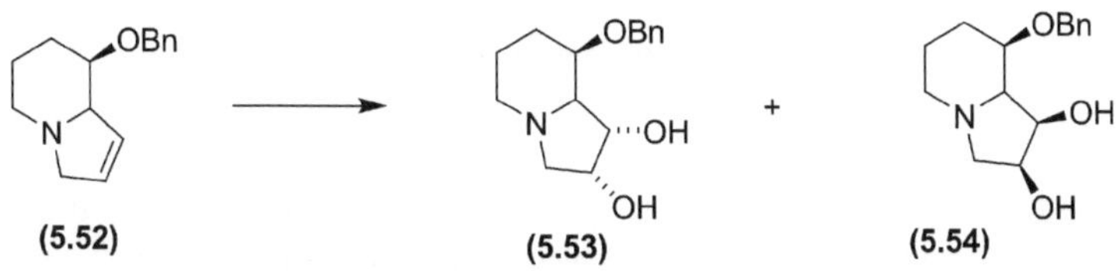

(5.52)   **(5.53)**   **(5.54)**

(5.53) : **(5.54)**
No ligand   2 : 1
(DHQD)₂   95 : 5
(DHQ)₂PHAL 98 : 2

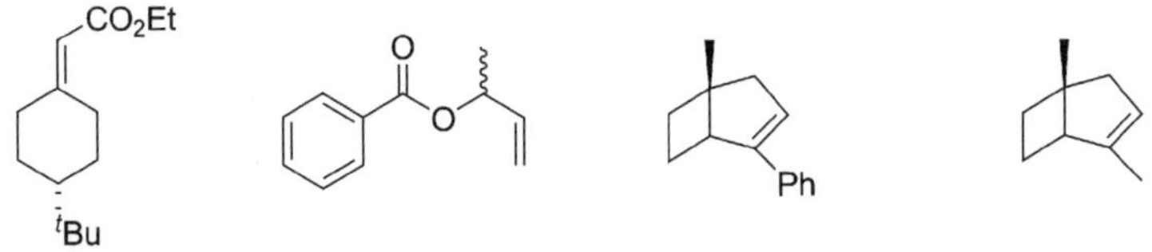

(5.55)   **(5.56)**   **(5.58)**   **(5.59)**
using (DHQD)₂-PHAL   using **(5.57)**   using (DHQD)₂-PHAL   using (DHQD)₂-PHAL

K_{rel} = 32 K_{rel} = 79 K_{rel} = 26 K_{rel} = 1.8

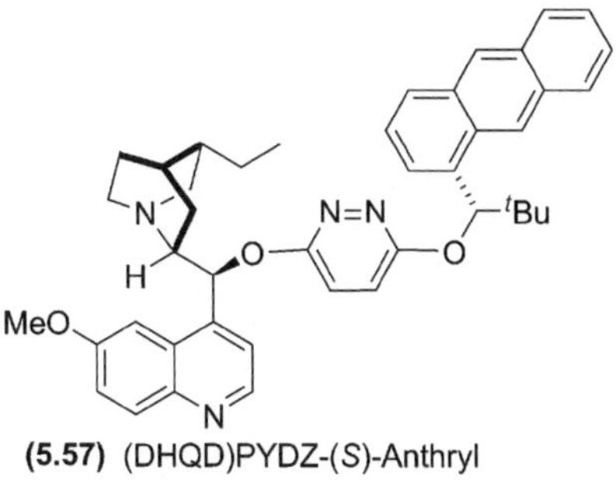

(5.57) (DHQD)PYDZ-(*S*)-Anthryl

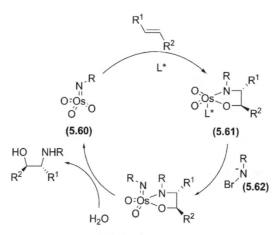

Figure 5.4 Catalytic cycle of aminohydroxylation

have been used, giving rise to amino alcohol derivatives with high enantioselectivity. N-halocarbamates,[46] sulfonamides, such as chloramine-M (MeSO$_2$NClNa)[47] and ^{t}BuSO$_2$NClNa,[48] and amides, such as N-bromoacetamide (+ base),[49] have all been employed to give high enantioselectivity in the AA reaction.

In general, N-halosulfonamides can be used to effect the AA reaction of α,β-enoates while N-halocarbamates and amides show wider scope and can also be used in the AA reaction of styrenes. The key intermediate is the imido-osmium(VIII) compound **(5.60)**, which undergoes addition to the alkene in a [2+2] or [3+2] process to give a ligand-bound osmium azaglycolate **(5.61)**. Reoxidation of the azaglycolate by reaction with N-halo species **(5.62)** followed by hydrolysis releases product and regenerates **(5.60)** (Figure 5.4).

The mnemonic device used to predict the sense of enantioselectivity in the AD reaction can also be used in the AA process. Typical examples include the asymmetric aminohydroxylation of alkenes **(5.63–5.67)**, all with excellent enantioselectivity. Heterocyclic groups are tolerated in the AA reaction and high ees have been obtained for the aminohydroxylation of furanoyl acrylates such as **(5.65)**.[50] In common with the AD reaction, pyrrolyl- and pyridyl-substituted olefins are difficult substrates and blocking of the nitrogen is required for enantioselective aminohydroxylation. However, indoles such as **(5.66)** undergo aminohydroxylation with good ee. The AA reaction has also been applied to the desymmetrisation of dienylsilane **(5.67)** by Landais and coworkers.[51] Whilst the enantioselectivity is not perfect, the reaction is still remarkably regio- and diastereoselective.

The regioselectivity of the AA reaction depends on the substitution pattern and, to some extent, the electron distribution of the alkene substrates. While many

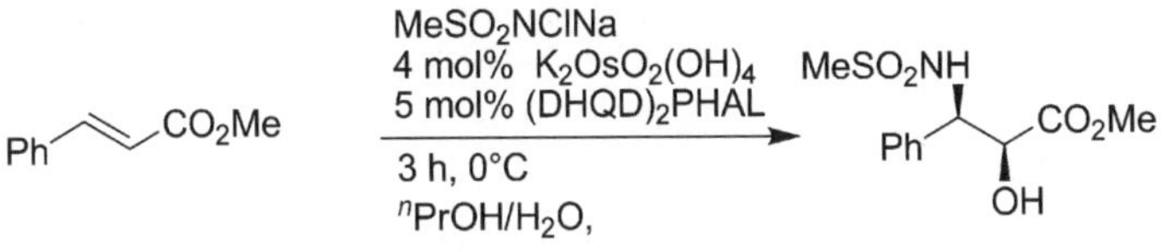

Ph—CH=CH—CO$_2$Me

(5.63)

MeSO$_2$NClNa
4 mol% K$_2$OsO$_2$(OH)$_4$
5 mol% (DHQD)$_2$PHAL
3 h, 0°C
^{n}PrOH/H$_2$O,

$\longrightarrow$

MeSO$_2$NH

Ph—CH(NHSO$_2$Me)—CH(OH)—CO$_2$Me

(5.68) 95% ee

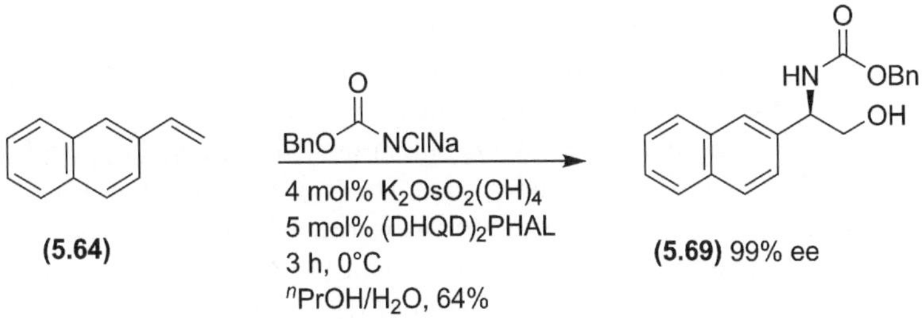

(5.64)

BnO—C(=O)—NClNa
4 mol% K$_2$OsO$_2$(OH)$_4$
5 mol% (DHQD)$_2$PHAL
3 h, 0°C
^{n}PrOH/H$_2$O, 64%

$\longrightarrow$

HN—C(=O)—OBn

(5.69) 99% ee

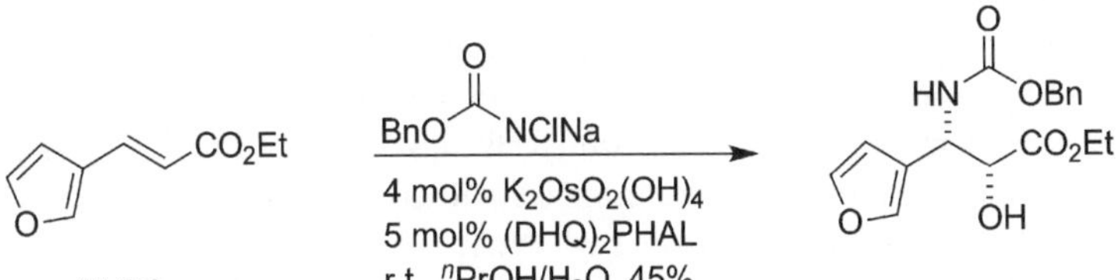

(5.65)

BnO—C(=O)—NClNa
4 mol% K$_2$OsO$_2$(OH)$_4$
5 mol% (DHQ)$_2$PHAL
r.t., ^{n}PrOH/H$_2$O, 45%

$\longrightarrow$

HN—C(=O)—OBn

CO$_2$Et
OH

(5.70) 99% ee
>15:1 regioselectivity

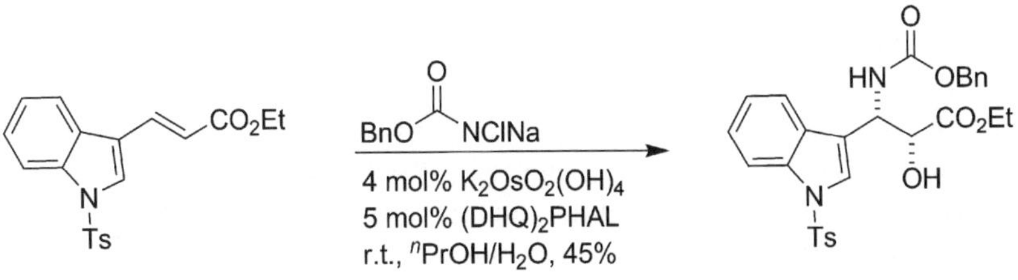

(5.66)

BnO—C(=O)—NClNa
4 mol% K$_2$OsO$_2$(OH)$_4$
5 mol% (DHQ)$_2$PHAL
r.t., ^{n}PrOH/H$_2$O, 45%

$\longrightarrow$

HN—C(=O)—OBn

CO$_2$Et
OH
Ts

(5.71) 93% ee
>15:1 regioselectivity

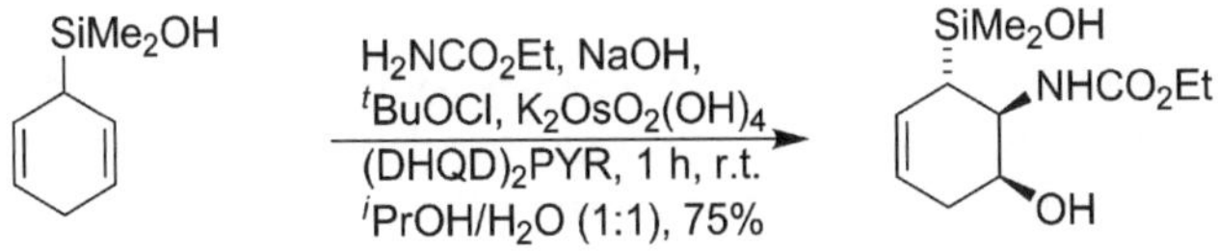

(5.67) SiMe$_2$OH

H$_2$NCO$_2$Et, NaOH,
^{t}BuOCl, K$_2$OsO$_2$(OH)$_4$
(DHQD)$_2$PYR, 1 h, r.t.
^{i}PrOH/H$_2$O (1:1), 75%

$\longrightarrow$

SiMe$_2$OH
NHCO$_2$Et
OH

(5.72) 68% ee, >98% de
>98% regioselectivity
>98% de

alkenes undergo amination at the most substituted carbon, the aminohydroxylation of styrenes gives primarily the benzylic amino derivative as the major regioisomer of product. Thus, *p*-bromostyrene **(5.73)** affords the regiosiomeric products **(5.74)** and **(5.75)** in an 80:20 ratio. These products can be converted into diamines[52] and amino acids.[53] As only the primary alcohol can be converted into the acid **(5.76)**, which is readily separated, the amino acid synthesis is effective using the regioisomeric mixture. It has been found that the regioselectivity of the AA reaction of styrenes is pH dependant and can be reversed if the pH of the reaction mixture is kept between 7.5 and 7.7 using phosphate buffer to give the 1-aryl-2-aminoethanol as the major product.[54]

The most electrophilic carbon is generally aminated preferentially and thus α,β-unsaturated esters and ketones undergo amination at the β-position. The use of the alternative ligand (DHQD)$_2$AQN **(5.39)** can overturn the normal regiochemistry of the AA of α,β-enoates.[55] For example, cinnamate **(5.63)** affords the amino alcohols **(5.77)** and **(5.78)**, with the benzylic alcohol regioisomer **(5.77)** predominating using this ligand. An in-depth study of the regioselectivity of the PHAL-mediated AA reaction based on an examination of the U-shaped ligand binding pocket postulated by Corey and coworkers[15] reveals that, while electronic factors do play some part, amination of *trans*-disubstituted alkenes occurs primarily on the alkene carbon bearing any aromatic-containing substituents, but β- to highly sterically demanding moieties such as bulky silyl ethers.[56] Thus, while enoate **(5.79)** is aminated primarily at the β-carbon, a complete reversal in selectivity is observed using β-silyloxyenoate **(5.80)** as substrate.

5.3　α-Heterofunctionalisation of Aldehydes and Ketones

The α-heterofunctionalisation of aldehydes and ketones can be achieved via the reaction of chiral nonracemic metal-enolates with suitable electrophiles or by activation of the electrophilic heteroatom source with an enantiomerically pure Lewis acid catalyst. Alternatively, metal-free α-functionalisation can be achieved via the reversible formation of enamines and this approach has proved a successful method for the synthesis of a range of enantioenriched α-heteroaldehydes and ketones.

5.3.1　α-Oxidation

Chiral nonracemic α-hydoxylated ketones are commonly accessed by asymmetric epoxidation or dihydroxylation of enol ethers and this methodology is discussed in the relevant sections of this book. Another general method for the enantioselective α-oxygenation of ketones and aldehydes is by reaction of an electrophilic source of oxygen with chiral nonracemic enamines or enolates or in the presence of Lewis acids.[57]

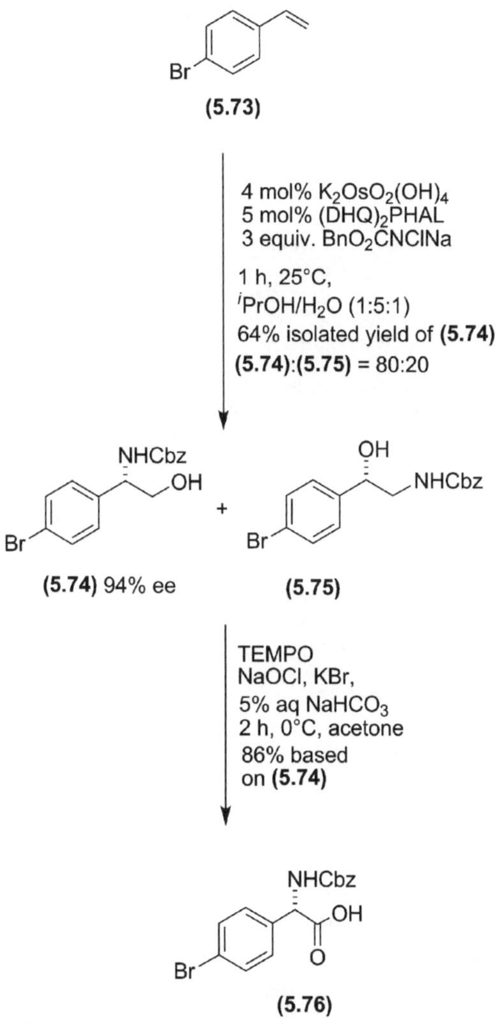

(5.73)

4 mol% $K_2OsO_2(OH)_4$
5 mol% $(DHQ)_2PHAL$
3 equiv. $BnO_2CNClNa$

1 h, 25°C,
$^iPrOH/H_2O$ (1:5:1)
64% isolated yield of **(5.74)**
(5.74):**(5.75)** = 80:20

NHCbz
OH

Br

(5.74) 94% ee

+

OH
NHCbz

Br

(5.75)

TEMPO
NaOCl, KBr,
5% aq $NaHCO_3$
2 h, 0°C, acetone
86% based
on **(5.74)**

NHCbz
OH
O

Br

(5.76)

TEMPO = (2,2,6,6-tetramethyl-1-piperidinyloxy, free radical)

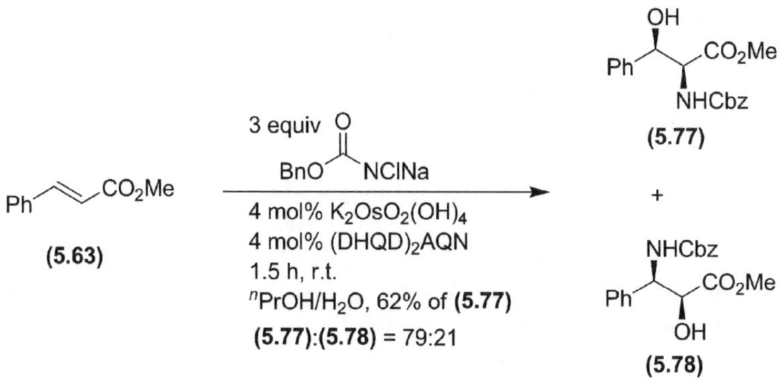

Ph
CO_2Me

(5.63)

3 equiv
O

BnO
NClNa

4 mol% $K_2OsO_2(OH)_4$
4 mol% $(DHQD)_2AQN$
1.5 h, r.t.
$^nPrOH/H_2O$, 62% of **(5.77)**
(5.77):**(5.78)** = 79:21

OH
CO_2Me
Ph
NHCbz

(5.77)

+

NHCbz
CO_2Me
Ph
OH

(5.78)

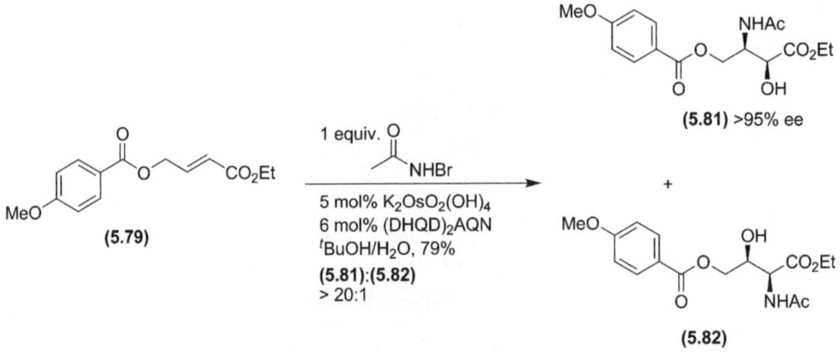

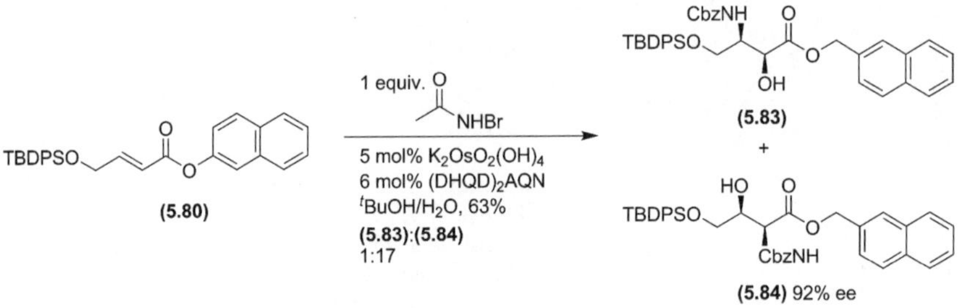

(5.81) >95% ee

(5.82)

(5.83)

(5.84) 92% ee

Yamamoto and coworkers have developed an asymmetric α-aminoxylation of cyclic tributyltin enolates using nitrosobenzene (5.85) as oxygen source in the presence of Lewis acidic 1:1 silver-BINAP complexes.[58] For example, the enolate (5.86) is converted into the α-aminoxyketone with high ee using this procedure. Cleavage of the N–O bond is then effected with no racemization, using catalytic copper sulfate in methanol.

Chiral nonracemic enamines formed *in situ* by reaction of aldehydes and ketones with proline-based organocatalysts also undergo stereoselective α-oxygenation.[57c] MacMillan and coworkers have achieved enantioselective α-oxidation of a range of aliphatic aldehydes such as (5.87) with 97–99% ee using proline in combination with nitrosobenzene (5.85). This method can be applied to the aminoxylation of six-membered cyclic ketones such as 1,4-cyclohexanedione monoethylene ketal (5.88).[59]

The organocatalysed oxygenation is thought to proceed via a chair-like transition state and enantioselectivity arises from hydrogen-bonding between the acidic hydrogen of the proline and the oxidant guiding attack at the *Re* face (Figure 5.5). Hydrolysis of the resulting oxygenated iminium species by water formed *in situ* releases proline to reenter the catalytic cycle.

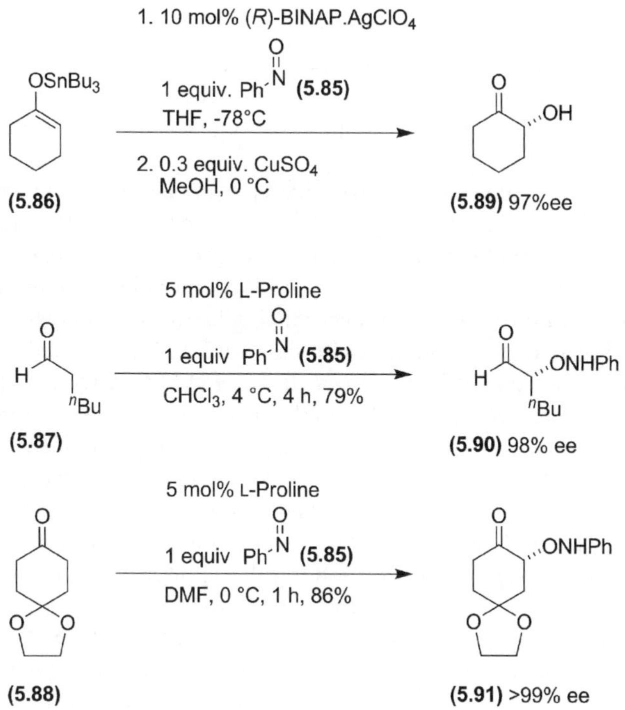

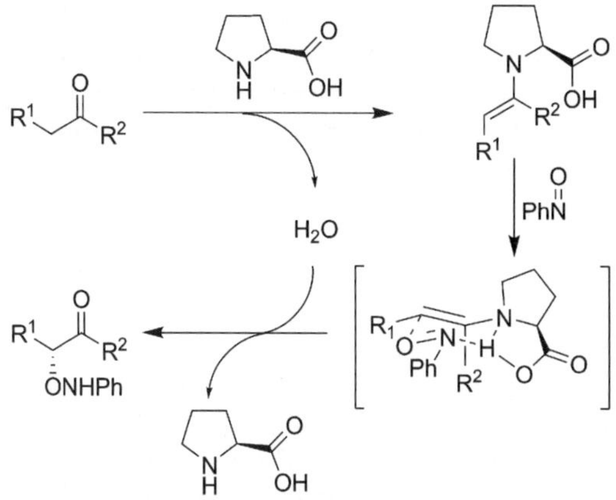

Figure 5.5 Mechanism of the proline-catalysed α-oxygenation of aldehydes and ketones

5.3.2 α-Amination

The direct α-amination of aldehydes and ketones provides a useful method for the synthesis of amino acids and, consequently, the asymmetric variant of this process using either metal-based complexes or organocatalysts has received much attention.

The α-amination of ketones can be achieved using Lewis acids in combination with electrophilic sources of nitrogen.[57b, 57c] As nitrosobenzene also functions as a nitrogen donor, modification of Yamomoto's protocol for α-oxygenation described above, by the use of a 2:1 silver:BINAP complex, leads to the synthesis of cyclic α-hydroxyaminoketones with ees ranging between 77 and 99%.[60] Copper bis-oxazoline-based Lewis acids have also shown utility as Lewis acids in the asymmetric amination. Aryl silyl enol ethers such as (**5.92**) undergo enantioselective amination with azocarboxylate (**5.93**) in the presence of the copper bis-oxazoline complex (**5.94**)[61] while 2-ketoesters such as (**5.95**) react with dibenzyl azodicarboxylate (**5.96**) using copper bis-oxazoline complex (**5.97**) to give the aminated product (**5.98**) with high ee.[62] In an alternate approach Evans and coworkers have achieved high ees in the amination of aryl N-acyloxazolidinones such as (**5.99**) by conversion to the magnesium enolate using the sulphonamide base (**5.100**) and quenching with di-*tert*-butylazadicarboxylate (**5.101**).[63]

Proline-derived enamines react stereoselectively with electrophilic sources of nitrogen in a similar fashion to the organocatalytic oxygenation depicted in Figure 5.5.[57c, 64] Both Jørgensen and List have developed an asymmetric L-proline catalysed amination of simple aliphatic aldehydes such as (**5.102**).[65,66] As the amino aldehyde product is prone to racemisation the products are isolated as the aminoalcohol. This proline-catalysed amination has also been applied to the chemo- and enantioselective amination of ketones.[67] The sense of enantioinduction in the organocatalytic amination can be reversed using proline catalysts with bulky aprotic side chains. For example, aldehyde (**5.102**) is converted into the (S)-enantiomer using proline (**5.103**) and diethylazodicarboxylate (DEAD).[68] In this case, the side chain shields the *Re* face from attack, forcing amination to occur from the opposite face. Cyclisation occurs on subsequent reduction to give corresponding (S)-oxazolidinone (**5.104**).

5.3.3 α-Halogenation

The asymmetric fluorination of β-ketoesters has been achieved in 62–90% ee using F-TEDA (Selectfluor) as fluorine source in the presence of 0.5 mol% of the chiral nonracemic titanium-based Lewis acid (**5.108**).[69] A greater range of β-ketoesters are fluorinated with higher ee using catalytic quantities of the palladium-BINAP complex (**5.109**) and N-fluorobenzenesulfonamide (NFSI).[70] In both cases the reaction proceeds through the intermediacy of a chiral enolate.

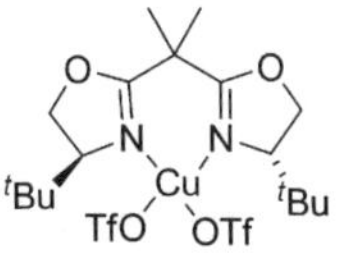

(5.94)

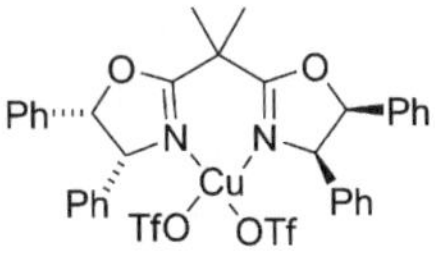

(5.97)

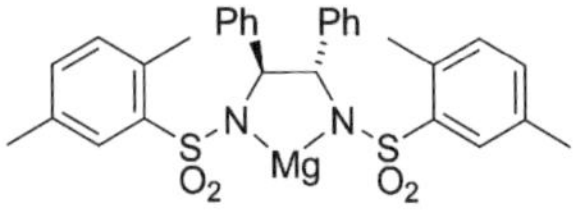

(5.100)

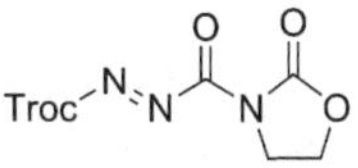

(5.93)
(Troc = CO₂CH₂CCl₃)

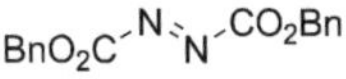

(5.96)

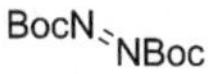

(5.101)

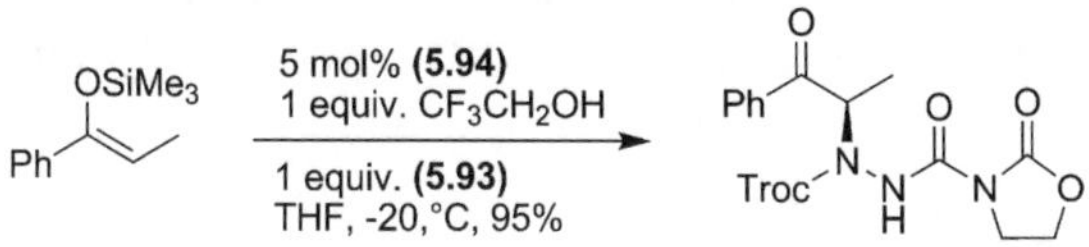

5 mol% **(5.94)**
1 equiv. CF₃CH₂OH
———————————→
1 equiv. **(5.93)**
THF, -20,°C, 95%

(5.92) → **(5.105)** 95% ee

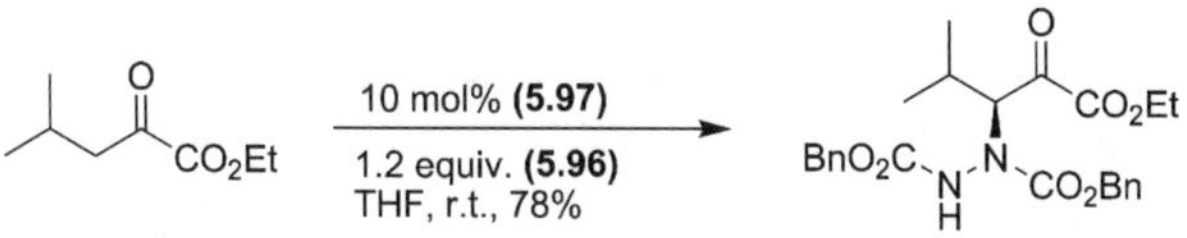

10 mol% **(5.97)**
———————————→
1.2 equiv. **(5.96)**
THF, r.t., 78%

(5.95) → **(5.98)** 95% ee

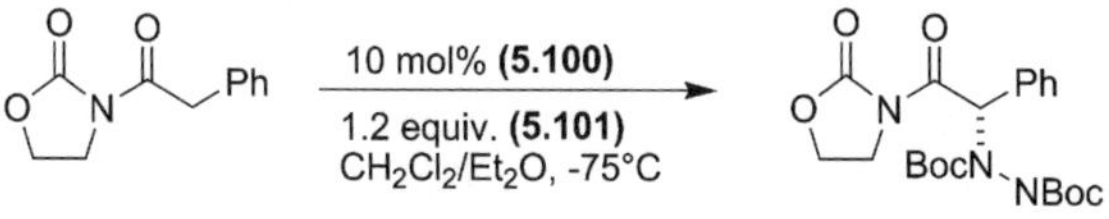

10 mol% **(5.100)**
———————————→
1.2 equiv. **(5.101)**
CH₂Cl₂/Et₂O, -75°C

(5.99) → **(5.106)** 86% ee

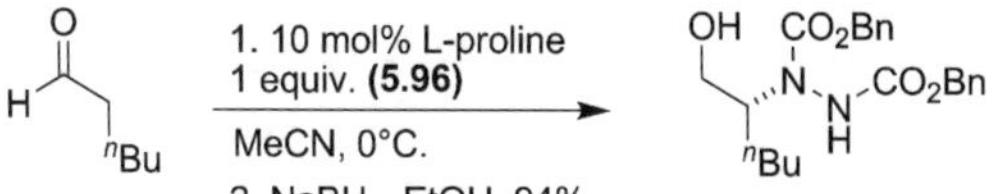

1. 10 mol% L-proline
1 equiv. **(5.96)**
———————————→
MeCN, 0°C.
2. NaBH₄, EtOH, 94%.

(5.102) → **(5.107)** 97% ee

1. 10 mol% **(5.103)**
1 equiv. DEAD
———————————→
CH₂Cl₂, r.t.
2. NaBH₄, EtOH, 83%.

(5.102) → **(5.104)** 97% ee

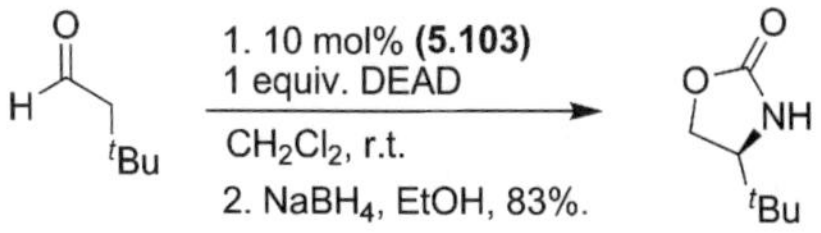

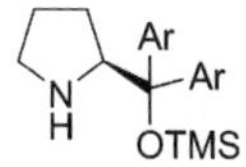

(5.103)
Ar = 3,5-(CF₃)₂-C₆H₃

The organocatalyst-based α-functionalisation strategy has been applied with much success to the asymmetric halogenation of aldehydes.[57c] The imidazolidinone salt (**5.110**) has been used by MacMillan and coworkers, in combination with NFSI, to effect enantioselective fluorination of a range of aldehydes, for example (**5.111**).[71]

The imidazolidinone salt (**5.110**) has also been used as a catalyst in the asymmetric chlorination of aldehydes. For example, octanal (**5.112**) is converted into the α-chloroaldehyde with high ee using the antipode of (**5.110**) and quinone (**5.113**) as the chlorine source.[72] Jørgensen and coworkers have achieved similar levels of enantioselectivity in this transformation using C_2-symmetric diphenylpyrrolidine (**5.114**) as the organocatalyst in combination with NCS.[73]

Organocatalysts also successfully mediate the enantioselective α-bromination of aldehydes.[68] For example, aldehyde (**5.102**) is brominated with high ee using catalyst (**5.103**) and bromoquinone (**5.115**).

5.4   Oxidation of C–H

The direct oxidation of unfunctionalised alkanes in an asymmetric fashion is a formidable challenge. However, oxidation of C–H bonds adjacent to suitable functional groups gives a handle on which to operate.[74] In particular, the allylic oxidation of cyclic alkenes utilising asymmetric variants of the Kharasch–Sosnovsky reaction has received considerable attention.[75] The reaction is catalysed by copper salts and requires a perester to give the allylic ester as product.

The highest selectivities to date have been achieved using copper-bisoxazolines such as (**5.119**)–(**5.121**) in combination with *tert*-butyl-*p*-nitrobenzoate (**5.122**). Cycloalkenes such as cyclopentene (**5.123**), cyclohexene (**5.124**) and cycloheptene (**5.125**) are particularly good substrates for this asymmetric transformation. Multiple equivalents of olefin are generally used and yields are based on the perester.

Propargylic oxidation of the acyclic alkyne (**5.126**) can be achieved with moderate enantioselectivity. It is noteworthy that in this case, the quoted yield is based on the starting material (rather than being based on the perester).[76a] One drawback with this procedure is the very slow reaction rate. It has been discovered that the rate of the oxidation can be enhanced using phenylhydrazine as an additive, to aid in the reduction of Cu(II) to Cu(I), and Cu-PYBOX ligand (**5.127**) as catalyst.[76b]

Catalytic asymmetric C–H oxidation has also been achieved using metal porphyrins and metal(salen) complexes. Benzylic oxidations using vaulted binaphthyl iron(III) porphyrins,[77] and manganese(III)(salen) complexes[78] have suffered from moderate yields and/or enantioselectivities. However, Miyafuji and Katsuki have obtained reasonable yield and good enantioselectivity for C–H

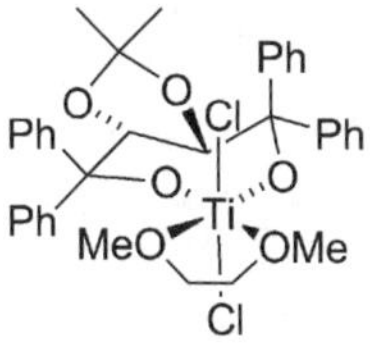
Ph
Ph
Ph
Ph
Cl
Ti
MeO
OMe
Cl

(5.108)

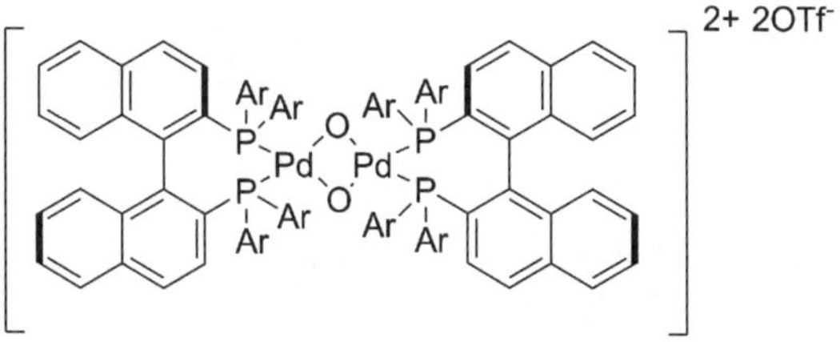
2+ 2OTf⁻
Ar
Pd
Pd
P

(5.109)

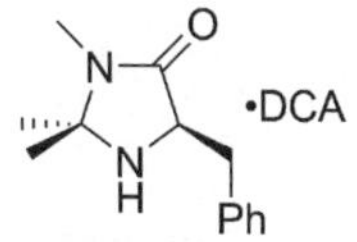
•DCA
Ph

(5.110)

DCA = dichloroacetic acid

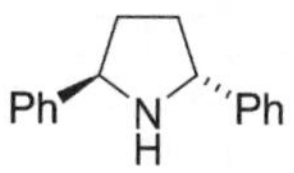
Ph
Ph

(5.114)

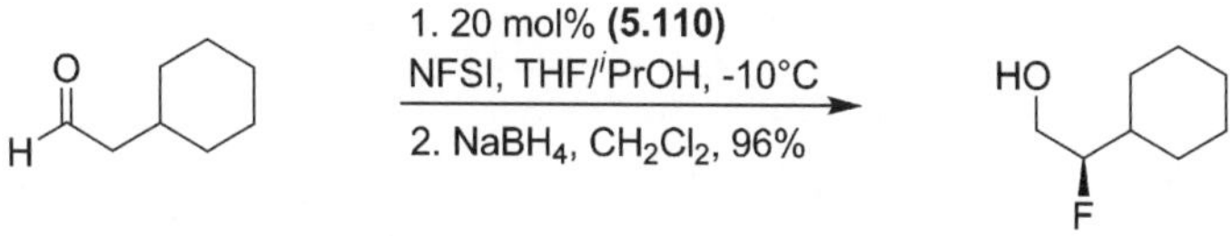
1. 20 mol% (5.110)
NFSI, THF/ⁱPrOH, -10°C
2. NaBH4, CH2Cl2, 96%
HO
F

(5.111)

(5.116) 99% ee

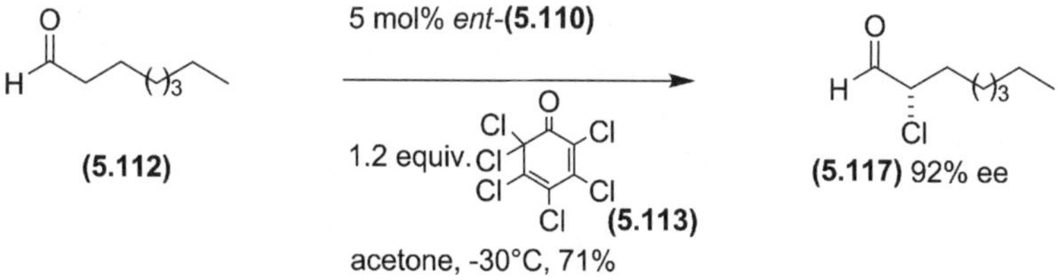
5 mol% ent-(5.110)
1.2 equiv.
(5.113)
acetone, -30°C, 71%
Cl

(5.112)

(5.117) 92% ee

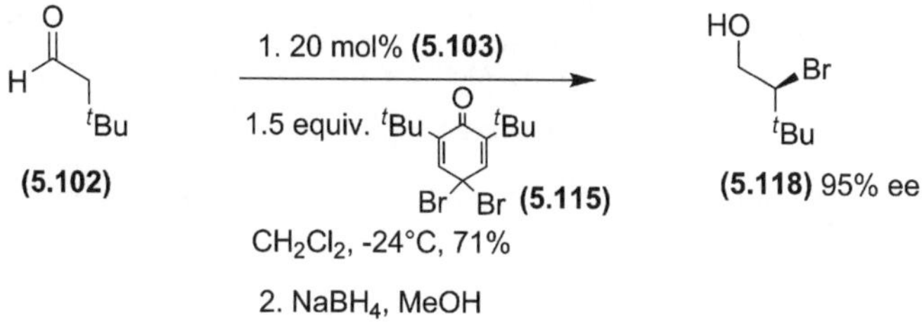
1. 20 mol% (5.103)
1.5 equiv.
(5.115)
CH2Cl2, -24°C, 71%
2. NaBH4, MeOH
HO
Br
ᵗBu

(5.102)

(5.118) 95% ee

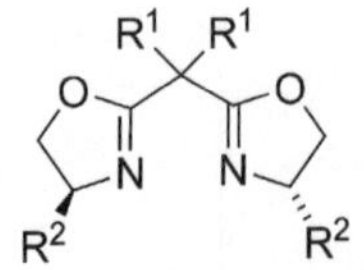
R1
R1
O
O
N
N
R2
R2

R^1 = Et, R^2 = Ph **(5.119)**
R^1 = Me, R^2 = Ph **(5.120)**
R^1 = Me, R^2 = ^{i}Pr **(5.121)**

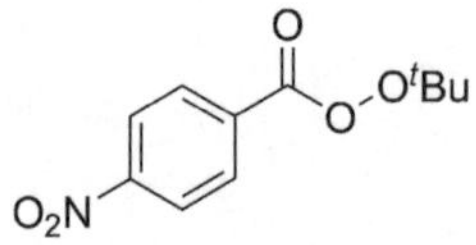
O
O
O^{t}Bu
O_2N

(5.123)

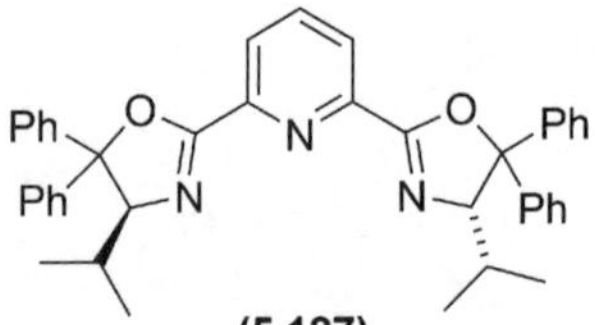
Ph
Ph
O
N
N
O
Ph
Ph
N

(5.127)

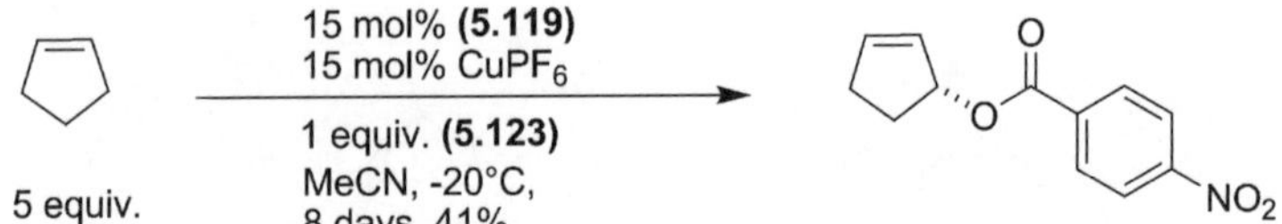
15 mol% (5.119)
15 mol% $CuPF_6$
1 equiv. (5.123)
MeCN, -20°C,
8 days, 41%
O
O
NO_2

5 equiv. **(5.123)**

(5.128) 99% ee

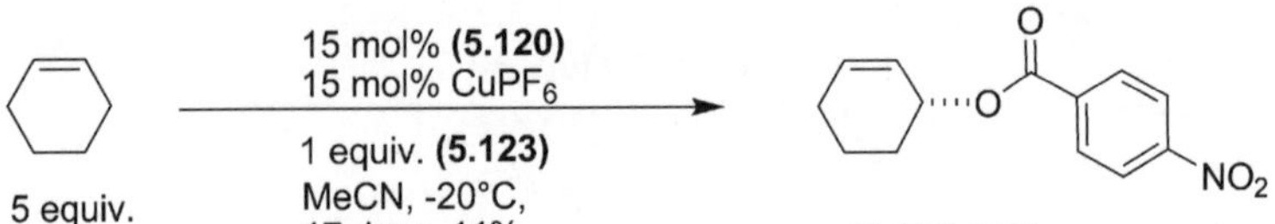
15 mol% (5.120)
15 mol% $CuPF_6$
1 equiv. (5.123)
MeCN, -20°C,
17 days, 44%
O
O
NO_2

5 equiv. **(5.124)**

(5.129) 96% ee

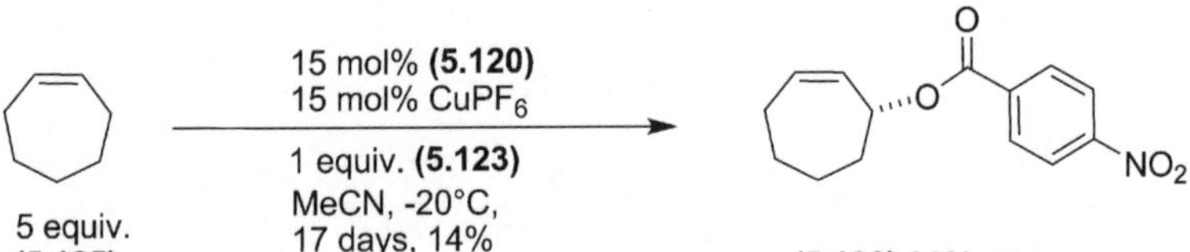
15 mol% (5.120)
15 mol% $CuPF_6$
1 equiv. (5.123)
MeCN, -20°C,
17 days, 14%
O
O
NO_2

5 equiv. **(5.125)**

(5.130) 99% ee

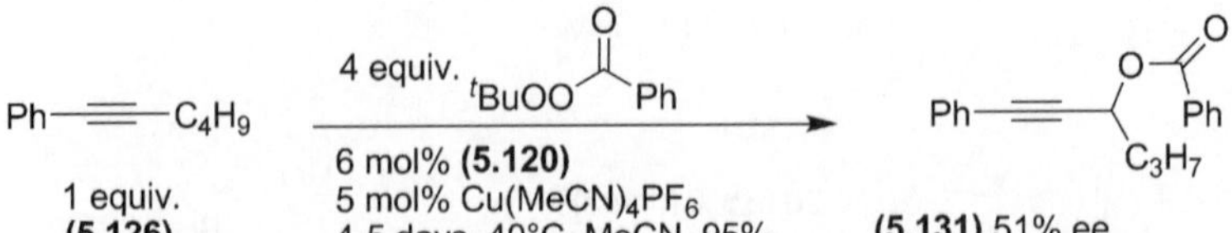
Ph
C_4H_9
4 equiv.
O
^{t}BuOO
Ph
6 mol% (5.120)
5 mol% $Cu(MeCN)_4PF_6$
4-5 days, 40°C, MeCN, 95%
Ph
O
O
Ph
C_3H_7

1 equiv. **(5.126)**

(5.131) 51% ee

oxidation adjacent to an ether in substrate (**5.132**), using manganese(salen) complex (**5.133**).[79]

5.5 Baeyer–Villiger Oxidation

The Baeyer–Villiger oxidation of ketones to esters can be achieved using transition metal catalysts, and an enantioselective variant using chiral nonracemic metal

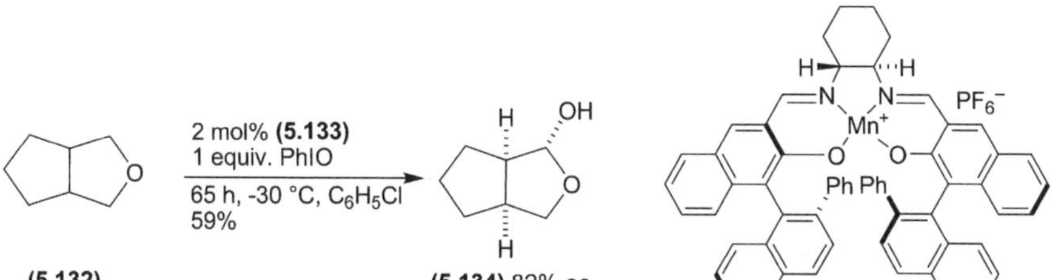

(5.132) → (5.134) 82% ee

2 mol% (5.133)
1 equiv. PhIO
65 h, -30 °C, C₆H₅Cl
59%

(5.133)

catalysts has been deployed in the desymmetrisation of prochiral ketones and the kinetic resolution of chiral, racemic ketones.[80] A number of catalyst systems have been investigated. Strukul and coworkers have reported the use of enantiomerically pure platinum complexes such as (5.135),[81] while Katsuki and coworkers have investigated the use of Zr(salen) complexes (5.136).[82] The group of Bolm has employed both copper complexes such as (5.137)[83] and aluminium-BINOL catalysts in this procedure.[84] 3-Phenylcyclobutanone (5.138) is a common test substrate for this reaction and is converted into the lactone (5.139) with up to 87% ee using Katsuki's Zr(salen complex) (5.136) and urea:hydrogen peroxide (UHP) as oxidant.[82]

Generally, the more highly substituted carbon centre preferentially migrates in the Baeyer–Villiger oxidation. Bond migration is under stereoelectronic control and only proceeds when the C–C bond is antiperiplanar to the peroxy moiety. Thus a chiral nonracemic catalyst that influences the conformation of the peroxy moiety can have a strong effect on the regioselectivity and even override the expected migratory aptitudes. Racemic fused bicyclic cyclobutanones are interesting substrates in this regard as each enantiomer has a different C–C bond in the antiperiplanar position in the substrate–ligand complex. As a result, oxidation of each enantiomer results in the formation of a different regioisomer. For example, bicyclooctanone (5.140) yields a 'normal' (5.141) and an 'abnormal' lactone (5.141) on oxidation using catalyst (5.137). This regiodivergence is also observed during enzyme-mediated oxidations of this type,[85] and biocatalysed Baeyer–Villiger oxidations are a particularly effective method for asymmetric induction.[80c, 86]

5.6 Oxidation of Sulfides

This asymmetric oxidation of sulfides has been achieved successfully using biotransformations.[87,88] However, a detailed discussion of these reactions is beyond the scope of the present book. A number of enatiomerically pure transition metal complexes in combination with terminal oxidants have been used to effect the asymmetric oxidations of sulfides to sulfoxides.[89]

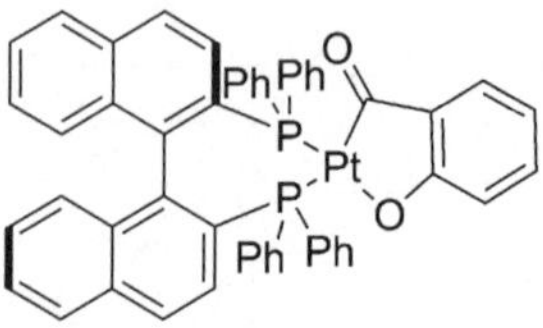

Ph Ph
P
Pt
P
O
O
Ph Ph

(5.135)

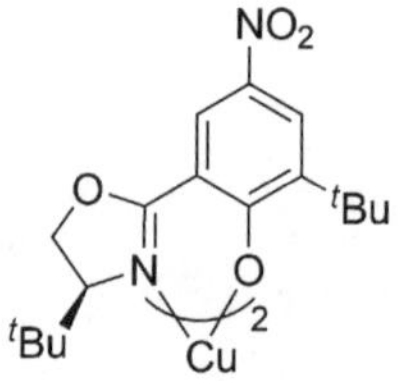

NO_2
^{t}Bu
O
N
O
^{t}Bu
Cu
2

(5.137)

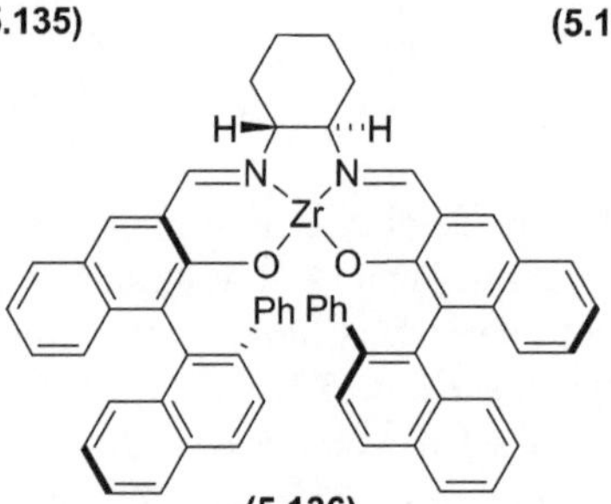

H
H
N
N
Zr
O
O
Ph Ph

(5.136)

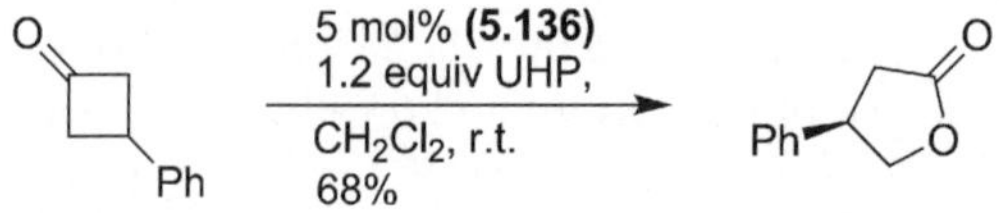

O
Ph
5 mol% (5.136)
1.2 equiv UHP,
CH_2Cl_2, r.t.
68%
Ph
O
O

(5.138) **(5.139)** 87% ee

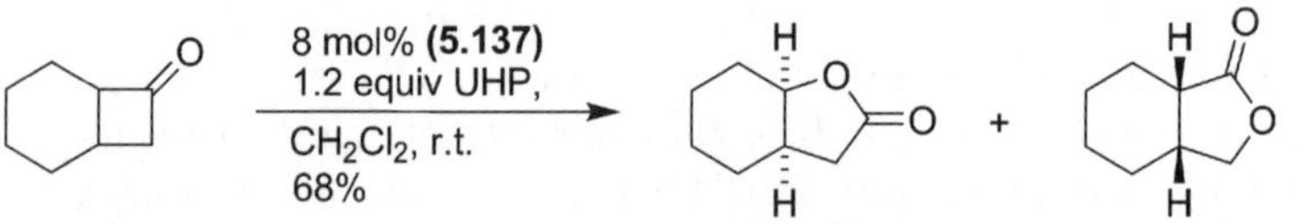

O
8 mol% (5.137)
1.2 equiv UHP,
CH_2Cl_2, r.t.
68%
H
O
O
H
+
H
O
O
H

(5.140) **(5.141)**, 54%, 82% ee **(5.142)**, 22% >99% ee

Kagan and Pitchen[90] and Modena and coworkers[91] independently reported the oxidation of sulfides to sulfoxides using modified Sharpless epoxidation catalyst (titanium/diethyl tartrate). By 1987,[92] Kagan had already reported a catalytic variation of the reaction and an improved catalytic system allows for the use of lower (10 mol%) loading of catalyst.[93] For example, sulfide (**5.143**) undergoes sulfoxidation with good enantioselectivity. An alternative catalyst based on $Ti(O^iPr)_4$ and BINOL is also effective for sulfoxidation, providing up to 96% ee.[94,95]

Jacobsen[96] and Katsuki[97,98] have both reported the use of manganese(III) (salen) catalysts for sulfide oxidation. These catalysts can be effective for the enantioselective oxidation of several arylmethylsulfides using iodosylbenzene as the stoichiometric oxidant. Additionally, titanium(salen) complexes function as efficient catalysts in this procedure, providing up to 94% ee in the oxidation of methyl phenyl sulfide using the more economical urea:hydrogen peroxide as oxidant.[99]

The vanadium complex formed *in situ* from VO(acac)$_2$ and salicylaldimine (**5.144**) is an effective sulfide oxidation catalyst in combination with H$_2$O$_2$ yielding a range of sulfoxides with ees between 53 and 70%, and up to 95% in the case of dithioacetal (**5.145**). Ellman and coworkers have adapted this procedure to the enantioselective oxidation of disulfides.[100] Salicylaldehyde-derived Schiff bases also function as effective ligands in the iron-catalysed asymmetric sulfoxidation of arylmethylsulfides with the highest ees obtained using ligand (**5.146**) and catalytic quantities of carboxylic acids such as (**5.147**).[101] For example, sulfide (**5.148**) is oxidised with high enantioselectivity using imine (**5.146**) in combination with Fe(acac)$_3$.

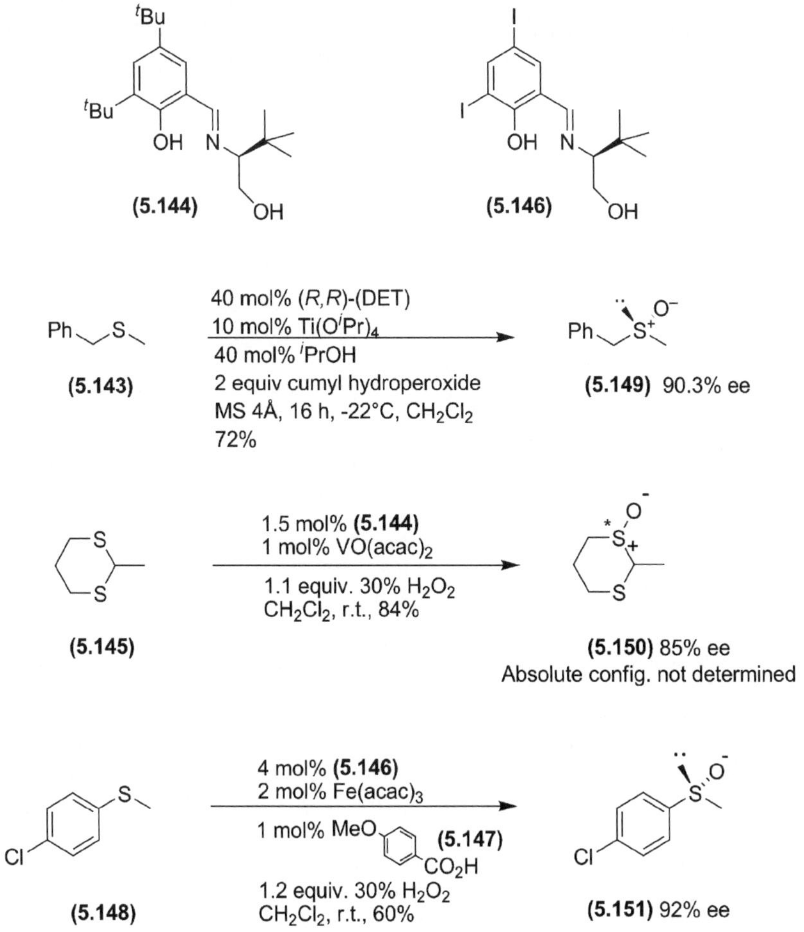

(**5.144**)   (**5.146**)

Ph⌒S⌒ (**5.143**)

40 mol% (*R*,*R*)-(DET)
10 mol% Ti(O*i*Pr)$_4$
40 mol% *i*PrOH
2 equiv cumyl hydroperoxide
MS 4Å, 16 h, -22°C, CH$_2$Cl$_2$
72%

Ph⌒S$^+$ (**5.149**) 90.3% ee

(**5.145**)

1.5 mol% (**5.144**)
1 mol% VO(acac)$_2$

1.1 equiv. 30% H$_2$O$_2$
CH$_2$Cl$_2$, r.t., 84%

(**5.150**) 85% ee
Absolute config. not determined

(**5.148**)

4 mol% (**5.146**)
2 mol% Fe(acac)$_3$

1 mol% MeO⌬CO$_2$H (**5.147**)

1.2 equiv. 30% H$_2$O$_2$
CH$_2$Cl$_2$, r.t., 60%

(**5.151**) 92% ee

Sulfides can usually be selectively oxidised to sulfoxides without (too much) over-oxidation to the corresponding sulfones. However, the conversion of sulfoxides into sulfones can be achieved under relatively mild catalytic conditions, if required. Uemura has demonstrated a kinetic resolution of sulfoxides by catalysed oxidation.[102] For example, the oxidation of racemic phenylmethylsulfoxide (**5.152**) affords the sulfone (**5.153**), but the unreacted starting material is recovered with very high enantiomeric excess. This kinetic resolution can be used to improve the enantioselectivity of the sulfoxidation process by further oxidation of the unwanted sulfoxide enantiomer. For example, Jackson and coworkers have discovered that an efficient kinetic resolution of the product sulfoxide occurs during sulfoxidation in chloroform at 0°C using the vanadium catalyst prepared from the antipode of (**5.146**).[103] This sulfide oxidation/sulfoxide kinetic resolution results in some of the highest ees to date in the sulfoxidation process. For example, 2-naphthyl methyl sulfide (**5.154**) is oxidised to the sulfoxide (**5.155**) with >99.5% ee and small amounts of the sulfone (**5.156**) by prolonged stirring with metal–ligand complex and oxidant at 0°C.

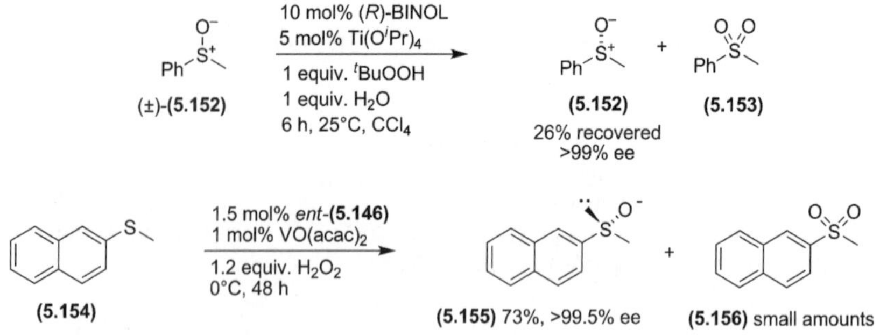

References

1. E. N. Jacobsen, I. Markó, W. S. Mungall, G. Schröder and K. B. Sharpless, *J. Am. Chem. Soc.*, **1988**, *110*, 1968.
2. For reviews on the AD reaction see: (a) B.B. Lohray, *Tetrahedron: Asymmetry*, **1992**, *3*, 1317. (b) R. A. Johnson and K. B. Sharpless, in *Catalytic Asymmetric Synthesis*, ed. I. Ojima, VCH, New York, **1993**, 227. (c) H. C. Kolb, M. S. VanNieuwenhze and K. B. Sharpless, *Chem. Rev.*, **1994**, *94*, 2483. (d) D. J. Berrisford, C. Bolm and K. B. Sharpless, *Angew. Chem., Int. Ed. Engl.*, **1995**, *34*, 1059. (e) I. E. Markó and J. S. Svendson, in *Comprehensive Asymmetric Catalysis*, Vol.2, ed. E. N. Jacobsen, A. Pfaltz and H. Yamamoto, Springer-Verlag, Berlin, **1999**, 713. (f) K. B. Sharpless, *Angew. Chem., Int. Ed.*, **2002**, *41*, 2024. (g) A. B. Zaitsev and H. Adolfsson, *Synthesis*, **2006**, 1725.

3. K. B. Sharpless, W. Amberg, M. Beller, H. Chen, J. Hartung, Y. Kawanami, D. Lübben, E. Manoury, Y. Ogino, T. Shibata and T. Ukita, *J. Org. Chem.*, **1991**, *56*, 4585.

4. K. B. Sharpless, W. Amberg, Y. L. Bennani, G. A. Crispino, J. Hartung, K.-S. Jeong, H.-L. Kwong, K. Morikawa, Z.-M. Wang, D. Xu and X.-L. Zhang, *J. Org. Chem.*, **1992**, *57*, 2768.

5. (a) B. B. Lohray, E. Nandanan and V. Bhushan, *Tetrahedron: Asymmetry*, **1996**, *7*, 2805. (b) I. Motorina and C. C. Crudden, *Org. Lett.*, **2001**, *3*, 2325. (c) B. M. Choudary, N. S. Chowdari, K. Jyothi and M. L. Kantam, *J. Am. Chem. Soc.*, **2002**, *124*, 5341.

6. (a) C. Bolm and A. Maischak, *Synlett*, **2001**, 93. (b) Y.-Q. Kuang, S.-Y. Zhang and L. L. Wei, *Tetrahedron Lett.*, **2001**, *42*, 5925. (c) A. Mandoli, D. Pini, M. Fiori and P. Salvadori, *Eur. J. Org. Chem.*, **2005**, 1271. (d) B. S. Lee, S. Mahajan and K. D. Janda, *Tetrahedron Lett.*, **2005**, *46*, 4491. (e) K. J. Kim, H. Y. Choi, S. H. Hwang, H. Soon, Y. S. Park, E. K. Kwueon, D. S. Choi, C. E. Song, *J. Chem. Soc., Chem Commun.*, **2005**, 3337.

7. (a) L. C. Branco and C. A. M. Afonso, *J. Chem. Soc., Chem Commun.*, **2002**, 3036. (b) A. Closson, M. Johansson and J. E. Bäckvall, *J. Chem. Soc., Chem Commun.*, **2004**, 1494. (c) Q. Yao, *Org. Lett.*, **2002**, *4*, 2197.

8. For reviews on the immobilised asymmetric dihydroxylation see: (a) C. Bolm and A. Gerlach, *Eur. J. Org. Chem.*, **1998**, 21. (b) P. Salvadori, D. Pini and A. Petri, *Synlett*, **1999**, 1181. (c) T. J. Dickerson, N. N. Read and K. D. Janda, *Chem. Rev.*, **2002**, *102*, 3325. (d) C. E. Song and S. Lee, *Chem. Rev.*, **2002**, *102*, 3495.

9. B. B. Lohray, V. Bhushan and E. Nandanan, *Tetrahedron Lett.*, **1994**, *35*, 4209.

10. T. Gobel and K. B. Sharpless, *Angew. Chem., Int. Ed. Engl.*, **1993**, *32*, 1329.

11. (a) U. Pidun, C. Boehme and G. Frenking, *Angew. Chem. Int. Ed.*, **1996**, 2817. (b). S. Dapprich, G. Ujaque, F. Maseras, A. Lledos, D. G. Musaev and K. Morokuma, *J. Am. Chem. Soc.*, **1996**, *118*, 11660. (c) M. Torrent, L. Deng, M. Duran, M. Sola and T. Ziegler, *Organometallics*, **1997**, *16*, 13.

12. A. J. DelMonte, J. Haller, K. N. Houk, K. B. Sharpless, D. A. Singleton, T. Strassner and A. A. Thomas, *J. Am. Chem. Soc.*, **1997**, *119*, 9907.

13. K. B. Sharpless, W. Amberg, Y. L. Bennani, G. A. Crispino, J. Hartung, K.-S. Jeong, H.-L. Kwong, K. Morikawa, Z.-M. Wang, D. Xu and X.-L. Zhang, *J. Org. Chem.*, **1992**, *57*, 2768.

14. G. M. Mehltretter, C. Dobler, U. Sundermeier and M. Beller, *Tetrahedron Lett.*, **2000**, *41*, 8083

15. E. J. Corey and M. C. Noe, *J. Am. Chem. Soc.*, **1996**, *118*, 11038.

16. (a) H. C. Kolb, P. G. Andersson and K. B. Sharpless, *J. Am. Chem. Soc.*, **1994**, *116*, 1278. (b) P.-O. Norrby, H. C. Kolb and J. B. Sharpless, *J. Am. Chem. Soc.*, **1994**, *116*, 8470.

17. H. C. Kolb, P. G. Andersson and K.B. Sharpless, *J. Am. Chem. Soc.*, **1994**, *116*, 1278.

18. K. J. Hale, S. Manaviazar and S. A. Peak, *Tetrahedron Lett.*, **1994**, *35*, 425.

19. Z.-M. Wang and K. B. Sharpless, *J. Org. Chem.*, **1994**, *59*, 8302.

20. K. P. M. Vanhessche, Z.-M. Wang and K. B. Sharpless, *Tetrahedron Lett.*, **1994**, *35*, 3469.

21. P. J. Walsh, P. T. Ho, S. B. King and K. B. Sharpless, *Tetrahedron Lett.*, **1994**, *35*, 5129.

22. (a) Z.-X. Feng and W.-S. Zhou, *Tetrahedron Lett.*, **2003**, *44*, 493. (b) C. Bonini, L. Chiummiento, M. De Bonis, M. Funicello, P. Lupattelli and R. Pandolfo, *Tetrahedron: Asymmetry*, **2006**, *17*, 2919.

23. D. Xu, C. Y. Park and K. B. Sharpless, *Tetrahedron Lett.*, **1994**, *35*, 2495.
24. Z.-M. Wang and K. B. Sharpless, *Tetrahedron Lett.*, **1993**, *34*, 8225.
25. E. J. Corey, A. Guzman-Perez and M. C. Noe, *Tetrahedron Lett.*, **1995**, *36*, 3481.
26. D. Xu, G. A. Crispino and K. B. Sharpless, *J. Am. Chem. Soc.*, **1992**, *114*, 7570.
27. K. Tani, Y. Sato, S. Okamoto and F. Sato, *Tetrahedron Lett.*, **1993**, *34*, 4975.
28. H. Becker, M. A. Soler and K. B. Sharpless, *Tetrahedron*, **1995**, *51*, 1345.
29. Z.-M. Wang, X.-L. Zhang, K. B. Sharpless, S. C. Sinha, A. Sinha-Bagchi and E. Keinan, *Tetrahedron Lett.*, **1992**, *33*, 6407.
30. Y. Miyazaki, H. Hotta and F. Sato, *Tetrahedron Lett.*, **1994**, *35*, 4389.
31. H. Becker, P. T. Ho, H. C. Kolb, S. Loren, P.-O. Norrby and K. B. Sharpless, *Tetrahedron Lett.*, **1994**, *35*, 7315.
32. T. Hashiyama, K. Morikawa and K. B. Sharpless, *J. Org. Chem.*, **1992**, *57*, 5067.
33. E. F. Marcune, S. Karady, P. J. Reider, R. A. Miller, M. Biba, L. DiMichele and R. A. Reamer, *J. Org. Chem.*, **2003**, *68*, 8088.
34. G. A. Crispino, K.-S. Jeong, H. C. Kolb, Z.-M. Wang, D. Xu and K. B. Sharpless, *J. Org. Chem.*, **1993**, *58*, 3785.
35. H. Becker and K. B. Sharpless, *Angew. Chem. Int. Ed,. Engl.* **1996**, *35*, 448.
36. X. Wang, M. Zak, M. Maddess, P. O'Sea, R. Tillyer, E. J. J. Grabowski and P. J. Reider, *Tetrahedron Lett.*, **2000**, *41*, 4865.
37. K. Morikawa, J. Park, P. G. Andersson, T. Hashiyama and K. B. Sharpless, *J. Am. Chem. Soc.*, **1993**, *115*, 8463.
38. L. Wang and K. B. Sharpless, *J. Am. Chem. Soc.*, **1992**, *114*, 7568.
39. L. Wang, K. Kakiuchi and K. B. Sharpless, *J. Org. Chem.*, **1994**, *59*, 6895.
40. N.-S. Kim, J.-R. Choi and J. K. Cha, *J. Org. Chem.*, **1993**, *58*, 7096.
41. K. B. Lindsay and S. G. Pyne, *J. Org. Chem.*, **2002**, *67*, 7774.
42. M. S. VanNieuwenhze and K. B. Sharpless, *J. Am. Chem. Soc.*, **1993**, *115*, 7864.
43. D. P. G. Hamon, K. L. Tuck and H. S. Christie, *Tetrahedron*, **2001**, *57*, 9499.
44. G. Li, H.-T. Chang and K. B. Sharpless, *Angew. Chem., Int. Ed. Engl.*, **1996**, *35*, 451.
45. For reviews on the AA reaction see: (a) O. Reiser, *Angew. Chem., Int. Ed. Engl.*, **1996**, *35*, 1308. (b) P. O'Brien, *Angew. Chem. Int. Ed.*, **1999**, *38*, 326. (c) J. A. Bodkin and M. D. McLeod, *J. Chem. Soc., Perkin Trans. 1*, **2002**, 2733. (d) K. B. Sharpless, *Angew. Chem. Int. Ed.*, **2002**, *41*, 2024. (e) K. Muñiz, *Chem. Soc. Rev.*, **2004**, *33*, 166.
46. (a) G. Li, H. H. Angert and K. B. Sharpless, *Angew. Chem., Int. Ed. Engl.*, **1996**, *35*, 2813. (b) K. L. Reddy, K. R. Dress and K. B. Sharpless, *Tetrahedron Lett.*, **1998**, *39*, 3667. (c) K. L. Reddy and K. B. Sharpless, *J. Am. Chem. Soc.*, **1998**, *120*, 1207. (d) P. O'Brien, S. A. Osbourne and D. D. Parker, *J. Chem. Soc., Perkin Trans 1*, **1998**, 2519.
47. J. Rudolph, P. C. Sennhenn, C. P. Vlaar and K. B. Sharpless, *Angew. Chem., Int. Ed. Engl.*, **1996**, *35*, 2810.
48. A. V. Gontcharov, H. Liu and K. B. Sharpless, *Org. Lett.*, **1999**, *1*, 783.
49. (a) M. Bruncko, G. Schlingloff and K. B. Sharpless, *Angew. Chem., Int. Ed. Engl.*, **1997**, *36*, 1483. (b) Z. P. Demko, M. Bartsch and K. B. Sharpless, *Org. Lett.*, **2000**, *2*, 2221.
50. (a) D. Raatz, C. Innertsberger and O. Reiser, *Synlett*, **1999**, 1907. (b) M. H. Haukaas and G. O'Doherty, *Org. Lett.*, **2001**, *3*, 401.
51. R. Angelaud, Y. Landais and K. Schenk *Tetrahedron Lett.*, **1997**, *38*, 1407.
52. P. O'Brien, S. A. Osborne and D. D. Parker, *Tetrahedron Lett.*, **1998**, *39*, 4099.

53. K. L. Reddy and K. B. Sharpless, *J. Am. Chem. Soc.*, **1998**, *120*, 1207.

54. V. Nesterenko, J. T. Byers, and P. J. Hergenrother, *Org. Lett.*, **2003**, *5*, 281.

55. (a) B. Tao, G. Schlingloff and K.B. Sharpless, *Tetrahedron Lett.*, **1998**, *39*, 2507. (b) A. J. Morgan, C. E. Masse and J. S. Panek, *Org. Lett.*, **1999**, *1*, 1949.

56. H. Han, C.-W. Cho and K. D. Janda, *Chem. Eur. J.*, **1999**, *5*, 1565.

57. (a) B. Pleitker, *Tetrahdron: Asymmetry*, **2005**, *16*, 3453. (b) J. M. Janey, *Angew. Chem. Int. Ed.*, **2005**, *44*, 4292. (c) M. Marigo and K. A. Jørgensen, *J. Chem. Soc., Chem Commun.*, **2006**, 2001.

58. N. Momiyama, H. Yamamoto, *J. Am. Chem. Soc*, **2003**, *125*, 6038.

59. (a) Y. Hayashi, J. Yamaguchi, T. Sumiya and M. Shoji, *Angew. Chem. Int. Ed.*, **2004**, *43*, 1112. (b) Y. Hayashi, J. Yamaguchi, T. Sumiya, K. Hibino and M. Shoji, *J. Org. Chem.*, **2004**, *69*, 5966.

60. N. Momiyama and H. Yamamoto, *J. Am. Chem. Soc.*, **2004**, *126*, 5360.

61. D. A. Evans and D. S. Johnson, *Org. Lett.*, **1999**, *1*, 595.

62. K. Juhl and K. A. Jørgensen, *J. Am. Chem. Soc.*, **2002**, *124*, 2420.

63. D. A. Evans and S. G. Nelson, *J. Am. Chem. Soc.*, **1997**, *119*, 6452.

64. R. O. Duthaler, *Angew. Chem. Int. Ed.*, **2003**, *42*, 975.

65. B. List, *J. Am. Chem. Soc.*, **2002**, *124*, 5656.

66. A. Bøgvig, K. Juhl, N. Kumaragurubaran, W. Zhuang and K. A. Jørgensen, *Angew. Chem. Int. Ed.*, **2002**, *41*, 1790.

67. N. Kumaragurubaran, K. Juhl, W. Zhuang, A. Bøgvig and K. A. Jørgensen, *J. Am. Chem. Soc.*, **2002**, *124*, 6254.

68. J. Franzen, M. Marigo, D. Fielenbach, T. C. Wabnitz, A. Kjaersgaard and K. A. Jørgensen, *J. Am. Chem. Soc.*, **2005**, *127*, 18296.

69. L. Hintermann and A. Togni, *Angew. Chem. Int. Ed.*, **2000**, *39*, 4359.

70. Y. Hamashima, K. Yagi, A. Takano, L. Tamás and M. Sodeoka, *J. Am. Chem. Soc.*, **2002**, *124*, 14530.

71. T. D. Beeson and D. C. MacMillan, *J. Am. Chem. Soc.*, **2005**, *127*, 8826.

72. M. P. Brochu, S. P. Brown and D. C. MacMillan, *J. Am. Chem. Soc.*, **2004**, *126*, 4108.

73. N. Halland, A. Braunton, S. Bachmann, M. Marigo and K. A. Jørgensen, *J. Chem. Soc.*, **2004**, *126*, 4790.

74. T. Katsuki, in *Comprehensive Asymmetric Catalysis*, Vol.2, ed. E. N. Jacobsen, A. Pfaltz and H. Yamamoto, Springer-Verlag, Berlin, **1999**, 791.

75. (a) A. Levina and J. Muzart, *Tetrahedron:Asymmetry*, **1995**, *6*, 147. (b) A. Levina and J. Muzart, *Synth. Commun.*, **1995**, *25*, 1789. (c) M. T. Rispens, C. Zondervan and B. L. Feringa, *Tetrahedron: Asymmetry*, **1995**, *6*, 661. (d) A. S. Gokhale, A .B. E. Minidis and A. Pfaltz, *Tetrahedron Lett.*, **1995**, *36*, 1831.(e) M. B. Andrus, A. B. Argade, X. Chen and M. G. Pamment, *Tetrahedron Lett.*, **1995**, *36*, 2945. (f) K. Kawasaki, S. Tsumura and T. Katsuki, *Synlett*, **1995**, 1245. (g) M. K. Södergren and P. G. Andersson, *Tetrahedron Lett.*, **1996**, *37*, 7577. (h) G. Sekar, A. DattaGupta and V. K. Singh, *J. Org. Chem.*, **1998**, *63*, 2961. (i) J. S. Clark, K. F. Tolhurst, M. Taylor and S. Swallow, *J. Chem. Soc., Perkin Trans 1.*, **1998**, 1167. (j) J. Eames and M. Watkinson, *Angew. Chem. Int. Ed.*, **2001**, *40*, 3567. (k) M. P. A. Lyle and P. D. Wilson, *Org. Biomol. Chem.*, **2006**, *4*, 41.

76a. J. S. Clark, K. F. Tolhurst, M. Taylor and S. Swallow, *Tetrahedron Lett.*, **1998**, *39*, 4913.

76b. S. K. Ginotra and V. K. Singh, *Tetrahedron*, **2006**, *62*, 3573.

77. J. T. Groves and P. Viski, *J. Org. Chem.*, **1990**, *55*, 3628.

78. K. Hamachi, R. Irie and T. Katsuki, *Tetrahedron Lett.*, **1996**, *37*, 4979.

79. A. Miyafuji and T. Katsuki, *Synlett*, **1997**, 836.

80. For reviews, see; (a) G. Strukul, *Angew. Chem., Int. Ed. Engl.*, **1998**, *37*, 1198. (b) C. Bolm and O. Beckmann, in *Comprehensive Asymmetric Catalysis*, Vol.2, ed. E. N. Jacobsen, A. Pfaltz and H. Yamamoto, Springer-Verlag, Berlin, **1999**, 803. (c) M. D. Mihovvilovic, F. Rudroff and B. Grötzl, *Curr. Org. Chem.*, **2004**, *8*, 1057.

81. (a) A. Gusso, C. Baccin, F. Pinna and G. Strukul, *Organometallics*, **1994**, *13*, 3442. (b) M. Colladon, A. Scarso and G. Strukul, *Synlett*, **2006**, 3515.

82. (a) A. Watanabe, T. Uchida, K. Ito and T. Katsuki, *Tetrahedron Lett.*, **2002**, *43*, 4481. (b) A. Watanabe, T. Uchida and T. Katsuki, *Proc. Natl. Acad. Sci. USA*, **2004**, *101*, 5737.

83. (a) C. Bolm, G. Schlingloff and K. Weickhardt, *Angew. Chem. Int. Ed. Engl.*, **1994**, *33*, 1848. (b) Bolm and G. Schlingloff, *J. Chem. Soc., Chem. Commun.*, **1995**, 1247.

84. J.-C. Frison, C. Palazzi and C. Bolm, *Tetrahedron*, **2006**, *62*, 6700.

85. D. R. Kelly, C. J. Knowles, J. G. Mahdi, I. N. Taylor, M. A. Wright, D. E. Hibbs, M. B. Hurstinghouse, A. K. Mish'al, S. M. Roberts, P. W. H. Wan, G. Grogan and A. J. Willets, *J. Chem. Soc., Perkin Trans.1*, **1995**, 2057.

86. C.-H. Wong and G. M. Whitesides, *Enzymes in Synthetic Organic Chemistry*, Tetrahedron Organic Chemistry Series, Vol. 12, Pergamon, Oxford, **1994**, 169.

87. H. L. Holland, *Chem. Rev.*, **1988**, *88*, 473.

88. G. Ottolina, P. Pasta, G. Carrea, S. Colonna, S. Dallavalle and H. L. Holland, *Tetrahedron: Asymmetry*, **1995**, *6*, 1375.

89. C. Bolm, K. Muñiz and J. P. Hildebrand, in *Comprehensive Asymmetric Catalysis*, Vol. 2, ed. E. N. Jacobsen, A. Pfaltz and H. Yamamoto, Springer-Verlag, Berlin, **1999**, 697.

90. (a) P. Pitchen and H. B. Kagan, *Tetrahedron Lett.*, **1984**, *25*, 1049. (b) J. M. Brunel, P. Diter, M. Duetsch and H. B. Kagan, *J. Org. Chem.*, **1995**, *60*, 8086.

91. S. H. Di Furia, G. Modena and G. Seraglia, *Synthesis*, **1984**, 325.

92. S. Zhao, O. Samuel and H. B. Kagan, *Tetrahedron*, **1987**, *43*, 5135.

93. J. M. Brunel and H. B. Kagan, *Synlett*, **1995**, 404.

94. N. Komatsu, Y. Nishibayashi, T. Sugita and S. Uemura, *Tetrahedron Lett.*, **1992**, *33*, 5391.

95. N. Komatsu, M. Hashizume, T. Sugita and S. Uemura, *J. Org. Chem.*, **1993**, *58*, 4529.

96. M. Palucki, P. Hanson and E. N. Jacobsen, *Tetrahedron Lett.*, **1992**, *33*, 7111.

97. K. Noda, N. Hosoya, K. Yanai, R. Irie and T. Katsuki, *Tetrahedron Lett.*, **1994**, *35*, 1887.

98. K. Noda, N. Hosoya, R. Irie, Y. Yamashita and T. Katsuki, *Tetrahedron*, **1994**, *50*, 9609.

99. B. Saito and T. Katsuki, *Tetrahedron Lett.*, **2001**, *42*, 3873.

100. (a) D.A. Cogan, G. Liu, K. Kim, B. J. Backes and J. A. Ellman, *J. Am. Chem. Soc.*, **1998**, *120*, 8011. (b) S. Blum, R. G. Bergman and J. A. Ellman, *J. Org. Chem.*, **2003**, *68*, 150.

101. (a) J. Legros and C. Bolm, *Angew. Chem. Int. Ed.*, **2003**, *42*, 5487. (b) J. Legros and C. Bolm, *Angew. Chem. Int.*, **2004**, *43*, 4225. (c) J. Legros and C. Bolm, *Chem. Eur. J.*, **2005**, *11*, 1086.

102. N. Komatsu, M. Hashizume, T. Sugita and S. Uemura, *J. Org. Chem.*, **1993**, *58*, 7624.

103. (a) C. Drago, L. Caggiano and R. F. W. Jackson, *Angew. Chem. Int. Ed.*, **2005**, *44*, 7221. (b) I. Mohammadpoor-Baltork, M. Hill, L. Caggiano and R. F. W. Jackson, *Synlett*, **2006**, 3540.

Chapter 6
Nucleophilic Addition to Carbonyl Compounds

The addition of nucleophiles to carbonyl groups is a fundamental process in organic synthesis. The addition of diethylzinc to aldehydes occurs with high ee in the presence of a wide range of aminoalcohol ligands and also titanium-based Lewis acids. This methodology has recently been extended to the enantioselective addition of alkenyl, alkynyl and arylzincs and also to the more challenging addition to ketones.

The addition of cyanide to aldehydes and some ketones has also been achieved in a highly enantioselective manner using a wide variety of catalyst systems including chiral nonracemic Lewis acids, Lewis bases, bifunctional catalysts and even small peptides.

The catalytic asymmetric allylation of aldehydes is another reaction that has received a great deal of attention. Both allylstannes and the less reactive allylsilanes undergo addition to aldehydes with high ee in the presence of enantiomerically pure Lewis acids and Lewis bases and asymmetric versions of the chromium-catalysed Kishi–Nozaki–Hiyama reaction utilising allyl halides have recently been developed.

The hydrophosphonylation of aldehydes can also proceed enantioselectively in the presence of chiral nonracemic Lewis acids and some bifunctional catalysts and a small subchapter discusses recent advances in this area.

The nucleophiles discussed above can also react with the C=N group and the final section of this chapter describes the many recent advances in the catalytic asymmetric additions to imines using both metal complexes and also organocatalysts.

6.1   Addition of Organozincs to Carbonyl Compounds

In general, rather forcing conditions are required for the alkylation of aldehydes by dialkylzinc reagents. However, the addition of small amounts of aminoalcohol

Catalysis in Asymmetric Synthesis 2e   © 2009 Vittorio Caprio and Jonathan M.J. Williams

(or other additives) catalyses the reaction. As long ago as 1984, the use of catalytic enantiomerically pure aminoalcohols as additives was shown to give an asymmetric reaction.[1] Since then, several hundred ligands, mostly aminoalcohols, have been examined for their efficiency in catalysing the enantioselective addition of dialkylzinc reagents to aldehydes, and much of this work has been reviewed.[2]

A few of these ligands are illustrated, including the camphor-derived ligand (6.03) (which was the first one to give high enantioselectivies),[3] the norephedrine-derived aminoalcohol (6.04),[4] the praline-derived ligand (6.05),[5] aminoalcohols with axial chirality such as binaphthyl (6.06)[6] and ferrocene-based aminoalcohols such as (6.07).[7]

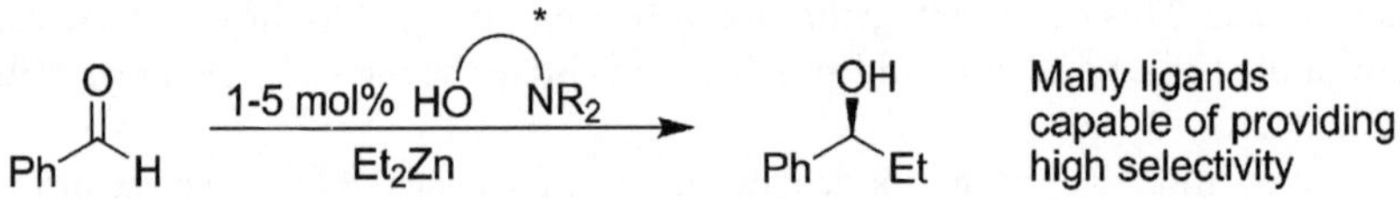

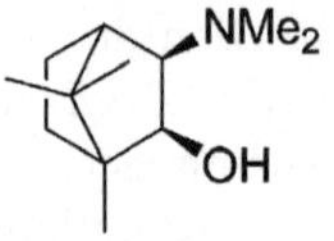

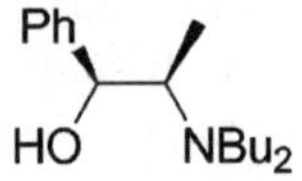

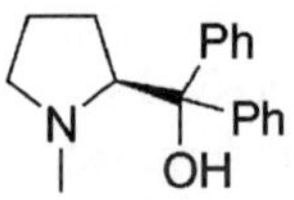

Whilst much of the published work in this field focusses on the addition of diethylzinc to simple arylaldehydes, many other possibilities have been reported, as demonstrated by the examples given in the schemes. Aldehydes (6.08–6.11) are all converted in good yields into the enantiomerically enriched alcohols (6.12–6.15) by catalysed addition of simple dialkylzinc reagents.

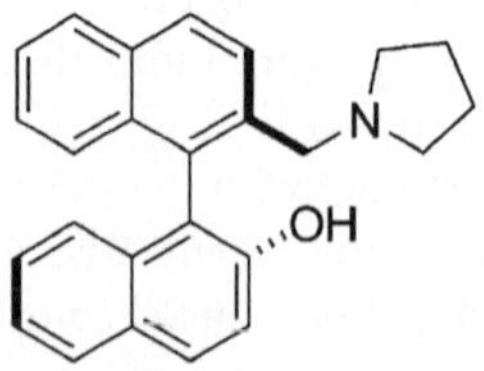

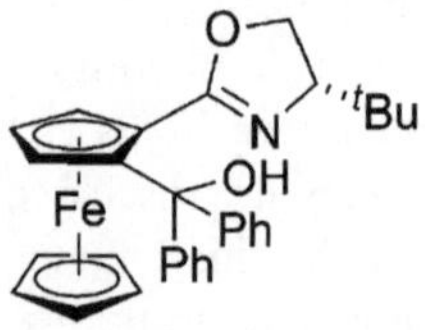

The mechanism of the aminoalcohol-promoted alkylation is generally considered to proceed via formation of zinc complex (6.16). This complex acts as both

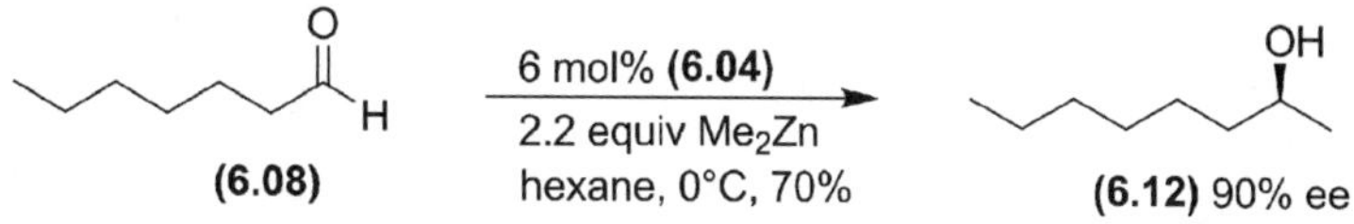

(6.08)

6 mol% **(6.04)**
2.2 equiv Me₂Zn
hexane, 0°C, 70%

(6.12) 90% ee

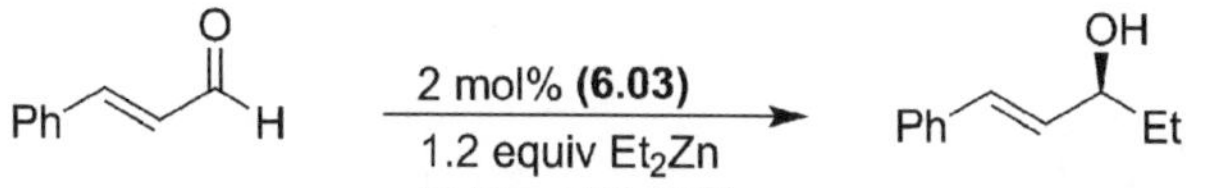

(6.09)

2 mol% **(6.03)**
1.2 equiv Et₂Zn
PhMe, 0°C, 81%

(6.13) 96% ee

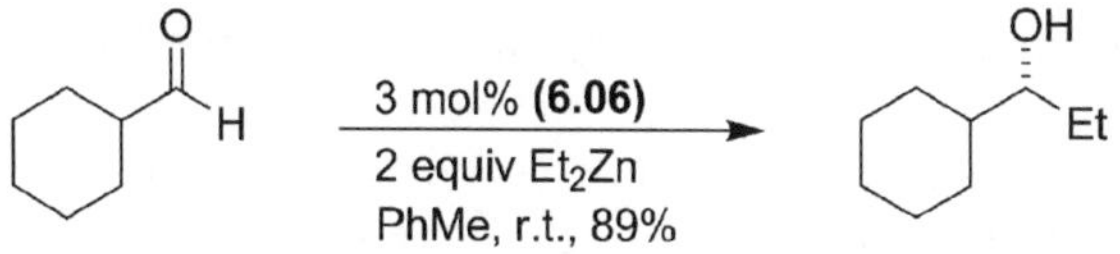

(6.10)

3 mol% **(6.06)**
2 equiv Et₂Zn
PhMe, r.t., 89%

(6.14) 98% ee

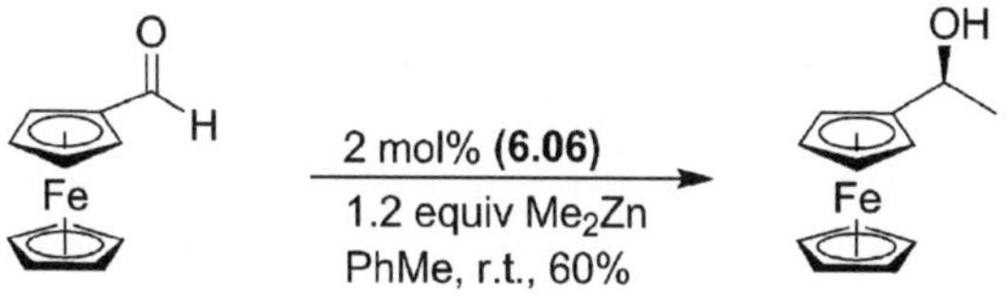

(6.11)

2 mol% **(6.06)**
1.2 equiv Me₂Zn
PhMe, r.t., 60%

(6.15) 81% ee

a Lewis acid to activate the aldehyde and also a Lewis base by coordination to another molecule of dialkylzinc to give active catalyst **(6.17)** (Figure 6.1). The alkyl transfer occurs as shown to the *si* face of the aldehyde to give the stereochemistry observed. [8]

One of the most remarkable features of the aminoalcohol-catalysed alkylation reaction is the observed chiral amplification. Thus, the alkylation of benzaldehyde with diethylzinc occurs with 95% ee, even when the ligand employed has a low enantiomeric excess (just 15% ee). [9] On first considerations, it would be reasonable to assume that the enantiomeric excess of the product would bear a direct linear correlation with enantiomeric excess of the ligand. However, the remarkable positive linear effect can lead to high selectivities being observed in the product using ligands of low enantiomeric excess. This is illustrated in Figure 6.2, where deviations from the expected linear relationship between enantiomeric excess of the ligand and the product are shown by positive nonlinear effects (+NLE) and negative nonlinear effects (−NLE).

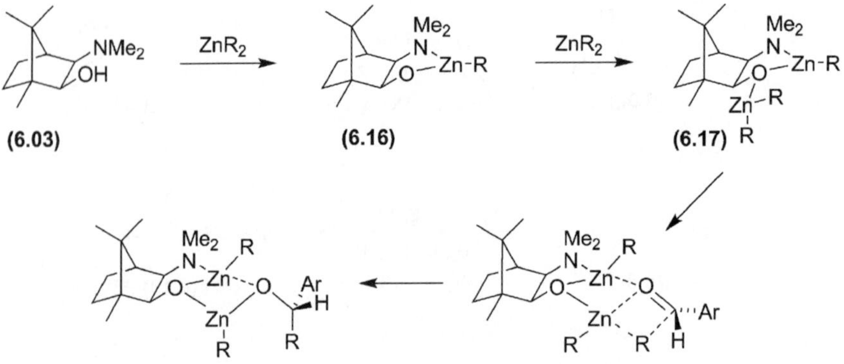

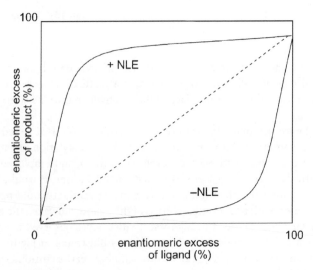

Figure 6.1 Mechanism of catalysed dialkylzinc addition to aldehyde

The topic of chiral amplification has been reviewed, [10] but is most well under-stood in the alkylation of aldehydes with dialkylzinc reagents. The origin of the remarkable positive nonlinear effect is due to dimerisation of alkylzinc intermediates. In order for reaction to occur, the aldehyde must be able to coordinate to the zinc atom of the monomer (**6.19**). However, the monomer (**6.19**) is in equilibrium with the dimer (**6.18**). When the dimer is comprised of the two different enantiomers of aminoalcohol (heterochiral) it is particularly stable and reaction does not proceed through it. However, the dimer comprised of aminoalcohols of the

Figure 6.2 Variation of enantiomeric excess of product and ligand. Abbreviations: + NLE, positive non-linear excess; −NLE, negative non-linear excess.

same chirality (homochiral) is more prone to dissociation, and hence the reaction pathway will preferentially use this material. For example, in the situation where the enantiomeric excess of the ligand is 20% (i.e. 60% (*R*) and 40% (*S*)), the more stable heterodimer will form selectively (using up 40% (*R*) and 40% (*S*)), leaving the (*R*)-enantiomer behind, which although it can form the homodimer, is much more prone to dissociation to the monomer, and is hence catalytically active.

(6.18)   (6.19)

An alternate approach to the asymmetric alkylation of aldehydes with dialkylzinc compounds uses chiral nonracemic titanium-based Lewis acid catalysts. This has been primarily achieved using complexes prepared from stoichiometric amounts of Ti(O^{i}Pr)$_4$ and enantiomerically enriched ligands. [2c-e] A variety of BINOL derivatives have been investigated as ligands in this process. For example, 3-substituted BINOL (**6.20**) can be used at relatively low catalyst loadings to effect alkylation of aromatic and α,β-unsaturated aldehydes with high ee in the presence of 1.4 equivalents of Ti(O^{i}Pr)$_4$. [11] Seebach and coworkers have achieved the enantioselective addition of diethylzinc to aldehydes in the presence of catalytic amounts of the spirotitanate (**6.21**). [12] This reaction can also be performed using dialkylzinc species generated by transmetallation of alkyl Grignard reagents such as butylmagnesium bromide (**6.22**) with ZnCl$_2$. [13] Knochel and coworkers have extended the range of nucleophiles to include functionalised organozinc reagents, prepared from the corresponding iodides. [14] For example, the functionalised aldehydes (**6.23**) and (**6.24**) undergo reaction with the functionalised zinc reagents (**6.25**) and (**6.26**) in the presence of the titanium catalyst prepared from ligand (**6.27**). The functionalised products which are obtained from these reactions are formed with good enantioselectivity. Clearly, these types of structure offer considerable synthetic opportunities.

Phenyl, alkenyl and alkynyl zinc species also undergo enantioselective addition to aldehydes. The enantioselective addition of diphenylzinc to aldehydes is challenging owing to the relatively rapid uncatalysed background addition. This problem can be overcome using a mixture of diphenylzinc and diethylzinc to give mixed ethylphenylzinc [15] and the enantioselective addition of diphenylzinc has been achieved using a variety of catalyst systems. Diarylmethanols can be synthesised with high ee using ferrocene-based catalysts such as (**6.07**). [15] Higher ees can be achieved with lower catalyst loadings using cyclopentadienylrhenium complex (**6.31**) and aromatic aldehydes such as (**6.32**) as substrate. [16] A number

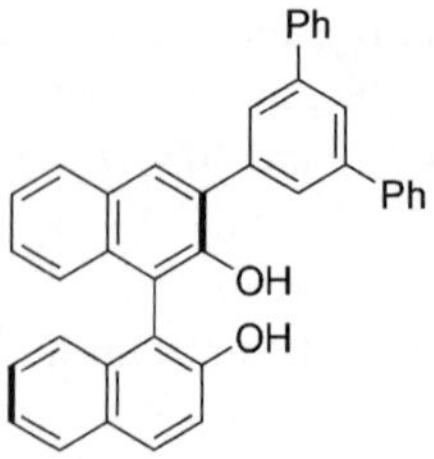

Ph
Ph
OH
OH

(6.20)

Ph
Ph
Ph
Ph
O
O
Ti
O
O
Ph
Ph
Ph
Ph

(6.21)

NHTf
NHTf

(6.27)

MgBr
(6.22)
1. ZnCl2, Et2O,
1,4-dioxane, r.t.
2. 14 mol% (6.21),
1.2 equiv Ti(OiPr)4,
PhCHO, -30°C, 70%
OH
Ph
(6.28) 98% ee

O
Me3Si
H
I
(6.23)
2 equiv. Ti(OiPr)4
8 mol% (6.27)
Zn
Tf
N
Bn
2
(6.25)
2-12 h, PhMe,
-55°C to -5°C, 62%
OH
Bn
N
Tf
Me3Si
I
(6.29) 82% ee

O
H
TBDPSOCH2
(6.24)
2 equiv Ti(OiPr)4/Ti(OtBu)4 (1:1)
8 mol% (6.27)
Zn
OPiv
2
(6.26)
12-24 h, PhMe, -35°C to -25°C, 76%
OH
OPiv
TBDPSOCH2
(6.30) 94% ee

of BINOL-based catalysts show high enantioselectivity in the diphenylzinc addition to aromatic aldehydes without the need for diethylzinc.[6,17] The partially hydrogenated BINOL (**6.33**) has been found to exhibit the greatest scope to date and can be used to effect selective addition of diphenylzinc to both aromatic and linear aliphatic aldehydes such as (**6.34**)- usually challenging substrates for this asymmetric transformation.

BINOL is an effective ligand in the titanium-catalysed asymmetric addition of alkynylzinc reagents to aldehydes.[2,18e] Aromatic, aliphatic and α,β-unsaturated aldehydes such as (**6.35**) are converted into the propargylic alcohol with 91–99% ee using the alkynylzinc, generated from phenylacetylene and diethylzinc. Carriera and coworkers have developed an enantioselective alkynylzinc addition that is also

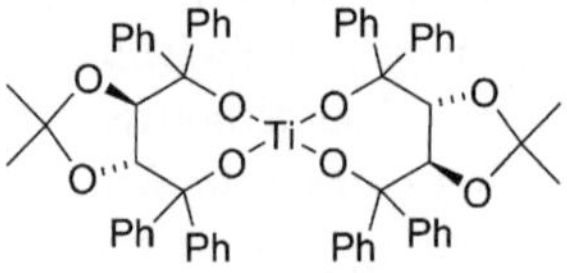

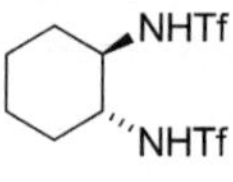

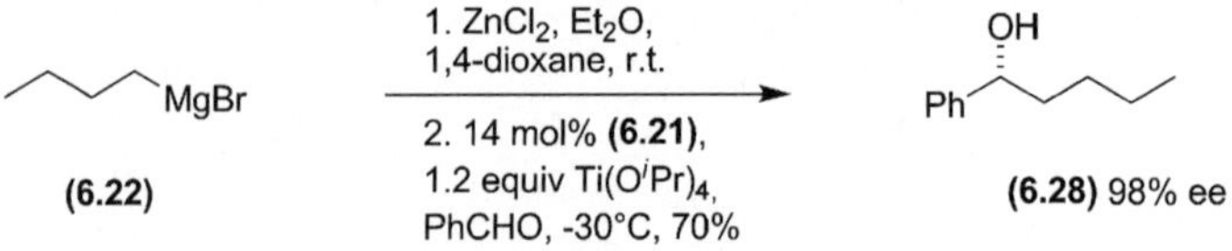

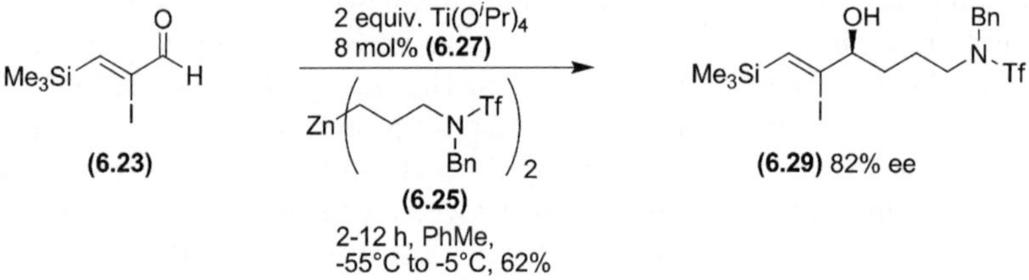

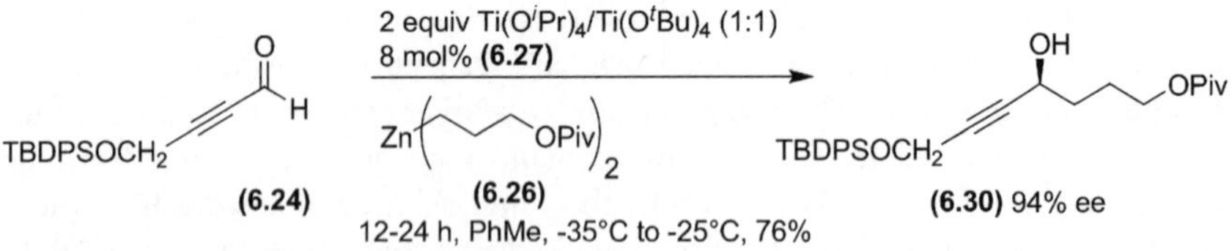

catalytic in metal- utilising only 20 mol% zinc triflate in combination with catalytic quantities of aminoalcohol (6.36).[19] Highly enantioselective additions of a variety of terminal acetylenes to aliphatic aldehydes are achieved using this method, but aromatic aldehydes cannot be used as substrates.

Alkenylzinc reagents are readily prepared by hydroboration followed by boron to zinc transmetallation and the enantioselective addition to aromatic and aliphatic aldehydes has been achieved using ligand (6.03).[20] High enantioselectivities have also been reported using aminothiol ligands such as (6.37)[21] and (6.38)[22] and the paracyclophane-based aminoalcohol (6.39).[23] The latter ligand exhibits good scope and has been successfully applied in the alkenylzinc addition to both aromatic aldehydes and aliphatic aldehydes such as (6.40).

The reduced reactivity of ketones makes them more challenging substrates for the asymmetric addition of organozinc species. Nevertheless, particularly reactive catalyst systems do effect asymmetric alkylation, alkynylation and arylation of ketones with high ee.[24] As diarylzincs are more reactive than the alkyl counterparts, most success in the addition to ketones has been achieved using diphenylzinc.[25] For example, addition of phenylzinc to ketones such as (6.45) proceeds with good ee in the presence of ligand (6.03).[26] The enantioselective addition of diethylzinc to ketones has been achieved using titanium catalysts formed from Ti(O^{i}Pr)$_4$ and sulfonamide ligands.[27] Walsh and coworkers used ligands such as (6.46), derived from 1,2-diaminocyclohexane and camphorsulfonyl chloride, in the enantioselective addition of diethylzinc to aryl ketones such as acetophenone and valerophenone[27c] and also cyclic α,β-enones such as (6.47).[27d] A variety of catalysts has been used to effect enantioselective addition of alkynylzinc species to ketones.[24] High ees have obtained using camphorsulfonamide ligands and acetophenone derivatives as substrates[28] and also using the BINOL/Ti(O^{i}Pr)$_4$ combination.[29] An asymmetric alkynylation of wide scope has been developed by Katsuki and coworkers utilising zinc(salen) complexes prepared from (6.48) as catalysts.[30] This method has been used to good effect in the alkynyation of aryl alkyl ketones and also aliphatic ketones such as 3,3-dimethyl-2-butanone (6.49) with phenylacetylene.

6.2 Addition of Cyanide to Aldehydes and Ketones

The conversion of aldehydes (6.53) into the corresponding cyanohydrins (6.54) is an appealing synthetic transformation, since cyanohydrins are readily converted into a number of important functional groups such as α-hydroxy acids (6.55) and β-amino alcohols (6.56).

The most popular and selective catalysts for the asymmetric transformation of aldehydes into cyanohydrins are oxynitrilase enzymes.[31] However, as the emphasis of this book is on chemically catalysed transformations, a discussion of biocatalysed

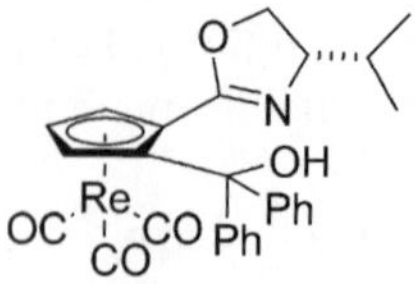

(6.31)

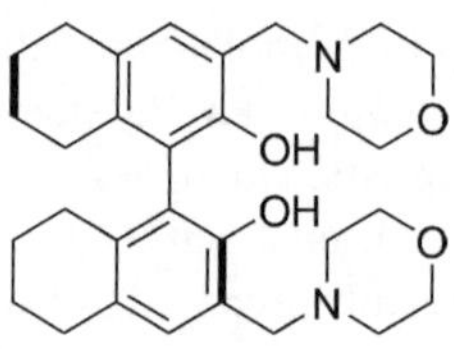

(6.33)

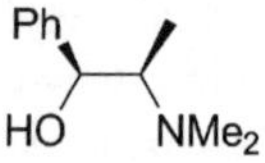

(6.36)

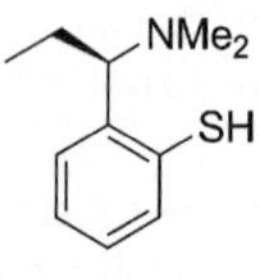

(6.37)

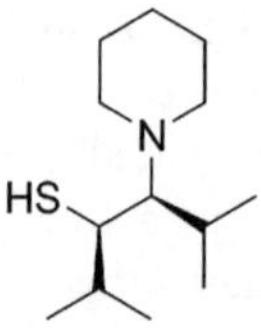

(6.38)

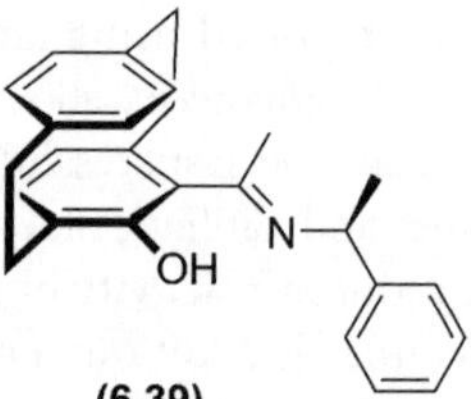

(6.39)

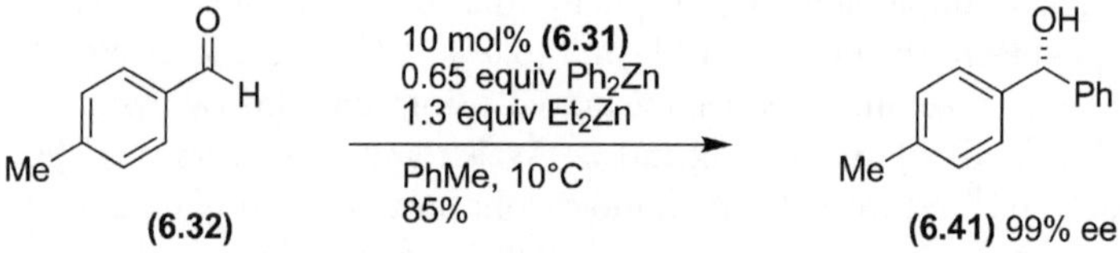

(6.32)

10 mol% **(6.31)**
0.65 equiv Ph_2Zn
1.3 equiv Et_2Zn

PhMe, 10°C
85%

(6.41) 99% ee

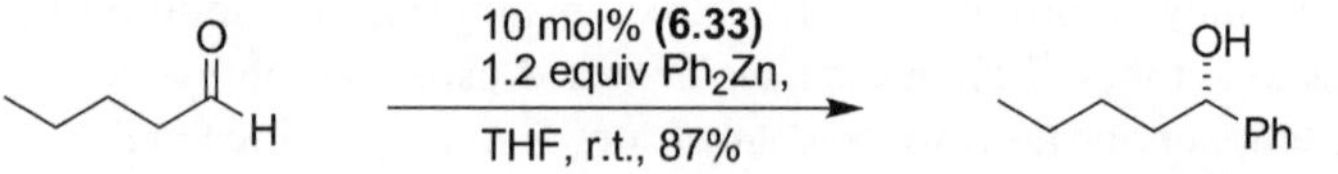

(6.34)

10 mol% **(6.33)**
1.2 equiv Ph_2Zn,

THF, r.t., 87%

(6.42) 92% ee

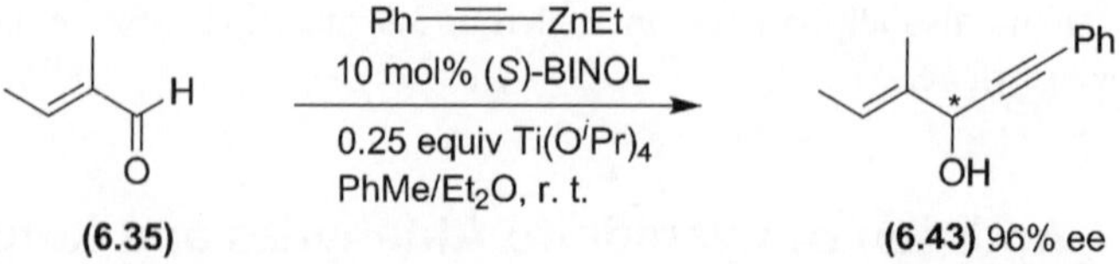

(6.35)

Ph———ZnEt
10 mol% (S)-BINOL

0.25 equiv $Ti(O^iPr)_4$
PhMe/Et_2O, r. t.

(6.43) 96% ee
absolute config.
not determined

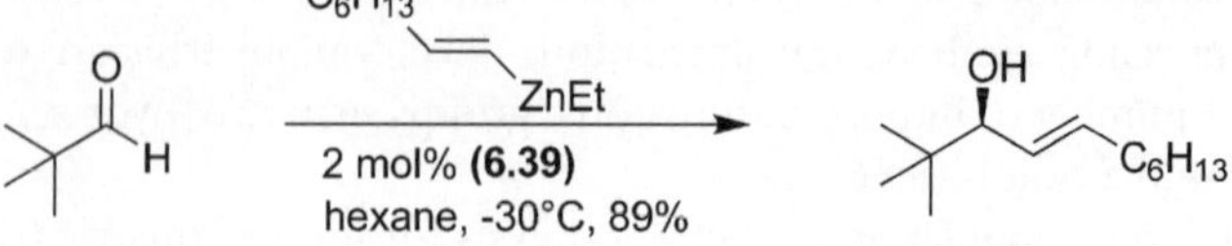

(6.40)

C_6H_{13}

ZnEt

2 mol% **(6.39)**
hexane, -30°C, 89%

(6.44) 98% ee

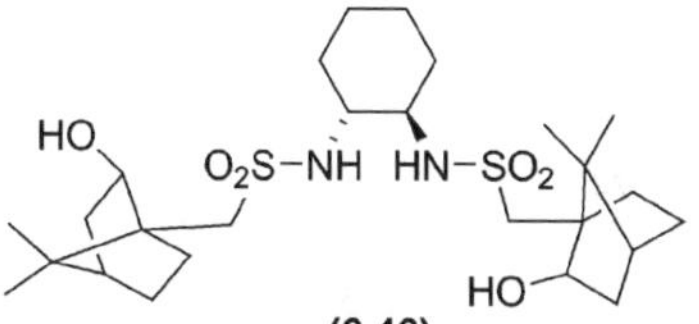

(6.46)

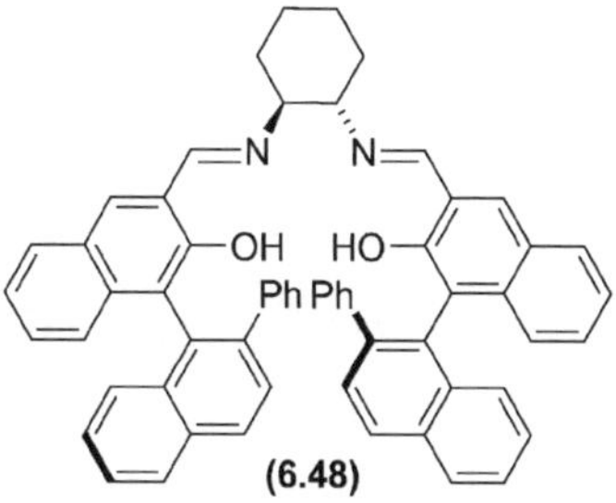

(6.48)

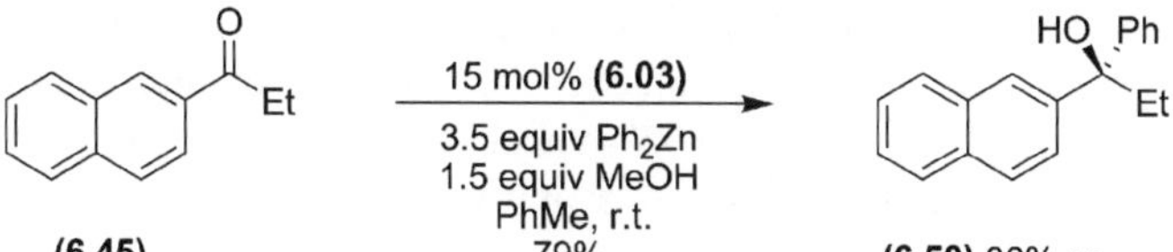

(6.45) → 15 mol% (6.03), 3.5 equiv Ph$_2$Zn, 1.5 equiv MeOH, PhMe, r.t., 79% → (6.50) 86% ee

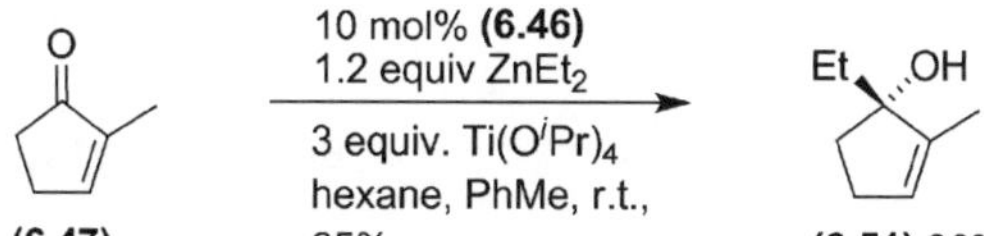

(6.47) → 10 mol% (6.46), 1.2 equiv ZnEt$_2$, 3 equiv. Ti(O^{i}Pr)$_4$, hexane, PhMe, r.t., 65% → (6.51) 96% ee

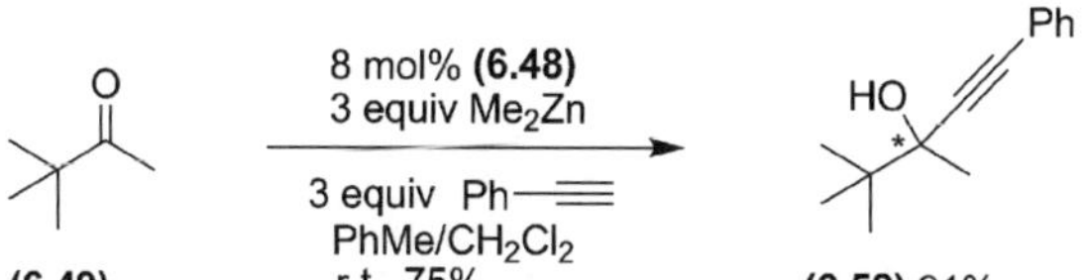

(6.49) → 8 mol% (6.48), 3 equiv Me$_2$Zn, 3 equiv Ph—≡, PhMe/CH$_2$Cl$_2$, r.t., 75% → (6.52) 91% ee, absolute config. not determined

cyanohydrin formation is beyond the scope of this section. A variety of metal-based Lewis acid catalysts and recently some bifunctional Lewis acid/Lewis base complexes have seen application as catalysts in the catalytic asymmetric cyanation of aldehydes.[31b,c, 32]

The Lewis acid-catalysed addition of trimethylsilyl cyanide to aldehydes has been reported using several different catalysts. Titanium-based Lewis acids have proved to be particularly popular.[2e, 33] In a typical reaction, benzaldehyde (6.01) is converted into the cyanohydrin (6.57) usually after removal of the trimethylsilyl group by hydrolysis. A wide variety of ligands have been used for this reaction and

(6.53) (6.54) (6.55) (6.56)

those exhibiting the greatest selectivity (**6.58–6.62**) are illustrated. Typically these ligands contain oxygen, nitrogen and /or sulfur donor atoms.

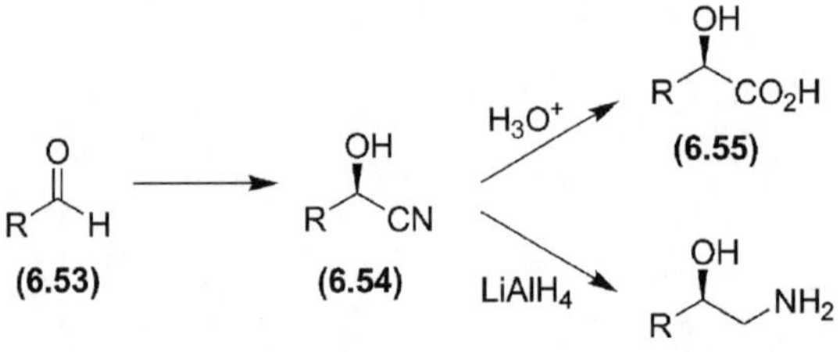

Some of the highest ees are obtained with ligand (**6.61**)[33g] and (**6.62**).[33f,h] The former also shows good scope and catalyses the cyanation of aliphatic aldehydes such as isobutyraldehyde with up to 95% ee. Ligand (**6.60**) is notable for its high catalytic efficiency in this procedure, since only 0.1 mol% is required.

(6.58)

(6.59)

(6.60) + $TiCl_4$

(6.61)

(6.62)

(6.01) **(6.57)**

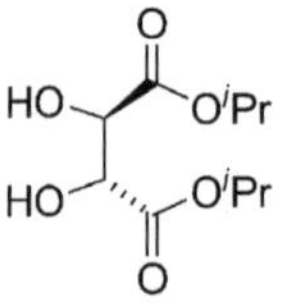

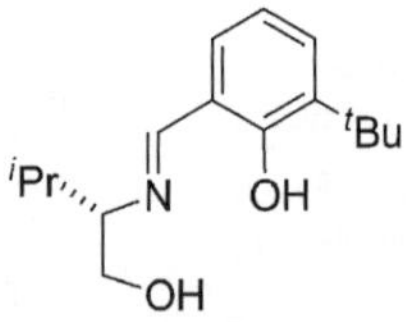

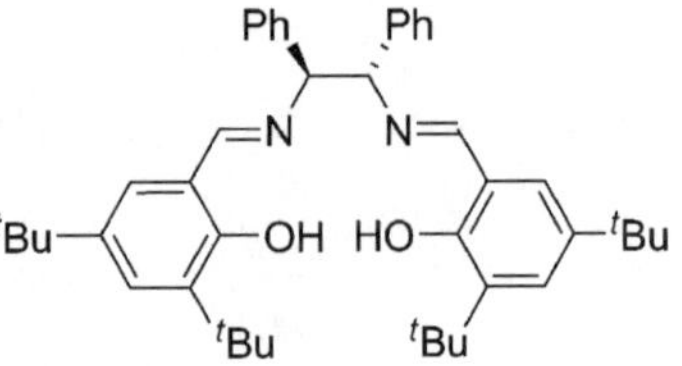

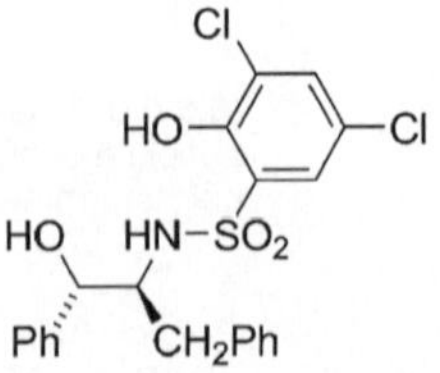

ligand	loading	yield	ee	conditions
(6.58)	20 mol%	84%	91%	0°C, 18 h
(6.59)	20 mol%	67%	85%	-80°C, 36 h
(6.60)	0.1 mol%	100%	86%	r.t., 24h
(6.61)	5 mol%	90%	94%	-65°C, 48 h
(6.62)	16.5 mol%	79%	94%	-78°C, 48h

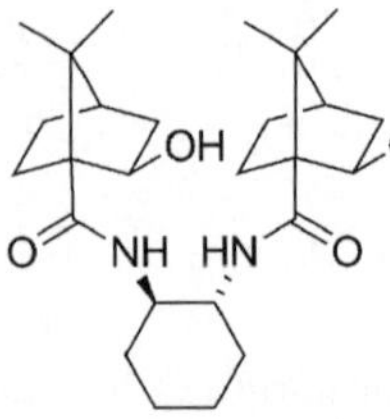

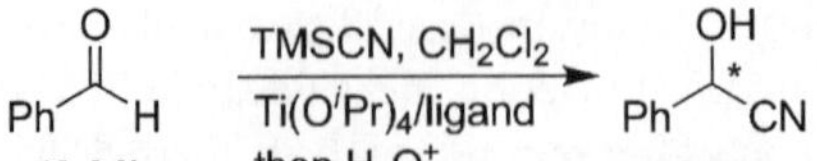

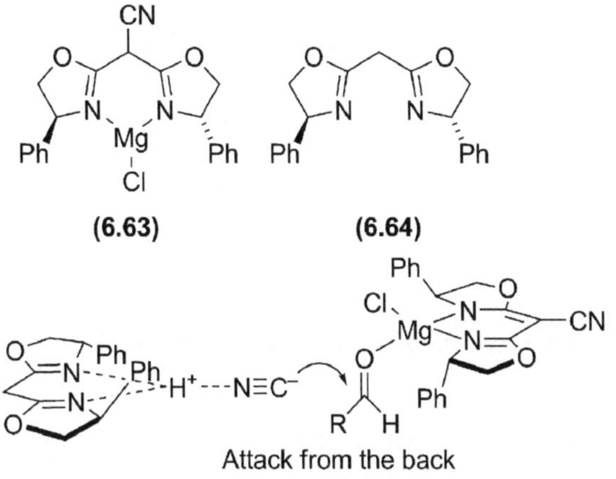

(6.63)   **(6.64)**

Attack from the back

Figure 6.3 The use of two bis-oxazolines in HCN addition to aldehydes.

Some of the highest ees have been obtained using bifunctional Lewis acid/Lewis base catalysts that activate both the aldehyde and the nucleophile. Corey and Wang have reported an enantioselective magnesium bis-oxazoline-catalysed trimethylsilylcyanation.[34] The reaction uses the magnesium complex (**6.63**), which activates the carbonyl group, in addition to the bis-oxazoline (**6.64**), which is believed to provide a asymmetric environment for the HCN, formed from TMSCN *in situ*. The favoured catalytic pathway may proceed as shown in Figure 6.3.

Aluminium-BINOL-based complexes have also been shown to be highly selective bifunctional asymmetric cyanation catalysts.[35] For example the bisphosphine oxide (**6.65**) developed by Shibasaki and coworkers catalyses the cyanation of both aromatic and aliphatic aldehydes with ees ranging from 83–98%.[35a]

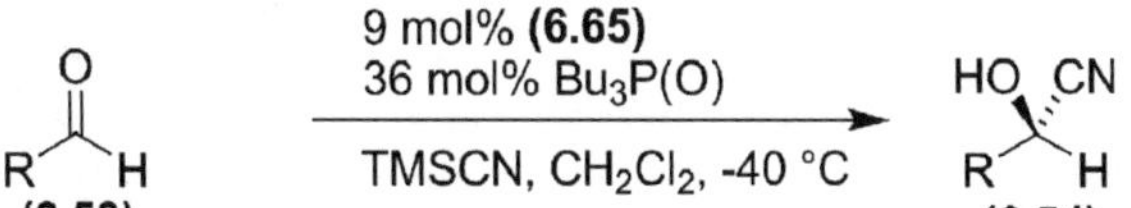

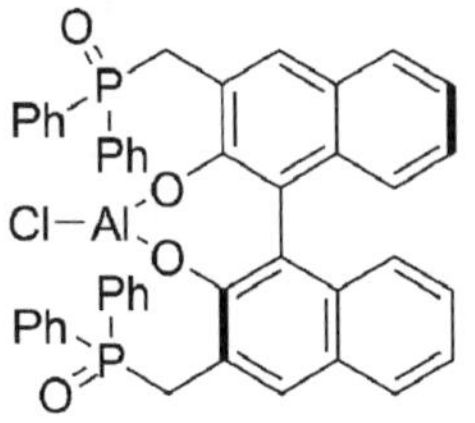

R	Yield	ee
$Ph(CH_2)_2$	97%	97%
$(CH_3CH_2)_2CH$	98%	83%
$PhCH=CH$	99%	98%
Ph	98%	96%

(6.65)

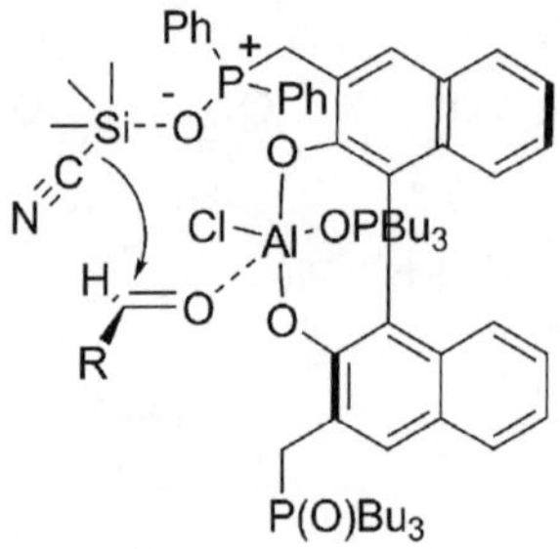

Figure 6.4 Model for asymmetric induction by complex (6.65)

In this procedure the nucleophile is activated by chelation with phosphorus while the aldehyde is coordinated to aluminium and oriented such that nucleophilic attack occurs from the *re*-face as depicted in Figure 6.4.

In the benzaldehyde 'test-bed' reaction, the unusual catalysts (6.66)[36] and (6.67)[37] have provided 84% ee (*R*) and 90% ee (*S*) respectively in the cyanohydrin formation reaction, and were shown to work well over a range of electron-rich arylaldehydes. Both of these catalytic systems use fairly low catalyst loadings.

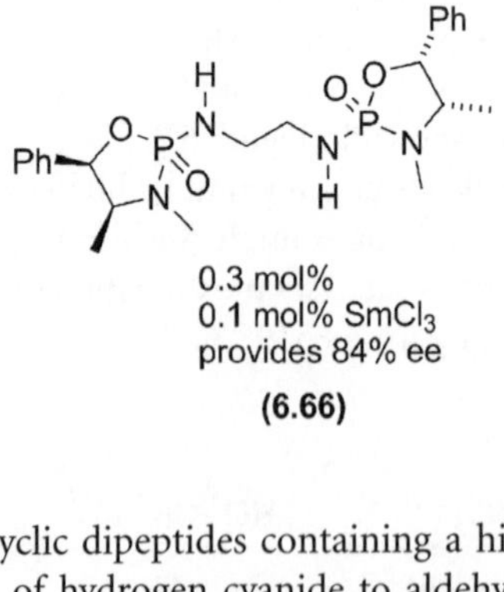

0.3 mol%
0.1 mol% SmCl$_3$
provides 84% ee

(6.66)

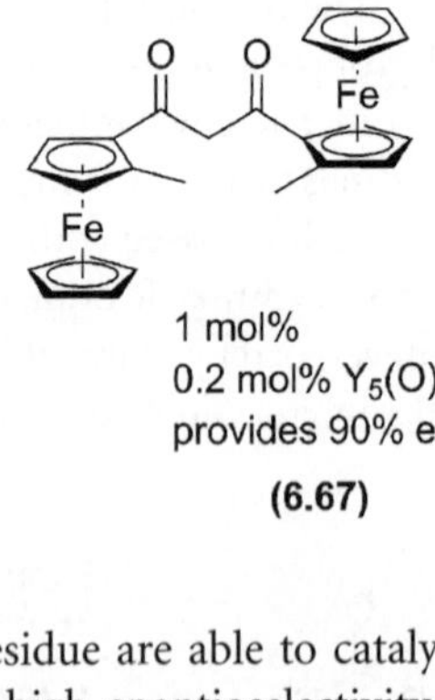

1 mol%
0.2 mol% Y$_5$(O)(O^{i}Pr)$_{13}$
provides 90% ee

(6.67)

Cyclic dipeptides containing a histidine residue are able to catalyse the addition of hydrogen cyanide to aldehydes with high enantioselectivity.[31b,c, 32c, 38,39] In particular, cyclo-[(*S*)-Phe-(*S*)-His] (6.68) is an effective catalyst, especially for arylaldehydes without an electron-withdrawing group, although some aliphatic aldehydes also undergo reaction with reasonable selectivity.[40] An assembly of reagents where the carbonyl group of the substrate is hydrogen bonded to the catalyst has been suggested, and is represented by structure (6.69). The imidazole group is protonated to generate cyanide ion which then attacks the carbonyl as indicated. It has been discovered that autoinduction occurs in this reaction as ees increase as the reaction proceeds. Furthermore, addition of small quantities of an enantiopure cyanohydrin at the beginning of the reaction leads to an enhancement in enantioselectivity. A new model has been proposed to account for these observations wherein a possibly dimeric cyclic peptide–cyanohydrin complex is the most active species.[41]

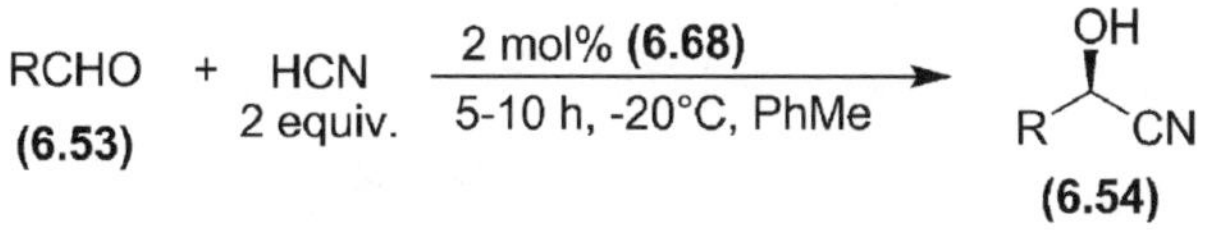

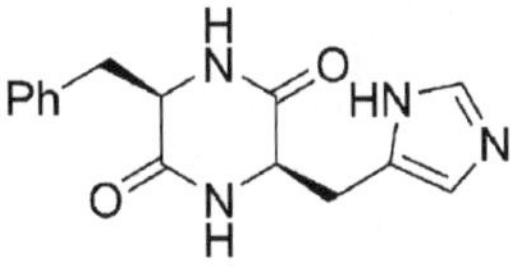

R = Ph	97% conversion	97% ee
R = 2-Naphth	61% conversion	91% ee
R = ^{t}Bu	60% conversion	58% ee

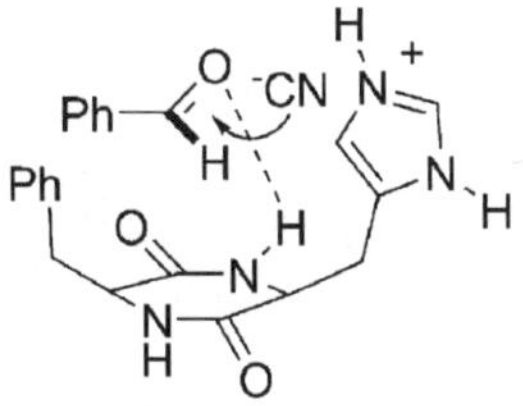

Ketones are not usually good substrates for catalysed cyanohydrin formation, since they are less reactive than aldehydes and in the past rather forcing conditions, utilising high pressure, were required to effect the catalytic asymmetric cyanation of ketones. However, recently, a diverse range of metal-based and organic catalysts systems have been developed which achieve this transformation under milder conditions with high ee under ambient and subambient temperature.[32b] The oxazaborolidinium salt (6.70) catalyses the asymmetric cyanosilylation of aromatic and some aliphatic methyl ketones via coordination to the carbonyl oxygen and the bifunctional titanium complex formed from the D-glucose-derived ligand (6.71) effects cyanosilylation of aromatic, aliphatic and cyclic ketones with good ee.[42] Interestingly, while the titanium complex provides the (R)-product use of lanthanide complexes of (6.71) results in a reversal in the mode of enantioselection giving the (S)-silyloxynitrile.[43] Asymmetric cyanosilylation of ketones can also be achieved using metal-free organocatalysts possessing Lewis basic sites.[32c] For instance amino acid salt (6.72) functions as an effective catalyst for the enantioselective cyanosilylation of aromatic and α,β−unsaturated ketones at low temperature.[44] Modified cinchona alkaloids such as (DHQD)$_2$AQN used in the Sharpless dihydroxylation (see Section 5.1) are also effective Lewis base catalysts providing up to 97% ee in the cyanosilylation of a range of cyclic and hindered aliphatic ketones.[45] Thiourea (6.73) is a bifunctional organocatalyst possessing a Lewis basic amine moiety and a Brønsted acidic thiourea group and, in a similar fashion to (6.72) catalyses the asymmetric cyanosilylation of aromatic and α,β-unsaturated ketones, but at lower catalysts loadings and temperatures.[46]

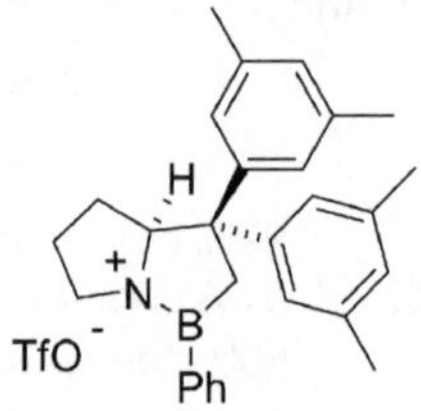

(6.70)

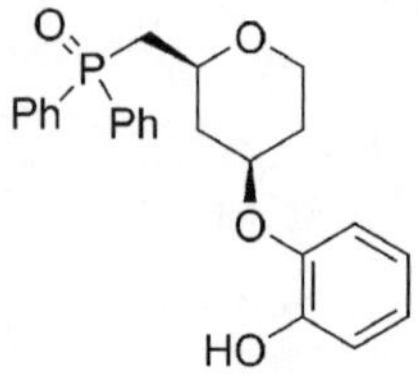

(6.71)

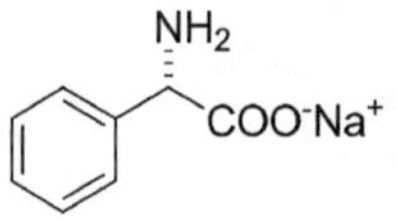

(6.72)

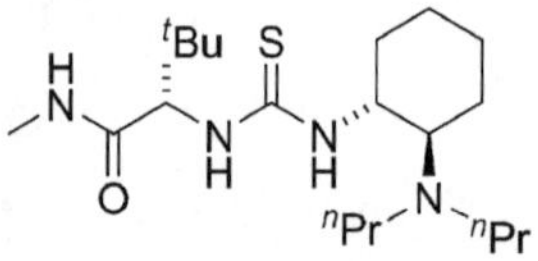

(6.73)

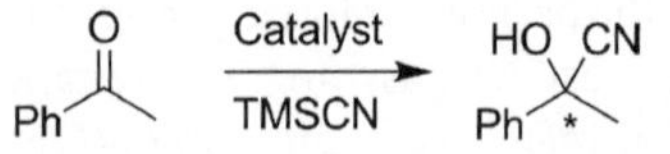

Catalyst system	Yield	Config./ee
10 mol% (6.70), 25°C	49%	(R) 65%
10 mol% (6.71)/Ti(O^{i}Pr)$_4$ -30°C	85%	(R) 92%
5 mol% (6.71)/Gd(O^{i}Pr)$_3$ -40°C	92%	(S) 92%
30 mol% (6.72), -45°C	96%	(S) 94%
5 mol% (6.73), -78°C	96%	(S) 97%

6.3   Allylation of Aldehydes

The allylation of aldehydes provides a useful route to homoallylic alcohols. The most popular methods for achieving this in an asymmetric fashion are by the use of metal-based Lewis acid catalysts or using Lewis basic organocatalysts.[47]

A variety of metal complexes have been investigated as catalysts in this procedure with boron- and titanium-based systems[2e] being the most popular. In 1991, H. Yamamoto reported that the allylation of aldehydes with allylsilanes (the Sakurai–Hosomi reaction) could be achieved with the enantiomerically pure boron catalyst (**6.74**).[48,49] Whilst the parent allylsilane (**6.75**) was not reactive enough to give the product (**6.76**) efficiently, alkyl substitution of the allylsilane (**6.77**) increases the reactivity of the silane towards benzaldehyde (**6.01**) and affords high levels of enantioselectivity and diastereoselectivity in the product (**6.78**). The corresponding reactions with aliphatic aldehydes are less satisfactory.

Marshall extended the scope of the reaction to the use of allylstannane reagents, which are more reactive nucleophiles.[50] The addition of triflic anhydride as a promoter overcame a problem with catalyst deactivation and allowed the reaction of allylstannane (**6.79**) to take place efficiently and with fairly good stereocontrol in the product (**6.80**).

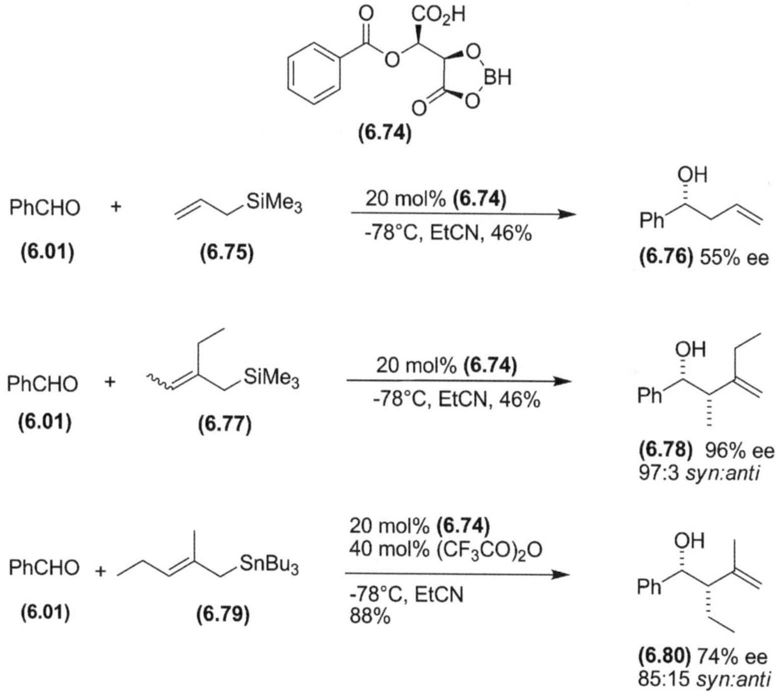

(**6.74**)

PhCHO + ⌁SiMe₃ → 20 mol% (**6.74**), -78°C, EtCN, 46% → Ph OH

(**6.01**)   (**6.75**)   (**6.76**) 55% ee

PhCHO + ⌁SiMe₃ → 20 mol% (**6.74**), -78°C, EtCN, 46% → Ph OH

(**6.01**)   (**6.77**)   (**6.78**) 96% ee
97:3 *syn:anti*

PhCHO + ⌁SnBu₃ → 20 mol% (**6.74**), 40 mol% (CF₃CO)₂O, -78°C, EtCN, 88% → Ph OH

(**6.01**)   (**6.79**)   (**6.80**) 74% ee
85:15 *syn:anti*

The relatively weaker Lewis acidic titanium complexes require the use of a stronger nucleophile than allylsilanes, and tributylallyltin (**6.81**) is the most common allylating agent employed when using titanium-based catalyst systems. In 1993, Umani-Ronchi[51] and Keck[52,53] published related results using BINOL/titanium derived catalysts. In the Umani-Ronchi system, BINOL is employed, in combination with TiCl₂(OⁱPr)₂ and shown to work well with aliphatic aldehydes, including octanal (**6.82**). The Keck system uses BINOL with titanium tetraisopropoxide in a ratio of either 2:1 or 1:1. This combination is effective for both aliphatic and aromatic aldehydes, including cyclohexane carboxaldehyde (**6.83**). Some of the highest ees have been obtained using bis(naphthoxy)titanium complex (**6.86**). This catalyst exhibits a wide scope and effects allylation of a variety of aldehydes with ees ranging from 95 to 99%.[54] The BINOL:Ti catalysed allylation of aldehydes has been shown to be subject to chiral amplification[54,55] and chiral poisoning.[56]

In an effort to improve catalyst efficiency, Yu and coworkers have shown that it is possible to use 'molecular accelerators' in these reactions.[57] They showed that the addition of the alkylthioborane $^{i}PrSBEt_3$ or the alkylthiotrimethylsilane $^{i}PrSSiMe_3$ allows the use of lower catalyst loadings (typically 1 mol%) in the addition of tributylallyltin to aldehydes.

(**6.86**)

R	Yield	ee
$CH_3(CH_2)_6$	84%	99%
PhCH=CH	70%	95%
Ph	90%	96%

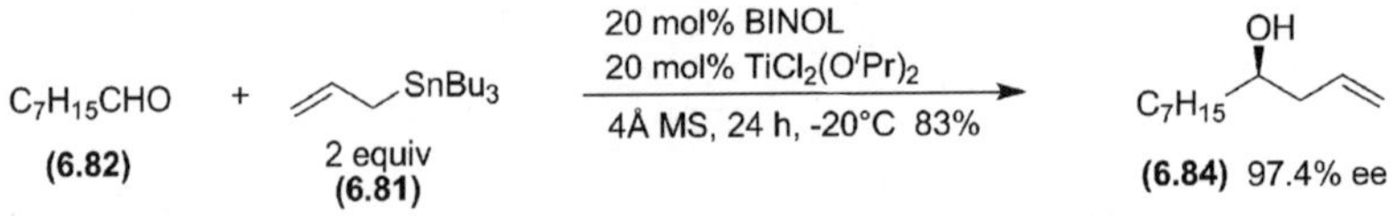

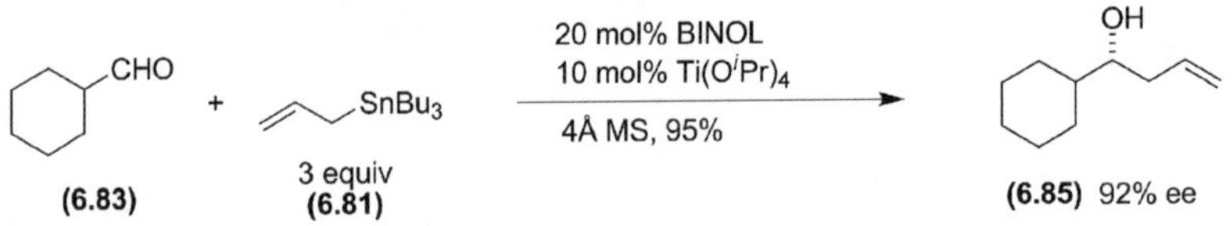

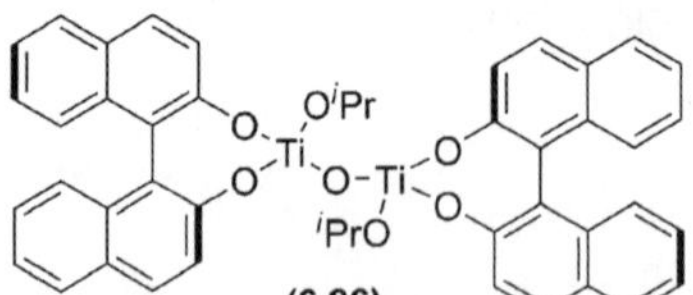

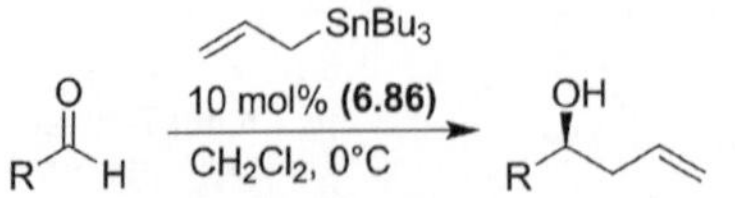

Silver catalysts have not been widely investigated for asymmetric synthesis. However, Yamamoto has shown that the combination of silver triflate with BINAP (**6.87**) is effective for the allylation of aldehydes with tributylallyltin.[58] The reaction is represented by the conversion of furfural (**6.88**) into the addition product (**6.89**) with high yield and enantioselectivity. The use of crotylstannanes (**6.90**) was found to lead to the *anti*-adducts (**6.91**) selectively, as shown in the reaction with benzaldehyde (**6.01**).[59] Enantiomerically pure zirconium BINOL complexes[60] and zinc bis-oxazoline complexes[61] have also been used to catalyse the asymmetric allylation of aldehydes with allylstannanes and some high ees in this transformation have been obtained using indium BINOL[62] and PYBOX complexes.[63]

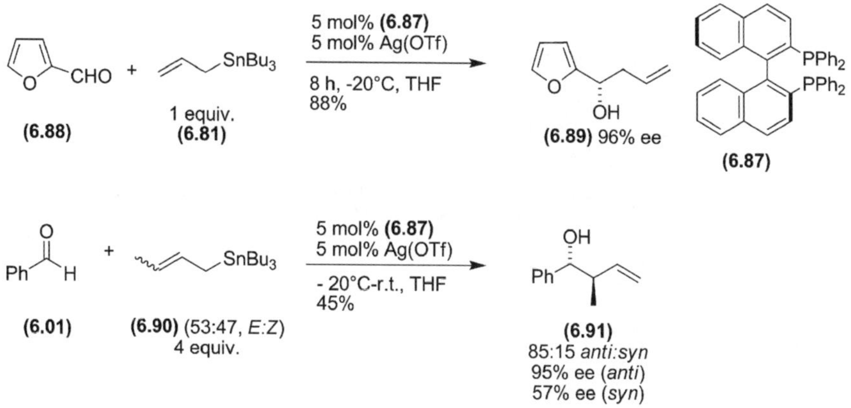

Although allyltributylstannane has been the most widely used allylating agent, the weaker nucleophile allyltrimethylsilane is cheaper and less toxic. Gauthier and Carreira have shown that allyltrimethylsilane is an effective allylating agent in the presence of a catalyst derived from the stronger Lewis acid titanium tetrafluoride and BINOL.[64] Pivaldehyde (**6.92**) undergoes allylation to afford the homoallylic alcohol (**6.93**) in good enantioselectivity and yield (after hydrolysis of the trimethylsilyl ether).

While allylstannanes and allylsilanes have been the most common allyl donors used in these studies the allylation of aldehydes with cheaper, more readily available allyl halides (Kishi–Nozaki–Hiyama reaction) can now also be performed in an asymmetric fashion. In this transformation the allyl halide is coupled to an aldehyde via the intermediacy of an allylchromium(III) species. This reaction can be performed using catalytic amounts of chromium(II) salts in the presence of stoichiometric manganese as a reductant and TMSCl as a scavenger, and a number of research groups have investigated the use of chiral nonracemic chromium catalysts. Moderate to good enantioselectivities have been obtained using chromium(salen) complexes[65] and the best ees to date have been achieved using the tridentate bis-oxazoline ligand (**6.94**), which is selective over a wide range of aldehydes.[66]

As well as allylation reactions catalysed by Lewis acids, it has been shown that the reaction can be catalysed by Lewis bases.[47b] In 1994, Denmark showed that allylation of benzaldehyde (**6.01**) with allyltrichlorosilane (**6.95**) could be achieved using phosphoramide (**6.96**) as a catalyst.[67,68] A range of such phosphoramides has

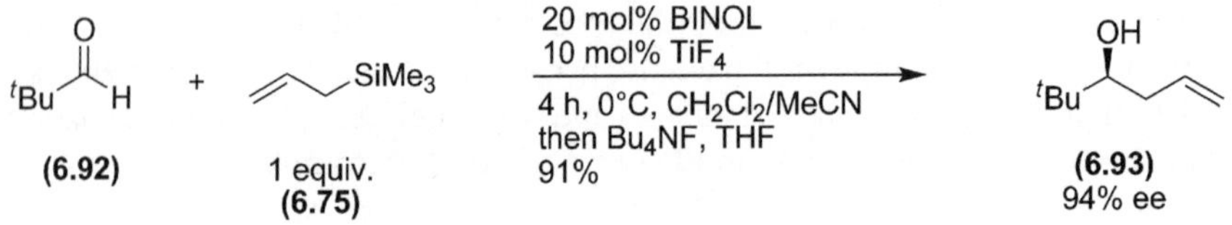

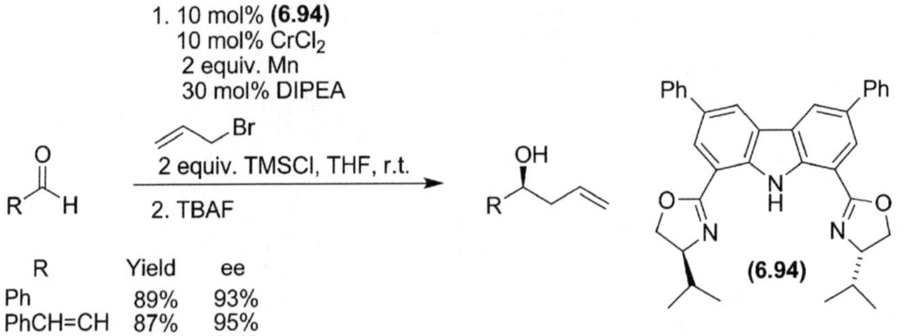

(6.92) + (6.75)
1 equiv.
(6.75)

20 mol% BINOL
10 mol% TiF$_4$
————————————→
4 h, 0°C, CH$_2$Cl$_2$/MeCN
then Bu$_4$NF, THF
91%

(6.93)
94% ee

1. 10 mol% (6.94)
   10 mol% CrCl$_2$
   2 equiv. Mn
   30 mol% DIPEA

   Br

   2 equiv. TMSCl, THF, r.t.
   ————————————→
2. TBAF

R	Yield	ee
Ph	89%	93%
PhCH=CH	87%	95%
c-C$_6$H$_{11}$	95%	94%

(6.94)

been synthesised by varying the substituents on nitrogen but ees have remained modest.[69] Improvements in the enantioselectivity of allylation have been observed by Iseki and coworkers using the proline-derived phosphoramide (6.97).[70] This Lewis base-catalysed allylation is postulated to proceed via a closed chair transition state with an octahedral cationic silicon incorporating two phosphoramide ligands as shown in Figure 6.5.[69,71] The Lewis base thus plays a dual role in activating the allyl group by coordination to silicon and also by enhancing the Lewis acidity of silicon by ionisation. The addition of crotylsilanes such as (6.98) proceeds with high diastereoselectivites and in a predictable manner owing to the closed, chair-like nature of this transition state. The mechanistic studies have led to the design of bisphosphoramides such as (6.99) which catalyses the allylation and crotylation of aromatic and α,β-unsaturated aldehydes with high ee.[72]

Enantiomerically pure formamides,[73] diamines[74] and ureas[75] have also been used to catalyse the asymmetric allylation.

N-oxides are the other most important class of Lewis basic catalysts for asymmetric allylation. A variety of catalysts have been investigated and some of these providing the highest ees are illustrated.[76] These ligands (6.101–6.104) typically contain heteroaromatic N-oxides moieties and catalyse the allylation of benzaldehyde and derivatives with good to high ee. While Lewis basic organocatalysts do not require the use of potentially toxic transition metal salts or allyltin compounds, these catalysts are limited in scope to the allylation of aromatic aldehydes and, to date, rather poor enantioselectivites have been obtained using aliphatic substrates.

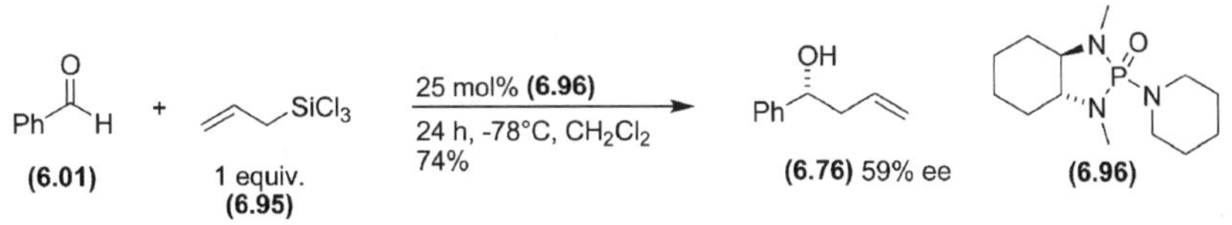

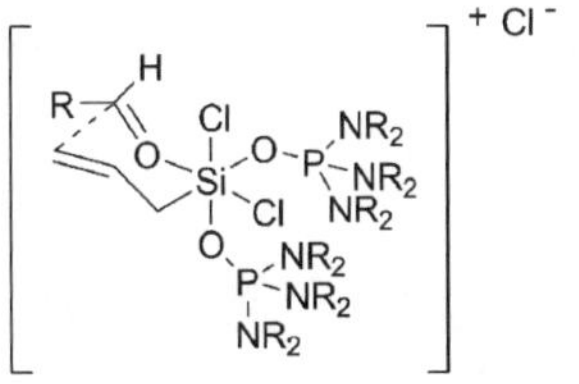

Figure 6.5 Transition state phosphoramide catalysed allylation

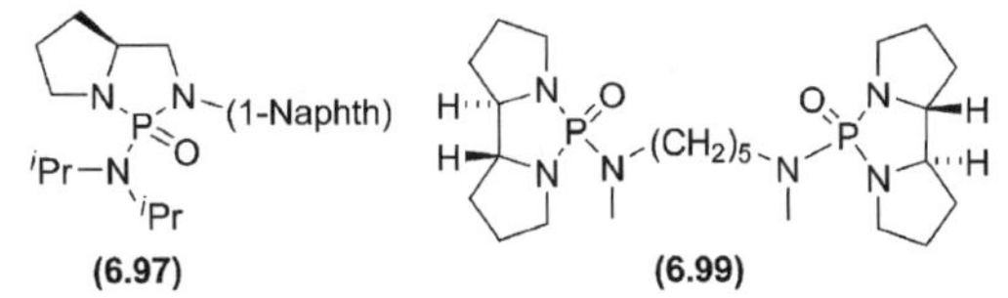

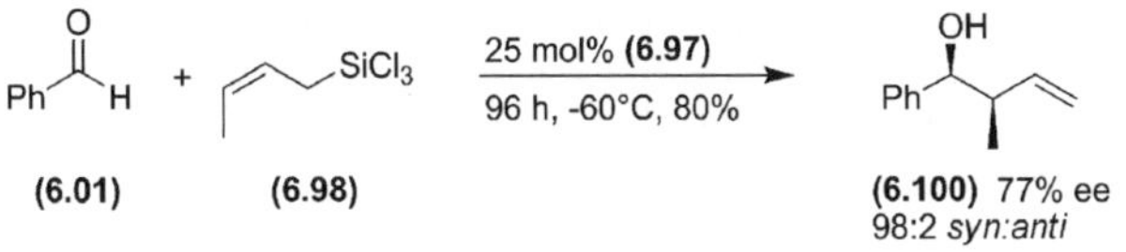

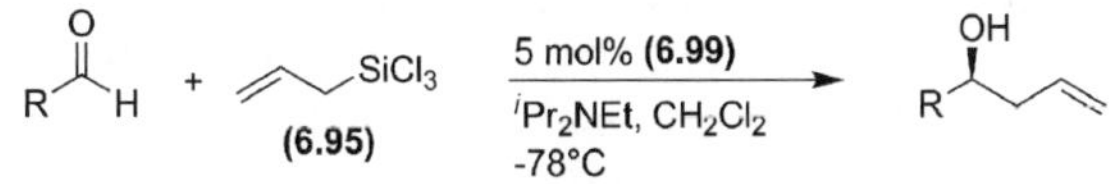

		R	Yield	ee
		Ph	85%	87%
		2-furyl	86%	81%
		(*E*)-C₆H₅CH=CH	59%	81%

6.4 Hydrophosphonylation of Aldehydes

The catalytic enantioselective hydrophosphonylation of aldehydes[77] was first re-ported in 1993 by Shibuya and coworkers.[78] These early results showed that the addition of diethylphosphite (**6.105**) to benzaldehyde (**6.01**) could be catalysed by the titanium complex formed from ligand (**6.58**) with moderate enantioselectivity in the α-hydroxyphosphonate product (**6.106**). The titanium catalyst serves to facilitate conversion of the phosphite into the tautomer HOP(OEt)₂ and activates the aldehyde towards addition.

LaLi₃tris(binaphthoxide) catalysts (LLB catalyst, see Section 7.1) have also been used for this reaction,[79,80] although the method of preparation of the catalyst is important for very high selectivities.[81] The LLB catalyst (**6.107**) was prepared by mixing LaCl₃.7H₂O (1 equiv.) BINOL dilithium salt (2.7 equiv.) and sodium

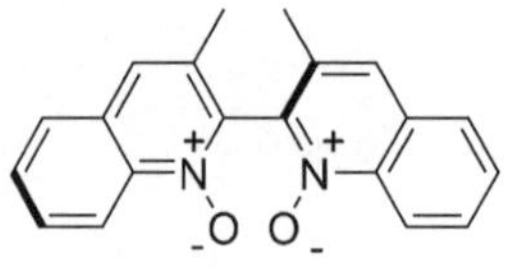

(6.101)

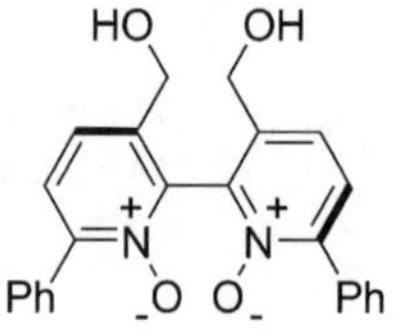

(6.102)

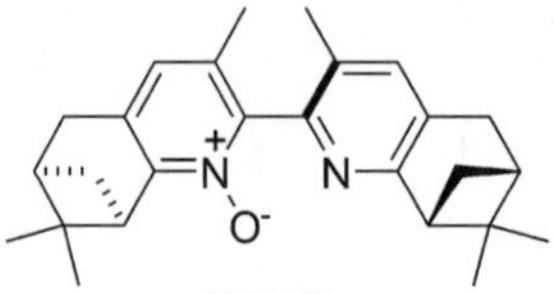

(6.103)

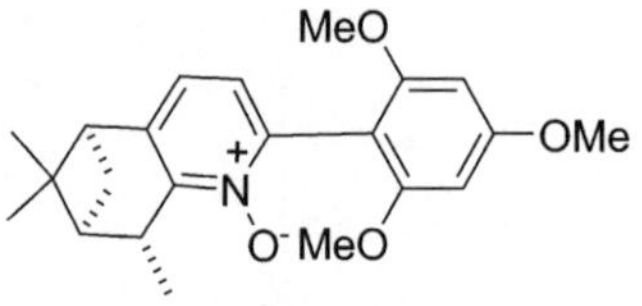

(6.104)

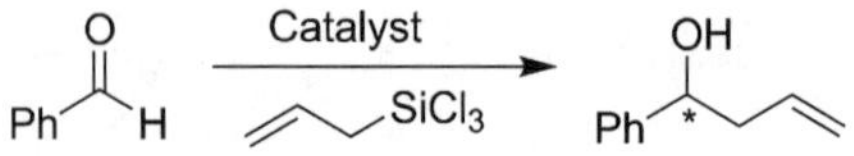

Catalyst	Yield	ee
10 mol% **(6.101)**, -78°C	85%	(R) 88%
0.1 mol% **(6.102)**, -45°C	95%	(S) 82%
7 mol% **(6.103)**, -60°C	72%	(S) 98%
10 mol% **(6.104)**, -20°C	95%	(S) 96%

tert-butoxide (0.3 equiv.) in THF at 50°C. The catalyst prepared in this way was effective for the hydrophosphonylation of aldehydes **(6.53)** using dimethyl phosphite **(6.108)** to give α-hydroxyphosphonates **(6.109)**. As well as La-Li-BINOL complexes, Shibasaki has shown that Al-Li-BINOL complexes catalyse hydrophosphonylation of aldehydes with good yields and enantioselectivities [82] and some of the highest ees to date have been obtained in the phosphonylation of a range of aldehydes by Katsuki and coworkers using aluminium(salalen) complex **(6.110)** in combination with dimethyl phosphite **(6.108)**. [83]

6.5   Nucleophilic Additions to Imines

Nucleophilic additions to imines are generally more challenging than additions to carbonyls, owing to the lower electrophilicity of the imine and the greater propensity for α-deprotonation. Furthermore, stereoselective processes may be complicated by the presence of *syn/anti* imine isomers. Nevertheless, the direct, enantioselective synthesis of substituted amines is an attractive goal and, while catalytic asymmetric additions to imines are less developed than the corresponding

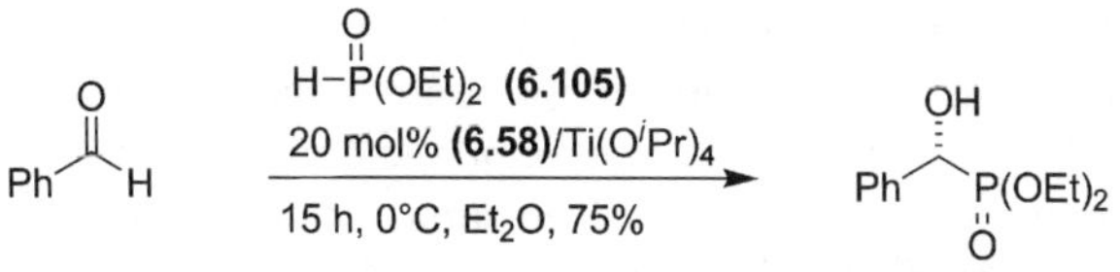

Ph-CHO (6.01) + H–P(OEt)$_2$ (6.105), 20 mol% (6.58)/Ti(O^{i}Pr)$_4$, 15 h, 0°C, Et$_2$O, 75% → Ph–CH(OH)–P(OEt)$_2$ (6.106) 53% ee

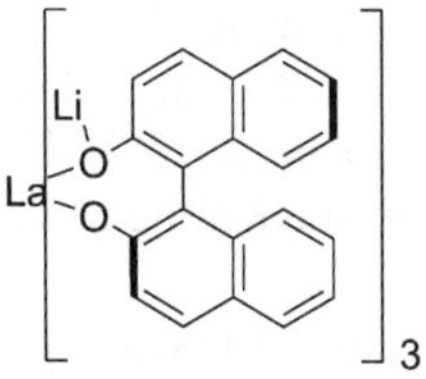

(6.107)

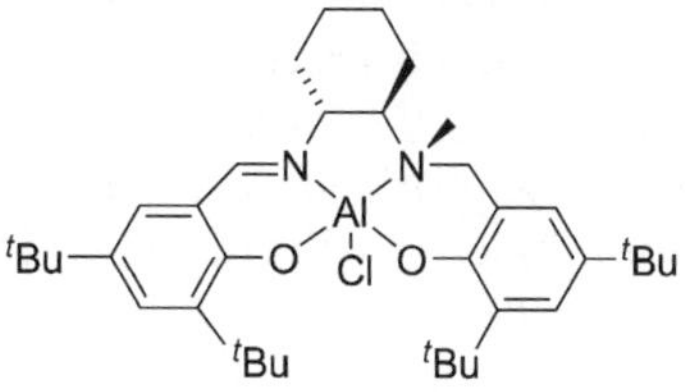

(6.110)

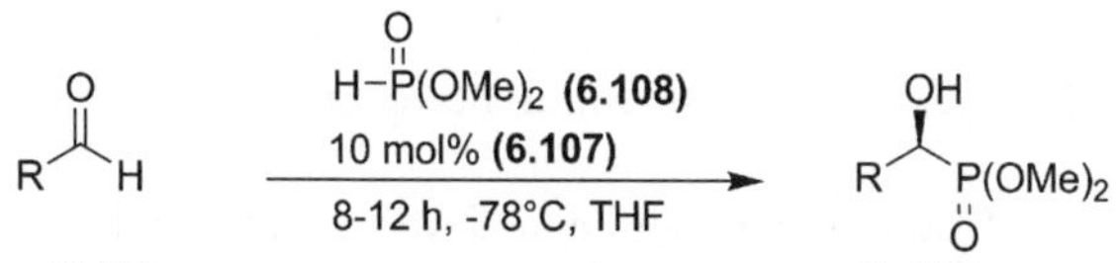

R-CHO (6.53) + H–P(OMe)$_2$ (6.108), 10 mol% (6.107), 8-12 h, -78°C, THF → R–CH(OH)–P(OMe)$_2$ (6.109)

R	yield	ee
Ph-	88%	79%
p-MeC$_6$H$_4$-	87%	93%
PhCH=CH-	90%	84%
C$_5$H$_{11}$-	88%	61%

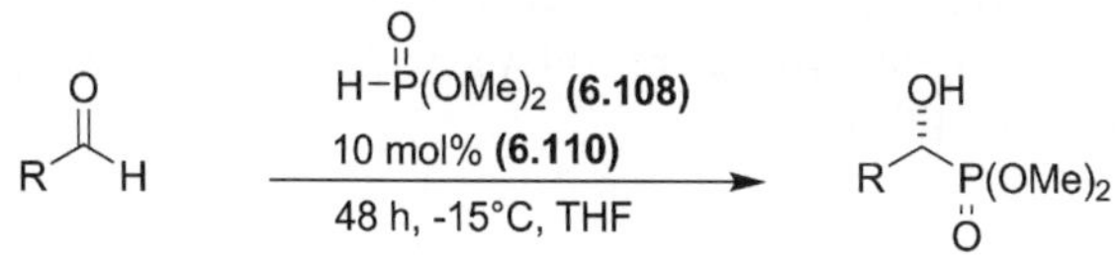

R-CHO (6.53) + H–P(OMe)$_2$ (6.108), 10 mol% (6.110), 48 h, -15°C, THF → R–CH(OH)–P(OMe)$_2$ (6.109)

R	yield	ee
Ph-	87%	90%
p-O$_2$NC$_6$H$_4$-	95%	94%
PhCH=CH-	77%	83%
CH$_3$CH$_2$-	61%	89%

enantioselective additions to carbonyls, a number of catalyst systems have been developed recently that achieve these transformations with good to high ee.[32a, 84] In general, enantioinduction in this process is achieved by activation of the imine within an asymmetric environment using a chiral nonracemic metal-based Lewis acid or organic Brønsted acid.

6.5.1 Alkylation, Allylation and Arylation

The catalytic asymmetric alkylation of imines has been achieved using both organolithium and organozinc reagents as nucleophiles in the presence of chiral nonracemic ligands.[84] While enantioselectivities remain moderate using organolithium species, some high ees have been obtained using dialkylzincs. Several ligands have been investigated, with (6.111–6.113) giving ees above 90% in the dialkylzinc addition to activated imines. For example, the copper(I) complex prepared from binaphthylthiophosphoramide (**6.111**) has been used in the enantioselective addition of diethylzinc to aromatic N-tosylimines such as (**6.114**), giving ees in the range 80–93%.[85] The alkylation of aliphatic imines can also be achieved in high ee. The copper complex prepared from amidophosphine (**6.112**) and Cu(OTf)$_2$ is effective in the diethylzinc addition to aryl and aliphatic N-tosylimines yielding ees ranging from 86 to 96%.[86] A range of aliphatic imines are alkylated with high ee in the presence of the zirconium complex prepared from ligand (**6.113**).[87] In this procedure the isolation of potentially unstable aliphatic imines is avoided by the alkylation of *in situ*-derived imines. Thus, treatment of a mixture of aliphatic aldehyde (**6.53**), o-anisidene (**6.115**), diethylzinc and Zr(^{i}OPr)$_4$/(**6.113**) leads to the isolation of the o-anisidylamine (**6.116**) in moderate yield and excellent ee.

Several catalyst systems have been studied in the asymmetric addition of allylsilanes and allylstannanes to imines.[84b] Amongst these the π-allylpalladium complex (**6.117**) developed by Yamamoto and coworkers has proven to be particularly successful using aromatic imines as substrates.[88] For example, N-benzylimine (**6.118**) is reduced with good ee using this catalyst in the presence of trimethylallylsilane and TBAF.

The asymmetric arylation of imines can also be achieved using chiral metal catalysts.[84b] For example, N-arylsulfonyl aromatic imines such as (**6.119**) are arylated with high ee in the presence of the rhodium catalyst prepared from MOP ligand (**6.120**).[89] The highest ees in this process are achieved with substrates bearing electron-withdrawing groups on nitrogen.

6.5.2 Cyanation of Imines

The addition of cyanide to imines forms the basis of the Strecker reaction, and can be used in the synthesis of amino acid derivatives by hydrolysis of the nitrile to the acid. The asymmetric variant of this reaction can be achieved using both metal-based catalysts and organocatalysts.[32a, 84b,c, 90]

Jacobsen and Sigman have developed an aluminium(salen) complex (**6.124**), which has been shown to provide high enantioselectivities in the addition of HCN to imines, including the conversion of imine (**6.125**) into an aminonitrile, which is isolated as the trifluoroacetamide (**6.126**).[91] Titanium complexes of dipeptide-based Schiff bases have been used by Snapper, Hoveyda and coworkers in the asymmetric cyanation of aromatic aldimines with TMSCN. In this procedure, the use of ^{i}PrOH as an additive leads to the isolation of almost all products in high yield with >99% ee after one recrystallisation.[92] An asymmetric cyanation of relatively wide scope has been developed by Kobayashi and coworkers utilising binuclear zirconium complex (**6.127**) as catalyst in combination with tributyltin cyanide or HCN.[93] Both aromatic and aliphatic aldimines (**6.128**) undergo asymmetric cyanation with good ee using this process.

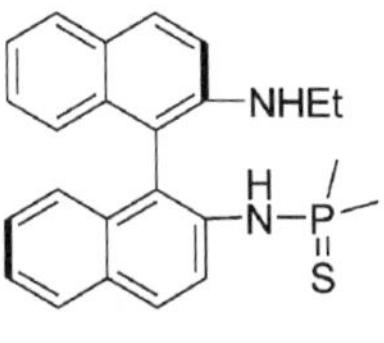

(6.111) 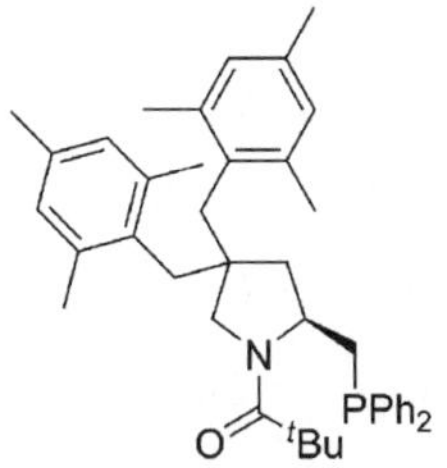　　　　**(6.112)** 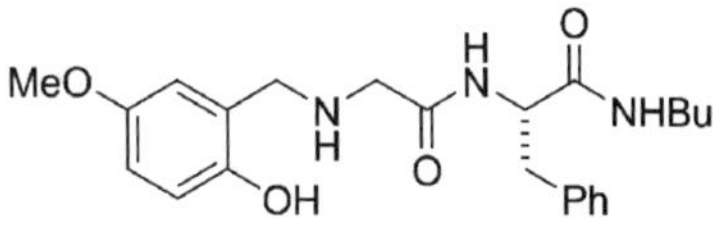　　　　**(6.113)**

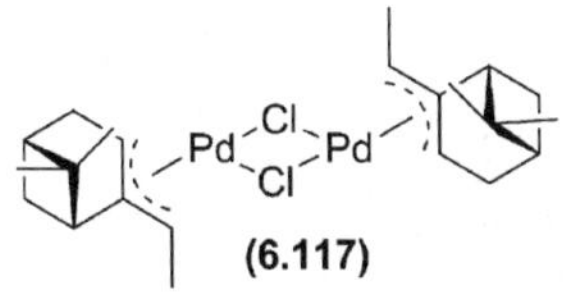

(6.117)

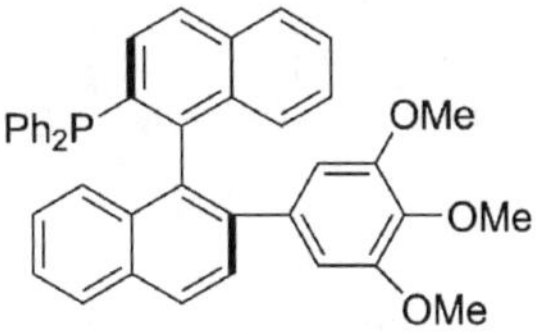

(6.120)

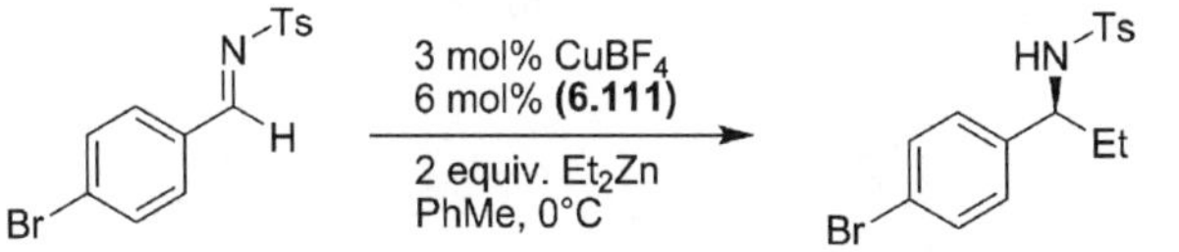

3 mol% CuBF$_4$
6 mol% **(6.111)**

2 equiv. Et$_2$Zn
PhMe, 0°C

(6.114)　　　　**(6.121)** 92% ee

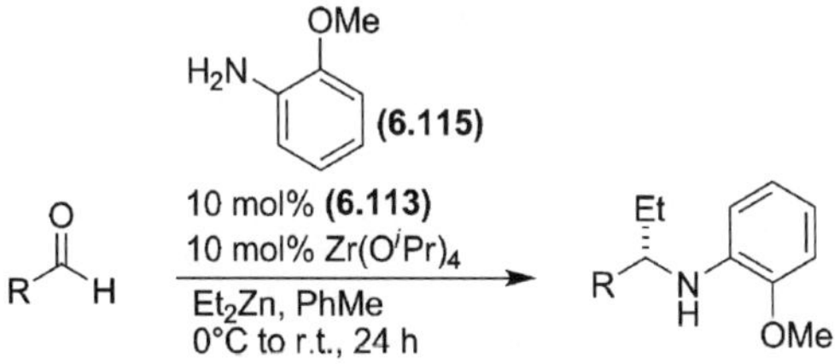

(6.115)

10 mol% **(6.113)**
10 mol% Zr(O^{i}Pr)$_4$

Et$_2$Zn, PhMe
0°C to r.t., 24 h

(6.53)　　　　**(6.116)**

R	Yield	ee
C$_4$H$_9$	69%	97%
HOCH$_2$	48%	>98%
PhCH$_2$CH$_2$	94%	94%
(CH$_3$)$_2$CHCH$_2$	58%	95%

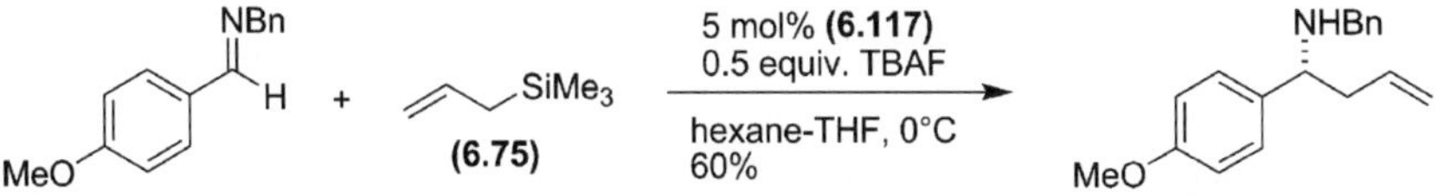

(6.75)

5 mol% **(6.117)**
0.5 equiv. TBAF

hexane-THF, 0°C
60%

(6.118)　　　　**(6.122)** 84% ee

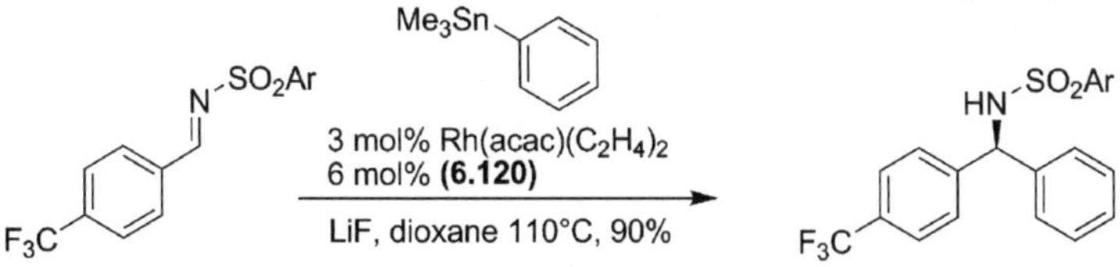

Me$_3$Sn

3 mol% Rh(acac)(C$_2$H$_4$)$_2$
6 mol% **(6.120)**

LiF, dioxane 110°C, 90%

(6.119)　　　　**(6.123)** 96% ee
Ar = *p*-C$_6$H$_4$NO$_2$

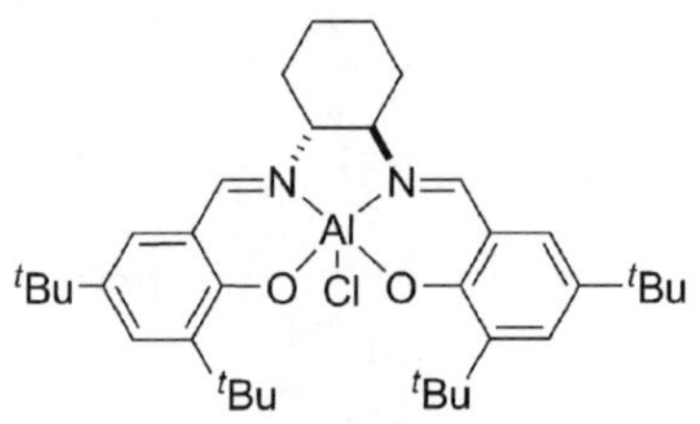

(6.124)

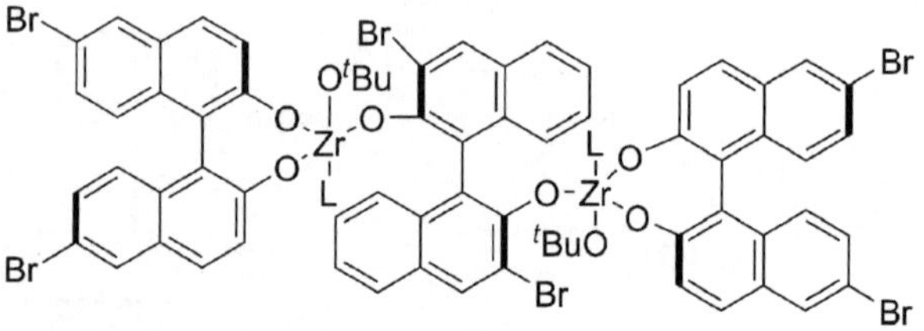

L = N-methylimidazole   (6.127)

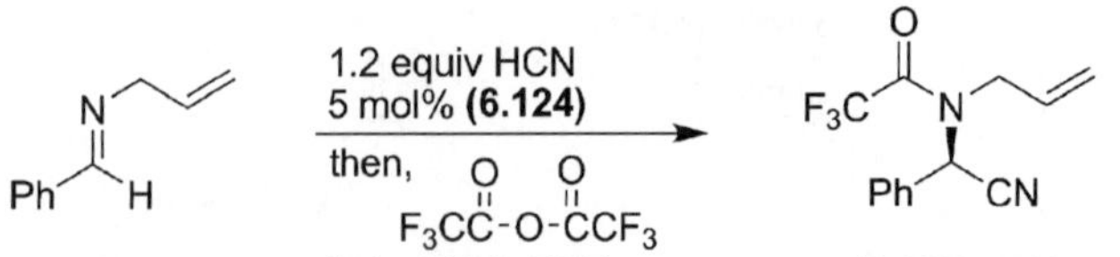

(6.125)   1.2 equiv HCN
5 mol% (6.124)

then,
$F_3CC\text{-}O\text{-}CCF_3$
18 h, -70°C, PhMe
91%

(6.126) 95% ee

(6.128)

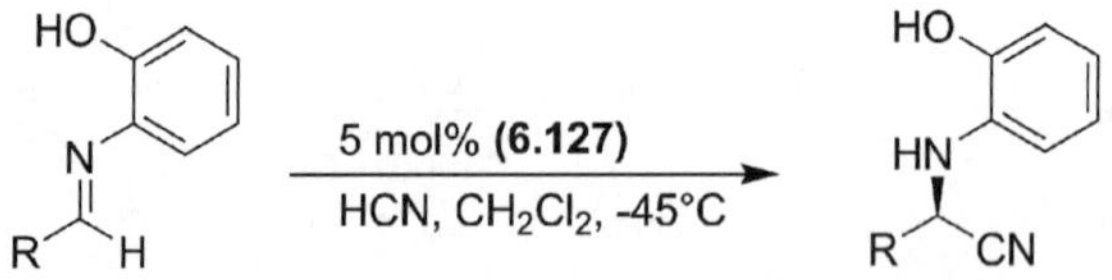

5 mol% (6.127)

HCN, CH$_2$Cl$_2$, -45°C

(6.129)

R	Yield	ee
Ph	80%	86%
^{i}Bu	99%	94%
c-C$_6$H$_{11}$	95%	94%

The bifunctional catalysts developed by Shibasaki and coworkers effective in the asymmetric cyanation of aldehydes and ketones (see Section 6.2) have been applied to good effect in the cyanation of imines.[90c] For instance, aluminium BINOL (6.65) catalyses the cyanation of aromatic and α,β-unsaturated N-fluorenylaldimines using TMSCN in good ee,[94] while gadolinium complexes of the glucose-derived ligand (6.71) and derivatives have been used in the enantioselective cyanation of ketimines.[95]

The asymmetric cyanation of imines can also be mediated by organocatalysts.[90d] The group of Jacobsen has developed a metal-free cyanation utilising thiourea (6.130)[96] and this catalyst has recently been used in an asymmetric three component Strecker reaction of an aldehyde, amine and acetyl cyanide.[97] This catalyst likely functions as a Brønsted acid via hydrogen bonding between the imine nitrogen and protons of the urea functional group. The sterically congested BINOL phosphate (**6.131**), used as a Brønsted acid catalyst in the transfer hydrogenation of imines (see Section 3.2), is also effective in the asymmetric Strecker reaction of aromatic imines. Use of this catalyst in combination with HCN yields the corresponding aminonitrile with 85–99% ee.[98] Corey and coworkers have investigated the C_2-symmetric guanidine (**6.132**) as an enantioselective organocatalyst in the Strecker reaction. While the enantioselectivities are generally not quite as high as those achieved using catalysts (**6.130**) and (**6.131**), both aromatic and aliphatic aldimines such as (**6.133**) can be used as substrates.[99] The mechanism of catalysis again centres on hydrogen bonding of the catalyst with the nitrogen atom of the aldimine. Potassium cyanide can be used in the asymmetric Strecker under phase transfer conditions. Use of quaternary ammonium salts, such as (**6.134**) as catalysts, results in the cyanation of a range of aliphatic *N*-arylsulfonylaldimines (**6.135**) with high ee using aqueous KCN.[100]

6.5.3 Hydrophosphonylation of Imines

Some of the metal-based catalysts used in the asymmetric hydrophosphonylation of aldehydes (see Section 6.4) can also be applied to the phosphonylation of imines. For instance, Shibasaki's heterobimetallic BINOL complexes work well for the catalytic asymmetric hydrophosphonylation of imines.[101] In this case lanthanum-potassium-BINOL complexes (**6.138**) have been found to provide the highest enantioselectivities for the hydrophosphonylation of acyclic imines (**6.139**). The hydrophosphonylation of cyclic imines using heterobimetallic lanthanoid complexes has been reported. Ytterbium and samarium complexes in combination with cyclic phosphites have shown the best results in the cases investigated so far.[102] For example, 3-thiazoline (**6.140**) is converted into the phosphonate (**6.141**) with 99% ee using ytterbium complex (**6.142**) and dimethyl phosphite (**6.108**). The aluminium(salalen) complex (**6.110**) developed by Katsuki and coworkers also functions as an effective catalyst for the hydrophosphonylation of both aromatic and aliphatic aldimines providing the resulting α-aminophosphonate with 81–91% ee.[83]

A number of Brønsted acidic organocatalysts have been applied to the asymmetric hydrophosphonylation of aldimines. Thiourea catalysts related to (**6.130**) catalyse the asymmetric hydrophosphonylation of a range of aliphatic and aromatic aldimines with high ee[103] and BINOL-derived phosphoric acid derivatives similar in structure to (**6.131**) are effective catalysts in the asymmetric phosphonylation of cinnamaldehyde-derived aldimines.[104] Asymmetric hydrophosphonylation of aromatic aldimines can also be achieved with high ee using cheap, commercially

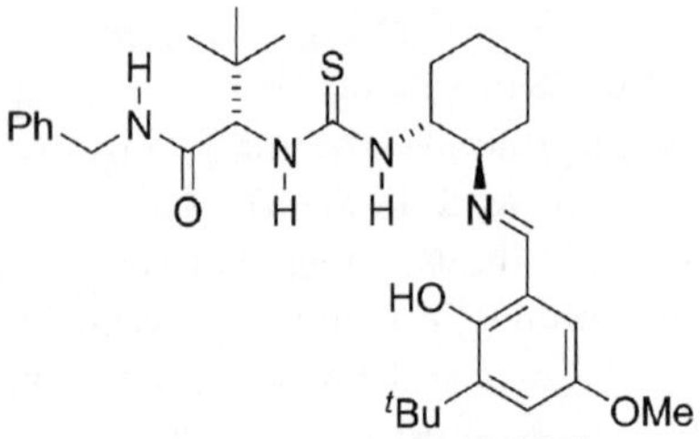

(6.130)

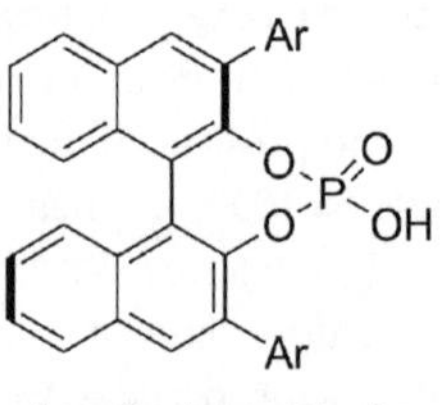

(6.131)

Ar = 9-phenanthryl

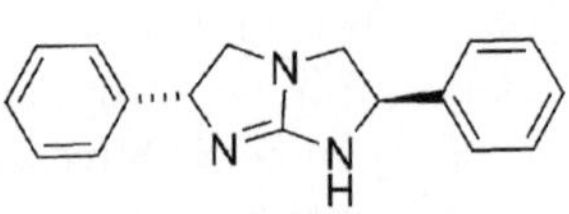

(6.132)

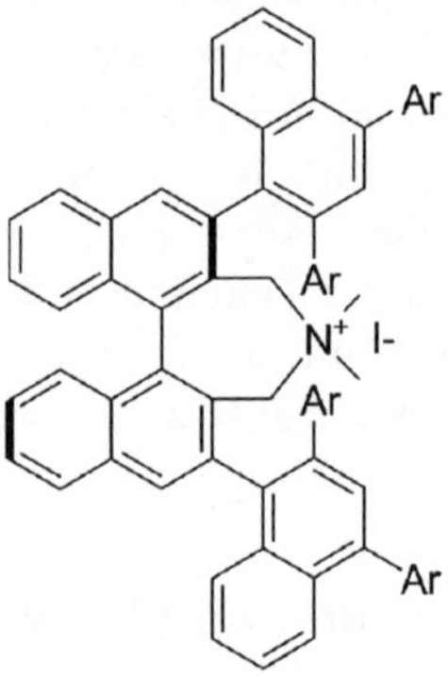

Ar = p-CF$_3$C$_6$H$_4$

(6.134)

(6.125)

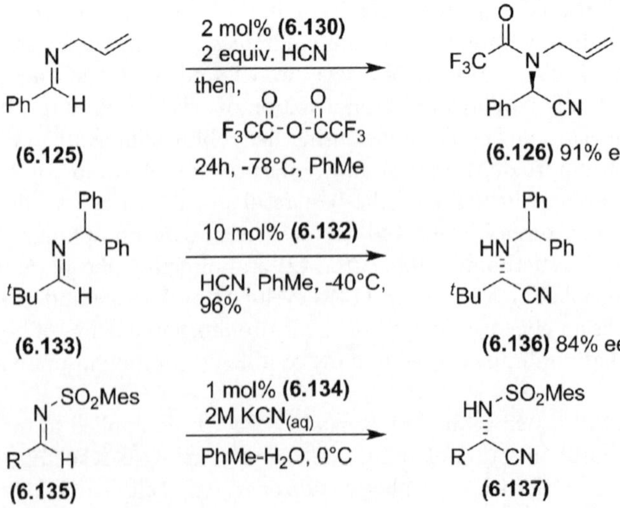

2 mol% **(6.130)**
2 equiv. HCN

then,

$$F_3CC\text{-}O\text{-}CCF_3$$

24h, -78°C, PhMe

(6.126) 91% ee

(6.133)

10 mol% **(6.132)**

HCN, PhMe, -40°C,
96%

(6.136) 84% ee

(6.135)

1 mol% **(6.134)**
2M KCN$_{(aq)}$

PhMe-H$_2$O, 0°C

(6.137)

R	Yield	ee
Ph(CH$_2$)$_2$	81%	90%
^{t}Bu	94%	94%
^{i}Pr	85%	93%

available quinine (**6.143**) as a catalyst. In this process it is postulated that the imine is activated by hydrogen bonding of the nitrogen atom with the hydroxyl group of quinine.[105]

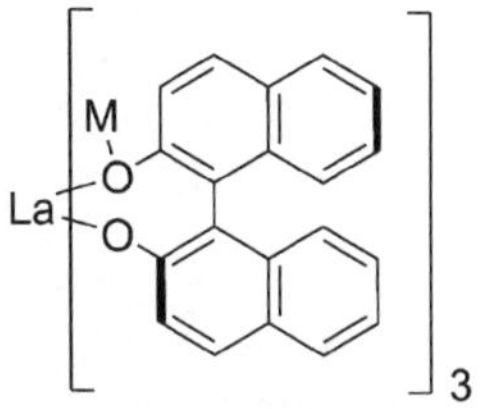

M = K (**6.138**)
M = Yb (**6.142**)

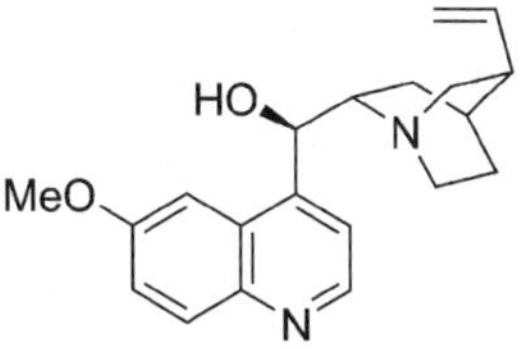

(**6.143**)

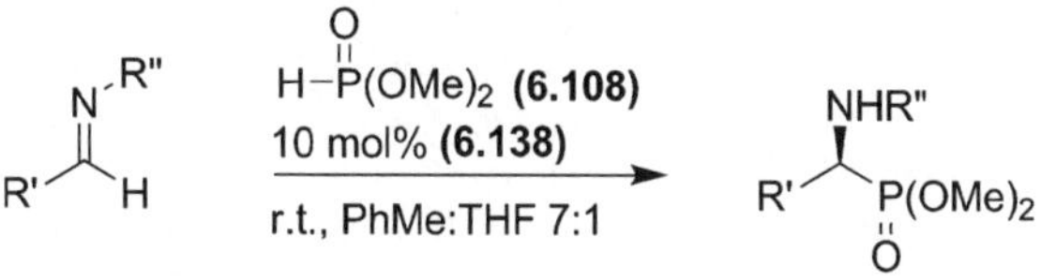

(**6.139**)

(**6.144**) up to 96% ee

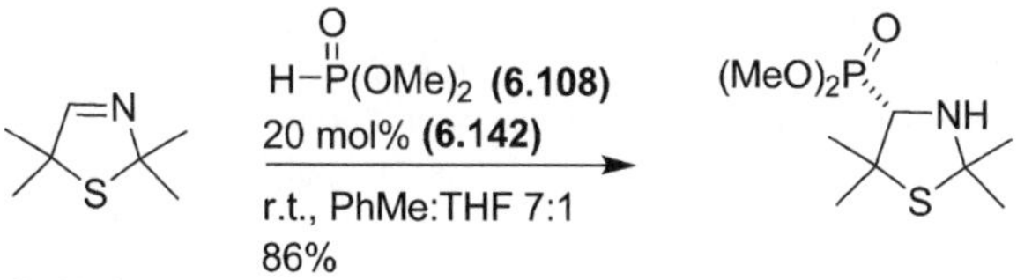

(**6.140**)

(**6.141**) 98% ee

References

1. N. Oguni and T. Omi, *Tetrahedron Lett.*, **1984**, *25*, 2823.
2. For reviews see: (a) R. Noyori and M. Kitamura, *Angew. Chem., Int. Ed. Engl.*, **1991**, *30*, 49. (b) K. Soai and S. Niwa, *Chem. Rev.*, **1992**, *92*, 833. (c) K. Soai and T. Shibata, in *Comprehensive Asymmetric Catalysis*, Vol. 2, ed. E. N. Jacobsen, A. Pfaltz and H. Yamamoto, Springer-Verlag, Berlin, **1999**, 911. (d) L. Pu and H.-B. Yu, *Chem Rev.*, **2001**, *101*, 757. (e) H. J. Zhu, J. X. Jiang, J. Ren, Y. M. Yan and C. U. Pittman, Jr., *Curr. Org. Syn.*, **2005**, *2*, 547. (e) D. J. Ramón and M. Yus, *Chem. Rev.*, **2006**, *106*, 2126.
3. (a) M. Kitamura, S. Suga, K. Kawai, and R. Noyori, *J. Am. Chem. Soc.*, **1986**, *108*, 6071. (b) R. Noyori, S. Suga, K. Kawai, M. Kitamura, N. Oguni, M. Hayashi, T. Kaneko and Y. Matsuda, *J. Organomet. Chem.*, **1990**, *382*, 19.

4. K. Soai, S. Yokoyama and T. Hayasaka, *J. Org. Chem.*, **1991**, *56*, 4264.
5. K. Soai, A. Ookawa, T. Kaba and K. Ogawa, *J. Am. Chem. Soc.*, **1987**, *109*, 7111.
6. D.-H. Ko, K. H. Kim and D.-C. Ha, *Org. Lett.*, **2002**, *4*, 3759.
7. C. Bolm, K. Muñiz-Fernández, A. Seger, G. Raabe and K. Günther, *J. Org. Chem.* **1998**, *63*, 7860.
8. For a detailed discussion of the various possibilities of bicyclic transition states, see: (a) R. Noyori, *Asymmetric Catalysis in Organic Synthesis*, John Wiley & Sons Inc., New York, **1994**, Chapter 5. (b) M. Yamakawa and R. Noyori, *J. Am. Chem. Soc.*, **1995**, *117*, 6327.
9. M. Kitamura, S. Okada, S. Suga and R. Noyori, *J. Am. Chem. Soc.*, **1989**, *111*, 4028.
10. (a) S. Mason, *Chem. Soc., Rev.*, **1988**, *17*, 347. (b) M. Avalos, R. Babiano, P. Cintas, J. L. Jiménez and J. C. Palacios, *Tetrahedron: Asymmetry*, **1997**, *8*, 2997. (c) K. Mikami and M. Yamanaka, *Chem. Rev.*, **2003**, *103*, 3369.
11. T. Harada and K. Kanda, *Org. Lett.*, **2006**, *8*, 3817.
12. B. Schmidt and D. Seebach, *Angew. Chem. Int. Ed., Eng.* **1991**, *30*, 99.
13. D. Seebach, L. Behrendt and D. Felix, *Angew. Chem. Int. Ed. Engl.*, **1991**, *30*, 1008.
14. (a) R. Ostwald, P.-V. Chavant, H. Stadtmüller and P. Knochel, *J. Org. Chem.*, **1994**, *59*, 4143. (b) C. Eisenberg and P. Knochel, *J. Org. Chem.*, **1994**, *59*, 3760. (c) H. Lütjens, S. Nowotny and P. Knochel, *Tetrahedron: Asymmetry*, **1995**, *6*, 2675.
15. C. Bolm, N. Hermanns, J. P. Hildebrand and K. Muniz, *Angew. Chem. Int. Ed.*, **2000**, *39*, 3465.
16. C. Bolm, M. Kesselgruber, N. Hermanns, J. P. Hildebrand and G. Raabe, *Angew. Chem. Int. Ed.*, **2001**, *40*, 1488.
17. (a) P. I. Dosa and G. C. Fu, *J. Am. Chem. Soc.*, **1998**, *120*, 445. (b) A.-S. Huang and L. Pu, *J. Org. Chem.*, **1999**, *64*, 4222.
18. (a) L. Pu, D. Moore, R.-G. Xie and L. Pu, *Org. Lett.*, **2002**, *4*, 4143. (b) G. Lu, X. Li, W. L. Chan and A. S. C. Chan, *J. Chem. Soc., Chem. Commun.*, **2002**, 172.
19. N. K. Anand and E. M. Carreira, *J. Am. Chem. Soc.*, **2001**, *123*, 9687.
20. (a) W. Oppolzer and R. N. Radinov, *Helv. Chim. Acta*, **1992**, *75*, 10 (b) W. Oppolzer, R. N. Radinov and E. El-Sayed, *J. Org. Chem.*, **2001**, *66*, 4766.
21. P. Wipf and S. Ribe, *J. Org. Chem.*, **1998**, *63*, 6454.
22. S.-L. Tseng and T.-K. Yang, *Tetrahedron Lett.*, **2005**, *16*, 773.
23. S. Dahmen and S. Bräse, *Org. Lett.*, **2001**, *3*, 4119.
24. C. García and V. S. Martín, *Curr. Org. Chem.*, **2006**, *10*, 1849.
25. (a) O. Prieto, D. J. Ramón and M. Yus, *Tetrahedron: Asymmetry*, **2003**, *14*, 1955. (b) C. García and P. J. Walsh, *Org. Lett.*, **2003**, *5*, 3641.
26. P. I. Dosa and G. C. Fu, *J. Am. Chem. Soc.*, **1998**, *120*, 445.
27. (a) D. J. Ramón and M. Yus, *Tetrahedron Lett.*, **1998**, *39*, 1239. (b) D. J. Ramón and M. Yus, *Tetrahedron*, **1998**, *54*, 5651. (c) C. García, L. K. LaRochelle and P. J. Walsh, *J. Am. Chem. Soc.*, **2002**, *124*, 10970. (d) S.-J. Jeon and P. J. Walsh, *J. Am. Chem. Soc.*, **2003**, *125*, 9544.
28. G. Lu, X. Li, X. Jia, W. L. Chan and A. S. C. Chan, *Angew. Chem. Int. Ed.*, **2003**, *42*, 5057.
29. Y. F. Zhou, R. Wang, Z. Q. Zu, W. J. Yan, L. Liu, Y. F. Kang and Z. J. Han, *Org. Lett.* **2004**, *6*, 4147.
30. B. Saito and T. Katsuki, *Synlett*, **2004**, 1557.

31. For reviews covering the enzyme-mediated cyanation of aldehydes see: (a) F. Effenberger, *Angew. Chem. Int. Ed. Engl.*, **1994**, *33*, 1555. (b) R. J. H. Gregory, *Chem. Rev.*, **1999**, *99*, 3649. (c) M. North, *Tetrahedron: Asymmetry*, **2003**, *14*, 147. (d) J. Sukumaran and U. Hanefeld, *Chem. Soc. Rev.*, **2005**, *34*, 530.

32. For reviews on chemically catalysed asymmetric cyanation see: (a) A. Mori and S. Inoue, in *Comprehensive Asymmetric Catalysis, Vol. 2*, ed. E. N. Jacobsen, A. Pfaltz and H. Yamamoto, Springer-Verlag, Berlin, **1999**, 983. (b) J. -M. Brunel and I. P. Holmes, *Angew. Chem. Int. Ed.*, **2004**, *43*, 2752. (c) A. Berkessel and H. Gröger, *Asymmetric Organocatalysis*, Wiley-VCH, Weinheim, **2005**, 130.

33. (a) M. Hayashi, T. Matsuda and N. Oguni, *J. Chem. Soc., Perkin Trans, 1*, **1992**, 3135. (b) D. Callant, D. Stanssens and J. G. de Vries, *Tetrahedron: Asymmetry*, **1993**, *4*, 185. (c) M. Hayashi, Y. Miyamoto, T. Inoue and N. Oguni, *J. Org. Chem*, **1993**, *58*, 1515. (d) W. Pan, X. Feng, L. Gong, W. Hu, Z. Li, A. Mi, Y. Jiang, *Synlett*, **1996**, 337. (e) Y. Belokon, M. Flego, N. Ikonnikov, M. Moscalenko, M. North, C. Orizu, V. Tararov and M. Tasinazzo, *J. Chem. Soc., Perkin Trans.*, *1*, **1997**, 1293. (f) C.-D. Hwang, D.-R. Hwanga and B.-J. Uang, *J. Org. Chem.* **1998**, *63*, 6762. (g) J.-S. You, H.-M. Gau and M. C. K. Choi, *J. Chem. Soc., Chem. Commun.* **2000**, 1963. (h) C.-W. Chang, C.-T. Yang, C.-D. Hwang and B. J. Uang, *J. Chem. Soc., Chem. Commun.*, **2002**, 54.

34. E. J. Corey and Z. Wang, *Tetrahedron Lett.*, **1993**, *34*, 4001.

35. (a) Y. Hamashima, D. Sawada, M. Kanai and M. Shibasaki, *J. Am. Chem. Soc.*, **1999**, *121*, 2641. (b) J. Casas, C. Nájera, J. Sansano and J. M. Saá, *Org. Lett.*, **2002**, *4*, 2589.

36. W.-B. Yang and J.-M. Fang, *J.Org. Chem.*, **1998**, *63*, 1356.

37. (a) A. Abiko and G.-Q. Wang, *J. Org. Chem.*, **1996**, *61*, 2264. (b) A. Abiko and G.-Q. Wang, *Tetrahedron*, **1998**, *54*, 11405.

38. J. Oku and S. Inoue, *J. Chem. Soc., Chem. Commun.*, **1981**, 229.

39. M. North, *Synlett*, **1993**, 807.

40. K. Tanaka, A. Mori and S. Inoue, *J. Org. Chem.*, **1990**, *55*, 181.

41. L. Xie, W. Hua, A. S. C. Chan and Y.-C. Leung, *Tetrahedron: Asymmetry*, **1999**, *10*, 4715.

42. Y. Hamashima, M. Kanai and M. Shibasaki, *J. Am. Chem. Soc.*, **2000**, *122*, 7412.

43. K. Yabu, S. Masumoto, S. Yamasaki, Y. Hamashima, M. Kanai, W. Du, D. P. Curran and M. Shibasaki, *J. Am. Chem. Soc.*, **2001**, *123*, 9908.

44. X. Liu, B. Qin, X. Zhou, B. He and X. Feng, *J. Am. Chem. Soc.*, **2005**, *127*, 12224.

45. S.-K. Tian and L. Deng, *J. Am. Chem. Soc.*, **2001**, *123*, 6195.

46. D. E. Fuerst and E. N. Jacobsen, *J. Am. Chem. Soc.*, **2005**, *127*, 8964.

47. (a) A. Yanigasawa, in *Comprehensive Asymmetric Catalysis*, Vol. 2, ed. E. N. Jacobsen, A. Pfaltz and H. Yamamoto, Springer-Verlag, Berlin, **1999**, 965. (b) S. E. Denmark and J. Fu, *Chem. Rev.*, **2003**, *103*, 2763.

48. K. Furuta, M. Mouri and H. Yamamoto, *Synlett*, **1991**, 561.

49. K. Ishihara, M. Mouri, Q. Gao, T. Maruyama, K. Furuta and H. Yamamoto, *J. Am. Chem. Soc.*, **1993**, *115*, 11490.

50. J. A. Marshall and Y. Tang, *Synlett*, **1992**, 653.

51. A. L. Costa, M.G. Piazza, E. Tagliavini, C. Trombini and A. Umani-Ronchi, *J. Am. Chem. Soc.*, **1993**, *115*, 7001.

52. G. E. Keck, K.H. Tarbet and L. S. Geraci, *J. Am. Chem. Soc.*, **1993**, *115*, 8467.

53. G. E. Keck and L.S. Geraci, *Tetrahedron Lett.*, **1993**, *34*, 7827.
54. H. Hanawa, T. Hashimoto and K. Maruoka, *J. Am. Chem. Soc.*, **2003**, *125*, 1708.
55. G. E. Keck, D. Krishnamurthy and M. C. Grier, *J. Org. Chem.*, **1993**, *58*, 6543.
56. J. W. Faller, D. W. I. Sams and X. Liu, *J. Am. Chem. Soc.*, **1996**, *118*, 1217.
57. (a) C.-M. Yu, H.-S. Choi, W.-H. Jung and S.-S. Lee, *Tetrahedron Lett.*, **1996**, *37*, 7095. (b) C.-M. Yu, H.-S. Choi, W.-H. Jung, H.-J. Kim and J. Shin, *J. Chem. Soc., Chem. Commun.*, **1997**, 761. (c) C.-M. Yu, H.-S. Choi, S.-K. Yoon and W.-H. Jung, *Synlett*, **1997**, 889.
58. A. Yanagisawa, H. Nakashima, A. Ishiba and H. Yamamoto, *J. Am. Chem. Soc.*, **1996**, *118*, 4723.
59. A. Yanagisawa, A. Ishiba, H. Nakashima, H. Yamamoto, *Synlett*, **1996**, 88.
60. P. Bedeschi, S. Casolari, A. L. Costa, E. Tagliavini and A. Umani-Ronchi, *Tetrahedron Lett.*, **1995**, *36*, 7897. (b) S. Casolari, P. G. Cozzi, P. A. Orioli, E. Tagliavini and A. Umani-Ronchi, *J. Chem. Soc., Chem. Commun.*, **1997**, 2123. (c) M. Kurosu, and M. Lorca, *Tetrahedron Lett.*, **2002**, *43*, 1765. (d) A. Yanagisawa, H. Nakashima, A. Ishiba and H. Yamamoto, *Synlett*, **1997**, 88.
61. (a) P. G. Cozzi, P. Orioli, E. Tagliavini and A. Umani-Ronchi, *Tetrahedron Lett.*, **1997**, *38*, 145. (b) Y. Imai, W. Zhang, T. Kida, Y. Nakatsuji and I. Ikeda, *J. Org. Chem.*, **2000**, *65*, 3326.
62. Y.-C. Teo, K.-T. Tan and T.-P. Loh, *J. Chem. Soc., Chem. Commun.*, **2005**, 1318.
63. J. Lu, S.-J. Ji, Y.-C. Teo and T.-P. Loh, *Org. Lett.*, **2005**, *7*, 159.
64. (a) D. R. Gauthier, Jr and E. M. Carreira, *Angew. Chem., Int. Ed. Engl.*, **1996**, *35*, 2363. (b) J. W. Bode, D. R. Gauthier Jr and E. M. Carreira, *J. Chem. Soc., Chem. Commun.*, **2001**, 2560.
65. (a) M. Bandini, P. G. Cozzi, P. Melchiorre and A. Umani-Ronchi, *Angew. Chem. Int. Ed.*, **1999**, *38*, 3357. (b) A. Berkessel, D. Menche, C. A. Sklorz, M. Schröder and I. Paterson, *Angew. Chem. Int. Ed.*, **2003**, *42*, 1032.
66. M. Inoue, T. Suzuki and M. Nakada, *J. Am. Chem. Soc.*, **2003**, *125*, 1140.
67. S. E. Denmark, D. M. Coe, N. E. Pratt and B. D. Griedel, *J. Org. Chem.* **1994**, *59*, 6161.
68. For a review of these studies see: S. E. Denmark and J. Fu, *J. Chem. Soc., Chem. Commun.*, **2003**, 167.
69. S. E. Denmark, J. Fu, D. M. Coe, X. Su, N. E. Pratt and B. D. Griedel, *J. Org. Chem.*, **2006**, *71*, 1513.
70. K. Iseki, Y. Kuroki, M. Takahashi, S. Kishimoto and Y. Kobayashi, *Tetrahedron*, **1997**, *53*, 3513.
71. S. E. Denmark and J. Fu, *J. Am. Chem. Soc.*, **2000**, *122*, 12021.
72. S. E. Denmark and J. Fu, *J. Am. Chem. Soc.*, **2001**, *123*, 9488.
73. (a) K. Iseki, S. Mizuno, Y. Kuroki and Y. Kobayashi, *Tetrahedron Lett.*, **1998**, *39*, 2767. (b) K. Iseki, S. Mizuno, Y. Kuroki and Y. Kobayashi, *Tetrahedron*, **1999**, *55*, 977
74. (a) S. Kobayashi and K. Nishio, *Tetrahedron Lett.*, **1995**, *36*, 6729. (b) R. M. Angell, A. G. M. Barret, D. C. Braddock, S. Swallow and B. D. Vickery, *J. Chem. Soc., Chem. Commun.*, **1997**, 919.
75. I. Chataigner, U. Piarulli and C. Gennari, *Tetrahedron Lett.*, **1999**, *40*, 3633.
76. (a) M. Nakajima, M. Saito, M. Shiro and S. I. Hashimoto, *J. Am. Chem. Soc.*, **1998**, *120*, 6419. (b) T. Shimada, A. Kina, S. Ikeda and T. Hayashi, *Org. Lett.*, **2002**, *4*, 2799. (c)

T. Shimada, A. Kina and T. Hayashi, *J. Org. Chem.*, **2003**, *68*, 6329. (d) A. v. Malkov, M. Orsini, D. Pernazza, K. W. Muir, V. Langer, P. Meghani and P. Kočovsky, *Org. Lett.*, **2002**, *4*, 1047. (e) A. V. Malkov, M. Bell, F. Castelluzzo and P. Kočovsky, *Org. Lett.*, **2005**, *7*, 3219.

77. For a review on the various enantioselective approaches to α−hydroxyphosphonates see: H. Gröger and B. Hammer, *Chem. Eur. J.*, **2000**, *6*, 943.

78. T. Yokomatsu, T. Yamagishi and S. Shibuya, *Tetrahedron: Asymmetry*, **1993**, *4*, 1779.

79. T. Yokomatsu, T. Yamagishi and S. Shibuya, *Tetrahedron: Asymmetry*, **1993**, *4*, 1783.

80. N. P. Rath and C.D. Spilling, *Tetrahedron Lett.*, **1994**, *35*, 227.

81. H. Sasai, M. Bougauchi, T. Arai and M. Shibasaki, *Tetrahedron Lett.*, **1997**, *38*, 2717.

82. T. Arai, M. Bougauchi, H. Sasai and M. Shibasaki, *J. Org. Chem.*, **1996**, *61*, 2926.

83. (a) B. Saito and T. Katsuki, *Angew. Chem. Int. Ed.*, **2005**, *44*, 4600. (b) B. Saito, H. Egami and T. Katsuki, *J. Am. Chem. Soc.*, **2007**, *129*, 1978.

84. For reviews of catalytic asymmetric additions to imines see: (a) S. E. Denmark and O. J.–C. Nicaise, in *Comprehensive Asymmetric Catalysis*, Vol. 2, ed. E. N. Jacobsen, A. Pfaltz and H. Yamamoto, Springer-Verlag, Berlin, **1999**, 924. (b) T. Vilaivan and W. Bhanthumnavin, *Curr. Org. Chem.*, **2005**, *9*, 1315. (c) G. K. Fristad and A. K. Mathies, *Tetrahedron*, **2007**, *63*, 2541.

85. C.-J. Wang and M. Shi, *J. Org. Chem.* **2003**, *68*, 6229.

86. T. Soeta, K. Nagai, H. Fujihara, M. Kuriyama and K. Tomioka, *J. Org. Chem.* **2003**, *68*, 9723.

87. J. R. Porter, J. F. Traverse, A. H. Hoveyda and M. L. Snapper, *J. Am. Chem. Soc.*, **2001**, *123*, 10409.

88. (a) H. Nakamura, K. Nakamura and Y. Yamamoto, *J. Am. Chem. Soc.*, **1998**, *120*, 4242. (b) K. Nakamura, H. Nakamura and Y. Yamamoto, *J. Org. Chem.* **1999**, *64*, 2614.

89. T. Hayashi and M. Ishigedani, *J. Am. Chem. Soc.*, **2000**, *122*, 976.

90. (a) L. Yet, *Angew. Chem. Int. Ed.*, **2001**, *40*, 875. (b) H. Gröger, *Chem. Rev.*, **2003**, *103*, 2795. (c) M. Kanai, N. Kato, E. Ichikawa and M. Shibasaki, *Pure Appl. Chem.*, **2005**, *77*, 2047. (d) A. Berkessel and H. Gröger, *Asymmetric Organocatalysis*, Wiley-VCH, Weinheim, **2005**, 85.

91. M. S. Sigman and E. N. Jacobsen, *J. Am. Chem. Soc.*, **1998**, *120*, 5315.

92. C. A. Krueger, K. W. Kuntz, C. D. Dzierba, W. G. Wirschun, J. D. Gleason, M. L. Snapper and A. H. Hoveyda, *J. Am. Chem. Soc.*, **1999**, *121*, 4284.

93. H. Ishitani, S. Komiyama, Y. Hasegawa and S. Kobayashi, *J. Am. Chem. Soc.*, **2000**, *122*, 762.

94. M. Takamura, Y. Hamashima, H. Usada, M. Kanai, and M. Shibasaki, *Angew. Chem. Int. Ed.*, **2000**, *112*, 1716.

95. S. Masumoto, H. Usuda, M. Suzuki, M. Kanai and M. Shibasaki, *J. Am. Chem. Soc.*, **2003**, *125*, 5634.

96. (a) M. S. Sigman and E. N. Jacobsen, *J. Am. Chem. Soc.*, **1998**, *120*, 4901. (b) P. Vachal and E. N. Jacobsen, *J. Am. Chem. Soc.*, **2002**, *124*, 10012.

97. S. C. Pan and B. List, *Org. Lett.*, **2007**, *9*, 1149.

98. M. Rueping, E. Sugiono and C. Azap, *Angew. Chem. Int. Ed.*, **2006**, *45*, 2617.

99. E. J. Corey and M. J. Grogan, *Org. Lett.*, **1999**, *1*, 157.

100. T. Ooi, Y. Uematsu and K. Maruoka, *J. Am. Chem. Soc.*, **2006**, *128*, 2548.

101. H. Sasai, S. Arai, Y. Tahara and M. Shibasaki, *J. Org. Chem.*, **1995**, *60*, 6656.
102. (a) H. Gröger, Y. Saida, H. Sasai, K. Yamaguchi, J. Martens and M. Shibasaki, *J. Am. Chem. Soc.*, **1998**, *120*, 3089. (b) I. Schlemminger, Y. Saida, H. Gröger, W. Maison, N. Durot, H. Sasai, M. Shibasaki and J. Martens, *J. Org. Chem.*, **2000**, *65*, 4818.
103. G. D. Joly and E. N. Jacobsen, *J. Am. Chem. Soc.*, **2004**, *126*, 4102.
104. T. Akiyama, H. Morita, J. Itoh and K. Fuchibe, *Org. Lett.*, **2005**, *7*, 2583.
105. D. Pettersen, M. Marcolini, L. Bernardi, F. Fini, R. P. Herrera, V. Sgarzani and A. Ricci, *J. Org. Chem.*, **2006**, *71*, 6269.

Chapter 7
The Aldol and Related Reactions

The standard aldol reaction involves the addition of an enolate to a ketone or an aldehyde. However, there are related processes and this chapter includes subsections on the isocyanide aldol, nitroaldol and Morita–Baylis–Hillmann reaction. In addition there are reactions involving additions of enolates to the C=N group and a large subsection is devoted to a discussion of the catalytic asymmetric Mannich reaction. As well as these mechanistically related processes, the carbonyl-ene reaction is also discussed here. Whilst the mechanism of the carbonyl-ene reaction is different from the aldol reaction, the synthetic result is rather similar, and perhaps fits most comfortably into this chapter.

Significant advances in the development of a catalytic asymmetric variant of the aldol reaction have been developed recently. Both enantiomerically pure Lewis acids and Lewis bases have been applied to the addition of silyl enol ethers to aldehydes and ketones and highly diastereoselective and enantioselective additions have been achieved. Often, the mode of diastereoselectivity can be rationalised from a consideration of the relevant open or closed transition states.

The direct asymmetric aldol reaction has also received much recent attention. High ees have been obtained using lanthanide- or zinc-based bifunctional catalysts bearing both Lewis acidic and Lewis basic sites. The most significant recent advance in this area is the discovery that cheap, readily available organic catalysts such as L-proline are also effective.

Some of the catalyst systems used in the asymmetric aldol reaction are also effective in related reactions. Thus, bifunctional catalysts and L-proline-based organocatalysts have been used to good effect in the nitroaldol reaction and Mannich reaction. The latter process is also effectively catalysed by enantiomerically pure Brønsted acids. Furthermore, much recent progress has been made in the development of a catalytic asymmetric Morita–Baylis–Hillman reaction using Lewis/Brønsted acid catalysts and bifunctional catalysts.

Catalysis in Asymmetric Synthesis 2e   © 2009 Vittorio Caprio and Jonathan M.J. Williams

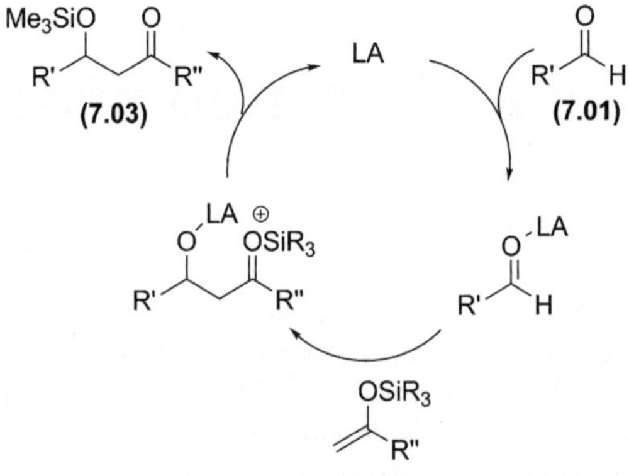

Figure 7.1 General catalytic cycle for aldol reactions of silyl enol ethers. Abbreviation: LA = Lewis acid

7.1 The Aldol Reaction

The aldol reaction and related processes have been of considerable importance in organic synthesis. The control of *syn/anti* diastereoselectivity, enantioselectivity and chemoselectivity has now reached impressive levels. The use of catalysts is a relatively recent addition to the story of the aldol reaction.[1] One of the most common approaches to the development of a catalytic asymmetric aldol reaction is based on the use of enantiomerically pure Lewis acids in the reaction of silyl enol ethers with aldehydes and ketones (the Mukaiyama reaction) and variants of this process have been developed for the synthesis of both *syn* and *anti* aldol adducts. A typical catalytic cycle is represented in Figure 7.1, where aldehyde (**7.01**) coordinates to the catalytic Lewis acid, which encourages addition of the silyl enol ether (**7.02**). Release of the Lewis acid affords the aldol product, often as the silyl ether (**7.03**).

Amongst the first catalytic aldol reactions were the tin triflate-catalysed reactions reported by Kobayashi and coworkers.[2,3] The tin complex of ligand (**7.04**) catalyses the addition of reactive silyl enol ethers such as compound (**7.05**) to give very good enantioselectivity in some cases. This methodology has been used in the preparation of natural products including sphingofungins. The initial asymmetric induction was achieved by coupling aldehyde (**7.07**) and ketene acetal (**7.08**).[4]

Enantiomerically pure boron-based Lewis acids have also been used successfully in catalytic aldol reactions. Corey's catalyst (**7.10a**) provides good enantioselectivity with ketone-derived silyl enol ethers, including compound (**7.11**).[5] Other oxazaborolidine complexes (**7.13**) derived from α,α-disubstituted α-amino acids give particularly high enantioselectivity,[6] especially with the disubstituted ketene

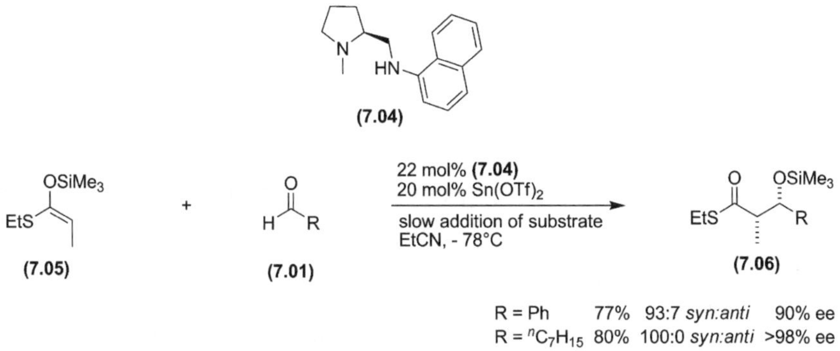

(7.04)

(7.05) + **(7.01)**

22 mol% **(7.04)**
20 mol% Sn(OTf)$_2$

slow addition of substrate
EtCN, - 78°C

(7.06)

R = Ph 77% 93:7 *syn:anti* 90% ee
R = $^nC_7H_{15}$ 80% 100:0 *syn:anti* >98% ee

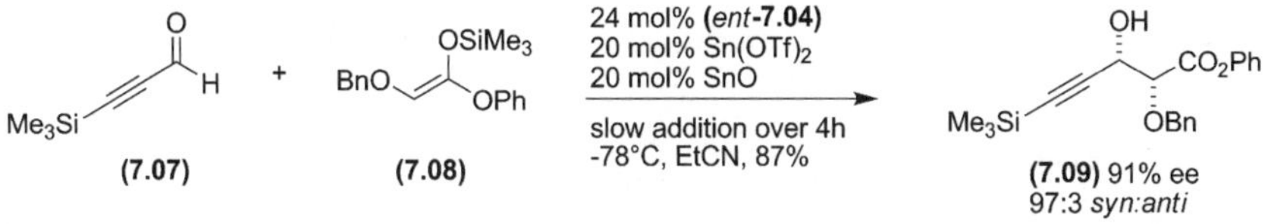

(7.07) + **(7.08)**

24 mol% (*ent*-**7.04**)
20 mol% Sn(OTf)$_2$
20 mol% SnO

slow addition over 4h
-78°C, EtCN, 87%

(7.09) 91% ee
97:3 *syn:anti*

acetal **(7.14)**. One drawback with these catalysts is the relatively high loadings required for good selectivity. The activity of Corey's catalyst can be improved by the incorporation of electron-withdrawing groups on boron. For instance, the aldol reaction of **(7.11)** with benzaldehyde proceeds with 91% ee using only 4 mol% of the trifluoromethylphenyl substituted oxazaborolidine **(7.10b)**.[7]

Yamamoto's CAB catalysts **(7.16)** (see Section 8.1) have also been used in catalytic aldol reactions. This reaction is stereoconvergent as either geometry of silyl enol ether **(7.18)** affords *syn* selectivity in the product **(7.19)** indicating that the reaction proceeds via an open anticlinal transition state with the minimum of steric interactions between the aldehyde substituent and α-substituent of the enol ether, as depicted in Figure 7.2.[8]

Titanium complexes are often encountered in Lewis acid-catalysed reactions. This is certainly true for catalysed aldol reactions. Mikami and Matsukawa demonstrated that titanium/BINOL complexes e.g. complex **(7.20)** afforded high yield and enantioselectivity in the aldol reactions of thioester ketene silylacetals with a variety of aldehydes.[9] In contrast to some of the aldol reactions described above, the stereochemistry of the adducts is dependant on the geometry of the enol ether. Thus, reaction of the (*E*)-enol ether **(7.21)** with aldehyde **(7.22)** yields the *syn*-aldol adduct **(7.23)** predominantly while the (*Z*)-enol ether **(7.24)** results in isolation of the *anti*-adduct **(7.25)** as the major product. The authors invoke a closed silatropic ene transition state (structure **(7.26)** for *syn*-transition state), substantiated by suitable crossover experiments, to explain the diastereoselectivities

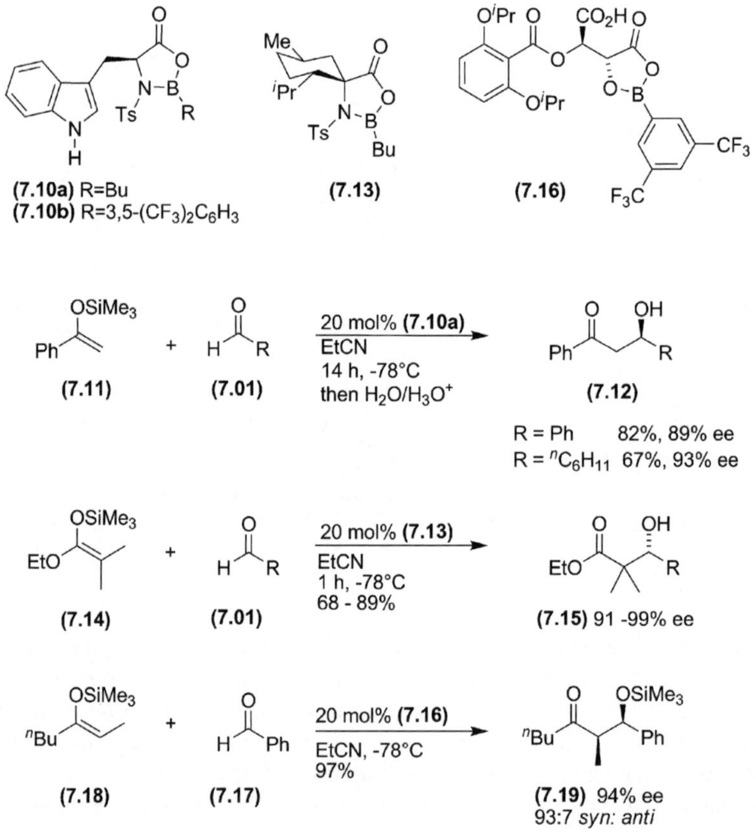

(7.10a) R=Bu
(7.10b) R=3,5-$(CF_3)_2C_6H_3$

(7.13)

(7.16)

(7.11) + **(7.01)**

20 mol% **(7.10a)**
EtCN
14 h, -78°C
then H_2O/H_3O^+

(7.12)

R = Ph   82%, 89% ee
R = $^nC_6H_{11}$ 67%, 93% ee

(7.14) + **(7.01)**

20 mol% **(7.13)**
EtCN
1 h, -78°C
68 - 89%

(7.15) 91 -99% ee

(7.18) + **(7.17)**

20 mol% **(7.16)**
EtCN, -78°C
97%

(7.19) 94% ee
93:7 *syn: anti*

observed. Further improvements to enantioselectivity were made by using improved silyl migrating groups (silacyclobutanes) and using pentafluorophenol as an achiral ligand associated with the titanium.[10] Keck and Krishnamurthy have also used a titanium/BINOL catalyst in aldol reactions to give high enantioselectivity.[11] The aldol reaction and 'normal' ene reaction are clearly related (the ene process is discussed in Section 7.7). Carreira and coworkers have reported a highly selective catalytic system for aldol reactions of silyl ketene acetals. Their catalyst **(7.27)** is derived from an amino-BINOL variant, along with an achiral salicylate ligand.[12,13] Very high enantioselectivities have been achieved using this catalyst for a range of aldehyde substrates **(7.01)** with enol ethers **(7.28)**, often using relatively low catalyst loading. The reaction has also been applied to aldol reactions of the dienolate **(7.29)**.[14] Aldehydes, including substrate **(7.30)** were used to give high enantioselectivities (80–94% ee) in the products **(7.31)**.

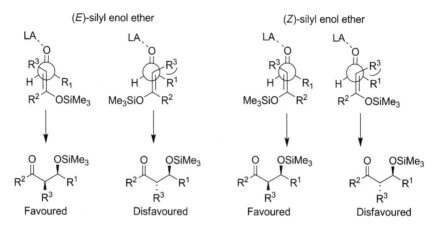

Figure 7.2 Open transition states for the Lewis acid catalysed Mukaiyama aldol reaction. Abbreviation: LA = Lewis acid

The air stable and storable zirconium catalyst, formed from $Zr(O^tBu)_4$, 3,3'-diiodo-1,1'-binaphthalene-2,2'-diol (3,3'-I_2-BINOL), *n*-propanol and water, with the putative dimeric structure (**7.33**) also catalyses *anti*-selective asymmetric aldol reactions.[15] While this process is believed to proceed through an acyclic transition state, as depicted in Figure 7.2, it is postulated that the greatest steric interaction is now between the silyl enol ether substituent R_3 and the bulky Lewis acid resulting in the formation of the *anti*-diastereomer predominantly.

Bis-oxazolines (BOX) have shown much utility as ligands in the asymmetric aldol reaction.[16] Evans has reported that copper(II) complexes of ligand (**7.34**) are very efficient catalysts for aldol reactions using benzyloxyacetaldehyde (**7.35**) as the electrophilic partner. This aldehyde is able to achieve two-point binding to the copper catalyst, and good enantioselectivities are obtained in the copper(II)-catalysed aldol reaction with silyl enol ethers such as (**7.36**).[17] Pyruvate esters such as (**7.38**) can also enjoy the advantages of two-point binding and have proven to be good substrates for the copper-catalysed reaction.[18] Interestingly, while use of the copper complex (**7.34**) results in formation of the *syn*-diastereomer (**7.40**), indicating operation of an open transition state (see Figure 7.2), use of the corresponding tin complex results in formation of the *anti*-adduct (**7.41**) with high ee.[19]

The nickel(II) (BOX) complex (**7.42**) is an effective catalyst in the direct aldol reaction of thiazolidinethione (**7.43**).[20] In this process, the enolate generated using 2,6-lutidine reacts with a range of aromatic and aliphatic aldehydes to give the *syn*-adduct (**7.44**) predominantly.

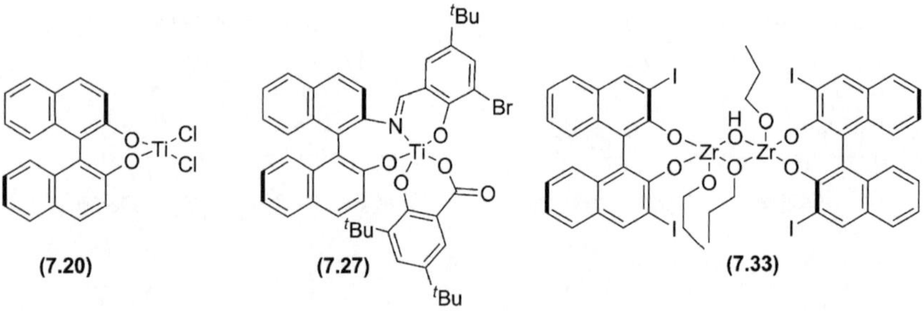

(7.20)   (7.27)   (7.33)

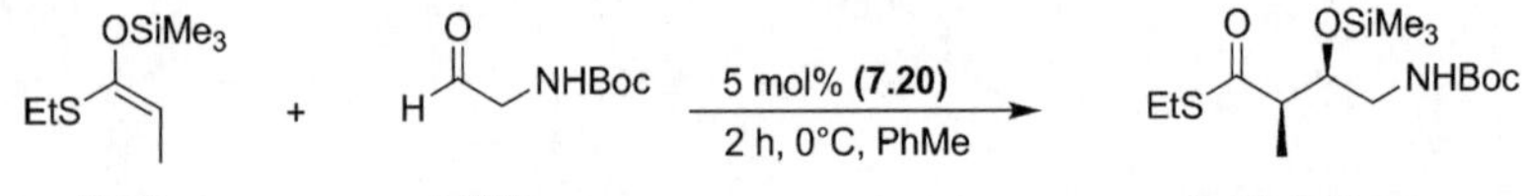

(7.21)   +   (7.22) $\xrightarrow[\text{2 h, 0°C, PhMe}]{\text{5 mol% (7.20)}}$ (7.23) 90% ee
                                                                                                  92:8 *syn:anti*

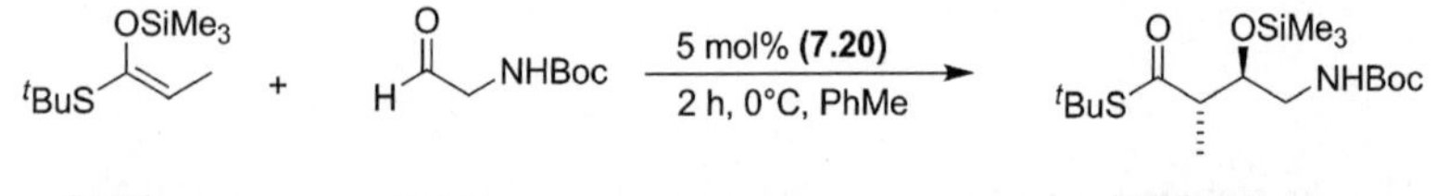

(7.24)   +   (7.22) $\xrightarrow[\text{2 h, 0°C, PhMe}]{\text{5 mol% (7.20)}}$ (7.25) 90% ee
                                                                                                  8:92 *syn:anti*

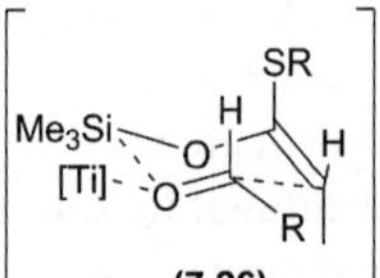

(7.26)

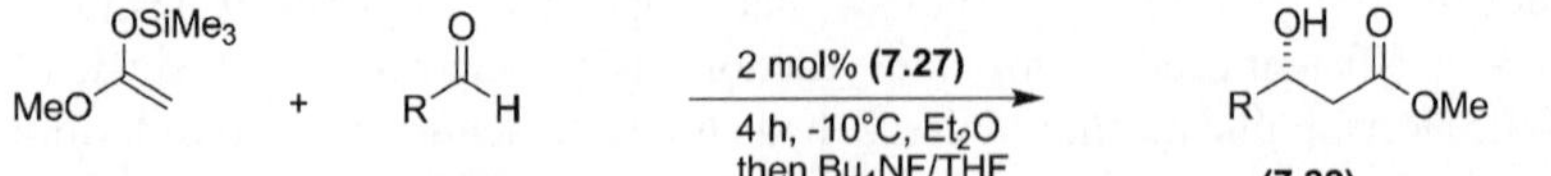

(7.28)   +   (7.01) $\xrightarrow[\substack{\text{4 h, -10°C, Et}_2\text{O}\\ \text{then Bu}_4\text{NF/THF}\\ \text{72-98\%}}]{\text{2 mol% (7.27)}}$ (7.32)

R= MeCH=CH-   97% ee
R = Ph-   96% ee
R = PhCH$_2$CH$_2$- 94% ee

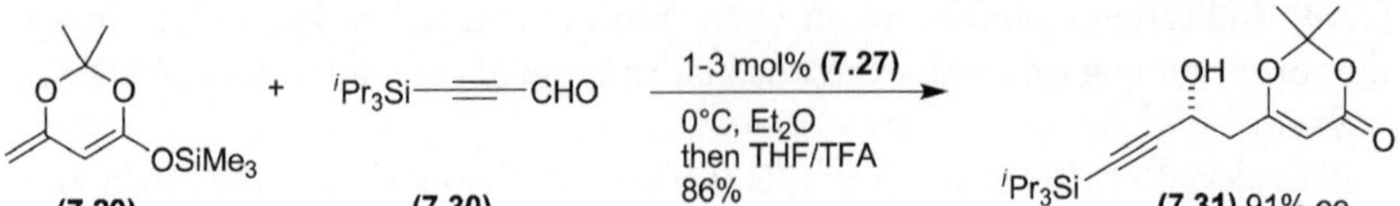

(7.29)   +   (7.30) ^{i}Pr$_3$Si——≡—CHO $\xrightarrow[\substack{\text{0°C, Et}_2\text{O}\\ \text{then THF/TFA}\\ \text{86\%}}]{\text{1-3 mol% (7.27)}}$ (7.31) 91% ee

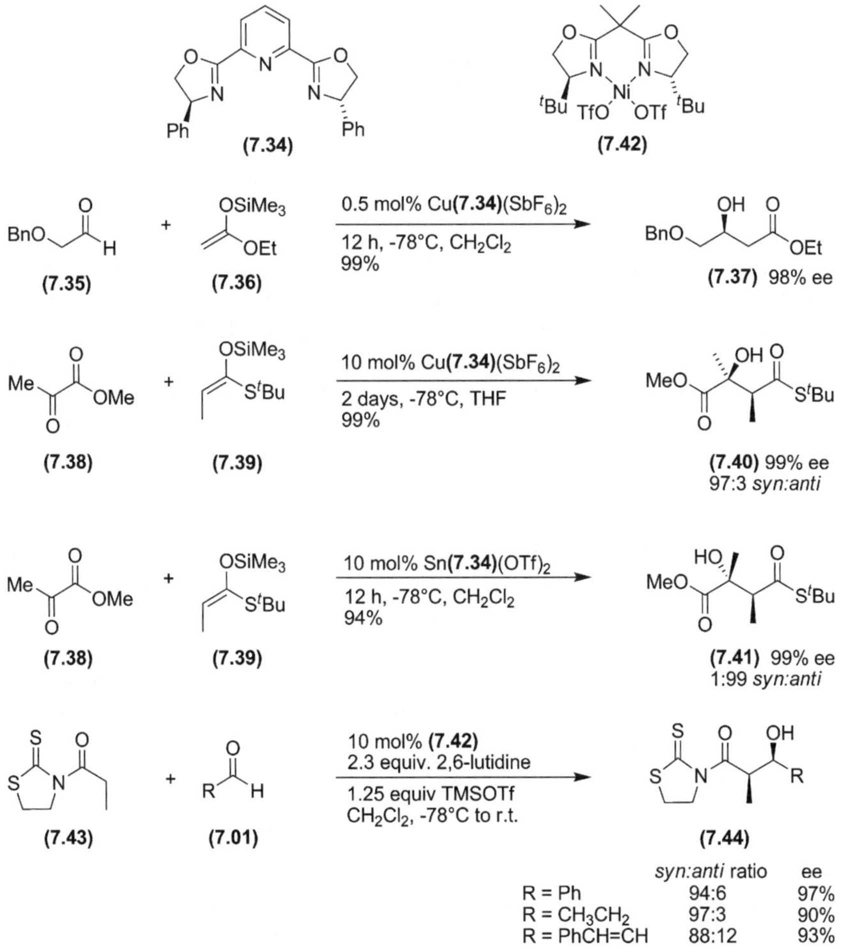

(7.34)

(7.42)

BnO—CHO (7.35) + OSiMe₃ / OEt (7.36)

0.5 mol% Cu(7.34)(SbF₆)₂

12 h, -78°C, CH₂Cl₂
99%

→ BnO—CH(OH)—CH₂—C(O)—OEt (7.37) 98% ee

Me—C(O)—C(O)—OMe (7.38) + OSiMe₃ / SᵗBu (7.39)

10 mol% Cu(7.34)(SbF₆)₂

2 days, -78°C, THF
99%

→ (7.40) 99% ee, 97:3 syn:anti

Me—C(O)—C(O)—OMe (7.38) + OSiMe₃ / SᵗBu (7.39)

10 mol% Sn(7.34)(OTf)₂

12 h, -78°C, CH₂Cl₂
94%

→ (7.41) 99% ee, 1:99 syn:anti

(7.43) + RCHO (7.01)

10 mol% (7.42)
2.3 equiv. 2,6-lutidine

1.25 equiv TMSOTf
CH₂Cl₂, -78°C to r.t.

→ (7.44)

	syn:anti ratio	ee
R = Ph	94:6	97%
R = CH₃CH₂	97:3	90%
R = PhCH=CH	88:12	93%

An alternative use of copper(II) catalysts has been reported by Carreira.[21] Using a fluoride counterion, they reason that the reaction proceeds via formation of a copper enolate rather than through Lewis acid-mediated activation of the aldehyde. Using TolBINAP as the ligand, the catalyst affords high enantioselectivities. The reaction worked well with most of the aldehydes that were reported (65–95% ee), including the reaction of benzaldehyde (7.17).

Palladium catalysts have also been reported to give good enantioselectivities in aldol reactions, again via catalytic formation of an enantiomerically pure enolate.[22]

An alternate strategy for catalysing the aldol reaction is by the use of Lewis bases that activate the donor. Denmark and coworkers have applied phosphoramide catalyts, effective in the Lewis base-catalysed addition of allylsilanes to aldehydes (see Section 6.3), to the asymmetric aldol reaction of trichlorosilyl enol ethers.[23] Good levels of enantioinduction have been observed in the aldol reaction of silyl enol ethers of ketones in the presence of phosphoramide (7.46). The diastereoselectivity of addition is dependant on the geometry of the enol ether indicating that a closed, chair-like transition state such as (7.47) incorporating a Lewis-acidic hypervalent cationic silicon is operative.[24] Thus, (E)-enol ether (7.48) reacts with benzaldehyde to give the *anti*-adduct (7.49) while (Z)-enol ether (7.50) yields the *syn*-adduct (7.51) as the major product. The Lewis base-catalysed procedure can also be applied to the aldol reaction of ketene acetal (7.52) with ketones. In this case the best ees are obtained with aromatic ketones such as acetophenone (7.53) and the bis-*N*-oxide (7.54).[25] Furthermore, this strategy has also been used to effect the challenging asymmetric cross aldol reaction of aldehyde-derived trichlorosilyl enol ethers with aldehydes using dimeric phosphoramide (7.56).[26] The group of Denmark has also developed a catalytic asymmetric Mukaiyama aldol reaction utilising stoichiometric quantities of the Lewis acid SiCl$_4$ activated by coordination with catalytic amounts of phosphoramide (7.56).[27] In this procedure, either geometry of the silyl enol ether leads to formation of the *anti*-aldol adduct indicating that an open transition state is operative.

The development of a catalytic asymmetric direct aldol reaction is an attractive concept as this does not require the prior generation of silyl enol ethers. Aldol reactions in living systems are catalysed using type I or type II aldolases. Type II aldolases are zinc-containing metalloenzymes possessing, Lewis acidic, Lewis basic and/or Brønsted basic sites and a number of enantiomerically pure, bifunctional polymetallic complexes have been designed that mimic the action of this enzyme type and catalyse the direct asymmetric aldol addition. Shibasaki and coworkers have developed LaLi$_3$-tris(binaphthoxide) (LLB) catalysts such as (7.57) possessing a Lewis acidic lanthanum and Brønsted basic lithium binaphthoxides capable of activating both aldol components in an asymmetric environment, as depicted in Figure 7.3.[28] The catalyst prepared by addition of KOH to (7.57) yields moderate to good ees in the aldol addition of a variety of aliphatic aldehydes, for instance (7.58), to a range of aromatic and aliphatic methyl ketones such as 2-butanone (7.59). The dinuclear zinc complex (7.61) developed by Trost and coworkers functions as an alternate bifunctional catalyst that is highly effective in the direct aldol reaction of aliphatic aldehydes with aryl methyl ketones[29] and has also been used

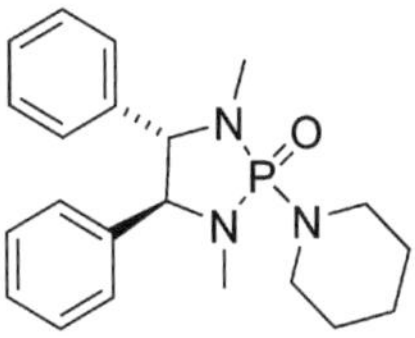

(7.46) 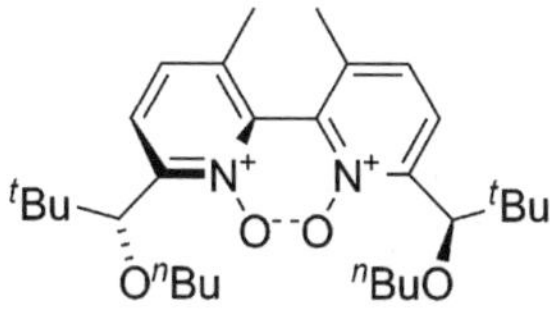 **(7.54)**

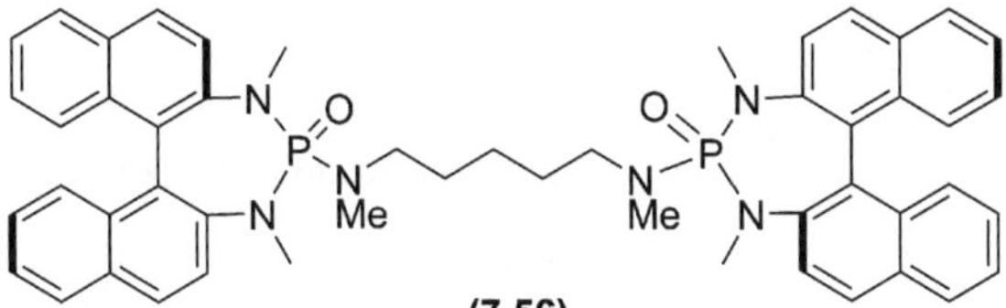

(7.56)

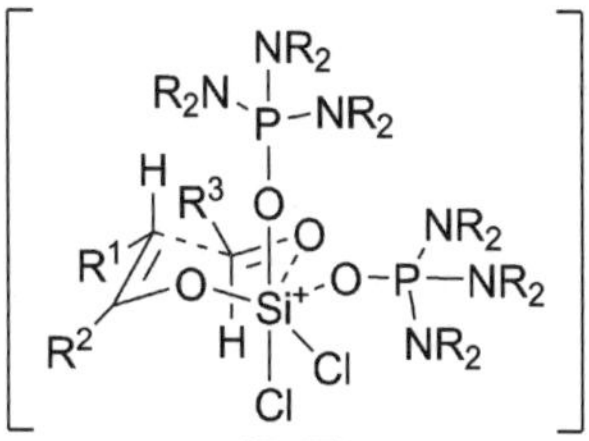

(7.47)

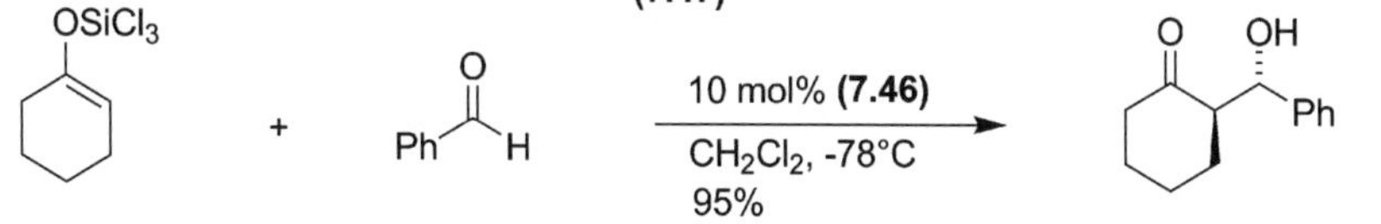

(7.48) + **(7.17)** → 10 mol% **(7.46)** / CH₂Cl₂, -78°C / 95% → **(7.49)** 93% ee / 1:49 *syn:anti*

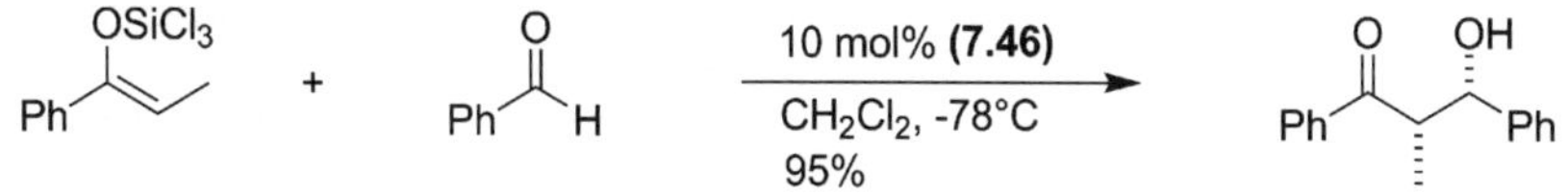

(7.50) + **(7.17)** → 10 mol% **(7.46)** / CH₂Cl₂, -78°C / 95% → **(7.51)** 95% ee

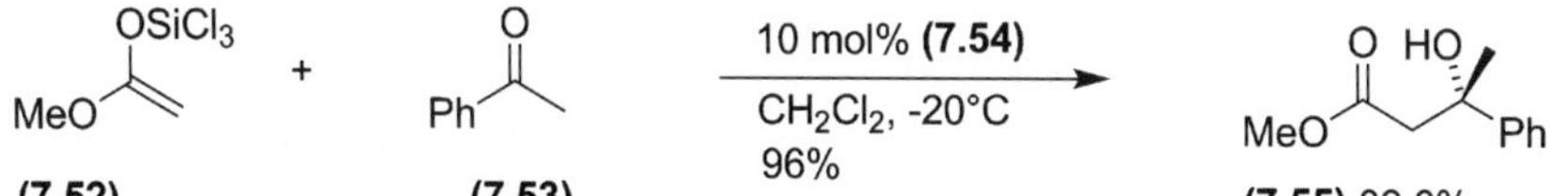

(7.52) + **(7.53)** → 10 mol% **(7.54)** / CH₂Cl₂, -20°C / 96% → **(7.55)** 82.6% ee

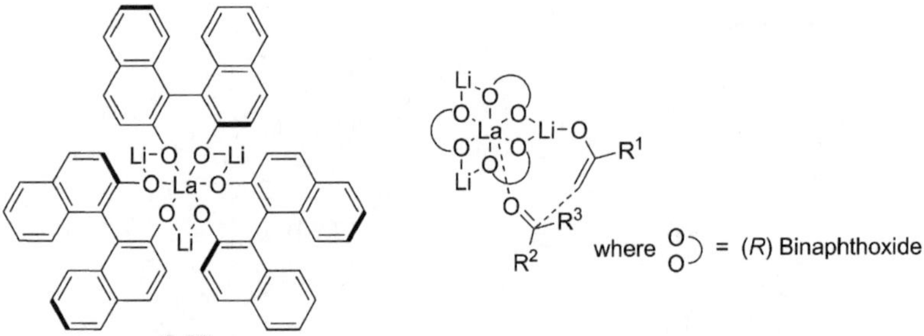

(7.57)

where $\begin{smallmatrix}O\\O\end{smallmatrix}$ = (R) Binaphthoxide

Figure 7.3 Heterobimetallic catalyst activation of the aldol reaction

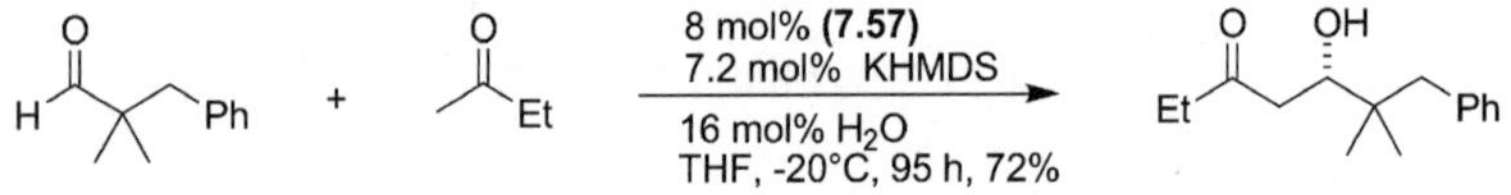

(7.58)   **(7.59)**

8 mol% **(7.57)**
7.2 mol% KHMDS
16 mol% H_2O
THF, -20°C, 95 h, 72%

(7.60) 88% ee

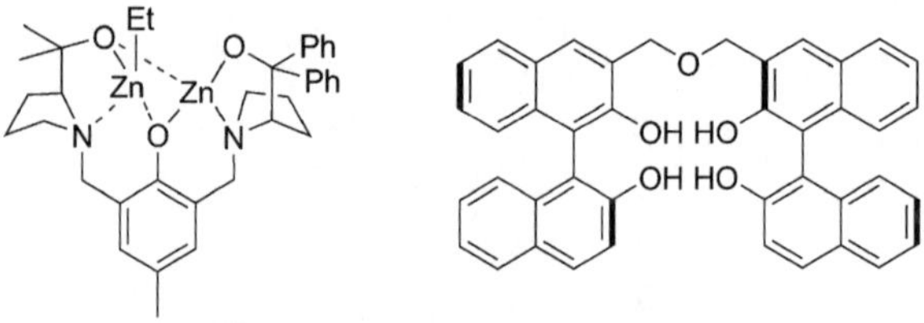

(7.61)   **(7.62)**

(7.63)   **(7.64)**

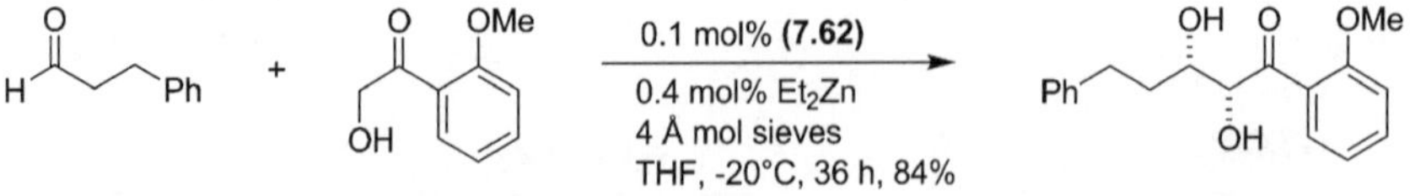

0.1 mol% **(7.62)**
0.4 mol% Et_2Zn
4 Å mol sieves
THF, -20°C, 36 h, 84%

(7.65) 92% ee
89:11 *syn:anti*

in the aldol reactions of methyl vinyl ketone[30] and methyl ynones.[31] In this example, one zinc atom reacts with the ketone to form a zinc enolate while the other zinc acts as a Lewis acid to activate the aldehyde. Zinc catalysts have also been applied to the asymmetric aldol reaction of α-hydroxyketones. Use of the catalyst prepared from *O*-linked BINOL **(7.62)** and either two or four equivalents of diethylzinc provides the aldol adduct with high *syn* selectivity. For example aldehyde **(7.63)** reacts with hydroxyketone **(7.64)** to give the adduct **(7.65)** with high ee at very low catalysts loadings.[32] Lanthanide complex (**7.57**) also catalyses the aldol reaction of α-hydroxyketones.[32] As the *anti*-1,2-diol is the major product in this case this reaction nicely complements the Sharpless AD process where the conversion of (*Z*)-olefins into *anti*-diols proceeds with poor selectivity (see Section 5.1).

Type I aldolases activate the aldol donor by the formation of enamines with active site amino acids and an alternate approach to the direct catalytic asymmetric aldol reaction centres on mimicking this process using proline-based organocatalysts. In fact, one of the earliest examples of asymmetric catalysis uses (*S*)-proline (**7.66**) as a catalyst for the intramolecular aldol reaction (the Hajos–Eder–Saeur–Wiechert reaction).[33] As an example the achiral triketone (**7.67**) cyclises to give the aldol product (**7.68**) with good enantioselectivity.

(**7.67**)   3 mol% (**7.66**)   CH$_3$CN   100%   (**7.68**) 93.4% ee   (**7.66**)

A major breakthrough in this area occurred in 2000 when List, Lerner and Barbas developed an intermolecular version of this process.[34] High ees are obtained using acetone (**7.69**) and hydroxyacetone (**7.70**) with branched aliphatic aldehydes such as (**7.71**) and (**7.72**). The latter reaction proceeds in all cases to give the *anti*-adduct predominantly. Aromatic aldehydes such as benzaldehyde (**7.17**) are also good substrates with the high ees in this case obtained using the 5,5-dimethyl thiazolidinium-4-carboxylate (DMTC) (**7.73**) as catalyst. The group of MacMillan has applied this methodology to the highly challenging asymmetric cross aldol reaction of aldehydes.[35] Slow addition of the donor allows the isolation of the *anti*- aldol adducts in high yield and ee with some of the best results obtained using α-thioacetals such as (**7.74**) as the acceptor.[36]

This proline-catalysed aldol proceeds via initial formation of an enamine with the donor component. This enamine then reacts with the *re*-face of the acceptor through a closed-transition state, as depicted in Figure 7.4, to give the *anti*-product as the major diastereomer. The proline also activates the acceptor by hydrogen bonding within this ensemble and thus plays a bifunctional role in the reaction. The resulting iminium ion is then hydrolysed to close the catalytic cycle.

There are some drawbacks associated with this methodology. L-Proline is poorly soluble in many organic solvents and requires the use of polar organic solvents such as DMSO or DMF. Unfortunately the use of water as reaction medium results in the formation of racemates. Furthermore, relatively high catalyst loadings are

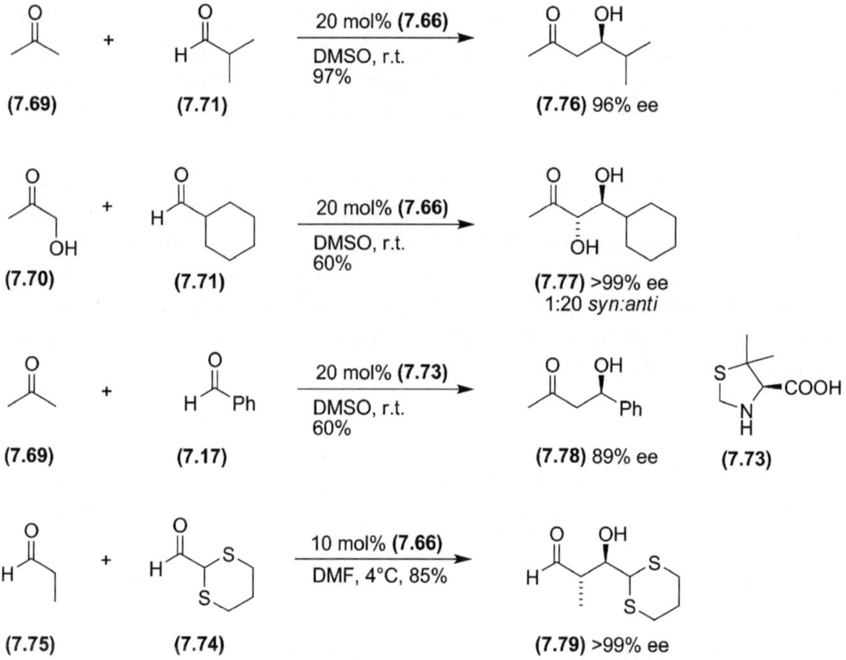

(7.69)
(7.71)
20 mol% (7.66)
DMSO, r.t.
97%
(7.76) 96% ee
(7.70)
(7.71)
20 mol% (7.66)
DMSO, r.t.
60%
(7.77) >99% ee
1:20 syn:anti
(7.69)
(7.17)
20 mol% (7.73)
DMSO, r.t.
60%
(7.78) 89% ee
COOH
(7.73)
(7.75)
(7.74)
10 mol% (7.66)
DMF, 4°C, 85%
(7.79) >99% ee

required to obtain good yields of aldol products, although this problem is partially remedied by the low cost of the catalyst. A large number of proline derivatives have been investigated as organocatalysts in the aldol reaction in an effort to alleviate some of these deficiencies and further improve on the scope of this process.[1f, 37] A representative set of such structures (**7.80–7.86**) is shown, all of which result in the formation of aldol adducts with high ee.[38] Replacement of the carboxylic acid moiety with a bioisosteric tetrazole results in a catalyst (**7.80**) that is both more reactive than L-proline (**7.66**) and more readily soluble in organic solvents such as THF.[38a,b] In a similar vein, acyl sulfonamides such as (**7.81**) give good enantioselectivities in the aldol reaction with aromatic aldehydes in organic solvents such as dichloromethane and acetone.[38a] The addition of stoichiometric amounts of water increases the activity of tetrazole (**7.80**) further and this allows the use of aldehydes such as chloral monohydrate (**7.87**) and formaldehyde, which have an affinity for water and are generally poor substrates for the catalytic asymmetric aldol reaction.[38c] Catalysts (**7.82**)[38e] and (**7.83**),[38d] with lipophilic substituents, provide high ees with water as the solvent. In these cases the aldol reaction occurs in a hydrophobic medium formed by association of the catalyst and substrates and the highest ees (up to 99%) are obtained using hydrophobic ketones such as cyclohexanone and aromatic aldehydes. Hydrazide (**7.84**) is active in toluene and

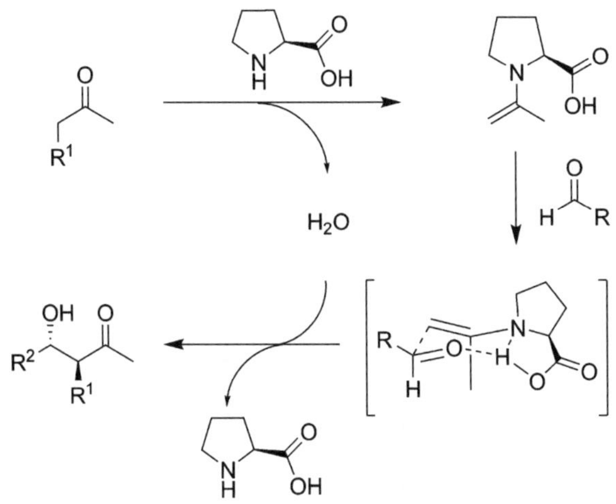

Figure 7.4 Catalytic cycle of the proline catalysed aldol reaction

catalyses the aldol reaction of acetone with aromatic aldehydes with 87–96% ee [38f] while β-aminoalcohol-derived catalyst (**7.85**) catalyses the asymmetric aldol reaction of a variety of aromatic and α-branched aliphatic aldehydes with acetone with ees in the range 96 to >99% ee. [38g] Diamide catalysts such as (**7.86**) are especially selective for the aldol reactions of heterocyclic ketones such as tetrahydropyranone (**7.89**) thiopyranone and piperidinones. [38h]

7.2 Isocyanide and Related Aldol Reactions

The asymmetric aldol reaction of α−isocyano-substituted carbonyls is a potentially useful route to chiral nonracemic β-hydroxy-α-amino acids. [39] In fact, isocyanoacetic esters, such as compound (**7.92**), were amongst the first nucleophiles to be employed in aldol-type reactions to give high enantioselectivity. [40] Although gold-catalysed reactions have not been widely investigated in asymmetric catalysis, they do work well in this reaction. The isonitrile group coordinates to the gold, whilst the side-arm in ligand (**7.93**) assists in deprotonation and stabilisation of the nucleophilic enolate. The reaction also works well with aliphatic aldehydes (85–97% ee, 85–100% *trans*). The reaction has been used with α-isocyano Weinreb amides, such as compound (**7.95**). This allows subsequent manipulation of the carbonyl compound in product (**7.97**) into aldehydes and ketones. [41] The aldehydes used gave very high enantioselectivities in the products (93–99% ee, 90–96% de) and an impressive range of substrates has been investigated in the reaction of related substrates. [39]

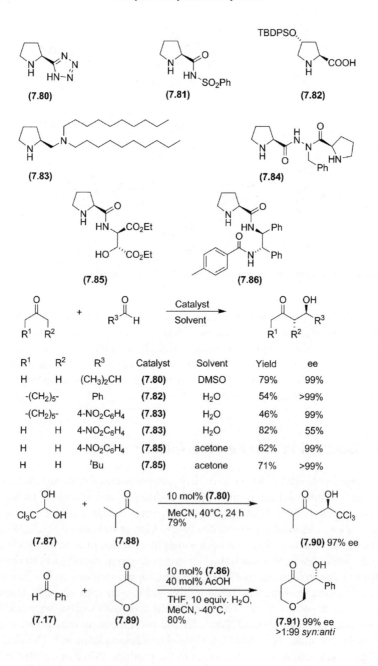

R^1	R^2	R^3	Catalyst	Solvent	Yield	ee
H	H	(CH$_3$)$_2$CH	**(7.80)**	DMSO	79%	99%
-(CH$_2$)$_5$-		Ph	**(7.82)**	H$_2$O	54%	>99%
-(CH$_2$)$_5$-		4-NO$_2$C$_6$H$_4$	**(7.83)**	H$_2$O	46%	99%
H	H	4-NO$_2$C$_6$H$_4$	**(7.83)**	H$_2$O	82%	55%
H	H	4-NO$_2$C$_6$H$_4$	**(7.85)**	acetone	62%	99%
H	H	^{t}Bu	**(7.85)**	acetone	71%	>99%

Cyanopropionates have also been employed in catalytic aldol reactions. The enolisation of the nucleophile (**7.98**) by the rhodium complex of TRAP ligand (**7.99**) is the basis for the catalysis.[42] The use of bulky esters affords high selectivity in the aldol reaction with formaldehyde (**7.100**), although only moderate *anti:syn* selectivity was observed when alternative aldehydes were employed.

Another approach to the synthesis of β-hydroxy-α-amino acids is by aldol reaction of imines derived from amino acids. The benzophenone imine of glycine (**7.102**) undergoes highly enantioselective aldol addition with a range of aliphatic aldehydes, including (**7.71**) under phase-transfer conditions in the presence of the bromide salt of phase-transfer catalyst (**7.103**).[43] A similar transformation is catalysed, in low to moderate ee, by the bimetallic catalysts developed by Shibasaki and coworkers.[44]

7.3   The Nitroaldol Reaction

The nitroaldol (Henry) reaction involves the addition of nitronates to aldehydes and ketones to give a β-nitroalcohol. These products are useful synthetic building blocks as the nitro group can be transformed into a range of other functional groups, and this has stimulated some recent research into the development of a catalytic asymmetric variant. Some of the catalyst systems used in the asymmetric aldol rection have been successfully employed in the catalytic asymmetric nitroaldol process.[45]

Silyl nitronates undergo enantioselective addition to aromatic aldehydes in the presence of an enantiomerically pure bifluoride derived from (**7.103**).[46] In this approach the *anti*-adduct is the major diastereomer formed. Thus silyl nitronate (**7.105**) undergoes addition to benzaldehyde to give adduct (**7.106**) with high ee. Alternately the coupling of silyl nitronates can be achieved with high ee using Lewis acids such as copper bis-oxazoline catalyst (**7.107**) in combination with a fluoride ion source.[47]

Copper bis-oxazoline catalysts have also been applied with success to the direct asymmetric nitroaldol reaction. The addition of nitromethane to α-ketoesters occurs with high ee using such catalysts in combination with triethylamine[48] and a base-free process is possible utilising copper bis-oxazoline catalysts such as (**7.108**)[49] and also copper-diamine catalysts[50] derived from Cu(OAc)$_2$. In these cases the acetate ligands act as mild Brønsted bases to effect deprotonation of the nitroalkane on binding to the catalysts. These complexes are active in the coupling of nitromethane (**7.109**) with a range of aromatic and aliphatic aldehydes including pivaldehyde (**7.110**). The heterobimetallic complexes such as (**7.57**) developed by the group of Shibasaki and Trost's bifunctional zinc-centred catalyst (**7.61**) active in the direct asymmetric aldol reaction are also effective in the direct nitroaldol

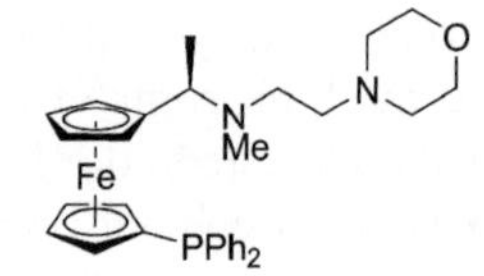

(7.93)

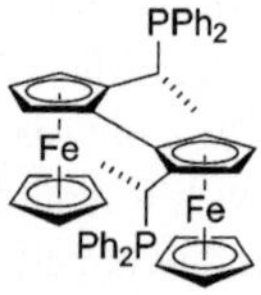

(7.99)

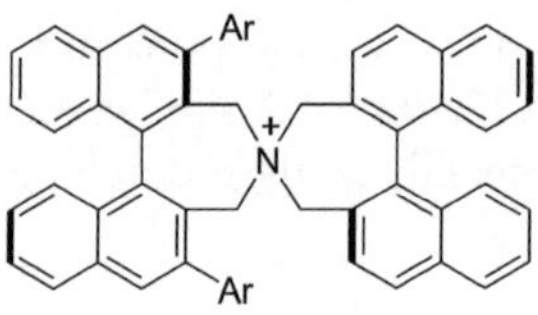

(7.103)

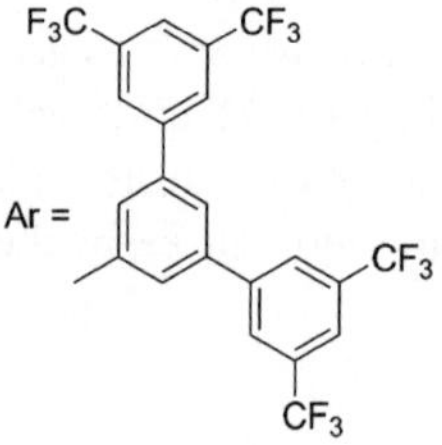

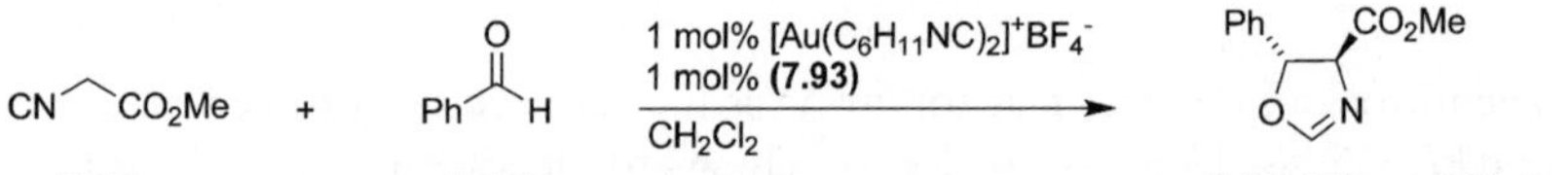

Ar =

CN〜CO₂Me + Ph—CHO →

(7.92)   (7.17)

1 mol% [Au(C₆H₁₁NC)₂]⁺BF₄⁻
1 mol% **(7.93)**
CH₂Cl₂

(7.94) 95% ee, 90% de

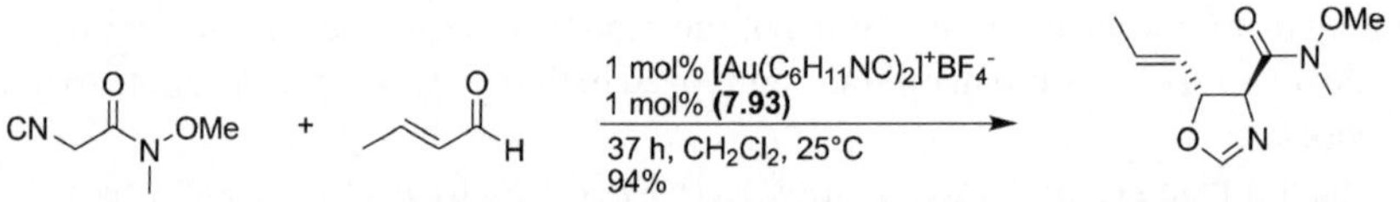

(7.95)   (7.96)

1 mol% [Au(C₆H₁₁NC)₂]⁺BF₄⁻
1 mol% **(7.93)**
37 h, CH₂Cl₂, 25°C
94%

(7.97) 99% ee, 94% de

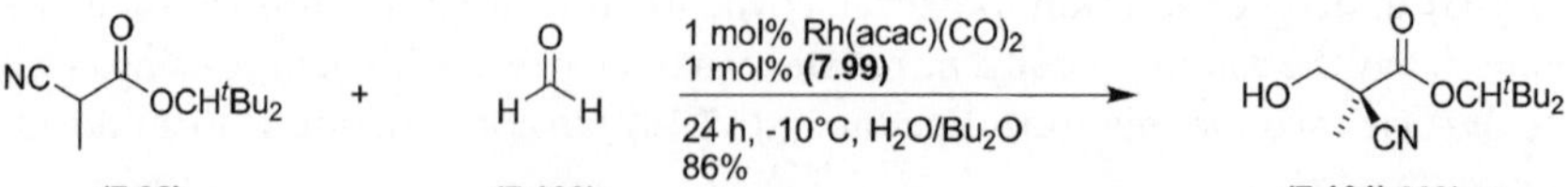

(7.98)   (7.100)

1 mol% Rh(acac)(CO)₂
1 mol% **(7.99)**
24 h, -10°C, H₂O/Bu₂O
86%

(7.101) 93% ee

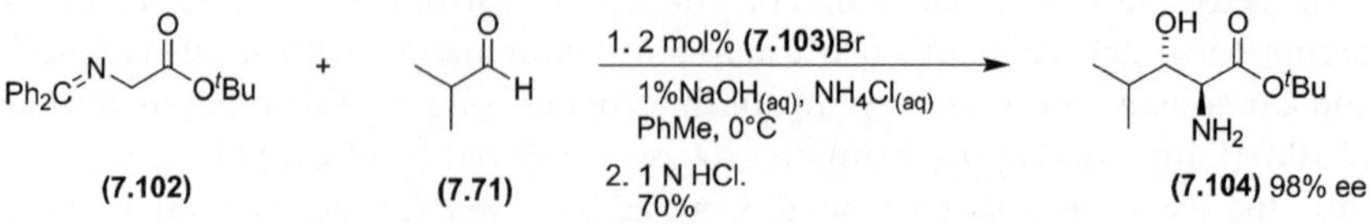

(7.102)   (7.71)

1. 2 mol% **(7.103)**Br
1%NaOH₍aq₎, NH₄Cl₍aq₎
PhMe, 0°C
2. 1 N HCl.
70%

(7.104) 98% ee

process affording high ees in the addition of nitroalkanes to aromatic and aliphatic aldehydes.[28c, 45a, 51] Organocatalysts have also been applied with some success to the asymmetric nitroaldol reaction.[37c,d, 45c] Some of the best ees have been obtained using catalysts (**7.112–7.114**). All of these compounds act as bifunctional catalysts to activate both the nitronate and aldehyde acceptor via hydrogen bonding interactions. Cinchona-derived catalyst (**7.112**) displays ees in the range 93–97% in the addition of nitromethane (**7.109**) to α-ketoesters,[52] while the structurally related thiourea-functionalised catalyst (**7.113**) displays good ees in the nitroaldol reaction with aromatic aldehydes.[53] The guanidine-thiourea-based catalyst (**7.114**) exhibits the widest scope.[54] This catalyst effects the enantioselective nitroaldol reaction of a variety of aliphatic aldehydes with high *syn* selectivity and is not restricted to the use of nitromethane as donor.

7.4   Addition of Enolates to Imines

The development of a catalytic asymmetric addition of enolates to imines (Mannich reaction) has only recently received attention and it was not until 1998 that there were reports of this process giving products in over 90% enantiomeric excess. Since then a large number of both metal-based catalysts and organocatalysts for the asymmetric Mannich reaction have been investigated.[55]

In common with the aldol reaction, the catalytic asymmetric Mannich process can be achieved by metal complex-catalysed addition of either silyl enol ethers or unmodified aldehydes and ketones to imines. Kobayashi has used the zirconium complex (**7.117**) to catalyse the addition of α-silyloxy ketene acetals (**7.118**) to imines.[56] High *syn* selectivity and enantiomeric excess were seen with arylimines (**7.119**). It has been discovered that catalysts of this type precipitate in hexane and the resulting white solid can be stored for up to six months with no loss of activity.[57] Sodeoka has employed the palladium TolBINAP complexes used in normal aldol reactions affording up to 90% ee in the formation of the β-aminoketone (**7.122**).[58] The use of various late transition metal BINAP complexes (**7.123–7.125**) affords high enantiomeric excesses in the imine/aldol reaction forming product (**7.127**), especially with copper(I) complexes.[59] The silver-catalysed asymmetric Mannich reaction can be performed using the silver complex formed from the isoleucine-derived phosphine (**7.128**) and AgOAc. This catalyst provides high enantioselectivities (up to 96% ee) in the reaction of aryl imines with silyl enol ethers derived from acetone or acetophenone.[60] The catalytic asymmetric Mannich reaction can even be performed in aqueous media if a stable imine substitute is used as acceptor. For instance, the asymmetric Mannich reaction of silyl enol ethers with acylhydrazono ester (**7.129**) and enolates such as (**7.130**) occurs with high ee in water using the zinc complex (**7.131**), formed *in situ* from zinc fluoride and the corresponding diamine and a surfactant such as cetyltrimethylammonium

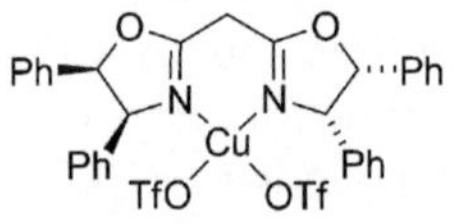

(7.107)

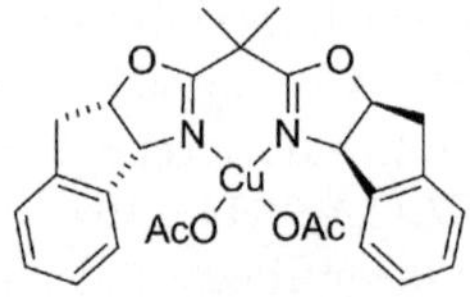

(7.108)

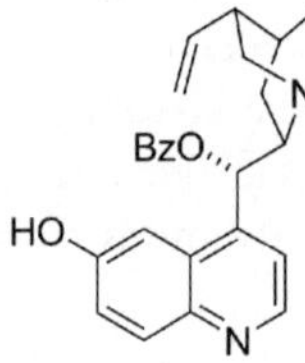

(7.112)

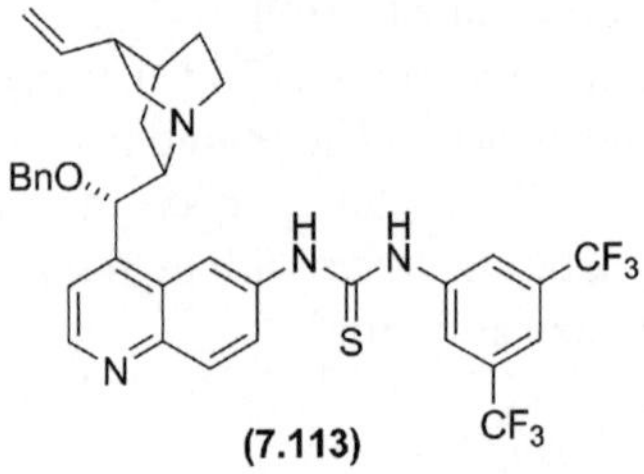

(7.113)

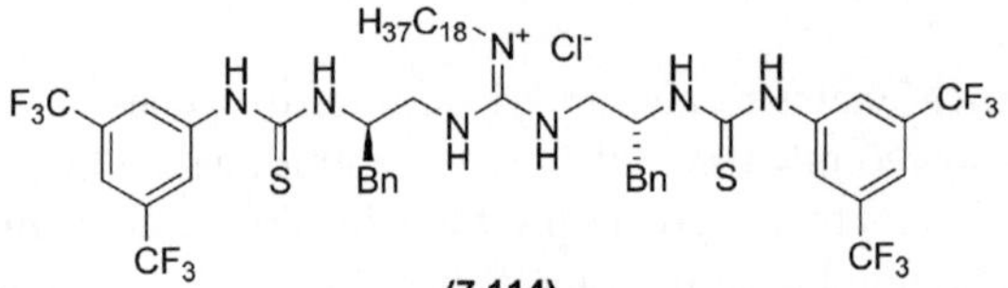

(7.114)

(7.105) **(7.17)** **(7.106)** 95% ee
8:92 *syn:anti*

(7.109) **(7.110)** **(7.111)**

(7.115) **(7.01)** **(7.116)**

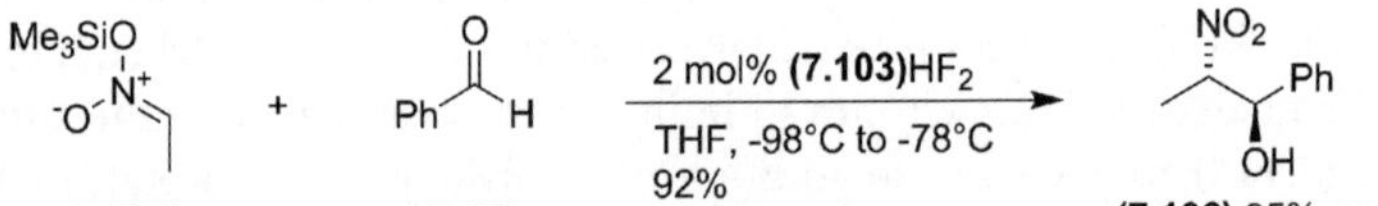

R^1	R^2	Yield	*syn:anti*	ee
$(CH_3)_2CH$	Et	50%	97:3	90%
$c\text{-}C_6H_{11}$	Et	77%	99:1	93%
$c\text{-}C_6H_{11}$	$CH_3(CH_2)_2$	61%	99:1	95%
$CH_3(CH_2)_2$	$PhCH_2$	70%	91:9	87%

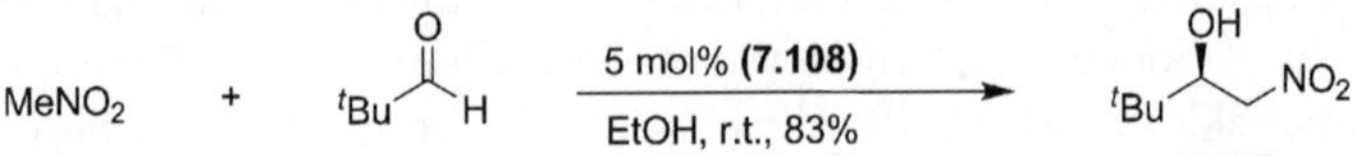

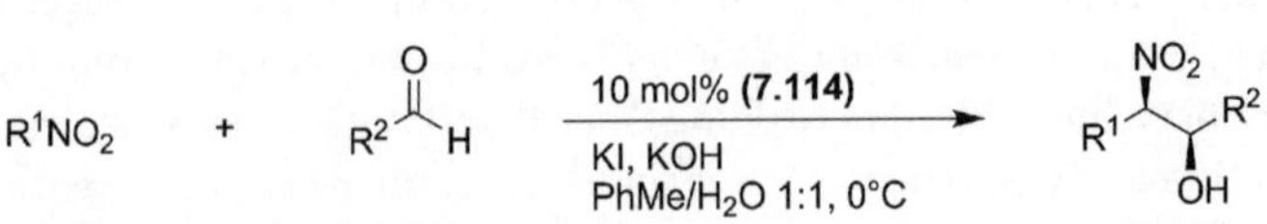

bromide (CTAB). In this process the catalyst acts primarily as a Lewis acid to activate the imine but also functions as a Lewis base by delivery of fluoride to the silyl enol ether.[61]

The direct catalytic asymmetric Mannich reaction can be achieved using the same metal-based catalysts deployed in the direct asymmetric aldol process (see Section 7.1). For instance, the zinc catalyst prepared from four equivalents of Et_2Zn and ligand (**7.62**) has been used to give high ee (98–99.5%) in the *anti*-selective Mannich reaction of 2-hydroxyacetophenones with nonenolisable imines.[62] The dinuclear zinc complex (**7.61**) developed by Trost and coworkers is also effective in the asymmetric Mannich reaction of 2-hydroxyacetophenone (**7.133**).[63] In this case both aromatic and aliphatic diphenylphosphinoylimines such as (**7.134**) are good substrates although the *anti*-selectivity is not as good as that obtained using ligand (**7.62**).

An alternate approach to the direct asymmetric Mannich reaction uses enantiomerically pure organocatalysts.[55b–d] L-Proline and derivatives, applied with much success to the catalytic asymmetric aldol reaction (see Section 7.1), also function as effective catalysts in the Mannich reaction. The mechanism of this process is similar to the L-proline-catalysed aldol reaction involving conversion of the donor into an enamine and proceeds via a closed six-membered transition state similar to that depicted in Figure 7.4. However, in contrast to the L-proline-catalysed aldol reaction, the *syn*-Mannich adduct is the major diastereomer formed and the *si* rather than the *re*-face of the acceptor undergoes attack, as depicted in Figure 7.5.

This reversal in selectivity is thought to arise from the reaction of an (*E*)-configured enamine aligned so as to minimise steric interactions of the nitrogen substituent with the pyrrolidine ring of the enamine.

As the intermediate enamine reacts faster with imines than aldehydes, a one-pot three component coupling of the donor ketone, aldehyde and amine is possible. List and coworkers have achieved high ees in this reaction utilising L-proline (**7.66**) and some aliphatic aldehydes and aromatic aldehydes such as (**7.136**) in combination with *p*-anisidene (**7.137**).[64] This catalyst system is also effective for the coupling of α-hydroxyketones. Use of the tetrazole-substituted proline (**7.80**) allows the reaction to be performed in dichloromethane rather than DMSO and high ees in the Mannich reaction between aliphatic ketones and imines derived from ethyl glyoxalate have been obtained under these reaction conditions.[38a]

Use of 3-pyrrolidine carboxylic acids (**7.139**) and (**7.140**) as catalysts results in formation of the *anti*-Mannich adduct as the major product.[65] This reversal in diastereoselectivity arises from reaction of the s-*cis* conformer of the enamine formed from (**7.139**) or (**7.140**), as depicted in Figure 7.6. Catalyst (**7.139**) provides high ees in the *anti*-Mannich reaction of aliphatic cyclic and acyclic ketones such as pentan-3-one (**7.141**)[65a] while the methyl-substituted analogue (**7.140**) catalyses the *anti*-addition of a range of aliphatic aldehydes in 98 to >99% ee.[65b]

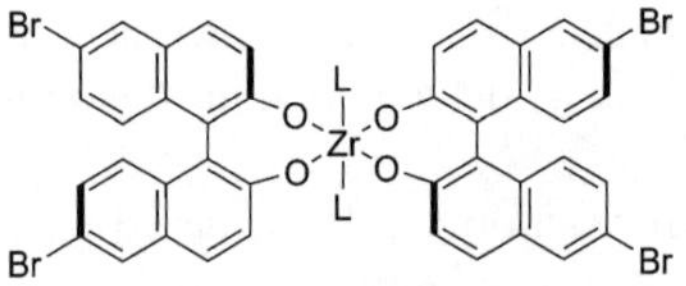

L = 1,2-dimethylimidazole
(7.117)

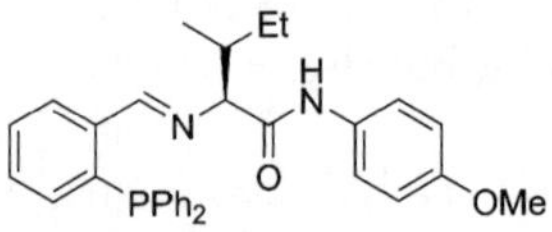

(7.128)

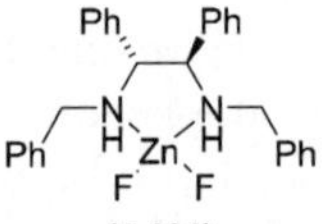

(7.131)

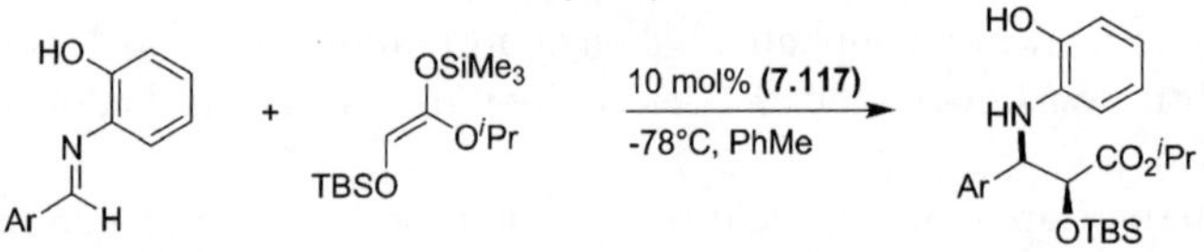

HO

(7.119)   +   (7.118) $\xrightarrow[\text{-78°C, PhMe}]{\text{10 mol% (7.117)}}$ (7.120)

Ar = Ph   100%, 96:4 *syn:anti* 95% ee
Ar = 1-Naphth 65%, >99:1 *syn:anti* 91% ee

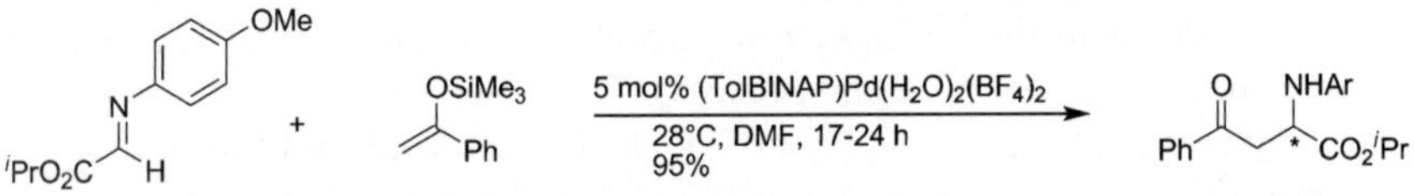

(7.121)   +   (7.11) $\xrightarrow[\substack{28°C, DMF, 17-24 h \\ 95\%}]{\text{5 mol% (TolBINAP)Pd(H}_2\text{O})_2\text{(BF}_4\text{)}_2}$ (7.122)

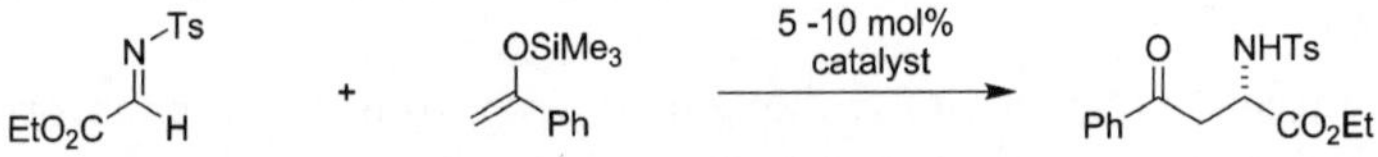

(7.126)   +   (7.11) $\xrightarrow[]{\substack{5\text{-}10 \text{ mol\%} \\ \text{catalyst}}}$ (7.127)

10 mol% [(R)-BINAP]AgSbF$_6$	**(7.123)**	THF, −80°C 95%, 90% ee
5 mol% [(R)-BINAP]Pd(ClO$_4$)$_2$	**(7.124)**	CH$_2$Cl$_2$, −80°C 91%, 80% ee
10 mol% [(R)-TolBINAP]CuClO$_4$	**(7.125)**	THF, 0°C 91%, 98% ee

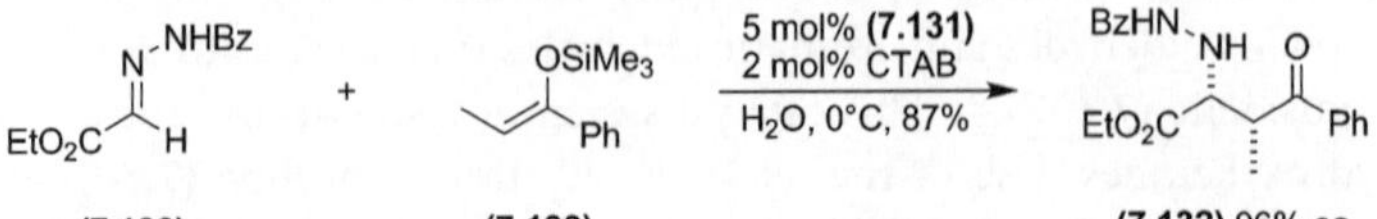

(7.129)   +   (7.130) $\xrightarrow[\text{H}_2\text{O, 0°C, 87\%}]{\substack{5 \text{ mol\% (7.131)} \\ 2 \text{ mol\% CTAB}}}$ (7.132) 96% ee
                                                                                93:7 *syn:anti*

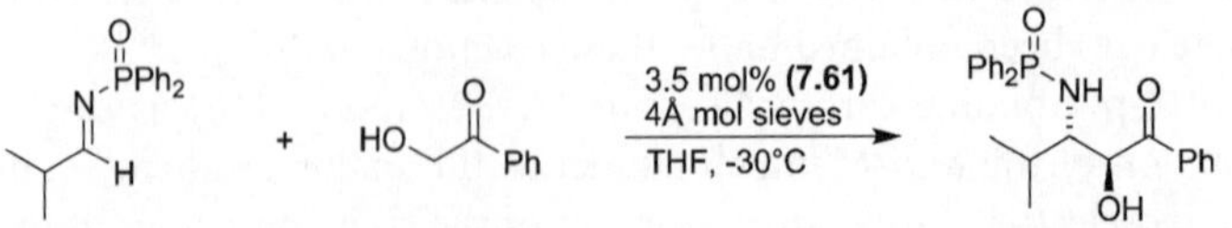

(7.134)   +   (7.133) $\xrightarrow[\text{THF, -30°C}]{\substack{3.5 \text{ mol\% (7.61)} \\ 4\text{Å mol sieves}}}$ (7.135) >99% ee
                                                                              1:6 *syn:anti*

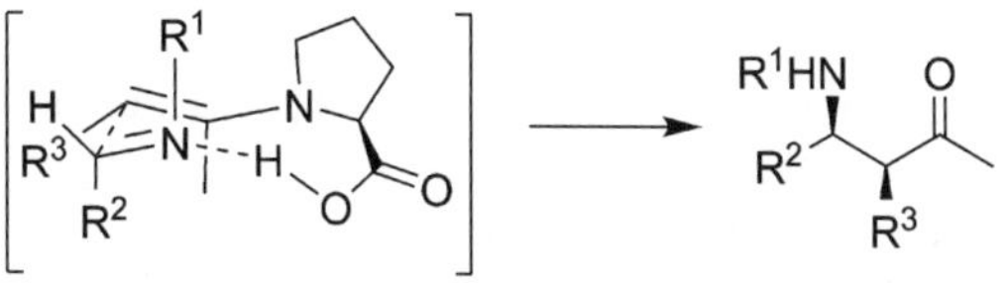

Figure 7.5 Transition state of the L-proline-catalysed Mannich Reaction

The enamine catalysis detailed above proceeds via activation of the Mannich donor. An alternate strategy to the catalysis of the Mannich reaction is by the use of Brønsted acids that activate the acceptor imine by protonation on nitrogen. Some of the most successful asymmetric variants of this process use BINOL-based phosphoric acids as catalysts. For instance Terada and coworkers used **(7.144)** to effect highly enantioselective addition of acetylacetone to a range of aryl aldimines[66]

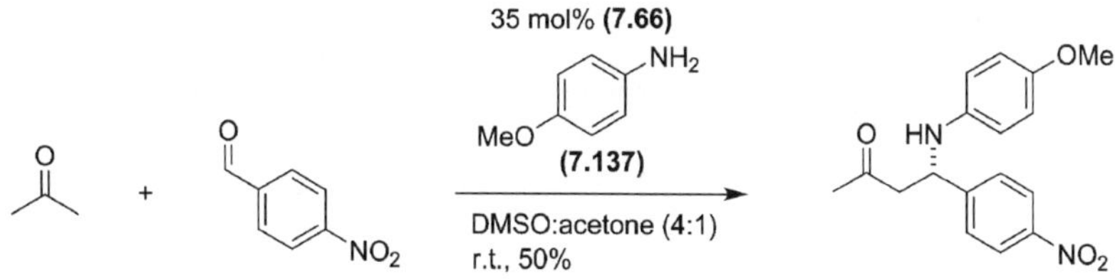

(7.69) **(7.136)** **(7.138)** 94% ee

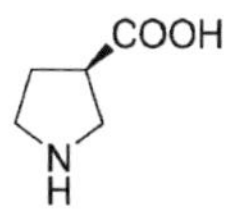

(7.139)

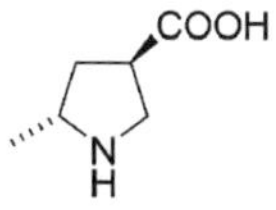

(7.140)

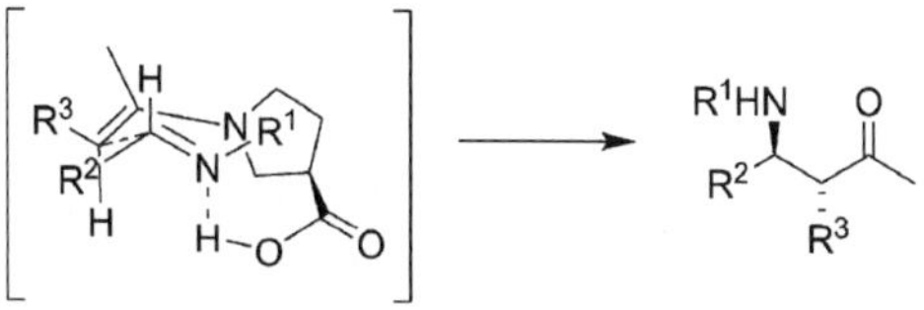

Figure 7.6 Transition state of the organocatalysed *anti*-Mannich reaction

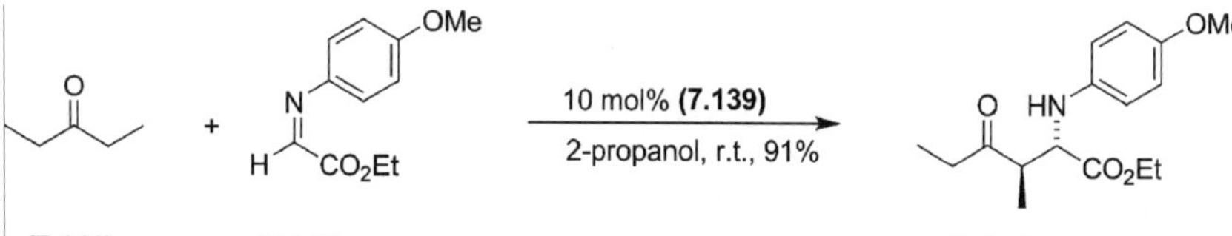

(7.141) **(7.142)** **(7.143)** 97%ee

while the group of Akiyama have applied acid (**7.145**) to the enantioselective *syn*-addition of silyl ketene acetals such as (**7.146**) to aromatic imines including (**7.147**).[67] The thiourea-base Brønsted acid catalysts developed by Jacobsen and coworkers, which are active in asymmetric additions in imines (see Section 6.5) are also effective in the Mannich reaction. For instance 5 mol% of urea (**7.148**) effects highly enantioselective addition of silyl ketene acetals to *N*-Boc-protected aryl aldimines.[68]

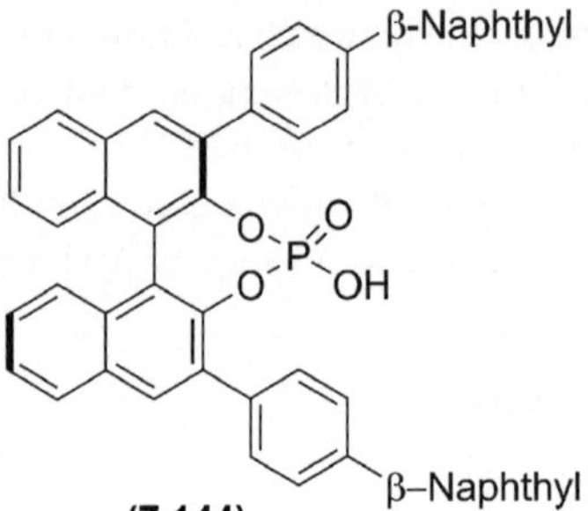

(**7.144**) 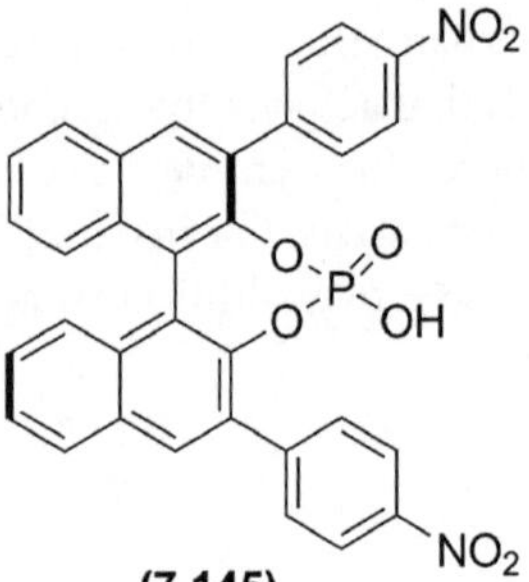　　　　　　(**7.145**)

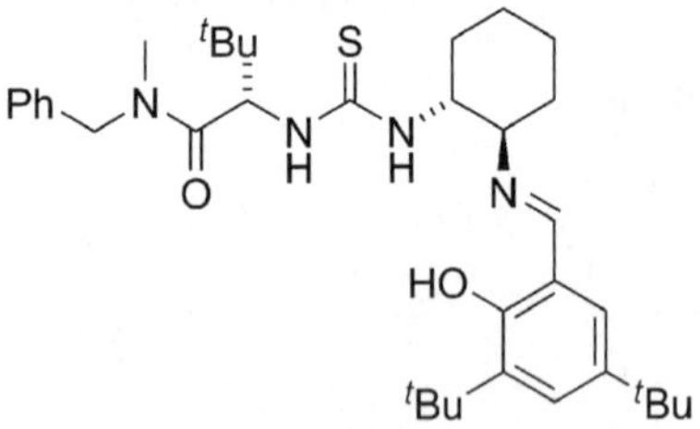

(**7.148**)

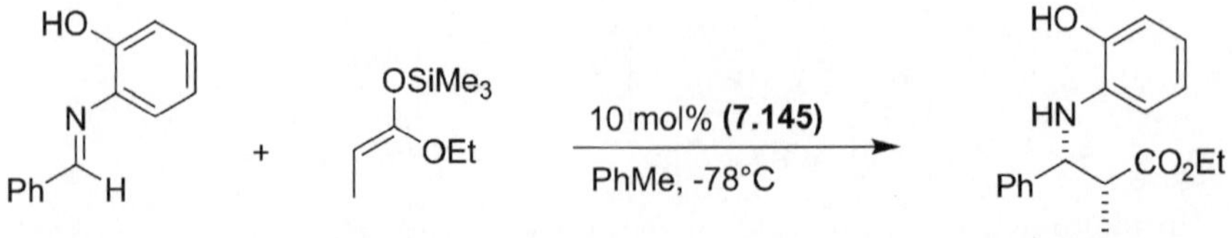

(**7.147**)　　　　(**7.146**)　　　10 mol% (**7.145**)　　　(**7.149**) 96% ee

PhMe, -78°C　　　　87:13 *syn:anti*

7.5　Darzens Condensation

The Darzens condensation is an aldol-like reaction in which the aldolate product closes to give an epoxy ketone. The reaction has been achieved with moderate enantiomeric excess using enantiomerically pure crown ethers[69,70] and, recently, using metal(salen) complexes in the presence of base.[71]

The highest ees have been obtained using enantiomerically pure ammonium salts under phase transfer conditions. The catalyst (**7.150**) has been used to give higher enantiomeric excesses with various aldehydes (42–79% ee), including the reaction of aliphatic aldehyde (**7.151**) with the α-chloroketone (**7.152**).[72] This type of catalyst has also been applied to the Darzens reaction of cyclic α-chloroketones,[73] and chloromethylsulfones.[74] Similar levels of enantioinduction have been achieved using the more easily prepared BINOL-based ammonium salt (**7.154**) and α-chloroamides as substrates.[75]

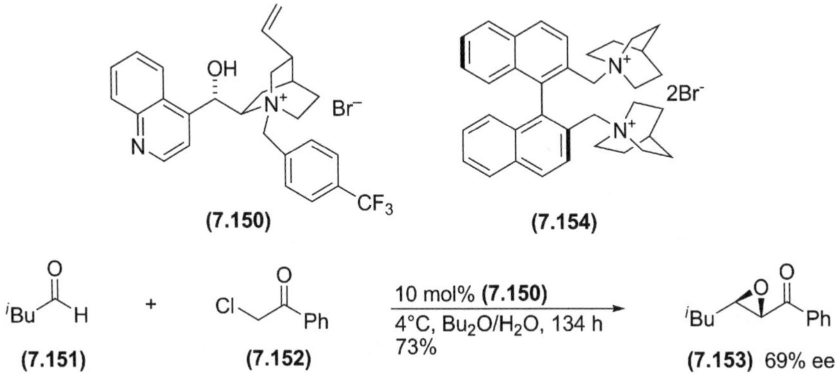

(7.150)   (7.154)

(7.151)   +   (7.152)   →   10 mol% (7.150) / 4°C, Bu₂O/H₂O, 134 h / 73%   →   (7.153) 69% ee

7.6 Morita–Baylis–Hillman Reaction

The Morita–Baylis–Hillman (MBH) reaction involves the conversion of an α,β-unsaturated carbonyl compound into an aldol-like adduct.[76] The reaction is catalysed by tertiary amines (**7.155**), which form an intermediate enolate (**7.157**) by conjugate addition (rather than by direct deprotonation of the α-proton). The enolate undergoes an aldol reaction with an aldehyde (**7.01**), followed by loss of the amine catalyst to provide the Baylis–Hillman adduct (**7.158**), as shown in Figure 7.7.

In addition to tertiary amines, triphenylphosphine is also an effective promoter and the use of enantiomerically pure amines or phosphines to catalyse the reaction is an interesting prospect, since the products would be synthetically useful. In addition there is also the potential for Lewis acid/Brønsted acid-catalysed asymmetric MBH reactions. While early attempts at the development of a catalytic asymmetric variant were only moderately successful, providing products with up to 50% ee,[77] some recent progress has been made in this area and high ees have been obtained in both the MBH and aza-Baylis–Hillman reaction of α,β-carbonyls with imines.

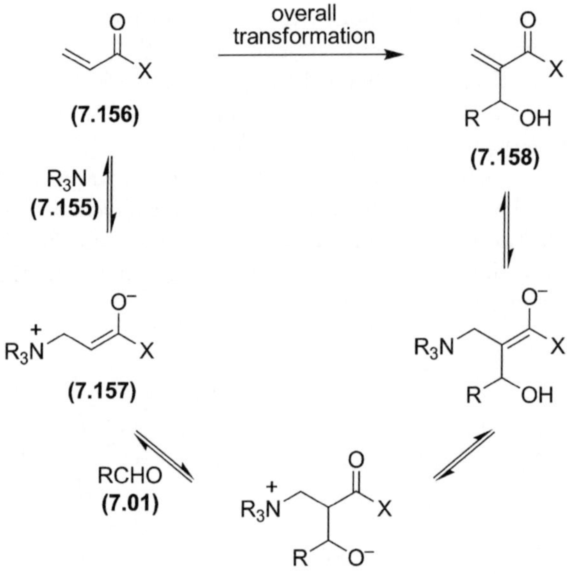

Figure 7.7 Catalytic cycle of the Baylis–Hillman reaction

The metal-based Lewis acid derived from the camphor-derived ligands such as (**7.159**) and La(OTf)$_3$ is effective in the MBH reaction of aromatic aldehydes with α,β-unsaturated aldehydes mediated by 1,4-diazabicyclo[2.2.2]octane (DABCO).[78] The best ees (up to 95%) are obtained using sterically bulky acrylates such as α-naphthyl acrylate. More success has been obtained using Brønsted acidic organocatalysts. The partially reduced BINOL (**7.160**) has been used to effect enantioselective MBH reaction of aliphatic aldehydes such as (**7.71**) with 2-cyclohexen-1-one (**7.161**) mediated by triethylphosphine,[79] while bis(thio)ureas such as (**7.163**) provide up to 96% ee in the coupling of this ketone with cyclohexanecarbaldehyde in the presence of DABCO.[80]

An alternate to the use of Lewis acids is the employment of amine catalysts in the MBH reaction. Hatakeyama and coworkers have used the quinidine (**7.164**) as a catalyst in the MBH reaction of both aromatic aldehydes such as benzaldehyde (**7.17**) and aliphatic aldehydes with the acrylate (**7.165**).[81] In all cases ees are high (91–99%), but yields are moderate. This amine has also been applied to the catalysis of the aza-Baylis–Hillman reaction of methylvinyl ketone (MVK) and methyl acrylate with *N*-tosylarylaldimines giving the product with high ee.[82]

It has been shown that L-proline in combination with quinidine (**7.164**) and imidazole leads to some enantioselection in the MBH reaction, presumably via the formation of a iminium species by reaction of L-proline with the α,β-unsaturated carbonyl acceptor.[83] Higher levels of enantioselectivity have been obtained in this procedure using the imidazole-containing octapeptide (**7.167**) in combination with L-proline, and some of the highest ees in the MVK-based MBH reaction have been obtained using this dual catalyst system.[84] For example MVK (**7.168**) reacts with aldehyde (**7.169**) to give the MBH adduct (**7.170**) with 81% ee under these conditions. Intramolecular variants of the MBH reaction can be carried out in a highly enantioselective fashion using this dual catalyst approach. In this case (S)-pipecolinic acid is employed in place of L-proline in combination with N-methylimidazole as the catalyst system.[85]

A number of BINOL-based bifunctional organocatalysts, for example (**7.171–7.173**), containing both Brønsted acidic and Lewis basic sites have been used to good effect in the asymmetric MBH reaction. The amine-thiourea (**7.171**) promotes the MBH reaction of aliphatic aldehydes with 2-cyclohexenone with ees ranging from 80 to 94%[86] while both the (pyridinylaminomethyl)BINOL (**7.172**)[87] and phosphine (**7.173**)[88] catalyse the aza-Baylis–Hillman reaction of simple α,β-carbonyls such as MVK and phenyl acrylate with N-tosyl arylaldmines with similar levels of enantioselectivity.

Barrett and Kamimura have developed a stepwise asymmetric catalytic variant of the Baylis–Hillman reaction.[89] A stoichiometric amount of nucleophile, Me₃Si-SPh or Me₃Si-SePh (**7.174**) is added, but only a catalytic amount of Lewis acid (**7.175**) in the reaction between methyl vinyl ketone (**7.168**) and acetaldehyde (**7.176**). The aldol adduct (**7.177**) was subjected to oxidation and selenoxide elimination to give the product (**7.178**), which is the same as the product from a Baylis–Hillman reaction.

7.7 Carbonyl-Ene Reactions

The ene reaction can be accelerated by catalysis. In particular the carbonyl-ene reaction, represented as shown in Figure 7.8, can be accelerated by Lewis acids. The reaction can be synthetically equivalent to an aldol reaction (when the ene component is a vinyl ether), and is considered in this section at the end of aldol reactions, and before concerted reactions.

The catalysed carbonyl-ene reaction frequently employs reactive aldehydes, especially glyoxalate esters.[90] Mikami's group has studied the titanium/BINOL catalysed carbonyl-ene reaction in considerable detail.[91] Typically, the catalyst is prepared *in situ* from diisopropoxytitanium dihalide and BINOL in the presence of 4 Å molecular sieves. Thus, alkenes (**7.179**) and (**7.180**) are converted into the homoallylic alcohols (**7.181**) and (**7.182**) with high enantioselectivity. Typical examples use up to 10 mol% of catalysts, but variation in the catalyst preparation allows the use of only 0.2 mol%.[92]

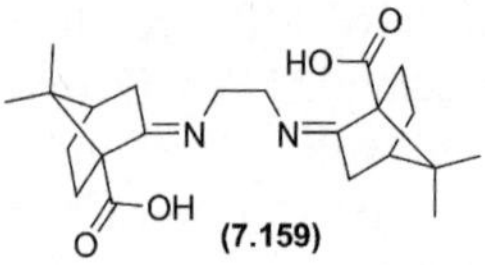

(7.159)

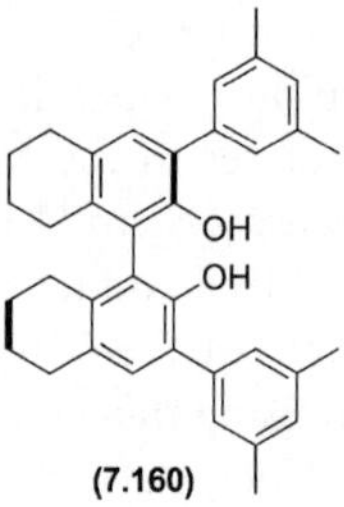

(7.160)

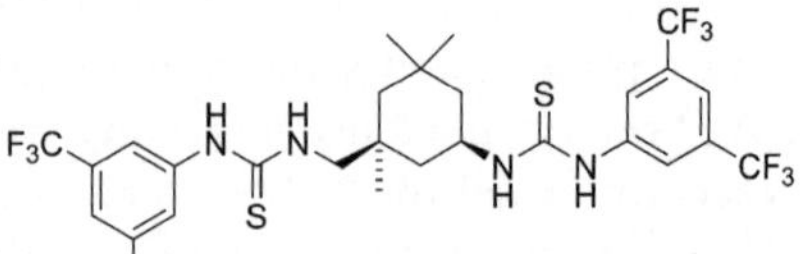

(7.163)

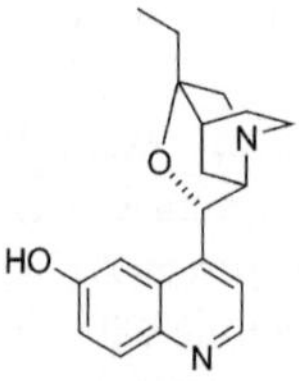

(7.164)

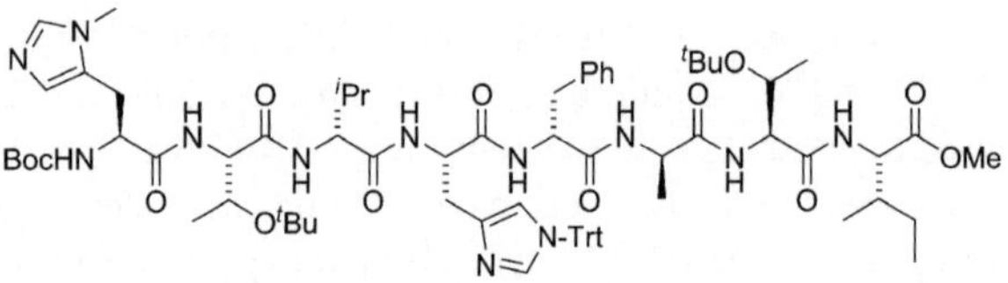

(7.167)

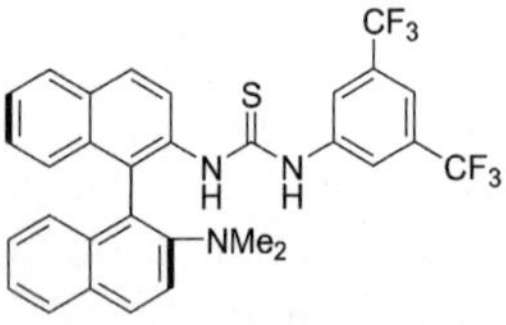

(7.171)

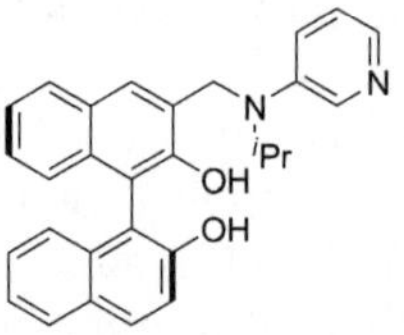

(7.172)

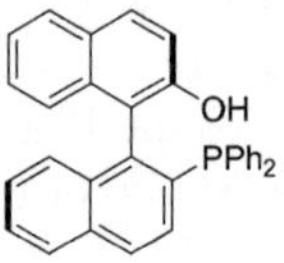

(7.173)

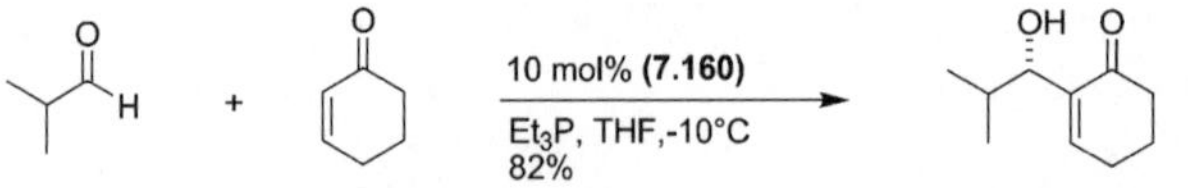

(7.71) + (7.161) → 10 mol% **(7.160)**, Et₃P, THF, -10°C, 82% → **(7.162)** 95% ee

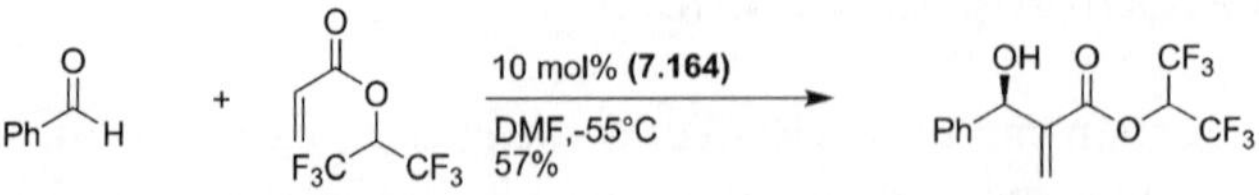

(7.17) + (7.165) → 10 mol% **(7.164)**, DMF, -55°C, 57% → **(7.166)** 95% ee

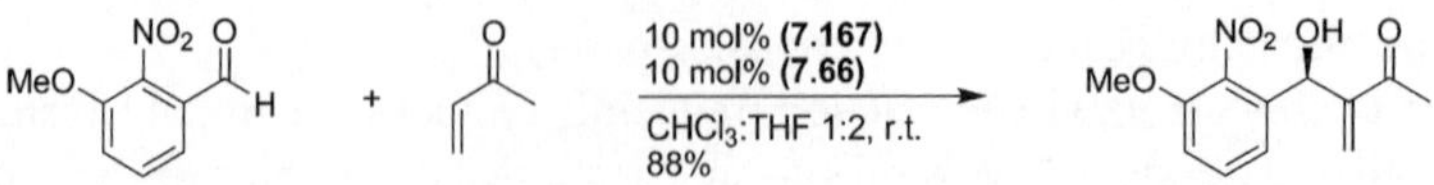

(7.169) + (7.168) → 10 mol% **(7.167)**, 10 mol% **(7.66)**, CHCl₃:THF 1:2, r.t., 88% → **(7.170)** 81% ee

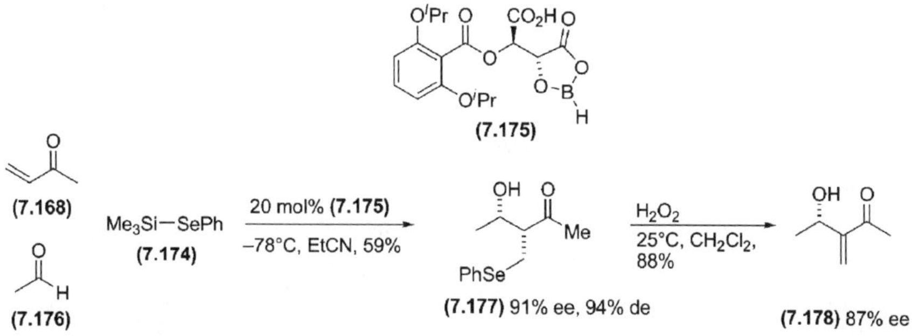

(7.175)

The titanium/BINOL catalysed ene reaction is subject to a strong positive non-linear effect (see Section 6.1). Thus, the use of BINOL of only 33% ee still provides product (**7.86**) with 91% ee. Higher enantioselectivites (91–99%) in this reaction have been obtained using as little as 0.1 mol% of catalysts prepared from 2:1 molar ratios of BINOL:titanium under nearly solvent-free conditions.[93]

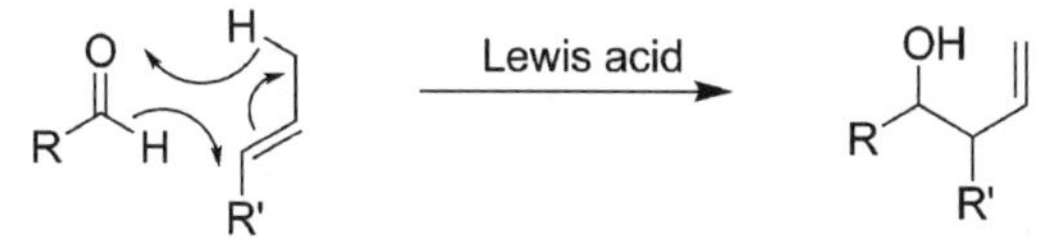

Figure 7.8 General mechanism of the carbonyl-ene reaction

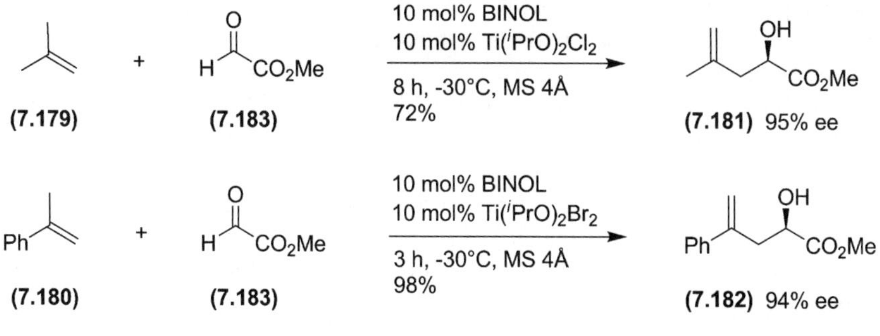

Normally, silyl enol ethers are considered to react with aldol substrates via an aldol mechanism, but Mikami and coworkers, in their examples, showed that the reaction involves an ene mechanism.[94] This is clear from the regiochemistry of the product (**7.186**) that is isolated from the reaction of silyl enol ether (**7.184**) and aldehyde (**7.185**) before hydrolysis. The same catalyst system has been used for the asymmetric desymmetrisation of the diene (**7.187**) with aldehyde (**7.183**). The

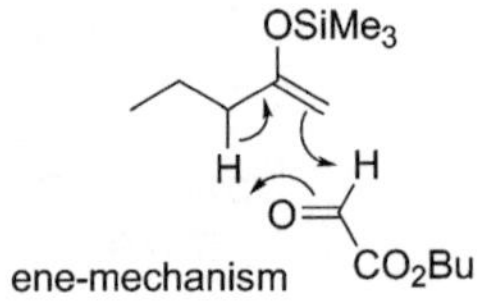

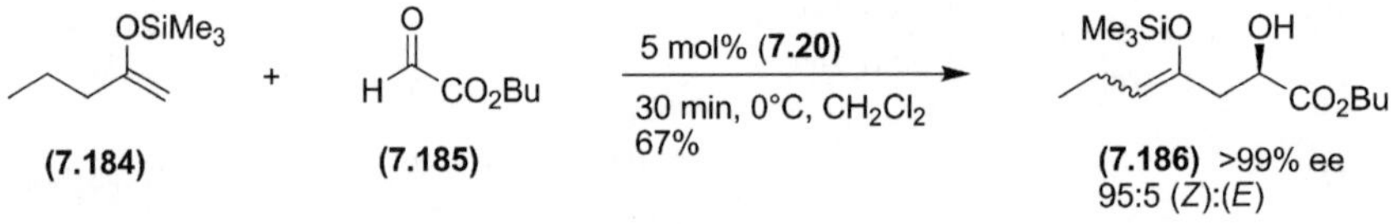

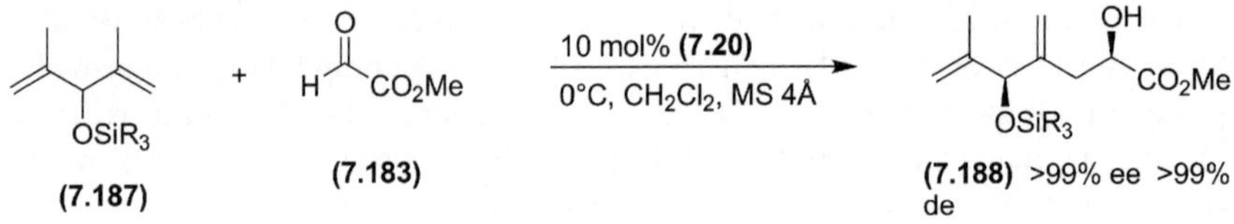

OSiMe3
H
H
O
ene-mechanism
CO2Bu

OSiMe3
(7.184)
+
O
H CO2Bu
(7.185)
5 mol% (7.20)
30 min, 0°C, CH2Cl2
67%
Me3SiO OH
CO2Bu
(7.186) >99% ee
95:5 (Z):(E)

OSiR3
(7.187)
+
O
H CO2Me
(7.183)
10 mol% (7.20)
0°C, CH2Cl2, MS 4Å
OH
CO2Me
OSiR3
(7.188) >99% ee >99% de

product (**7.188**) is obtained with superb diastereocontrol and enantiocontrol.[95] A transition state model rationalising this remarkable selectivity has been proposed by Corey and coworkers.[96]

As well as glyoxylate substrates, Mikami has employed fluoral (CF_3CHO) as a reactive aldehyde component of a carbonyl-ene reaction, still with very high selectivities.[97]

Carriera and coworkers employed titanium Schiff base catalysts such as (**7.189**) (see Section 7.1) in an ene reaction of 2-methoxypropene (**7.190**) with various aldehydes (**7.01**) (not especially reactive ones).[98] The overall synthetic strategy of using 2-methoxypropene (**7.190**) has significant merit in comparison with aldol reactions, because it is a cheap starting material, and means that a silyl enol ether does not have to be prepared prior to an aldol reaction. The Schiff base chromium complexes such as (**7.192**) developed by Jacobsen and coworkers is effective with both aliphatic and aromatic aldehydes providing up to 96% ee in the reaction of the latter type of substrate with 2-methoxypropene (**7.190**). This complex also catalyses the ene reaction with 2-silyloxypropene with high ee.[99]

The Evans copper(II) bis-oxazoline (BOX) catalysts (**7.193**) and (**7.194**) (see Section 7.1) have also been used effectively for glyoxalate ene reactions.[100] Even monosubstituted alkenes can be used as the ene component, where the alkene (**7.195**) reacts with ethyl glyoxalate (**7.196**) to give the α-hydroxy ester (**7.197**) with very high enantioselectivity. Other alkenes were also effective, providing enantiomeric excesses of over 90%, including alkene (**7.198**), which is converted into the ene-product (**7.199**).

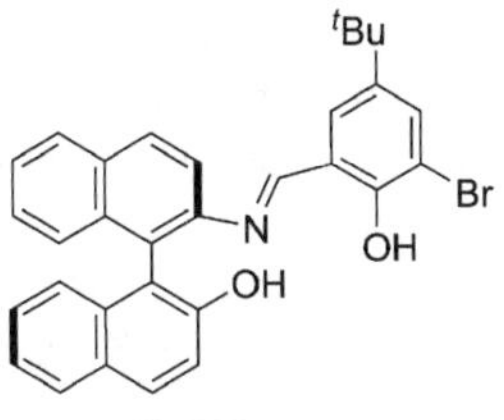

tBu
Br
N
OH
OH

(7.189)

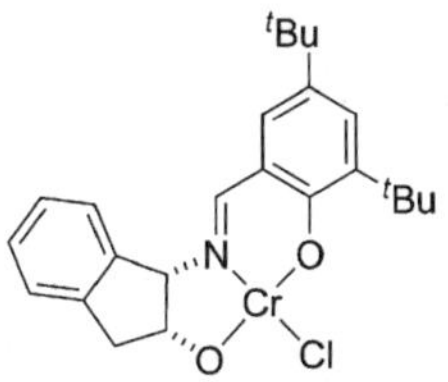

tBu
tBu
N
O
Cr
O
Cl

(7.192)

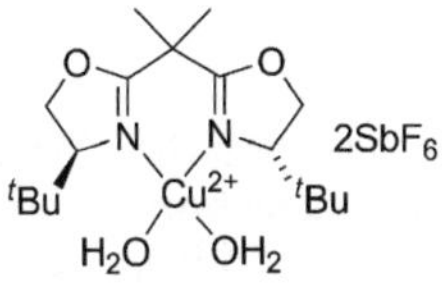

2SbF6
tBu
Cu2+
tBu
H2O
OH2

(7.193)

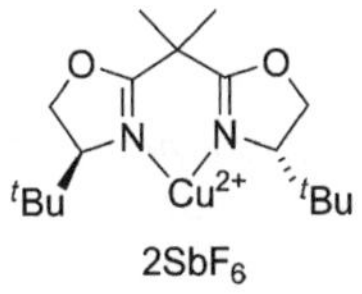

tBu
Cu2+
tBu
2SbF6

(7.194)

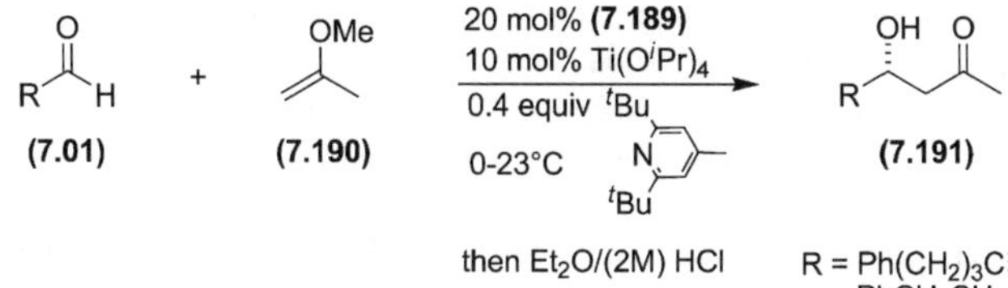

R
H
OMe
20 mol% (7.189)
10 mol% Ti(OiPr)4
0.4 equiv tBu
0-23°C
tBu
then Et2O/(2M) HCl
OH O
R

(7.01) **(7.190)** **(7.191)**

R = Ph(CH$_2$)$_3$C≡C 99%, 98% ee
PhCH$_2$CH$_2$- 98%, 90% ee
Ph- 83%, 66% ee

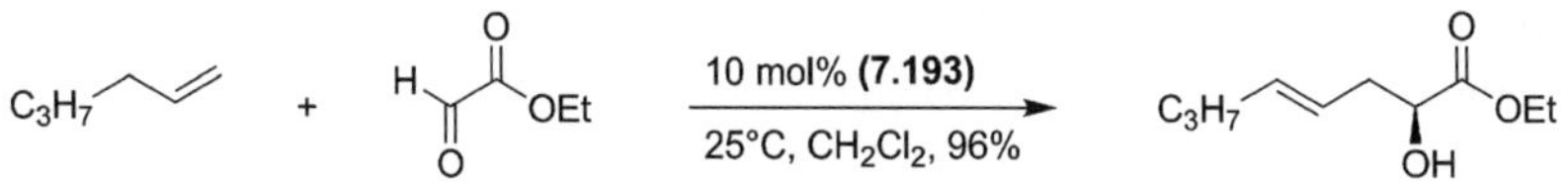

C3H7
H
OEt
10 mol% (7.193)
25°C, CH2Cl2, 96%
C3H7
OEt
OH

(7.195) **(7.196)** **(7.197)** 97% ee
96:4 *E:Z*

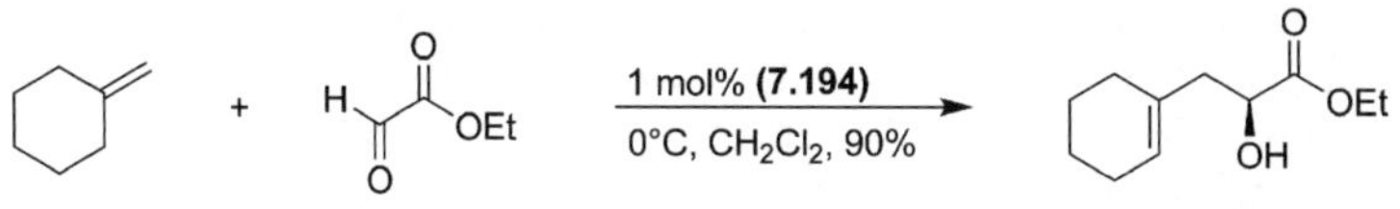

H
OEt
1 mol% (7.194)
0°C, CH2Cl2, 90%
OEt
OH

(7.198) **(7.196)** **(7.199)** 97% ee

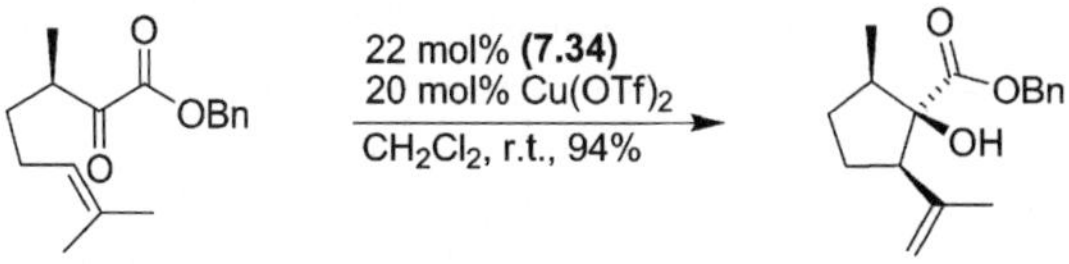

OBn
22 mol% (7.34)
20 mol% Cu(OTf)2
CH2Cl2, r.t., 94%
OBn
OH

(7.200) **(7.201)** 99.3% ee

The observed sense of asymmetric induction is consistent with a square planar copper(II)-glyoxylate complex with approach of the ene from the face opposite to the nearby *tert*-butyl substituent. Copper bis-oxazoline catalysts have also been used in the intramolecular carbonyl ene reaction. For instance, unsaturated α-ketoester (**7.200**) undergoes ene cyclisation to give the product (**7.201**) with high ee in the presence of the catalyst derived from BOX-ligand (**7.34**) and $Cu(OTf)_2$.[101]

References

1. For reviews on the catalytic asymmetric aldol reaction see: (a) S.G. Nelson, *Tetrahedron: Asymmetry*, **1998**, *9*, 357. (b) E. M. Carreira, in *Comprehensive Asymmetric Catalysis*, Vol. *3*, ed. E. N. Jacobsen, A. Pfaltz and H. Yamamoto, Springer-Verlag, Berlin, **1999**, 998. (c) T. D. Machajewski and C.-H. Wong, *Angew. Chem. Int. Ed.*, **2000**, *39*, 1352. (d) C. Palomo, M. Oiarbide and J. M. García, *Chem. Soc. Rev.*, **2004**, *33*, 65. (e) E. M. Carriera, A. Fettes and C. Marti, *Org. Reacts.*, **2006**, *67*, 1. (f) B. Scetter and R. Mahrwald, *Angew. Chem. Int. Ed.*, **2006**, 7506.
2. S. Kobayashi, Y. Fujishita and T. Mukaiyama, *Chem. Lett*, **1990**, 1455.
3. S. Kobayashi, M. Furuya, A. Ohtsubo and T. Mukaiyama, *Tetrahedron: Asymmetry*, **1991**, *2*, 635.
4. S. Kobayashi, T. Furuta, T. Hayashi, M. Nishijima and K. Hanada, *J. Am. Chem. Soc.*, **1998**, *120*, 908.
5. E. J. Corey, C. L. Cywin and T.D. Roper, *Tetrahedron Lett.*, **1992**, *33*, 6907.
6. E. R. Parmee, O. Tempkin, S. Masamune and A. Abiko, *J. Am. Chem. Soc.*, **1991**, *113*, 9365.
7. K. Ishihara, S. Kondo and H. Yamamoto, *J. Org. Chem.*, **2000**, *65*, 9125.
8. (a) K. Furuta, T. Maruyama and H. Yamamoto, *J. Am. Chem. Soc.*, **1991**, *113*, 1041. (b) S. Murata, M. Suzuki and R. Noyori, *J. Am. Chem. Soc.*, **1980**, *102*, 3248. (c) R. Mahrwald, *Chem. Rev.*, **1999**, 1095.
9. K. Mikami and S. Matsukawa, *J. Am. Chem. Soc.*, **1994**, *116*, 4077.
10. S. Matsukawa and K. Mikami, *Tetrahedron: Asymmetry*, **1995**, *6*, 2571.
11. G. E. Keck and D. Krishnamurthy, *J. Am. Chem. Soc.*, **1995**, *117*, 2363.
12. E. M. Carreira, R.A. Singer and W. Lee, *J. Am. Chem. Soc.*, **1994**, *116*, 8837.
13. R. A. Singer, M.S. Shephard and E. M. Carreira, *Tetrahedron*, **1998**, *54*, 7025.
14. R. A. Singer and E. M. Carreira, *J. Am. Chem. Soc.*, **1995**, *117*, 12360.
15. (a) H. Ishitani, Y. Yamashita, H. Shimizu and S. Kobayashi, *J. Am. Chem. Soc.*, **2000**, *122*, 5403. (b) Y. Yamashita, H. Ishitani, H. Shimizu and S. Kobayashi, *J. Am. Chem. Soc.*, **2002**, *124*, 3292. (c) K. Seki, M. Ueno and S. Kobayashi, *Org. Bimol. Chem.*, **2007**, *5*, 1347.
16. (a) G. Desimoni, G. Faita and K. A. Jorgensen, *Chem. Rev.*, **2006**, *106*, 3561. (b) G. Desimoni, G. Faita and P. Quadrelli, *Chem. Rev.*, **2003**, *103*, 3119. (c) J. S. Johnson and D. A. Evans, *Acc. Chem. Res.*, **2000**, *33*, 325.
17. (a) D.A. Evans, J.A. Murry and M.C. Kozlowski, *J. Am. Chem. Soc.*, **1996**, *118*, 5814.

18. D. A. Evans, M. C. Kozlowski, C. S. Burgey and D. W. C. MacMillan, *J. Am. Chem. Soc.*, **1997**, *119*, 7893. (b) D. A. Evans, C. S. Burgey, M. C. Kozlowski and S. W. Tregay, *J. Am. Chem. Soc.*, **1999**, *121*, 686.

19. D.A. Evans, D.W.C. MacMillan, K.R. Campos, *J. Am. Chem. Soc.*, **1997**, *119*, 10859.

20. D. A. Evans, C. W. Downey and J. L. Hubbs, *J. Am. Chem. Soc.*, **2003**, *125*, 8706.

21. J. Krüger and E. M. Carreira, *J. Am. Chem. Soc.*, **1998**, *120*, 837.

22. (a) M. Sodeoka, K. Ohrai and M. Shibasaki, *J. Org. Chem.*, **1995**, *60*, 2648. (b) M. Sodeoka, R. Tokunoh, F. Miyazaki, E. Hagiwara and M. Shibasaki, *Synlett*, **1997**, 463. (c) S.-I. Kiyooka, Y. Takeshita, Y. Tanaka, T. Higaki and Y. Wada, *Tetrahedron Lett.*, **2006**, *47*, 4453.

23. S. E. Denmark and R. A. Stavenger, *Acc. Chem. Res.*, **2000**, *33*, 432.

24. (a) S. E. Denmark, K.-T. Wong and R. A. Stavenger, *J. Am. Chem. Soc.*, **1997**, *119*, 2333. (b) S. E. Denmark, R. A. Stavenger and K.-T. Wong, *J. Org. Chem.*, **1998**, *63*, 918. (c) S. E. Denmark, R. A. Stavenger, K.-T. Wong and X. Su, *J. Am. Chem. Soc.*, **1999**, *121*, 4982.

25. S. E. Denmark, Y. Fan and M. D. Eastgate, *J. Org. Chem.*, **2005**, *70*, 5235.

26. S. E. Denamrk and S. K. Ghosh, *Angew. Chem. Int. Ed.*, **2001**, *40*, 4759.

27. (a) S. E. Denamark, T. Wynn and G. L. Beutner, *J. Am. Chem. Soc.*, **2002**, *124*, 13405. (b) S. E. Denamrk and T. Bui, *J. Org. Chem.*, **2005**, *70*, 10190.

28. (a) Y.M.A. Yamada, N. Yoshikawa, H. Sasai and M. Shibasaki, *Angew. Chem., Int. Ed. Engl.*, **1997**, *36*, 1871. (b) N. Yoshikawa, Y. M. A. Yamada, J. Das, H. Sasai and M. Shibasaki, *J. Am. Chem. Soc.*, **1999**, *121*, 4168. (c) M. Shibasaki and N. Yoshikawa, *Chem. Rev.*, **2002**, *102*, 2187.

29. B. M. Trost and H. Ito, *J. Am. Chem. Soc.*, **2000**, *122*, 12003.

30. B. M. Trost, S. Shin and J. A. Sciafani, *J. Am. Chem. Soc.*, **2005**, *127*, 8602.

31. B. M. Trost, A. Fettes and B. T. Shireman, *J. Am. Chem. Soc.*, **2004**, *126*, 2660.

32. N. Yoshikawa, N. Kumagai, S. Matsunaga, G. Moll, T. Ohshima, T. Suzuki and M. Shibasaki, *J. Am. Chem. Soc.*, **2001**, *123*, 2466.

33. (a) U. Eder, G. Sauer and R. Wiechert, *Angew. Chem., Int. Ed. Engl.*, **1971**, *10*, 496. (b) Z. G. Hajos and D. R. Parrish, *J. Org. Chem.*, **1974**, *39*, 1615.

34. (a) B. List, R. A. Lerner and C. F. Barbas III, *J. Am. Chem. Soc.*, **2000**, *122*, 2395. (b) K. Sakthivel, W. Notz, T. Bui and C. F. Barbas III, *J. Am. Chem. Soc.*, **2001**, *123*, 5260. (c) W. Notz, F. Tanaka and C. F. Barbas III, *Acc. Chem. Res.*, **2004**, *37*, 580.

35. A. B. Northrup and D. W. C. MacMillan, *J. Am. Chem. Soc.*, **2002**, *124*, 6798.

36. R. I. Storer and D. W. C. MacMillan, *Tetrahedron*, **2004**, *60*, 7705.

37. For reviews on the organocatalytic aldol reaction: see: (a) P. I. Dalko and L. Moisan, *Angew. Chem. Int. Ed.*, **2004**, *43*, 5138. (b) J. Seayad and B. List, *Org. Biomol. Chem.*, **2005**, *3*, 719. (c) A. Berkessel and H. Gröger, *Asymmetric Organocatalysis*, Wiley-VCH, Weinheim, **2005**, 130. (d) H. Pellisier, *Tetrahedron*, **2007**, *63*, 9267.

38. (a) A. J. A. Cobb, D. M. Shaw, D. A. Longbottom, J. B. Gold and S. V. Ley, *Org. Biomol. Chem.*, **2005**, *3*, 84. (b) A. Hartikka and P. I. Arvidsson, *Eur. J. Org. Chem.*, **2005**, 4287. (c) H. Torii, M. Nakadai, K. Ishihara, S. Saito and H. Yamamoto, *Angew. Chem. Int. Ed.*, **2004**, *43*, 1983. (d) N. Mase, Y. Nakai, N. Ohara, H. Yoda, K. Takabe, F. Tanaka and C. F. Barbas III, *J. Am. Chem. Soc.*, **2006**, *128*, 734. (e) Y. Hayashi, T. Sumiya, J. Takahashi, H. Gotoh, T. Urushima and M. Shoji, *Angew. Chem. Int. Ed.*, **2006**, *45*, 958.

(f) C. Cheng, J. Sun, C. Wang, Y. Zhang, S. Wei, F. Jiang and Y. Wu, *J. Chem. Soc., Chem. Commun.*, **2006**, 215. (g) Z. Tang, Z.-H. Yang, X.-H. Chen, L.-F. Cun, A.-Q. Mi, Y. Z. Jiang and L.-Z. Gong, *J. Am. Chem. Soc.*, **2005**, *127*, 9285. (h) J.-R. Chen, X.-Y. Li, X.-N. Xing and W.-J. Xiao, *J. Org. Chem.*, **2006**, *71*, 8198.

39. (a) M. Sawamura and Y. Ito, in *Catalytic Asymmetric Synthesis*, ed. I. Ojima, VCH, New York, **1993**, Chapter 7.2, 367. (b) R. Kuwano and Y. Ito, in *Comprehensive Asymmetric Catalysis*, Vol. 3, ed. E. N. Jacobsen, A. Pfaltz and H. Yamamoto, Springer-Verlag, Berlin, **1999**, 1067.

40. (a) Y. Ito, M. Sawamura and T. Hayashi, *J. Am. Chem. Soc.*, **1986**, *108*, 6405. (b) T. Hayashi, M. Sawamura and Y. Ito, *Tetrahedron*, **1992**, *48*, 1999.

41. M. Sawamura, Y. Nakayama, T. Kato and Y. Ito, *J. Org. Chem.*, **1995**, *60*, 1727.

42. R. Kuwano, H. Miyazaki and Y. Ito, *J. Chem. Soc.,Chem. Commun.*, **1998**, 71.

43. T. Ooi, M. Kaneda, M. Taniguchi and K. Maruoka, *J. Am. Chem. Soc.*, **2004**, *126*, 9685.

44. N. Yoshikawa and M. Shibasaki, *Tetrahedron*, **2002**, *58*, 8289.

45. For reviews on the catalytic asymmetric nitroaldol reaction see: (a) M. Shibasaki and H. Gröger, in *Comprehensive Asymmetric Catalysis*, Vol. 3, ed. E. N. Jacobsen, A. Pfaltz and H. Yamamoto, Springer-Verlag, Berlin, **1999**, 1076. (b) C. Palomo, M. Oiarbide and A. Mielgo, *Angew. Chem. Int. Ed.*, **2004**, *43*, 5442. (c) C. Palomo, M. Oiarbide and A. Laso, *Eur. J. Org. Chem.*, **2007**, 2561.

46. T. Oi, K. Doda and K. Maruoka, *J. Am. Chem. Soc.*, **2003**, *125*, 2054.

47. T. Risgaard, K. V. Gothelf and K. A. Jørgensen, *Org. Biomol. Chem.* **2003**, *1*, 153.

48. C. Christensen, K. Juhl, R. G. Hazell and K. A. Jørgensen, *J. Org. Chem.*, **2002**, *67*, 4875.

49. D. A. Evans, D. Seidel, M. Rueping, H. W. Lam, J. T. Shaw and C. W. Downey, *J. Am. Chem. Soc.*, **2003**, *125*, 12692.

50. M. Bandini, F. Piccinelli, S. Tommasi, A. Umani-Ronchi and C. Ventrici, *J. Chem. Soc., Chem. Commun.*, **2007**, 616.

51. (a) H. Sasai, T. Suzuki, S. Arai, T. Arai and M. Shibasaki, *J. Am. Chem. Soc.*, **1992**, *114*, 4418. (b) H. Sasai, T. Suzuki, N. Itoh, K. Tanaka, T. Date, K. Okamura and M. Shibasaki, *J. Am. Chem. Soc.*, **1993**, *115*, 10372. (c) T. Arai, Y. M. A. Yamada, N. Yamamoto, H. Sasai and M. Shibasaki, *Chem. Eur. J.*, **1996**, *2*, 1368. (d) B. M. Trost and V. S. C. Yeh, *Angew. Chem. Int. Ed.*, **2002**, *41*, 861.

52. H. Li, B. Wang and L. Deng, *J. Am. Chem. Soc.*, **2006**, *128*, 732.

53. T. Marcelli, R. N. S. van der Haas, J. H. van Maarseveen and H. Hiemstra, *Angew. Chem. Int. Ed.*, **2006**, *45*, 929.

54. Y. Sohtome, Y. Hashimoto and K. Nagasawa, *Eur. J. Org. Chem.*, **2006**, 2894.

55. For reviews on the catalytic asymmetric Mannich reaction see: (a) S. E. Denmark and O. J.-C. Nicaise in *Comprehensive Asymmetric Catalysis*, Vol. 2, ed. E. N. Jacobsen, A. Pfaltz and H. Yamamoto, Springer-Verlag, Berlin, **1999**, 954. (b) T. Vilaivan, W. Bhanthumnavin and Y. Sritana-Anant, *Curr. Org. Chem.*, **2005**, *9*, 1315. (c) A. Berkessel and H. Gröger, *Asymmetric Organocatalysis*, Wiley-VCH, Weinheim, **2005**, 85. (d) G. K. Friestad and A. K. Mathies, *Tetrahedron*, **2007**, *63*, 2541.

56. (a) S. Kobayashi, H. Ishitani and M. Ueno, *J. Am. Chem. Soc.*, **1998**, *120*, 431. (b) H. Ishitani, M. Ueno and S. Kobayashi, *J. Am. Chem. Soc.*, **1997**, *119*, 7153. (c) H. Ishitani, M. Ueno and S. Kobayashi, *J. Am. Chem. Soc.*, **2000**, *122*, 8180.

57. K. Saruhashi and S. Kobayashi, *J. Am. Chem. Soc.*, **2006**, *128*, 11232.
58. E. Hagiwar, A. Fujii and M. Sodeoka, *J. Am. Chem. Soc.*, **1998**, *120*, 2474.
59. D. Ferraris, B. Young, T. Dudding and T. Lectka, *J. Am. Chem. Soc.*, **1998**, *120*, 4548.
60. N. S. Josephsohn, M. L. Snapper and A. H. Hoveyda, *J. Am. Chem. Soc.*, **2004**, *126*, 3734.
61. (a) S. Kobayashi, T. Hamada and K. Manabe, *J. Am. Chem. Soc.*, **2002**, *124*, 5640. (b) T. Hamada, K. Manabe and S. Kobayashi, *J. Am. Chem. Soc.*, **2004**, *126*, 7768. (c) T. Hamada, K. Manabe and S. Kobayashi, *Chem. Eur. J.*, **2006**, *12*, 1205.
62. (a) S. Matsunaga, N. Kumagai, S. Harada and M. Shibasaki, *J. Am. Chem. Soc.*, **2003**, *125*, 4712. (b) S. Matsunaga, T. Yoshida, H. Morimoto, N. Kumagai and M. Shibasaki, *J. Am. Chem. Soc.*, **2004**, *126*, 8777.
63. B. M. Trost, J. Jaratjaroonphong and V. Reutrakul, *J. Am. Chem. Soc.*, **2006**, *128*, 2778.
64. (a) B. List, *J. Am. Chem. Soc.*, **2000**, *122*, 9336. (b) B. List, P. Pojarliev, W. T. Biller and H. J. Martin, *J. Am. Chem. Soc.*, **2002**, *124*, 827.
65. (a) H. Zhang, M. Mifsud, F. Tanaka and C. F. Barbas, III, *J. Am. Chem. Soc.*, **2006**, *128*, 9630. (b) S. Mitsumori, H. Zhang, P. H.-Y. Cheong, K. N. Houk, F. Tanaka and C. F. Barbas, III, *J. Am. Chem. Soc.*, **2006**, *128*, 1040.
66. D. Uraguchi and M. Terada, *J. Am. Chem. Soc.*, **2004**, *126*, 5356.
67. T. Akiyama, J. Itoh, K. Yokota and K. Fuchibe, *Angew. Chem. Int. Ed.*, **2004**, *43*, 1566.
68. A. G. Wenzel and E. N. Jacobsen, *J. Am. Chem. Soc.*, **2002**, *124*, 12964.
69. P. Bakó, Á. Szöllősy, P. Bombicz and L. Tőke, *Synlett*, **1997**, 291.
70. P. Bakó, K. Vizvárdi, Z. Bajor and L. Tőke, *J. Chem. Soc., Chem. Commun.*, **1998**, 1193.
71. T. J. R. Achard, Y. N. Belokon, M. Ilyin, M. Moskalenko, M. North and F. Pizzato, *Tetrahedron Lett.*, **2007**, *48*, 2965.
72. (a) S. Arai and T. Shioiri, *Tetrahedron Lett.*, **1998**, *39*, 2145. (b) S. Arai, Y. Shirai, T. Ishida and T. Shioiri, *Tetrahedron*, **1999**, *55*, 6375.
73. S. Arai, Y. Shirai, T. Ishida and T. Shioiri, *J. Chem. Soc., Chem. Commun.*, **1999**, 49.
74. S. Arai and T. Shioiri, *Tetrahedron*, **2002**, *58*, 1407.
75. S. Arai, K. Tokumara and T. Aoyama, *Tetrahedron Lett.*, **2004**, *45*, 1845.
76. (a) D. Basavaiah, P.D. Rao and R.S. Hyma, *Tetrahedron*, **1996**, *52*, 8001. (b) E. Ciganek, *Org. Reacts.*, **1997**, *51*, 201. (b) D. Basavaiah, A. J. Rao and T. Satyanarayana, *Chem. Rev.*, **2003**, *103*, 811. (c) Y.-L. Shi and M. Shi, *Eur. J. Org. Chem.*, **2007**, *18*, 2905.
77. (a) T. Oishi, H. Oguri and M. Hirama, *Tetrahedron: Asymmetry*, **1995**, *6*, 1241. (b) I.E. Markó, P. R. Giles and N. J. Hindley, *Tetrahedron*, **1997**, *53*, 1015. (c) T. Hayase, T. Shibata, K. Soai and Y. Wakatsuki, *J. Chem. Soc., Chem. Commun.*, **1998**, 1271.
78. K.-S. Yang, W.-D. Lee, J.-F. Pan and K. Chen, *J. Org. Chem.* **2003**, *68*, 915.
79. N. T. McDougal and S. E. Schaus, *J. Am. Chem. Soc.*, **2003**, *125*, 12094.
80. (a) Y. Sohtome, A. Tanatani, Y. Hashimoto and K. Nagasawa, *Tetrahedron Lett.*, **2004**, *45*, 5589. (b) A. Berkessel, K. Roland and J. M. Neudörfl, *Org. Lett.*, **2006**, *8*, 4195.
81. Y. Iwabuchi, M. Nakatani, N. Yokoyama and S. Hatakeyama, *J. Am. Chem. Soc.*, **1999**, *121*, 10219.
82. M. Shi and Y.-M. Xu, *Angew. Chem. Int. Ed.*, **2002**, *41*, 4507.
83. (a) M. Shi, J.-K. Jiang and C. Q-. Li, *Tetrahedron Lett.*, **2002**, *43*, 127. (b) M. Shi and J.-K. Jiang, *Tetrahedron: Asymmetry*, **2002**, *13*, 1941.

84. (a) J. E. Imbriglio, M. M. Vasbinder and S. J. Miller, *Org. Lett.*, **2003**, *5*, 3741. (b) M. M. Vasbinder, J. E. Imbriglio and S. J. Miller, *Tetrahedron*, **2006**, *62*, 11450.
85. C. E. Arroyan, M.M. Vasbinder and S. J. Miller, *Org. Lett.*, **2005**, *7*, 3849.
86. J. Wang, H. Li, X. Yu, L. Zu and W. Wang, *Org. Lett.*, **2005**, *7*, 4293.
87. K. Matsui, S. Takizawa and H. Sasai, *J. Am. Chem. Soc.*, **2005**, *127*, 3680.
88. (a) M. Shi and L.-H. Chen, *J. Chem. Soc., Chem. Commun.*, **2003**, 1310. (b) M. Shi, L.-H. Chen and C.-Q. Li, *J. Am. Chem. Soc.*, **2005**, *127*, 3790. (c) M. Shi and L.-H. Chen, *Pure Appl. Chem.*, **2005**, *7*, 2105.
89. A.G.M. Barrett and A. Kamimura, *J. Chem. Soc., Chem. Commun.*, **1995**, 1755.
90. (a) D. J. Berrisford and C. Bolm, *Angew. Chem., Int. Ed.*, **1995**, *34*, 1717. (b) K. Mikami and M. Terada, in *Comprehensive Asymmetric Catalysis*, Vol. 3, ed. E. N. Jacobsen, A. Pfaltz and H. Yamamoto, Springer-Verlag, Berlin, **1999**, 1144. (c) L. C. Dias, *Curr. Org. Chem.*, **2000**, *4*, 305.
91. (a) K. Mikami, M. Terada, S. Narisawa and T. Nakai, *Synlett*, **1992**, 255. (c) K. Mikami, *Pure Appl. Chem.*, **1996**, *68*, 639.
92. M. Terada and K. Mikami, *J. Chem. Soc., Chem. Commun.*, **1994**, 833.
93. Y. Yuan, X. Zhang and K. Ding, *Angew. Chem. Int. Ed.*, **2003**, *42*, 5478.
94. K. Mikami and S. Matsukawa, *J. Am. Chem. Soc.*, **1993**, *115*, 7039.
95. K. Mikami, S. Narisawa, M. Shimizu and M. Terada, *J. Am. Chem. Soc.*, **1992**, *114*, 6566.
96. E. J. Corey, D. Barnes-Seeman, T. W. Lee and S.N. Goodman, *Tetrahedron Lett.*, **1997**, *38*, 6513.
97. K. Mikami, T. Yajima, M. Terada, E. Kato and M. Maruta, *Tetrahedron: Asymmetry*, **1994**, *5*, 1087.
98. E. M. Carreira, W. Lee and R. A. Singer, *J. Am. Chem. Soc.*, **1995**, *117*, 3649.
99. (a) R. T. Ruck and E. N. Jacobsen, *J. Am. Chem. Soc.*, **2002**, *124*, 2882. (b) R. T. Ruck and E. N. Jacobsen, *Angew. Chem. Int. Ed.*, **2003**, *42*, 4771.
100. (a) D. A. Evans, C. S. Burgey, N. A. Paras, T. Vojkovsky and S. W. Tregay, *J. Am. Chem. Soc.*, **1998**, *120*, 5824. (b) D. A. Evans, S. W. Tregay, S. S. Burgey, N. A. Paras and T. Vojkovsky, *J. Am. Chem. Soc.*, **2000**, *122*, 7936.
101. D. Yang, M. Yang and N.-Y. Zhu, *Org. Lett.*, **2003**, *5*, 3749.

Chapter 8
Cycloadditions

The greater part of this chapter is concerned with the Diels–Alder and hetero-Diels–Alder reaction. The asymmetric version of both of these reactions can be catalysed with metal-based Lewis acids and also organocatalysts. The catalytic asymmetric 1,3-dipolar cycloaddition of nitrones and azomethine ylides is also discussed. Again, most success in this area has been achieved using metal-based Lewis acids and the use of organocatalysts is beginning to be explored. This chapter concludes with a brief account of recent research into the asymmetric [2+2]-cycloaddition, catalysed by enantiomerically pure Lewis acids and amine bases, and also the Pauson–Khand [2+2+1] cycloaddition mediated by titanium, rhodium and iridium complexes.

8.1   Diels–Alder Reactions

The Diels–Alder reaction is without doubt one of the most important methods for the generation of six-membered rings and it comes as no surprise that the development of a catalytic asymmetric variant has been an intense area of study.[1] One of the most common strategies is based on the use of enantiomerically pure metal-based Lewis acids and this section considers the different metals employed as Lewis acids in turn, including catalysts based on boron, aluminium and titanium, as well as a host of other transition metal complexes which give high ees in the Diels–Alder reaction with α-substituted enals and oxazolidinones. The use of Lewis acids further activated by Brønsted acids and other Lewis acids is an interesting new development in this area, which has led to a significant increase in the scope of this reaction. In addition, the Diels–Alder reaction is also catalysed by a range of organocatalysts. These catalysts activate the dienophile *in situ* either by conversion into a iminium ion or by hydrogen bonding.

8.1.1   Boron-Based Lewis Acids

Oxazaborolidines of the general type (**8.01**) were reported by Yamamoto[2] and by Helmchen[3] to give good asymmetric induction as catalysts in Diels–Alder

Catalysis in Asymmetric Synthesis 2e   © 2009 Vittorio Caprio and Jonathan M.J. Williams

reactions. However, Corey and Loh demonstrated that the *N*-tosyl tryptophan-derived catalyst (**8.02**) was particularly effective at providing high enantioselectivity.[1c, 4]

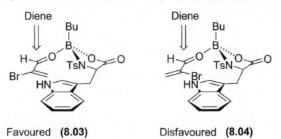

The indole unit, which is a π-base, is believed to have an attractive interaction with the dienophile, which is a π-acid. When the dienophile has associated to the Lewis acidic boron, the enal may adopt either the s-*cis* (**8.03**) or s-*trans* (**8.04**) conformation, as illustrated in Figure 8.1. Complex (**8.04**) (s-*trans*) undergoes an unfavourable interaction between the bromine atom and indole ring as the diene approaches, and the bromine atom is forced downwards. The reaction prefers to proceed through the s-*cis* conformation (**8.03**) and the model accounts for the very high selectivity that is observed.[5]

Bromoacrolein (**8.05**) is a good substrate for the enantioselective Diels–Alder reaction, and reacts with cyclopentadiene (**8.06**) to give the *exo*-Diels–Alder adduct (**8.07**) in good yield and with excellent selectivity. Corey subsequently used this catalyst and related catalysts, including oxazaborolidine (**8.08**), to provide precursors to various natural products.[6] For example, the reaction of bromoacrolein (**8.05**) with the elaborated cyclopentadiene (**8.09**) affords the product (**8.10**) with excellent diastereo- and enantiocontrol; the product was used in a synthesis of a gibberellic acid. The catalyst also lends itself well to Diels–Alder reactions with furan (**8.11**) as the dienophile;[7] the oxabicyclic product (**8.12**) is also a useful synthetic building block. High levels of stereocontrol in this process are partially dependant on selective binding of the Lewis acid to one carbonyl

Figure 8.1 The favoured s-*cis* transtion state with oxazaborolidine catalysts

lone pair. In aldehydes coordination typically occurs *syn* to the formyl proton. However, the lone pairs of ketone carbonyls are less distinct. Furthermore, α,β-enones are less electron-deficient dienophiles and a stronger Lewis acid may be required. Thus enantioselective Diels–Alder reactions of α,β-enones has proved

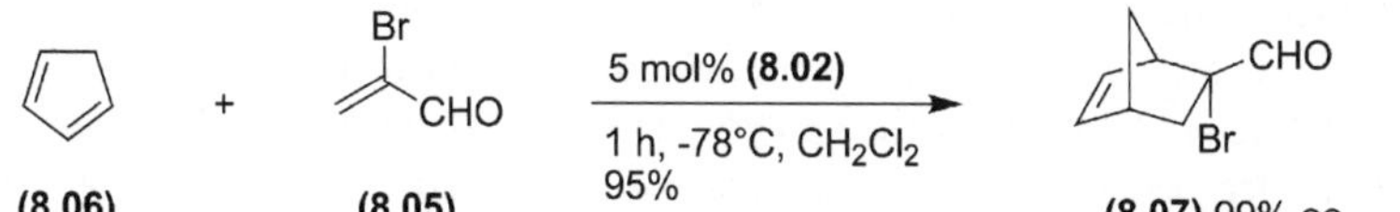

(8.06)   **(8.05)**

5 mol% **(8.02)**
1 h, -78°C, CH$_2$Cl$_2$
95%

(8.07) 99% ee
96:4 *exo* CHO:*endo*

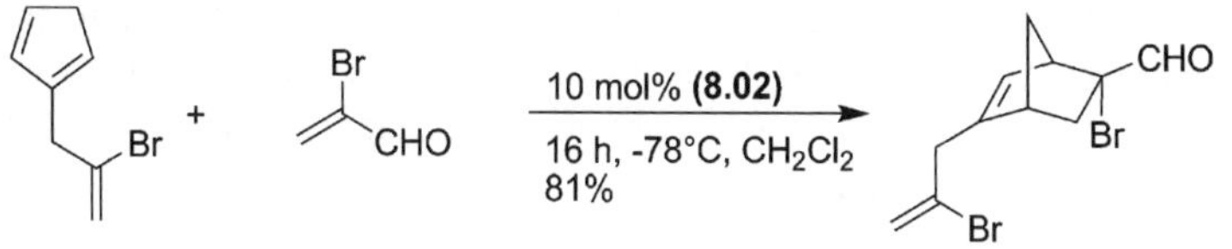

(8.09)   **(8.05)**

10 mol% **(8.02)**
16 h, -78°C, CH$_2$Cl$_2$
81%

(8.10) 99% ee
99:1 *exo* CHO:*endo*

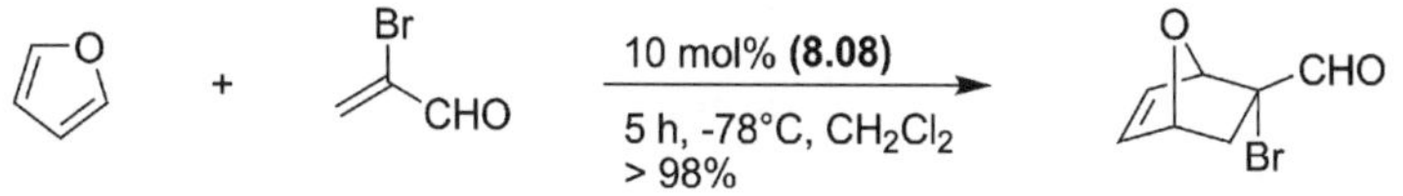

(8.11)   **(8.05)**

10 mol% **(8.08)**
5 h, -78°C, CH$_2$Cl$_2$
> 98%

(8.12) 92% ee
99:1 exo CHO:*endo*

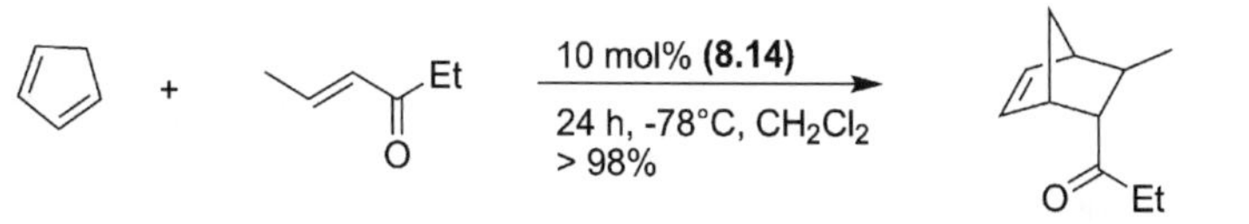

(8.06)   **(8.13)**

10 mol% **(8.14)**
24 h, -78°C, CH$_2$Cl$_2$
> 98%

(8.15) 94% ee
2:98 *exo* COEt:*endo*

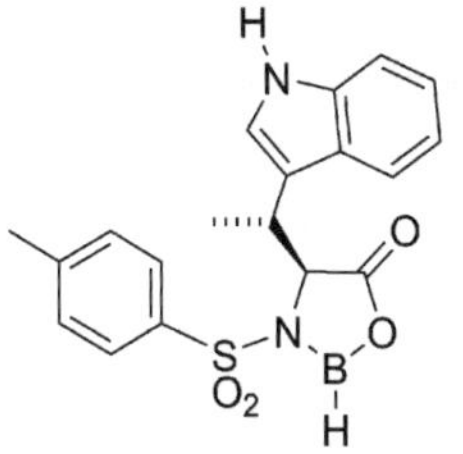

(8.08) 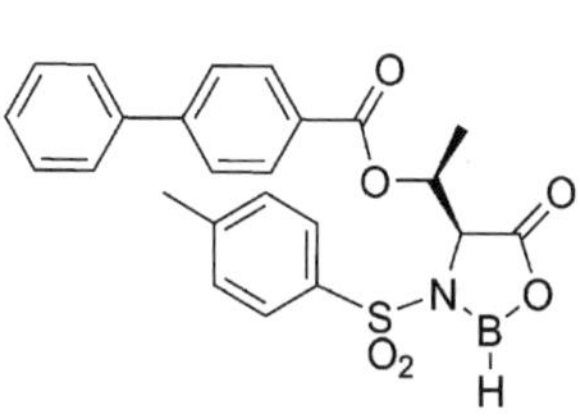   **(8.14)**

challenging. Nevertheless high ees in the Diels–Alder reaction of simple acyclic enones such as (8.13) with cyclopentadiene have been achieved using the *allo*-threonine-derived oxazaborolidinone (8.14) to give the *endo*-isomer (8.15) predominantly.

In 1988, Yamamoto and coworkers reported the use of enantiomerically pure (acyloxy)borane (CAB) catalysts (8.16) for the enantioselective Diels–Alder reaction.[8] These catalysts are derived from tartaric acid, and again, their reactions have been particularly selective with α-substituted enals as substrate.[9] The dienophile (8.17) undergoes a highly selective Diels–Alder reaction, affording the adduct (8.18), which contains four new stereocentres controlled by the catalysed reaction. The CAB catalysts were successfully applied to a cyclic example, where the acyclic starting material (8.19) affords a bicyclic product (8.20).[10]

The solution conformations of CAB complexes with methacrolein and crotonaldehyde were investigated by NOE spectroscopy.[11] The s-*trans* conformation was found to be favoured in most cases, although this doesn't necessarily prove that it is the most reactive conformation.

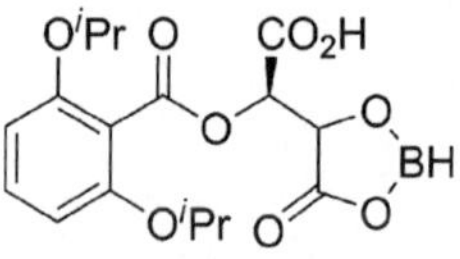

(8.16)

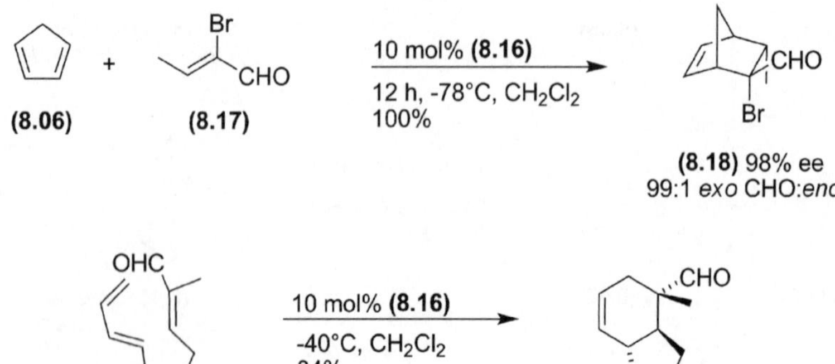

(8.06)   **(8.17)**

10 mol% **(8.16)**

12 h, -78°C, CH$_2$Cl$_2$
100%

(8.18) 98% ee
99:1 *exo* CHO:*endo*

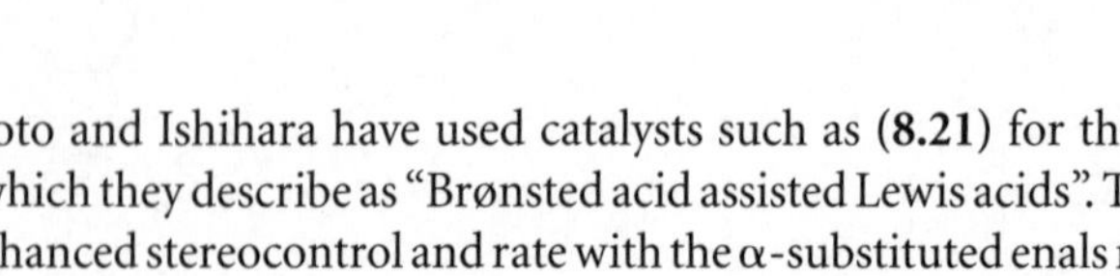

(8.19)

10 mol% **(8.16)**

-40°C, CH$_2$Cl$_2$
84%

(8.20) 92% ee
99:1 *endo:exo*

Yamamoto and Ishihara have used catalysts such as (8.21) for the Diels–Alder reaction, which they describe as "Brønsted acid assisted Lewis acids". These catalysts provide enhanced stereocontrol and rate with the α-substituted enals reported.[12] In

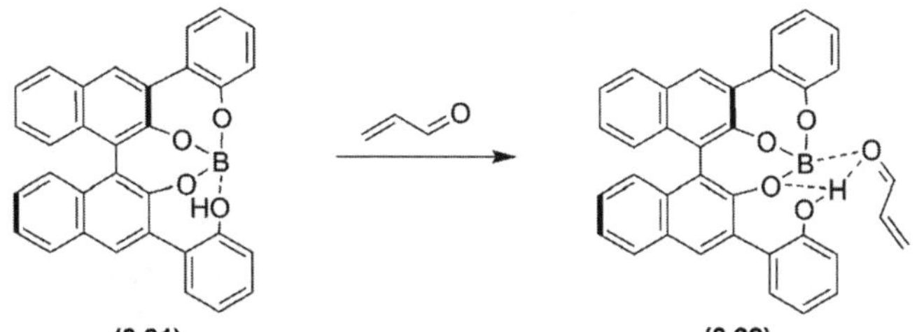
O
O
B
O
HO
O
O
B
O
H
O
O

(8.21) (8.22)

Figure 8.2 Coordination of dienophiles to catalyst **(8.21)**

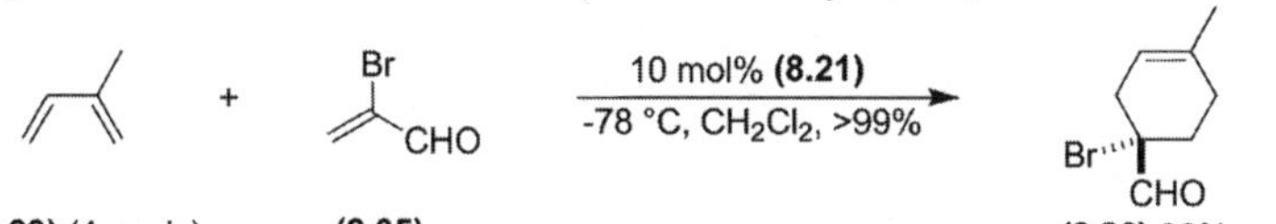
+
Br
CHO
10 mol% (8.21)
-78 °C, CH2Cl2, >99%
Br
CHO

(8.23) (4 equiv) **(8.05)** **(8.26)** 98% ee

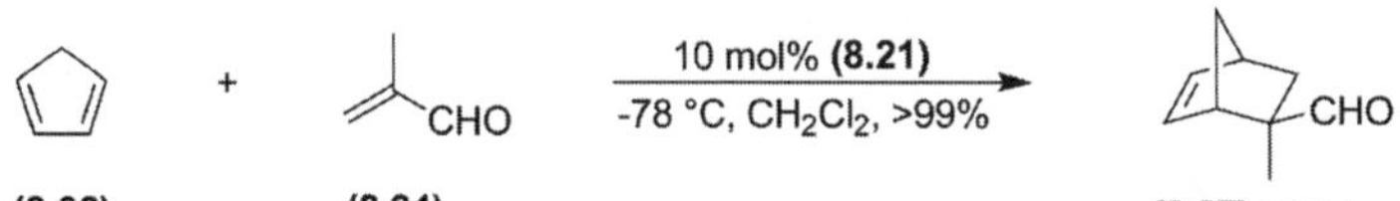
+
CHO
10 mol% (8.21)
-78 °C, CH2Cl2, >99%
CHO

(8.06) **(8.24)** **(8.27)** >99:1 *exo* CHO:*endo* 98% ee

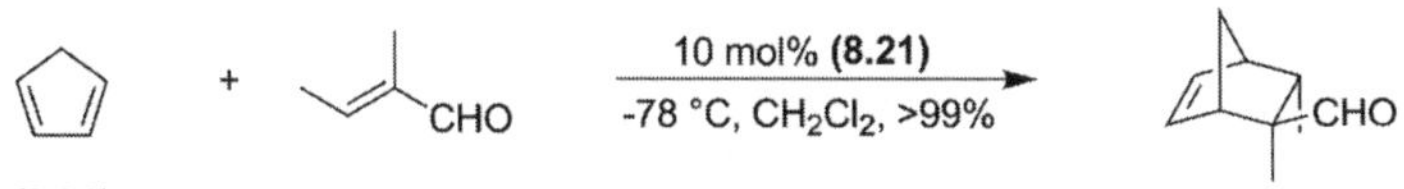
+
CHO
10 mol% (8.21)
-78 °C, CH2Cl2, >99%
CHO

(8.06) **(8.25)** **(8.28)** >99:1 *exo* CHO:*endo* 98% ee

the transition state assembly, one of the phenoxy groups is protonated, which allows the carbonyl group of the dienophile to coordinate to both the Lewis acidic boron atom and undergo hydrogen bonding with the catalyst. Further hydrogen bonding of the hydroxyl group to a B–O bond, as depicted in Figure 8.2, enhances the Lewis acidity of boron. Bromoacrolein **(8.05)**, methacrolein **(8.24)** and dienophile **(8.25)** all react with essentially complete stereoselectivity to give the Diels–Alder adducts **(8.26–8.28)**.

Cationic Lewis acids show improved acidity and hence activity. For instance, even at −94°C, relatively unreactive dienes such as butadiene **(8.29)** react with bromoacrolein **(8.05)** with excellent yields and good selectivities in the presence of the oxazaborinane **(8.30)**.[13] The Lewis acidity of oxazoborolidines can also be enhanced by protonation using strong Brønsted acids. The triflate salt **(8.32)** generated by protonation of the corresponding oxazaborolidine with triflic acid shows high activity and wide scope catalysing the Diels–Alder reaction of acyclic

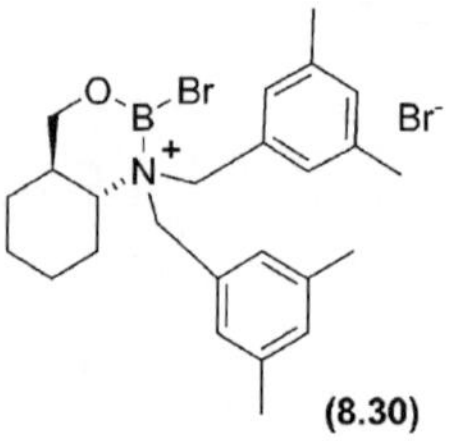

(8.30)

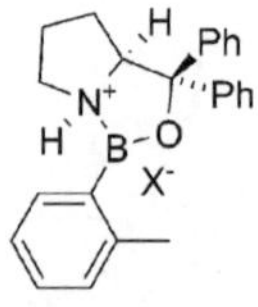

(8.32) X = ‾OTf
(8.35) X = ‾N(Tf)$_2$

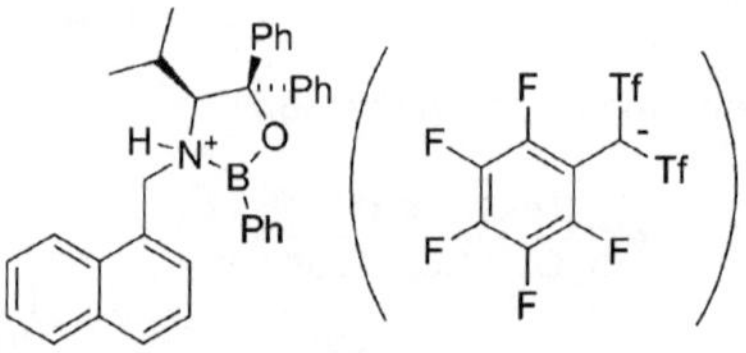

(8.36)

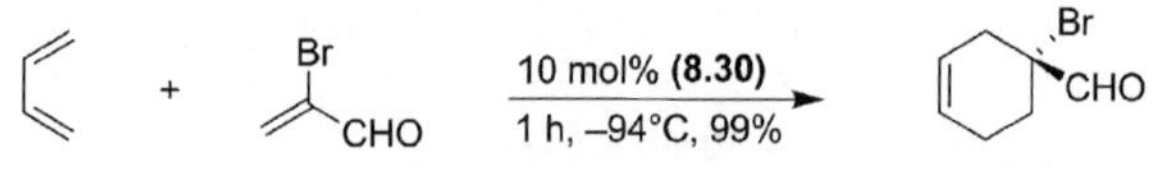

(8.29)   **(8.05)** $\xrightarrow[\text{1 h, –94°C, 99%}]{\text{10 mol% (8.30)}}$ **(8.31)** 94% ee

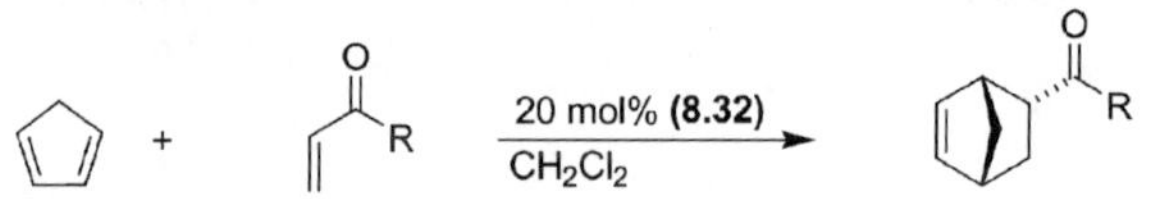

(8.06) $\xrightarrow[\text{CH}_2\text{Cl}_2]{\text{20 mol% (8.32)}}$

R	Conditions	endo:exo	Yield	ee
OH	-35°C, 1.5 h	95:5	99%	98%
Et	-20°C, 2 h	94:6	99%	97%
OEt	-20°C, 16 h	97:3	96%	>99%

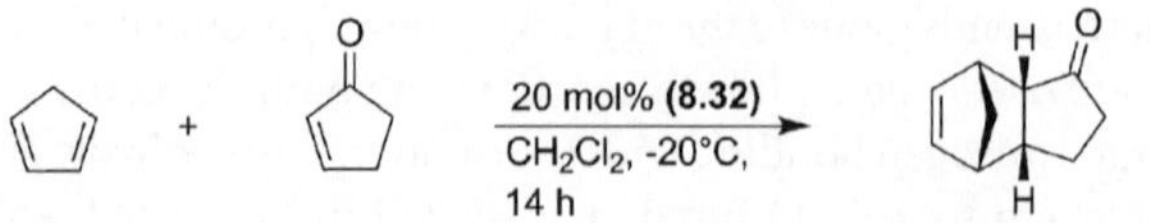

(8.06)   **(8.33)** $\xrightarrow[\substack{\text{CH}_2\text{Cl}_2, -20°C, \\ 14 \text{ h}}]{\text{20 mol% (8.32)}}$ **(8.34)** 5:95 *exo:endo*
92% ee

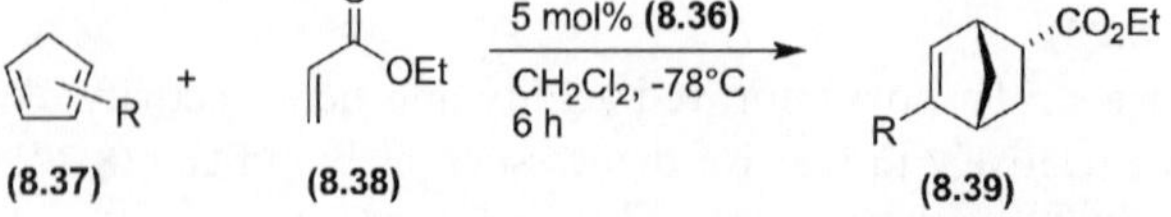

(8.37)   **(8.38)** $\xrightarrow[\substack{\text{CH}_2\text{Cl}_2, -78°C \\ 6 \text{ h}}]{\text{5 mol% (8.36)}}$ **(8.39)**
exo:endo >1:99

R = alkyl, allyl, benzyl
yields, 70-98%
ee's - 97-99%

and cyclic enals, enones, including cyclic ketones such as 2-cyclopentenone (**8.33**), and enoates.[14] While triflate (**8.32**) decomposes above 0°C, the triflimide salt (**8.35**) shows greater stability and can be used at room temperature to effect the enantioselective Diels–Alder reaction with less reactive dienes.[15] The activated oxazaborolidine (**8.36**) has been used as a catalyst in the asymmetric Diels–Alder reaction of a range of 2-substituted cyclopentadienes – challenging substrates owing to the presence of a 1:1 mixture of 1- and 2-substituted isomers at ambient temperature.[16] In this process, steric interactions between the catalyst and substrate favour the reaction of the 2-substituted cyclopentadiene (**8.37**) with ethyl acrylate (**8.38**) to give the cycloadduct (**8.39**) with high diastereoselectivity and ee. These oxazoborolidines can also be activated by coordination of other Lewis acids such as AlBr$_3$[17] and SnCl$_4$[18] to the basic nitrogen.

There are many other enantiomerically pure boron-based Lewis acid catalysts used in the Diels–Alder reaction,[19] Amongst these the dichloroborane (**8.40**) used by Hawkins is noteworthy because the substrates used were enoate esters, such as methyl acrylate (**8.41**)[20] α,β-unsaturated ketones and acid chlorides such as (**8.42**)[21] rather than the enals generally tested in Diels–Alder reactions.

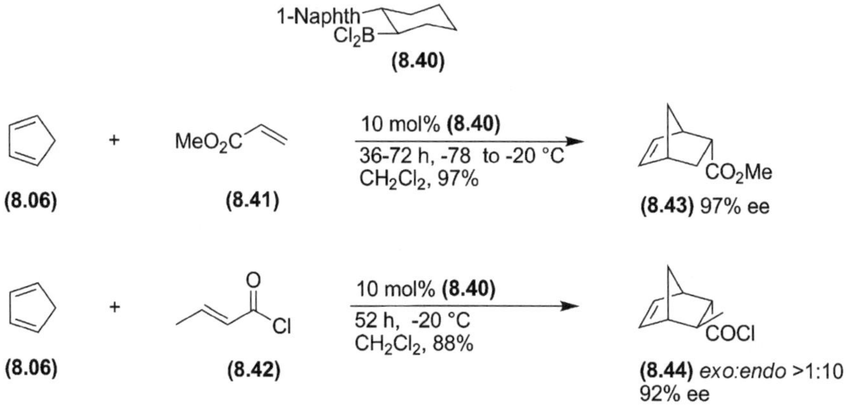

1-Naphth
Cl$_2$B
(**8.40**)

(**8.06**) + MeO$_2$C⟍ →[10 mol% (**8.40**)][36-72 h, -78 to -20 °C, CH$_2$Cl$_2$, 97%] (**8.43**) 97% ee, CO$_2$Me

(**8.06**) + (**8.42**) →[10 mol% (**8.40**)][52 h, -20 °C, CH$_2$Cl$_2$, 88%] (**8.44**) exo:endo >1:10, 92% ee, COCl

8.1.2 Aluminium-Based Lewis Acids

Some of the earliest work with catalytic asymmetric Diels–Alder reaction used aluminium catalysts.[22] Corey and coworkers have used aluminium catalyst (**8.45**) in the Diels–Alder reaction between a substituted cyclopentadiene (**8.46**) and the acryloyl oxazolidinone (**8.47**).[23] The cycloadduct (**8.48**) was obtained with good selectivity.

A particularly impressive example of the catalytic asymmetric Diels–Alder reaction has been provided by Wulff, who used the vaulted BINOL-aluminium complex (**8.49**) as a catalyst.[24] Not only is the reaction highly selective, but it uses an unusually low catalyst loading of just 0.5 mol%. Amongst Lewis acid-catalysed reactions in general and Diels–Alder reactions in particular, this is a remarkably small amount of catalyst.

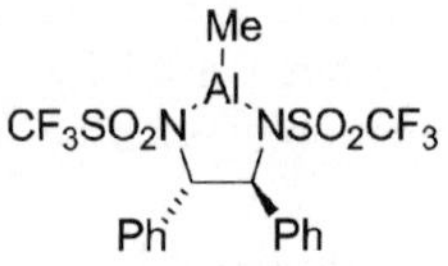

(**8.45**) 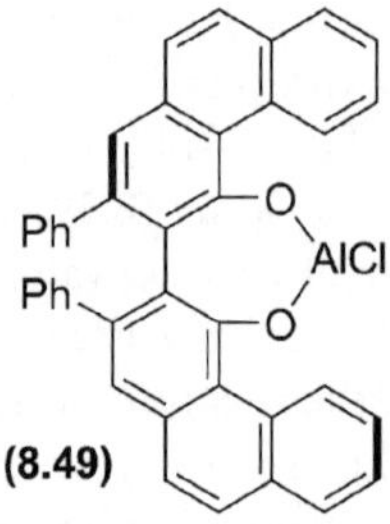   (**8.49**)

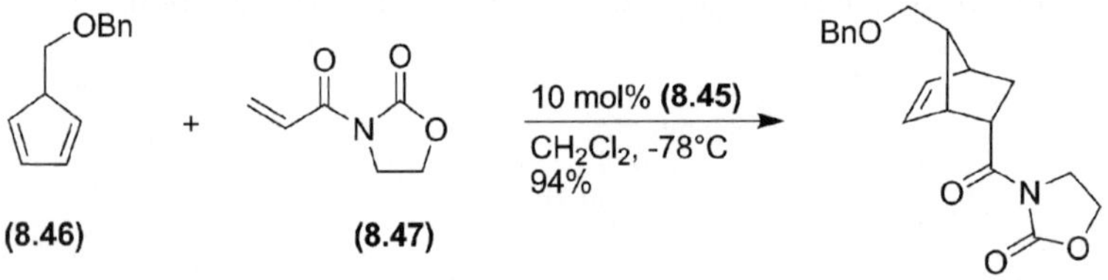

(**8.46**)   (**8.47**)

10 mol% (**8.45**)

CH$_2$Cl$_2$, -78°C
94%

(**8.48**) 94% ee

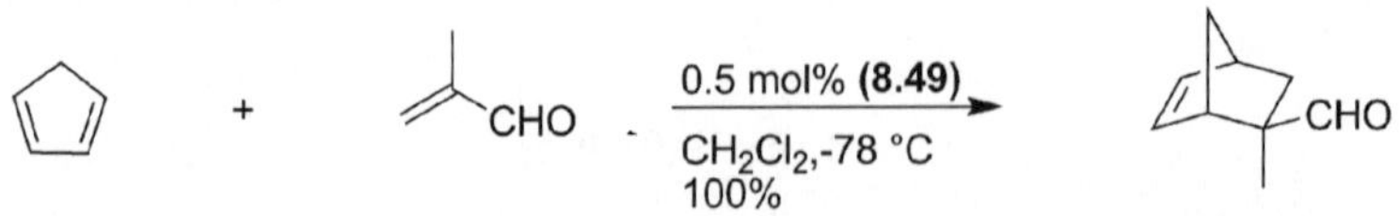

(**8.06**)   (**8.24**)

0.5 mol% (**8.49**)

CH$_2$Cl$_2$,-78 °C
100%

(**8.27**)
97:3 *exo* CHO: *endo*
97.7% ee

8.1.3 Titanium-Based Lewis Acids

There are several common Lewis acids based around titanium, including titanium tetrachloride and titanium tetraisopropoxide. Enantiomerically pure variants of these Lewis acids provide the basis for catalytic asymmetric reactions.[1, 25]

Narasaka and coworkers used the titanium complex of the TADDOL ($\alpha,\alpha,\alpha',\alpha'$-tetraaryl-1,3-dioxolane-4,5-dimethanol) ligand (**8.50**) to catalyse Diels–Alder reactions of acyloxazolidinones.[26] Thus, the crotonyl derivative (**8.51**) was converted into the Diels–Alder adduct (**8.52**) upon reaction with cyclopentadiene (**8.06**). The use of oxazolidinone substrates with titanium catalysts allows two-point binding as indicated by structure (**8.53**), and the bidentate nature of the interaction offers rigidity and hence good selectivity. Three papers were published independently in 1995 discussing the mechanism of Ti-TADDOLate-catalysed Diels–Alder reactions.[27] More recent studies indicate that enantioselectivity is dependant on the dienophile/catalyst ratio indicating the presence of a number of reactive intermediates and competitive mechanisms.[28] In general, the best results are obtained using conditions leading to formation of a 1:1 dienophile:catalyst complex.

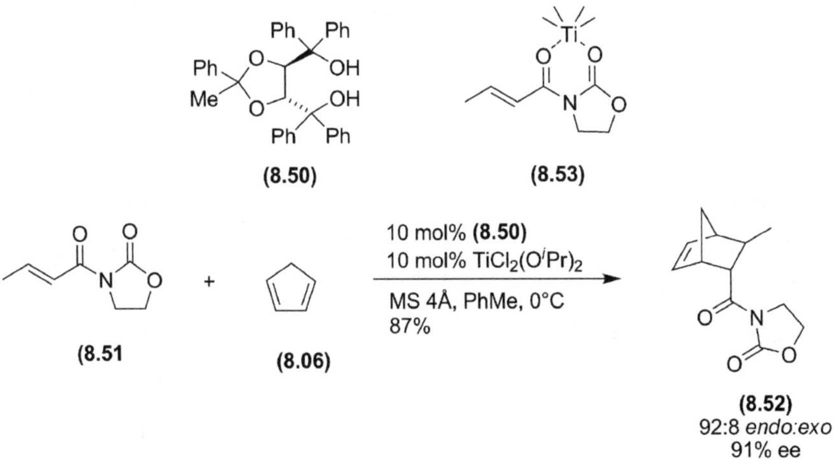

(**8.50**)   (**8.53**)

(**8.51**   (**8.06**)

10 mol% (**8.50**)
10 mol% TiCl$_2$(O^{i}Pr)$_2$

MS 4Å, PhMe, 0°C
87%

(**8.52**)
92:8 *endo:exo*
91% ee

BINOL/titanium complexes have also provided high enantioselectivities in Diels–Alder reactions.[29] High ees in the Diels–Alder reaction of quinones and quinone monoketals such as (**8.54**) have been obtained using the catalyst prepared from (*S*)-BINOL and Cl$_2$Ti(O^{i}Pr)$_2$.[30] The 6,6'-dibromoBINOL/titanium complex (**8.57**) gives slightly improved selectivities in some cases.[31] For example, the Diels–Alder reaction between diene (**8.58**) and methacrolein (**8.24**) is catalysed by this titanium complex, providing the product (**8.59**) with very good selectivity.

Other titanium complexes have been used to give good enantioselectivities in the catalysed Diels–Alder reaction, including the titanocene complex (**8.60**),[32] although the zirconocene analogue was slightly superior. Also, the elaborated BINOL ligand complex (**8.61**),[33] and the sulfonamide complex (**8.62**) have given good selectivities in the Diels–Alder reactions of cyclopentadiene with dienophiles (**8.47**), (**8.63**) and (**8.05**).[34]

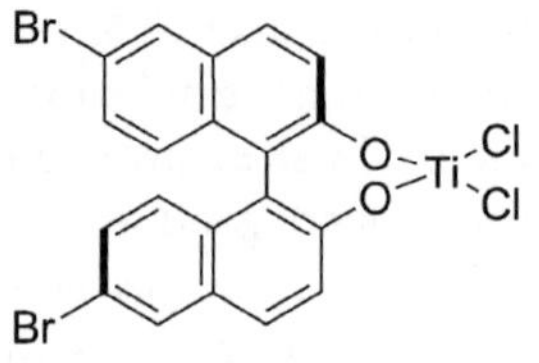

(8.57)

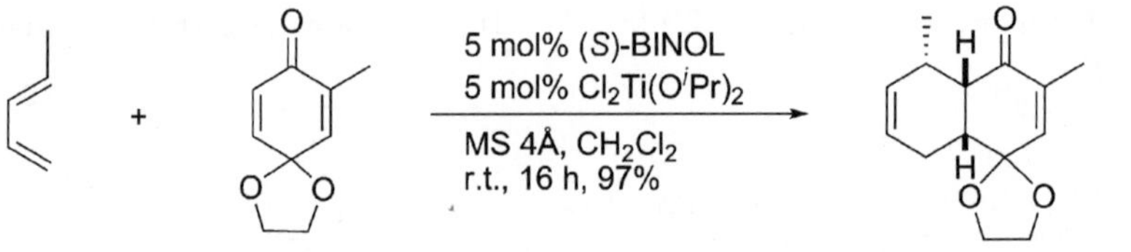

(8.55)   **(8.54)**

5 mol% (S)-BINOL
5 mol% Cl$_2$Ti(O^{i}Pr)$_2$
———————————
MS 4Å, CH$_2$Cl$_2$
r.t., 16 h, 97%

(8.56) 98% ee

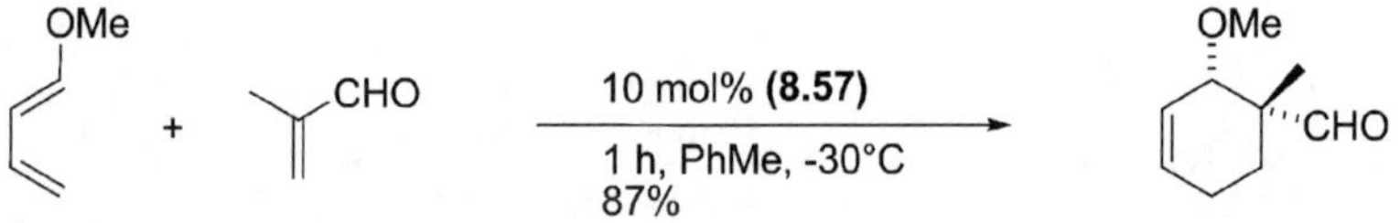

OMe

(8.58)   **(8.24)**

10 mol% **(8.57)**
———————————
1 h, PhMe, -30°C
87%

OMe

(8.59)
99:1 *endo:exo*
94% ee

8.1.4   Metal/Oxazoline Catalysts

Ligands based around the oxazoline unit have been successful in several metal-catalysed enantioselective processes.[35] They have certainly shown their value as ligands in the Diels–Alder reaction. The first report of a bis-oxazoline being used as a ligand in a Diels–Alder reaction was published by the Corey group in 1991.[36] They showed that an enantioselectivity of up to 86% ee with 99:1 *endo:exo* selectivity for the reaction between cyclopentadiene (**8.06**) and acryloyloxazolidinone (**8.47**) with a bis-oxazoline/iron(III) catalyst. The use of a similar magnesium complex of ligand (**8.67**) afforded slightly higher enantioselectivity.[37] The coordination of the substrate to the magnesium complex offers one face of the alkene selectively for reaction with cyclopentadiene, which preferentially approaches from the top face of the transition state assembly, as shown in Figure 8.3. Alternative magnesium/oxazoline complexes have also been reported by other researchers.[38, 39]

Evans and coworkers have examined the use of copper(II) complexes of bis-oxazoline (**8.68**) as catalysts for Diels–Alder reactions which can provide very high

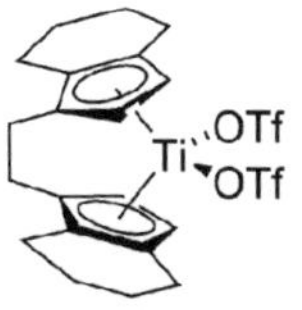

(8.60)

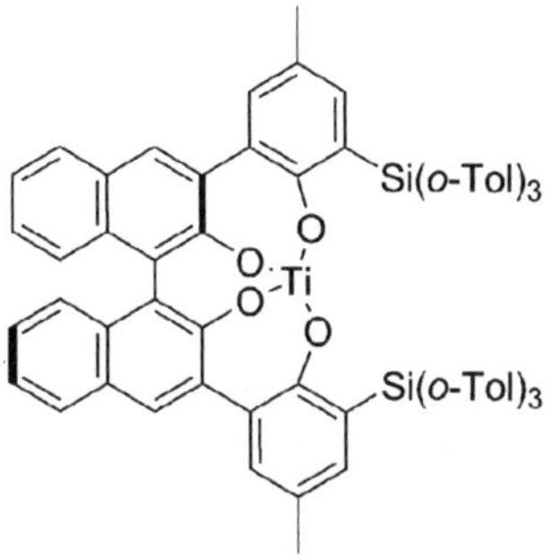

(8.61)

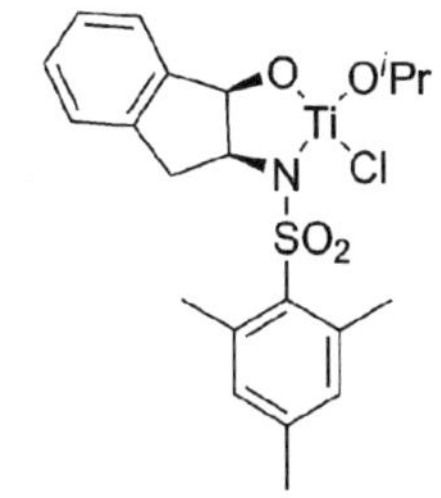

(8.62)

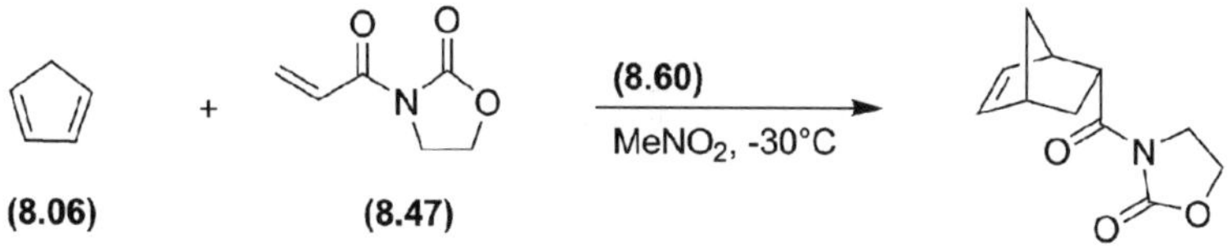

(8.06) + **(8.47)** → **(8.60)** / MeNO$_2$, -30°C → **(8.64)**
7:1 *endo:exo*
89% ee

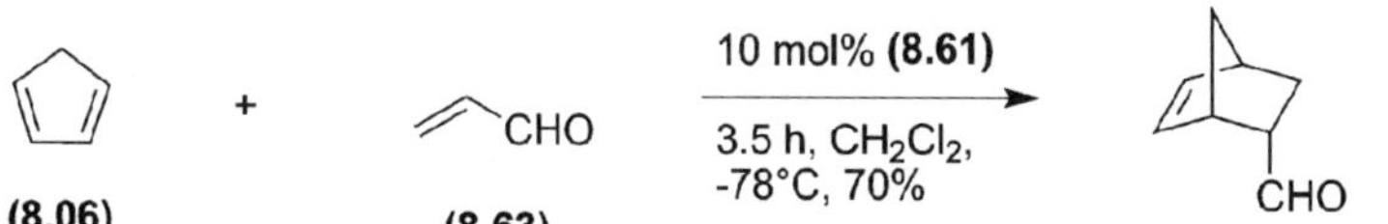

(8.06) + **(8.63)** CHO → 10 mol% **(8.61)** / 3.5 h, CH$_2$Cl$_2$, -78°C, 70% → **(8.65)** CHO
85:15 *endo:exo*
96% ee

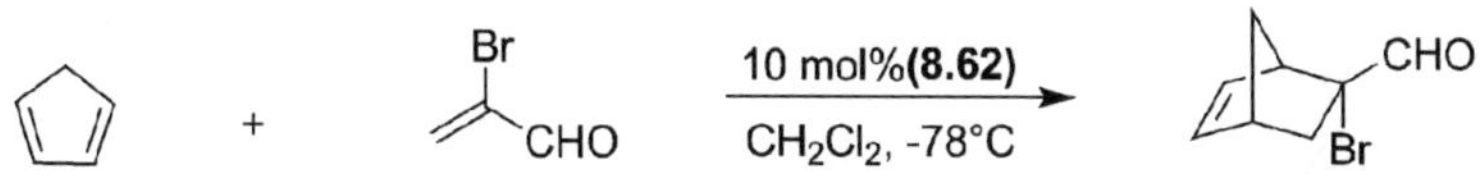

(8.06) + **(8.05)** → 10 mol% **(8.62)** / CH$_2$Cl$_2$, -78°C → **(8.66)**
67:1 *exo:endo* CHO
93% ee

asymmetric induction, as shown in the Diels–Alder reaction of the acryloyloxazo-lidinone (**8.47**) with cyclopentadiene (**8.06**).[40] In this case a square planar complex explains the stereochemical outcome, as shown in Figure 8.3.

The reaction has also been successfully applied to the use of furan (**8.69**) as the dienophile, where replacement of the triflate counterion with SbF$_6$$^-$

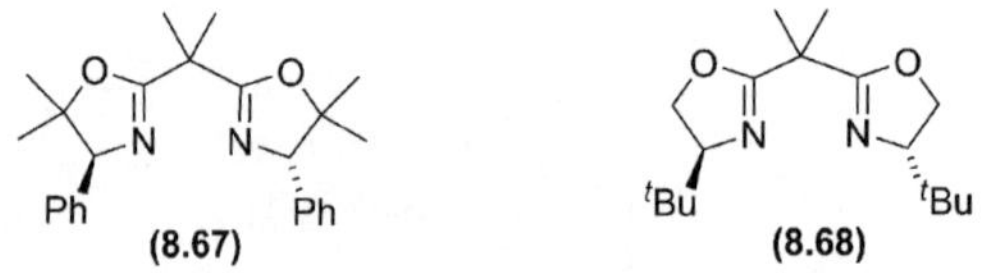

(8.67)

(8.68)

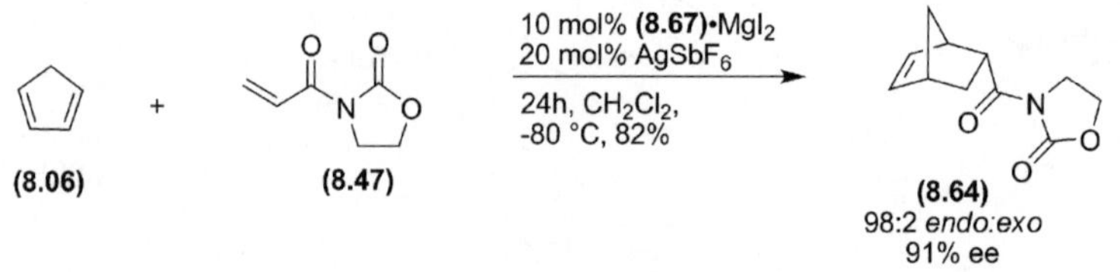

(8.06)   +   **(8.47)**

10 mol% **(8.67)**•MgI$_2$
20 mol% AgSbF$_6$

24h, CH$_2$Cl$_2$,
-80 °C, 82%

(8.64)
98:2 *endo:exo*
91% ee

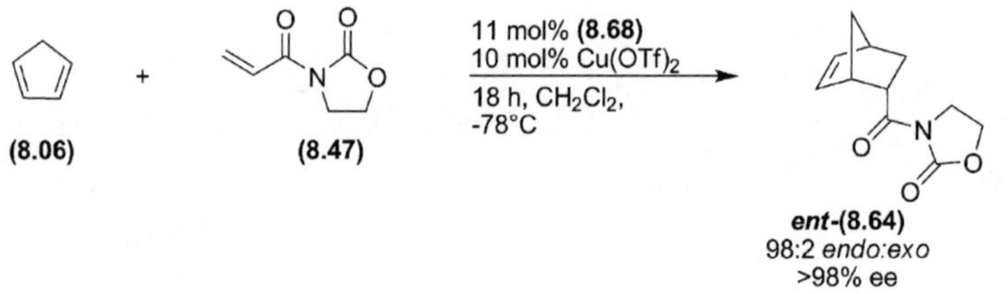

(8.06)   +   **(8.47)**

11 mol% **(8.68)**
10 mol% Cu(OTf)$_2$

18 h, CH$_2$Cl$_2$,
-78°C

***ent*-(8.64)**
98:2 *endo:exo*
>98% ee

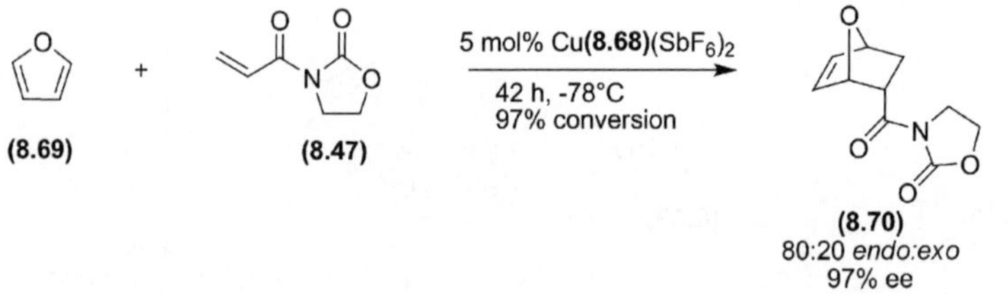

(8.69)   +   **(8.47)**

5 mol% Cu(**8.68**)(SbF$_6$)$_2$

42 h, -78°C
97% conversion

(8.70)
80:20 *endo:exo*
97% ee

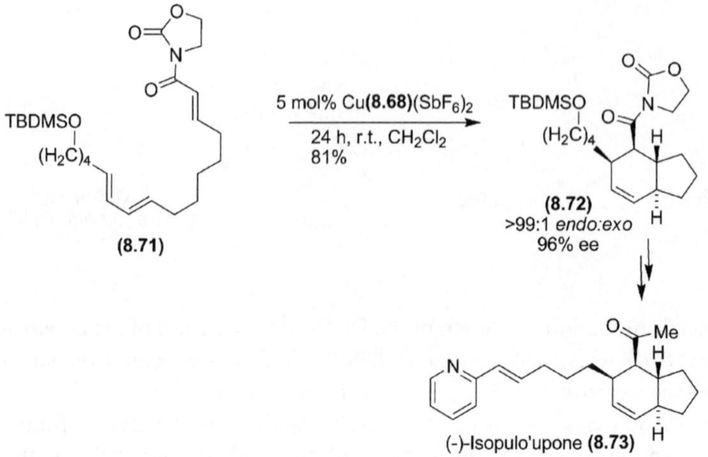

TBDMSO
(H$_2$C)$_4$

(8.71)

5 mol% Cu(**8.68**)(SbF$_6$)$_2$

24 h, r.t., CH$_2$Cl$_2$
81%

TBDMSO
(H$_2$C)$_4$

(8.72)
>99:1 *endo:exo*
96% ee

(-)-Isopulo'upone **(8.73)**

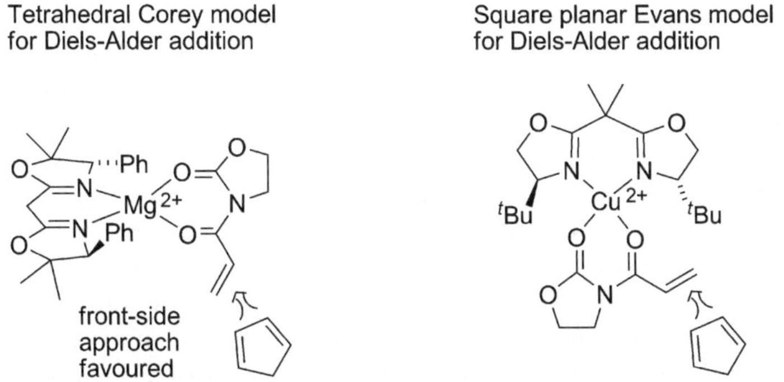

Tetrahedral Corey model
for Diels-Alder addition

Square planar Evans model
for Diels-Alder addition

front-side
approach
favoured

Figure 8.3 Corey and Evans models for asymmetric induction

provides a more reactive catalyst.[40c, 41] At higher temperatures ($-20°$C), the furan Diels–Alder adduct (**8.70**) was isolated with no enantiomeric excess after 24 h, although after 2–5 h a 59% ee was seen, indicating that racemisation of the product was occurring under these conditions. The lower temperature ($-78°$C) circumvented this problem.

The copper(II) catalyst has been applied to the intramolecular Diels–Alder reaction.[40c, 42] The precursor (**8.71**) undergoes an intramolecular Diels–Alder reaction (IMDA reaction) which proceeds with remarkable selectivity to give the product (**8.72**), which was subsequently converted into the marine toxin $(-)$-isopulo'upone (**8.73**).

The bis-oxazoline-catalysed Diels–Alder reaction is not restricted to the use of oxazolidinone-based dienophiles. For instance the α-sulfenylacrylate (**8.74**)[43] and α-hydroxyenones such as (**8.75**)[44] undergo enantioselective Diels–Alder reaction with cyclopentadiene in the presence of the copper complex of (**8.67**) and (**8.68**) respectively.

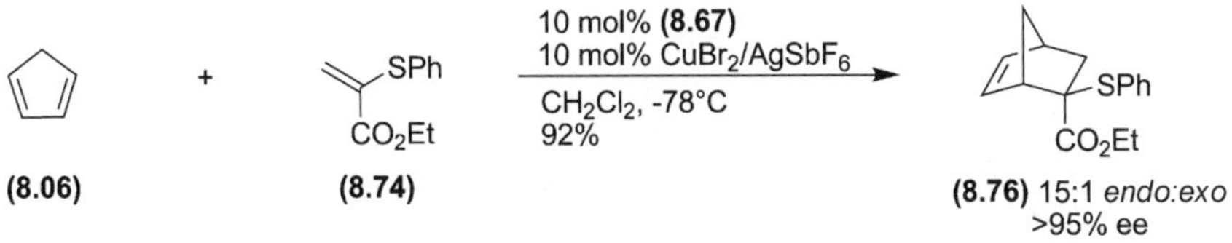

(8.06) + (8.74)

10 mol% (**8.67**)
10 mol% CuBr$_2$/AgSbF$_6$
CH$_2$Cl$_2$, -78°C
92%

(**8.76**) 15:1 *endo:exo*
>95% ee

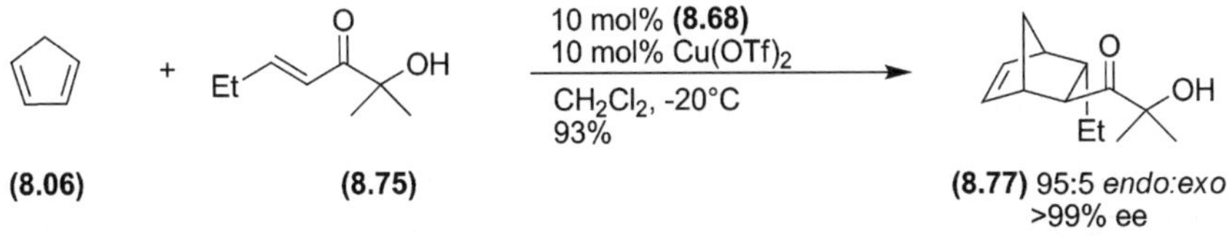

(8.06) + (8.75)

10 mol% (**8.68**)
10 mol% Cu(OTf)$_2$
CH$_2$Cl$_2$, -20°C
93%

(**8.77**) 95:5 *endo:exo*
>99% ee

Other oxazoline ligands have also been used successfully in copper-catalysed Diels–Alder reactions[45] including phosphino-oxazoline ligands.[46] Rhodium[47] and ruthenium[48] oxazoline complexes have been reported to give fairly good enantios-electivies and lanthanide complexes of pyridine bis-oxazolines (PYBOX) provide high ees in the Diels–Alder reaction with oxazolidinones[49] and quinones.[50] Ligand (**8.78**) provides an asymmetric environment for several metal salts for the catalytic Diels–Alder reaction.[51] In each case, the catalyst is prepared *in situ* by mixing of the ligand and metal salt. One of the key features of this bis-oxazoline is its ability to act as a *trans*-chelator in an octahedral environment. The change in coordination chemistry means that even though the bis-oxazolines (**8.68**) and (**8.78**) possess differing absolute configurations, they afford the same enantiomer of the Diels–Alder adduct (**8.64**).

In the nickel-catalysed reaction, the use of ligand with only 20% enantiomeric excess still produced a Diels–Alder adduct with up to 96% ee, which is a highly effective example of chiral amplification (see Section 6.1).

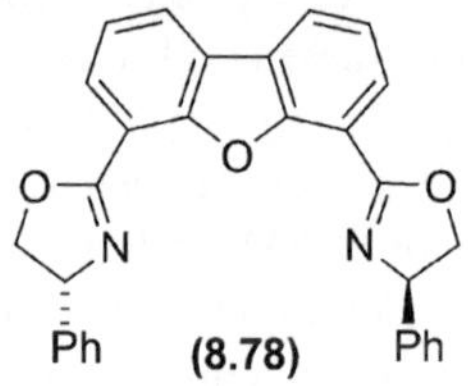

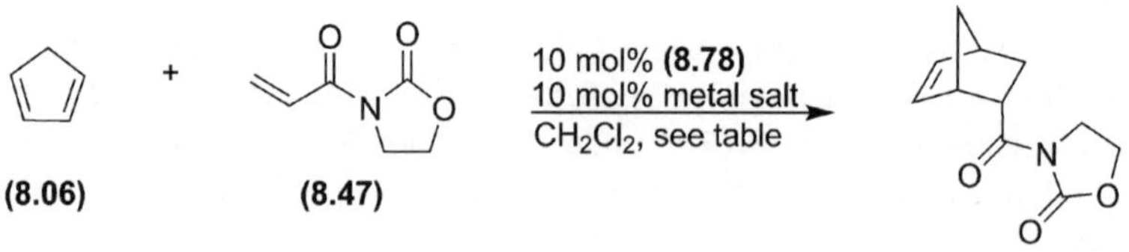

metal salt	conditions	yield (%)	endo:exo	ee (%)
Mg(ClO$_4$)$_2$	-40°C, 10 h	100	97:3	91
Ni(ClO$_4$)$_2$.6H$_2$O	-40°C, 14 h	96	97:3	>99
Mn(ClO$_4$)$_2$.6H$_2$O	-40°C, 96 h	96	97:3	83
Fe(ClO$_4$)$_2$	-40°C, 48 h	90	99:1	98
Co(ClO$_4$)$_2$.6H$_2$O	-40°C, 48 h	97	97:3	99
Cu(ClO$_4$)$_2$ + 3H$_2$O	-40°C, 15 h	99	97:3	96
Zn(ClO$_4$)$_2$	-40°C, 48 h	99	98:2	97

8.1.5   Other Metal Catalysts

There have been many other reports of catalytic asymmetric Diels–Alder reactions, including the heterobimetallic complexes developed by Shibasaki and coworkers,[52] and salen complexes. While Mn(salen) complexes have been used with some success,[53] much higher ees in the Diels–Alder reaction of 1-amino-1,3-dienes have been obtained using chromium and cobalt(salens).[54] The latter are active even at relatively low loadings. For instance, aminodiene (**8.79**) reacts with dienophile (**8.80**) to give adduct (**8.81**) with high ee using only 0.1 mol% of the enantiomerically pure cobalt(salen) (**8.82**).[54c]

Further examples also show particularly impressive selectivity. Kündig has used the cationic iron catalyst (**8.83**) in the Diels–Alder reaction to give cycloadduct (**8.85**).[55] The ruthenium (BINAP) catalyst (**8.86**) effects the Diels–Alder reaction of cyclopentadiene with 2-substituted acrylates with ees up to 99%.[56] Kobayashi has shown that lanthanide and scandium triflate complexes of BINOL with 1,2,6-trimethylpiperidine, formulated as complex (**8.87**), are effective with oxazolidinone-based substrates[57, 58, 59] and gives good selectivity in the formation of the Diels–Alder product (**8.52**).

Homo-Diels–Alder reactions have also been achieved with, in some cases, very high selectivity. In 1990, two groups independently reported achieving the homo-Diels–Alder reaction between norbornadiene (**8.88**) and phenylacetylene (**8.89**), affording the deltacyclene product (**8.90**).[60] When the phosphine used in this reaction was Norphos, the product was formed with remarkable selectivity.

8.1.6 Organocatalysts

In addition to Lewis acids the Diels–Alder reaction is also catalysed by amines. Recently a number of organocatalytic Diels–Alder reactions have been studied utilising enantiopure amines.[61] Initial work in this area, carried out by Kagan and coworkers, using cinchona alkaloids as catalysts, resulted in modest ees in the reaction of maleimide with anthrone.[62] A breakthrough was achieved in 2005 by the group of MacMillan using imidazolidinone salt (**8.91**) to catalyse the highly enantioselective Diels–Alder reaction of α,β–unsaturated aldehydes with a variety of dienes.[63] In this process the catalyst forms an (E)-iminium ion (**8.92**) by reaction with the enal, which is a more reactive dienophile. Molecular modelling reveals that the benzyl group of the catalyst shields the *re* face of the iminium ion. Cycloaddition followed by hydrolysis then provides the cycloadduct (**8.94**), as depicted in Figure 8.4.

High ees in the Diels–Alder cycloaddition of cyclopentadiene (**8.06**) and a range of enals (**8.95**) are obtained using (**8.91**), and similar levels of enantioselectivity are seen in the reaction of acrolein (**8.63**) and crotonaldehyde (**8.99**) with a variety of simple dienes such as (**8.98**). The related catalyst (**8.101**) has been applied with success to the intramolecular Diels–Alder reaction. For instance the trienal (**8.102**) undergoes cycloaddition, in the presence of this catalyst, to give decalin (**8.103**), which was converted in four further steps to the marine natural

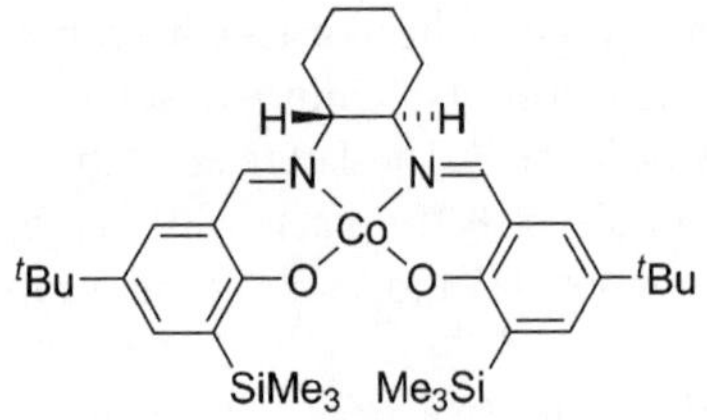

(8.82)

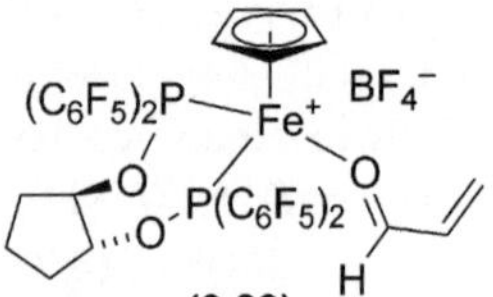

(8.83)

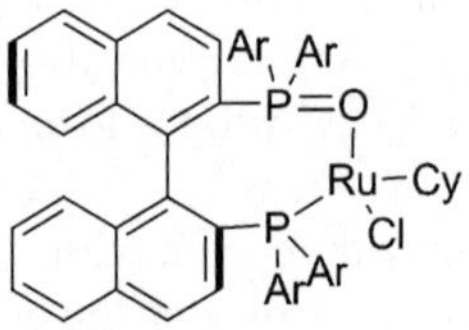

(8.86)

Ar = *p*-tolyl
Cy = cymene

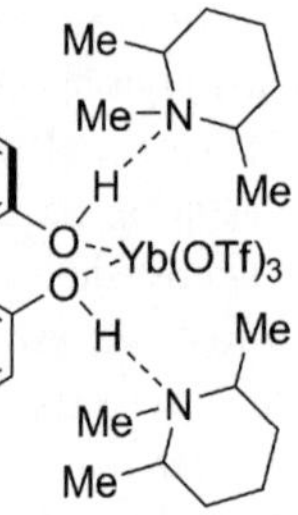

(8.87)

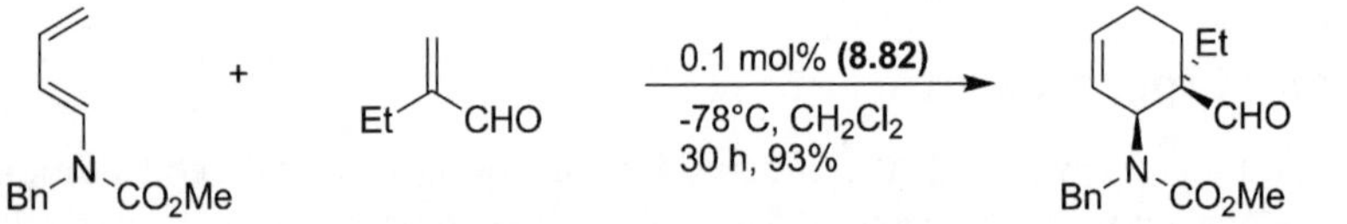

(8.79)   **(8.80)**

0.1 mol% **(8.82)**
-78°C, CH$_2$Cl$_2$
30 h, 93%

(8.81) >97% ee

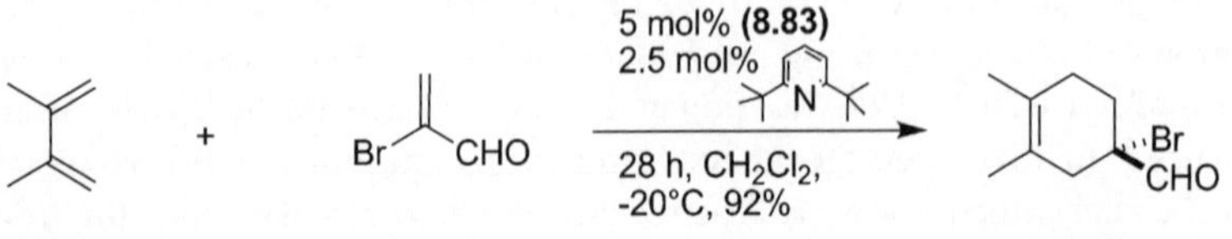

(8.84)   **(8.05)**

5 mol% **(8.83)**
2.5 mol%

28 h, CH$_2$Cl$_2$,
-20°C, 92%

(8.85)
97-98% ee

(8.06)   **(8.51)**

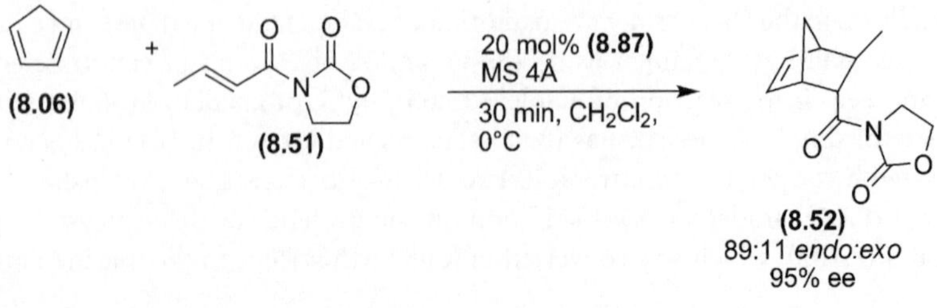

20 mol% **(8.87)**
MS 4Å

30 min, CH$_2$Cl$_2$,
0°C

(8.52)
89:11 *endo:exo*
95% ee

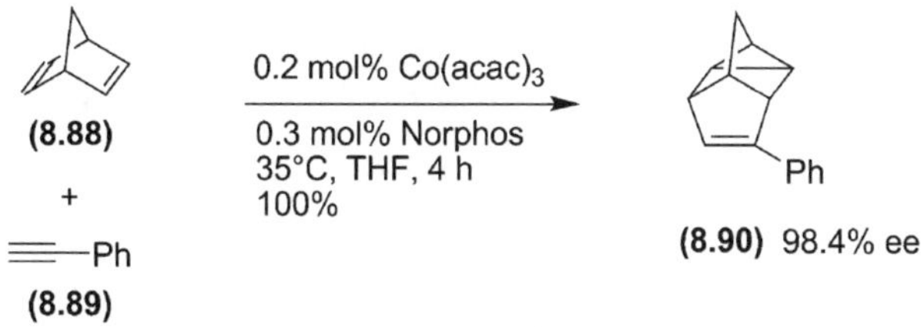

product solanapyrone.[64] Imidazolidinones of this type can also be applied to catalysis of the more challenging Diels–Alder reaction of α,β-enones. Cyclopentadiene and a range of substituted acyclic dienes including 1-methoxybutadiene and aminodiene (**8.104**) react with ketones such as ethyl vinyl ketone (**8.105**) with high *endo*-selectivity in the presence of the furyl-substituted imidazolidinone (**8.106**).[65] The organic catalysts (**8.108–8.110**) also provide high ees in the Diels–Alder reaction.[66] The hydrazide (**8.108**) functions effectively in water as solvent,[65a] while the BINAP-based diamine (**8.109**) provides some of the highest *exo*-selectivities (up to 20:1)[66b] in the cycloaddition of cyclopentadiene with α,β-enals. While both (**8.108**) and (**8.109**) display similar scope to the imidazolinone (**8.91**), the triamine (**8.110**) is effective in the Diels–Alder reaction of α-acyloxyacroleins such as (**8.111**).[66c]

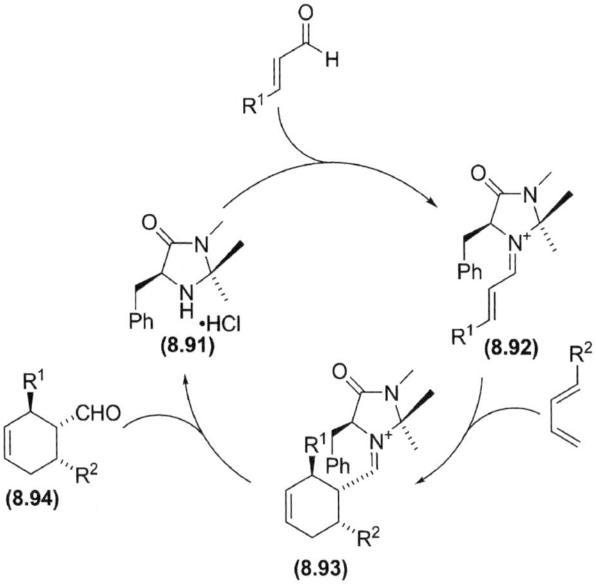

Figure 8.4 Catalytic cycle of imidazolidinone catalysed Diels–Alder reaction.

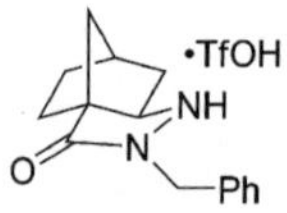

•TfOH
NH
N
O
Ph

(8.108)

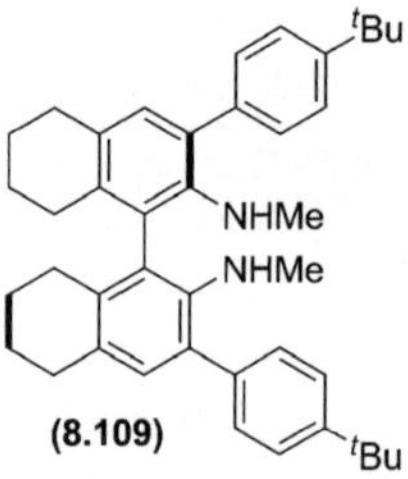

^{t}Bu
NHMe
NHMe
^{t}Bu

(8.109)

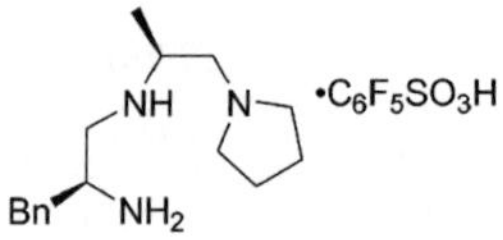

NH
N
•$C_6F_5SO_3H$
Bn
NH_2

(8.110)

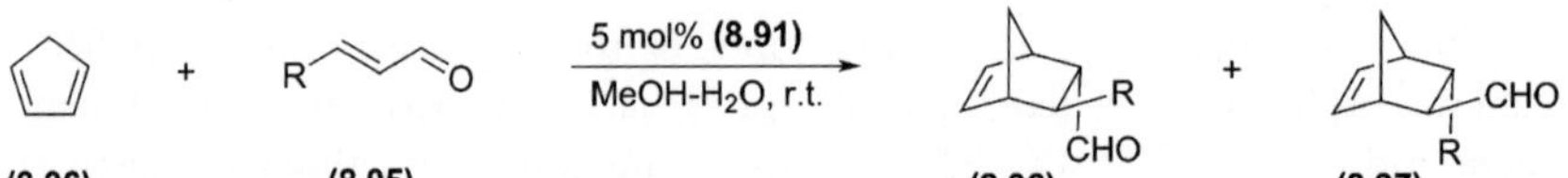

+
R
O
5 mol% (8.91)
MeOH-H_2O, r.t.
R
CHO
+
CHO
R

(8.06) **(8.95)** **(8.96)** *endo* **(8.97)** *exo*

R	Yield	*exo:endo*	ee (*exo*)	ee (*endo*)
Me	75%	1:1	86%	90%
Pr	92%	1:1	86%	90%
Ph	99%	1.3:1	93%	93%

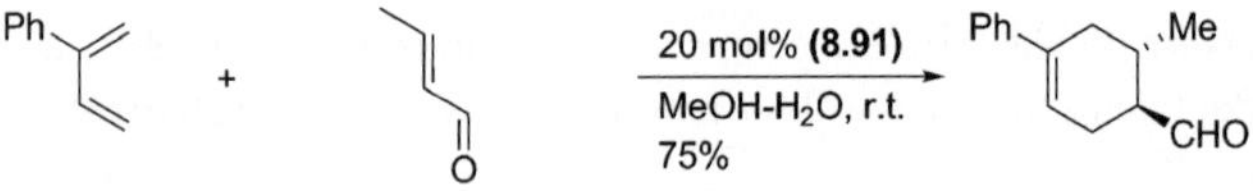

Ph
+
O
20 mol% (8.91)
MeOH-H_2O, r.t.
75%
Ph
Me
CHO

(8.98) **(8.99)** **(8.100)** 90% ee

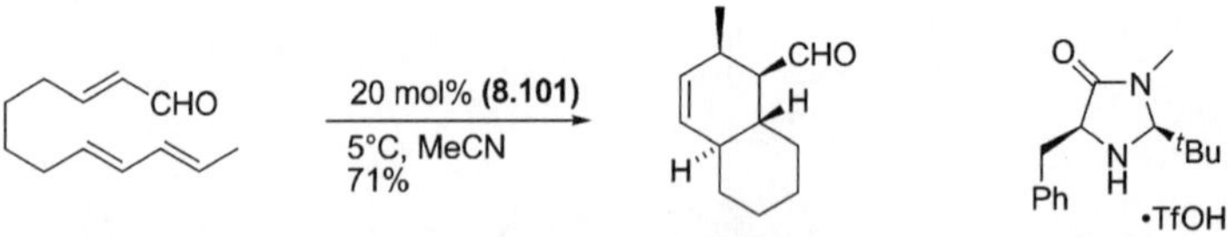

CHO
20 mol% (8.101)
5°C, MeCN
71%
CHO
H
H
O
N
N
H
^{t}Bu
Ph
•TfOH

(8.102) **(8.103)** 20:1 *dr* 90% ee **(8.101)**

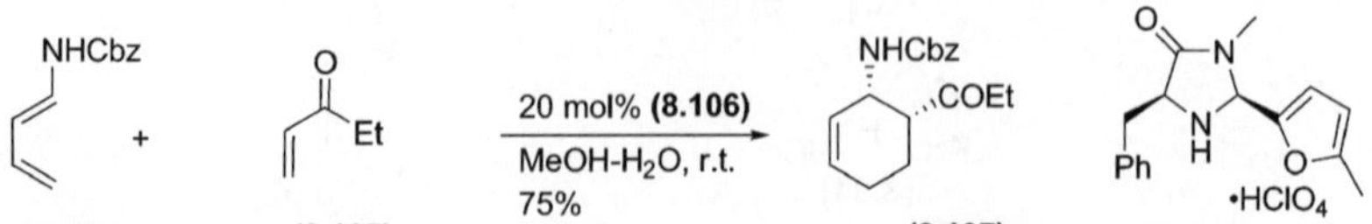

NHCbz
+
O
Et
20 mol% (8.106)
MeOH-H_2O, r.t.
75%
NHCbz
COEt
O
N
N
H
O
Ph
•$HClO_4$

(8.104) **(8.105)** **(8.107)** >100:1 *endo:exo* 98% ee **(8.106)**

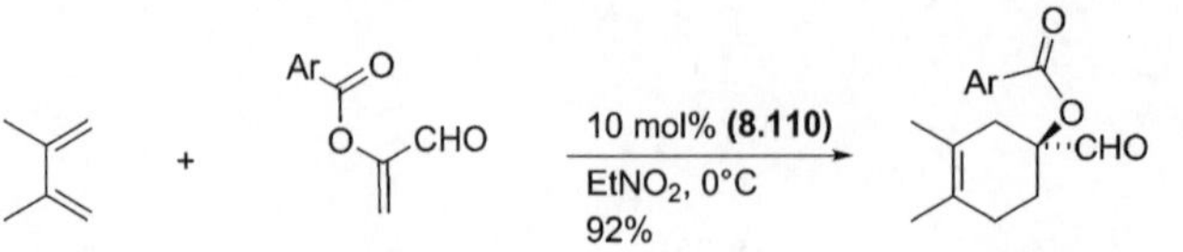

+
Ar
O
O
CHO
10 mol% (8.110)
$EtNO_2$, 0°C
92%
O
Ar
O
CHO

(8.84) **(8.111)** **(8.112)** 92% ee

The dienophile may also be activated by hydrogen bonding of the carbonyl oxygen with suitable protic molecules and it has been observed that Diels–Alder reactions are accelerated when performed in protic solvents such as 2-butanol. Thus the opportunity exists for the development of an enantioselective alcohol-catalysed asymmetric Diels–Alder reaction. Indeed, Rawal and coworkers have recently discovered that the use of catalytic amounts of the enantiomerically pure diol α,α,α′,α′-tetraaryl-1,3-dioxolane-4,5-dimethanol (TADDOL) (8.113) in the Diels–Alder reaction of aminosilyloxydiene (8.114) with 2-substituted acroleins such as (8.24), results in the formation of the cycloadduct (8.115) with high ee.[67] Similiarly, the strong Brønsted acid (8.116) effectively cataly-ses the Diels–Alder reaction of silyloxydienes such as (8.117) with α,β-enone (8.105).[68]

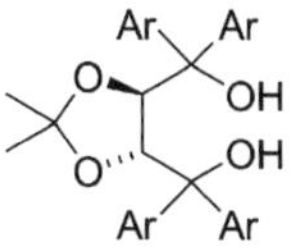

(8.113)

Ar = 1-naphthyl

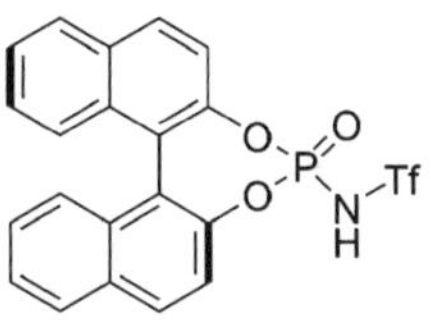

(8.116)

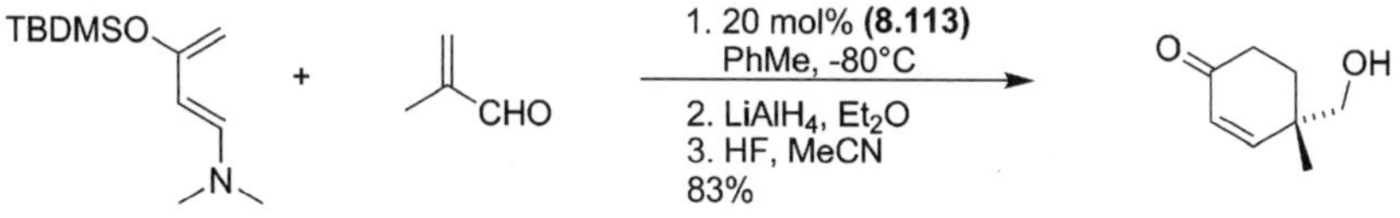

1. 20 mol% (8.113)
   PhMe, -80°C
2. LiAlH₄, Et₂O
3. HF, MeCN
83%

(8.114) (8.24) (8.115) 91% ee

20 mol% (8.116)

PhMe, -78°C

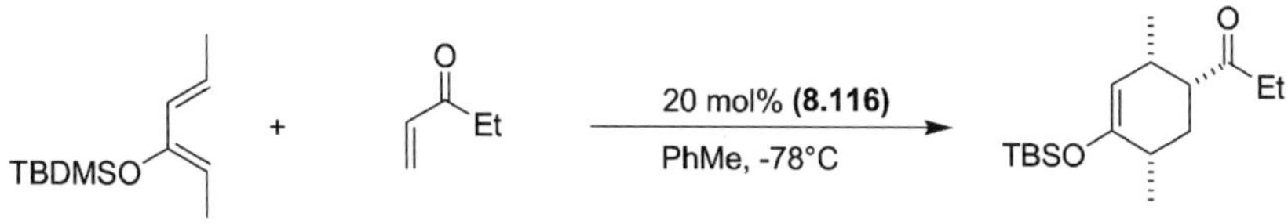

(8.117) (8.105) (8.118) 92% ee

8.2 Inverse Electron Demand Diels–Alder Reactions

Inverse electron demand Diels–Alder reactions involve a cycloaddition between an electron-rich dienophile and an electron-poor diene, the opposite electronic requirements from a normal Diels–Alder.

Markó and Evans have used ytterbium triflate complexes of BINOL and found that vinyl sulfide (**8.119**) provided the highest enantioselectivity in the inverse electron demand Diels–Alder reaction with the diene (**8.120**).[69]

Posner has used titanium complexes of BINOL to promote inverse electron demand Diels–Alder reactions between the same diene (**8.120**) and vinyl ethers.[70] Most of this work involved the use of stoichiometric Lewis acid to give 95–98% ee. Curiously, catalytic conditions (10 mol%) afforded the opposite enantiomer in 50–60% ee.

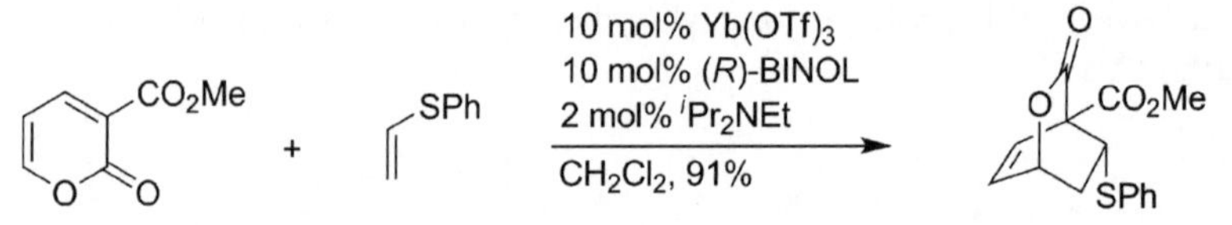

(**8.120**)   (**8.119**)   (**8.121**) >95% ee

8.3 Hetero-Diels–Alder Reactions

The hetero-Diels–Alder (HDA) reaction provides the opportunity to incorporate a heteroatom into the Diels–Alder product. Most commonly the catalytic asymmetric version of this reaction involves the reaction between an aldehyde (**8.122**) and a reactive diene (**8.123**) (typically with one or two oxygen substituents attached). Normally, the isolated products, after acidic work-up, are the enones (**8.124**). The products can either be formed by a direct cycloaddition or via a two step aldol-Michael sequence, according to Figure 8.5.

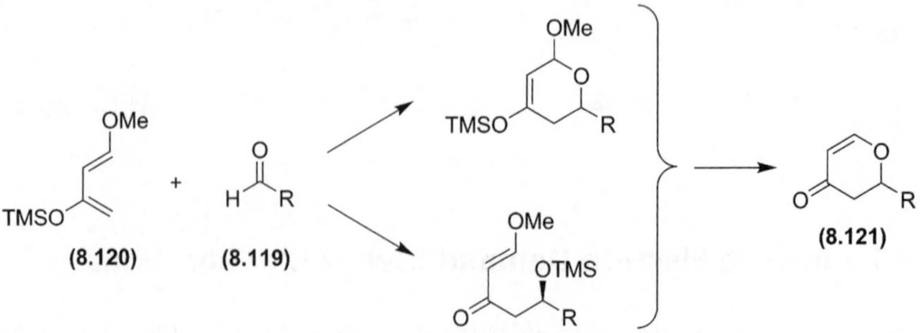

(**8.120**)   (**8.119**)   (**8.121**)

Figure 8.5 Alternative routes for the hetero-Diels–Alder reaction

The development of a catalytic enantioselective process has received much attention and, in common with the all-carbon Diels–Alder reaction, most success has been achieved using metal-based Lewis acids and, most recently, organocatalysts.[71] The earliest work on an enantioselective variant of this reaction was performed by Danishefsky.[72] The enantiomerically pure europium complex Eu(hfc)$_3$ provided moderate enantioselectivities for the hetero-Diels–Alder reaction between derivatives of diene (**8.123**) and benzaldehyde (**8.125**), performed under solvent-free conditions. Higher ees have been obtained using the hindered BINOL ligand (**8.126**), as its aluminium complex.[73] Yamamoto's CAB catalysts (see Section 8.1) such as (**8.129**), with the appropriate substituent on boron, have also been used to good effect.[74]

A range of transition metal-based Lewis acids has been investigated as catalysts in this reaction. Chromium(salen) complexes, which have shown to be useful catalysts in the all carbon Diels–Alder reaction, have been applied with success to the hetero Diels–Alder reaction.[75] For instance, the chromium complex (**8.130**), developed by Jacobsen and coworkers provides up to 93% ee in the cycloaddition of diene (**8.123**) with a range of aliphatic and aromatic aldehydes.[75a] The asymmetric HDA reaction is not limited to the use of Danishefsky's diene (**8.123**) and derivatives. The group of Jacobsen has also developed tridentate Schiff base chromium catalysts such as (**8.131**) and applied these to the HDA reaction of less activated, monooxygenated dienes such as (**8.58**), and aromatic and aliphatic aldehydes such as (**8.132**). Furthermore, the HDA reaction of Brassard's diene (**8.134**) with benzaldehyde (**8.125**) and derivatives proceeds with ees ranging from 90 to 99% in the presence of the catalyst prepared from Schiff base (**8.136**) and Ti(O^{i}Pr)$_4$.[76] Some exceptional ees at very low catalyst loadings have been obtained using transition metal catalysts in the HDA reaction. For instance, Ding and coworkers have discovered, using combinatorial techniques, that titanium catalysts prepared from two equivalents of H$_8$-BINOL and/or H$_4$-BINOL and one equivalent of Ti(O^{i}Pr)$_4$ catalyse the HDA reaction of Danishefsky's diene (**8.123**) with aromatic aldehydes with ee values ranging between 97 and >99%, at catalyst loadings as low as 0.005 mol%.[77] Moreover, the dirhodium catalyst (**8.137**) also provides high ees between 91 and 99% in the HDA reaction of (**8.123**) and monooxygenated dienes with aromatic and α,β-unsaturated aldehydes at 0.002 mol% loadings.[78]

Cu(II)-bis-oxazoline complexes have been reported to be useful catalysts for hetero-Diels–Alder reactions.[79] The choice of solvent was shown by Jørgensen to be important, with nitroalkanes providing the best results. The authors propose that the use of a polar solvent provides a higher degree of dissociation of the counterion from the metal. Under these conditions, the reaction of simple, unactivated dienes such as (**8.138**) with the activated aldehyde (**8.139**) affords the bicyclic adduct (**8.140**) with high selectivity.[80]

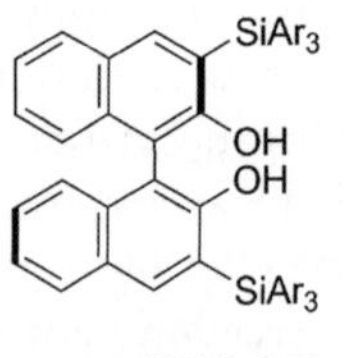

Ar = 3,5-(CH₃)₂C₆H₃-

(8.126)

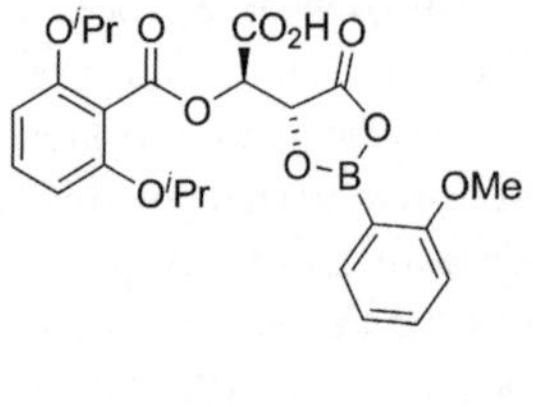

(8.129)

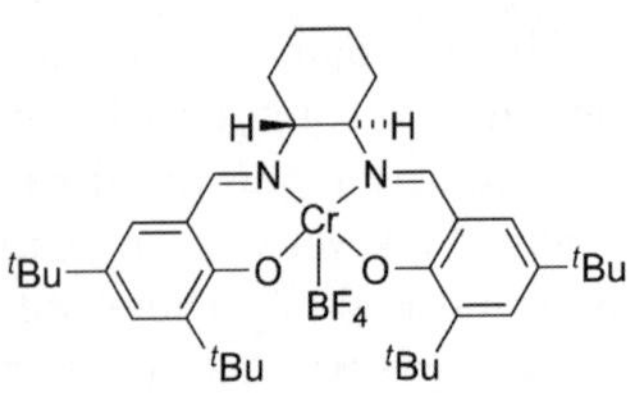

(8.130)

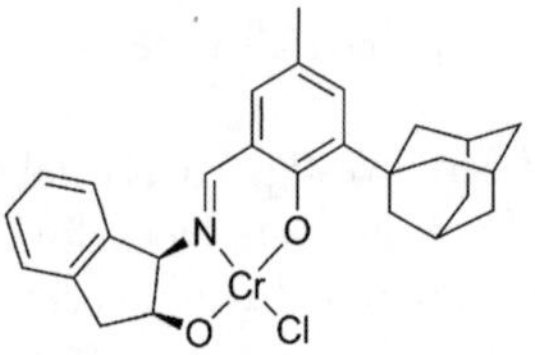

(8.131)

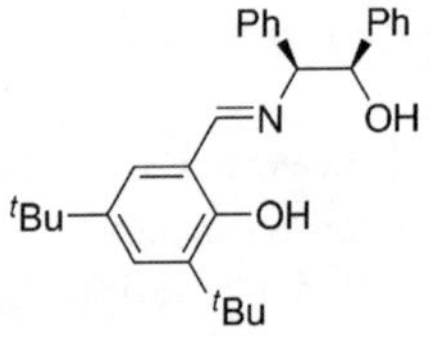

(8.136)

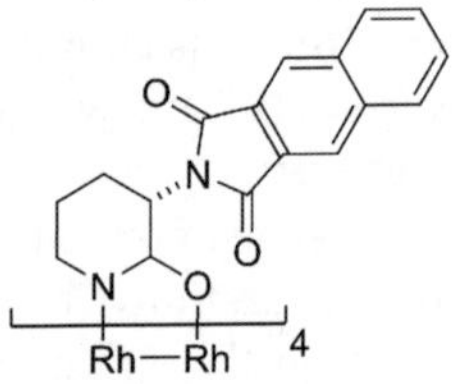

(8.137)

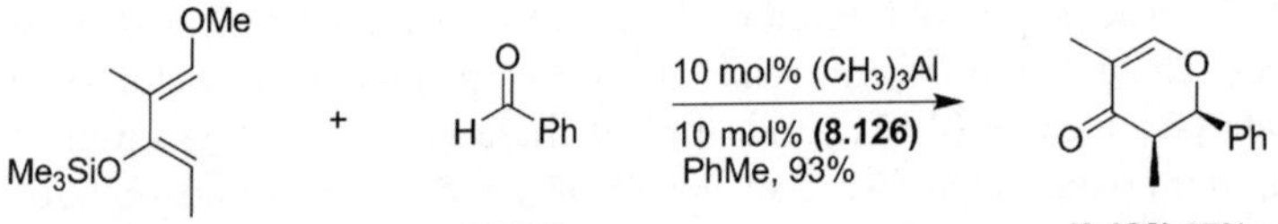

(8.127) + (8.123)

10 mol% (CH₃)₃Al
10 mol% **(8.126)**
PhMe, 93%

(8.128) 97% ee
30:1 cis/trans

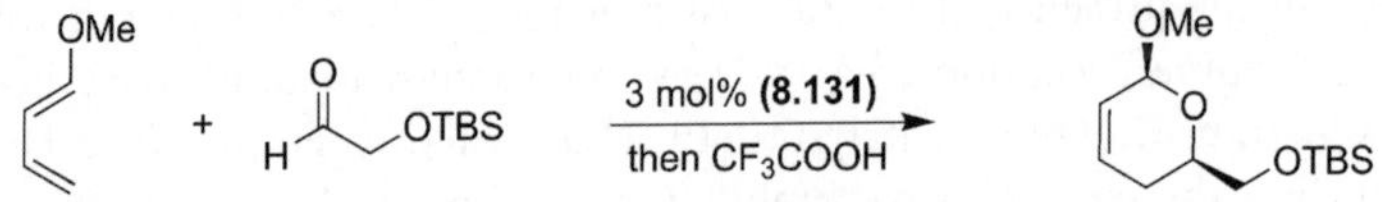

(8.58) + (8.132)

3 mol% **(8.131)**
then CF₃COOH

(8.133) >99% ee

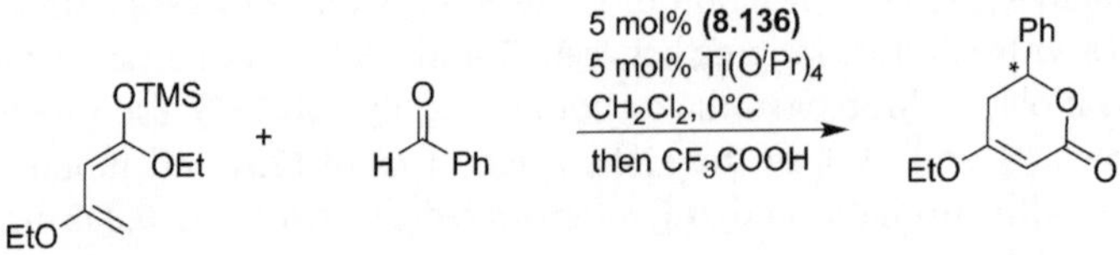

(8.134) + (8.123)

5 mol% **(8.136)**
5 mol% Ti(OⁱPr)₄
CH₂Cl₂, 0°C
then CF₃COOH

(8.135) 93% ee

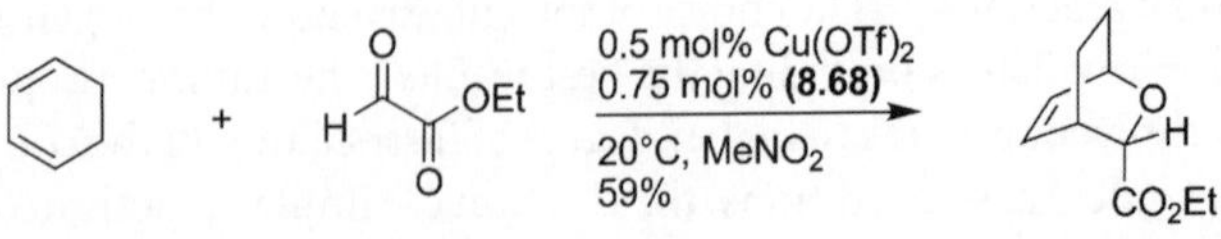

(8.138) + (8.139)

0.5 mol% Cu(OTf)₂
0.75 mol% **(8.68)**
20°C, MeNO₂
59%

(8.140) 95% ee
100% endo

The HDA reaction of activated ketones is also possible using copper bis-oxazoline complexes.[71c, 81] The highest ees in this procedure have been obtained using the copper complex prepared from the bis(sulfoximine) (**8.141**), which possesses chiral sulfur atoms and provides up to 98% ee in the HDA reaction of 1,3-cyclohexadiene (**8.138**) with substrate (**8.142**).[82]

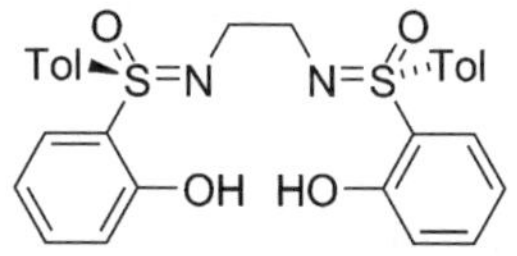

(8.141)

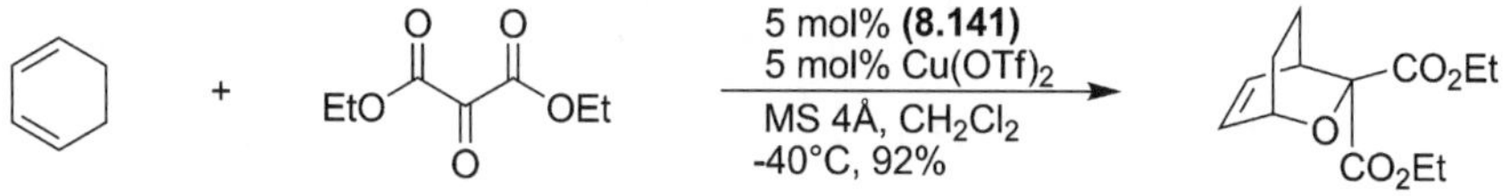

| (**8.138**) | (**8.142**) | (**8.143**) 98%ee |

The majority of metal-based Lewis acid catalysts used in the HDA reaction are moisture sensitive and are thus usually prepared *in situ*. The stable and storable zirconium-BINOL Lewis acids developed by Kobayashi and coworkers, effective in the aldol reaction (see Section 7.1) can also be used as asymmetric catalysts in the HDA reaction of aliphatic and aromatic aldehydes with dioxygenated dienes.[83]

Metal-free catalytic asymmetric HDA reactions have been developed, utilising enantiomerically pure protic molecules that activate the aldehyde component by hydrogen bonding to the carbonyl group.[61b] Rawal and coworkers have achieved up to 98% ee in the HDA reaction of aminodiene (**8.114**) with aromatic aldehydes in the presence of 20 mol% of TADDOL (**8.113**).[67, 84] The biaryl dimethanols (**8.144a**) and (**8.144b**) exhibit greater scope and can be used to effect highly enantioselective HDA reaction of aminodiene (**8.114**) with aromatic and aliphatic aldehydes (**8.122**).[85] Other classes of hydrogen bond catalysts for the asymmetric HDA reaction include the sulfonamido-substituted oxazoline (**8.145**) effective in the HDA reaction of aminodiene (**8.114**) with aromatic aldehydes[86] and bis-sulfonamides such as (**8.146**) have shown use in the cycloaddition of dioxygenated dienes such as (**8.147**) with pyruvates,[87] glyoxylates and phenyl glyoxal (**8.148**).[88]

The HDA reaction can also be performed using imines as dienophiles to give piperidinone derivatives as the cycloadducts. This aza-Diels–Alder process has been performed in an asymmetric fashion using a variety of enantiomerically pure metal-based Lewis acids.[89] Early work in this area was carried out by Kobayashi and coworkers using zirconium catalysts derived from BINOL (**8.150**) in the cycloaddition of diene (**8.123**) with aryl imines derived from 2-hydroxyaniline such as (**8.151**).[90] More recently aryl imines derived from 2-methoxyaniline (*o*-anisidine)

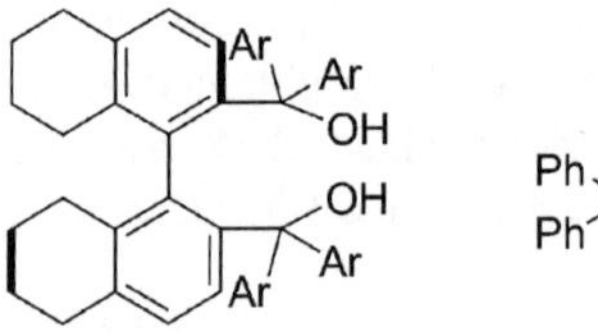

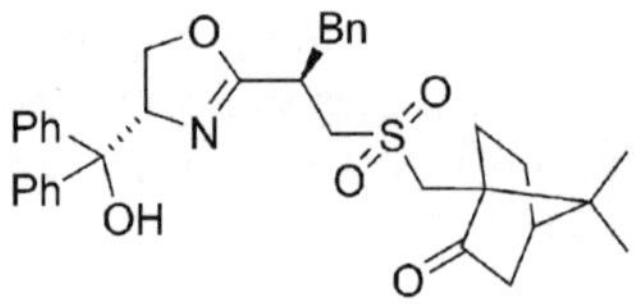

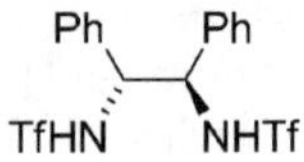

(8.144a) Ar = 4-F-3,5-Me$_2$C$_6$H$_5$ **(8.145)**   **(8.146)**
(8.144b) Ar = 4-F-3,5-Et$_2$C$_6$H$_5$

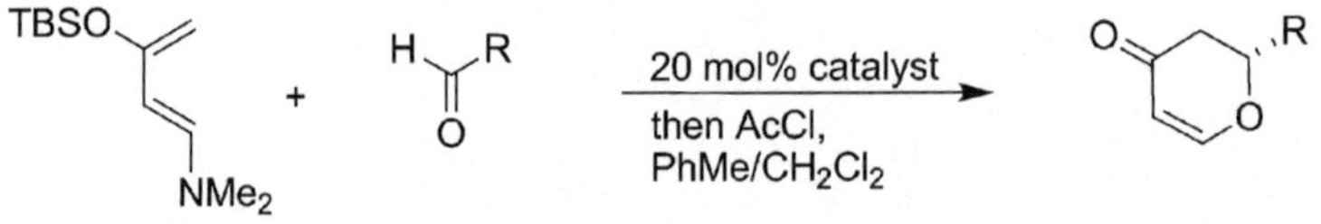

(8.114)   **(8.122)**   **(8.124)**

R	Catalyst	Yield	ee
Me	**(8.144b)**, -80°C	75%	97%
n-propyl	**(8.144a)**, -40°C	76%	94%
c-hexyl	**(8.144a)**, -40°C	99%	84%
Ph	**(8.144b)**, -80°C	84%	98%
2-furyl	**(8.144b)**, -80°C	96%	>99%

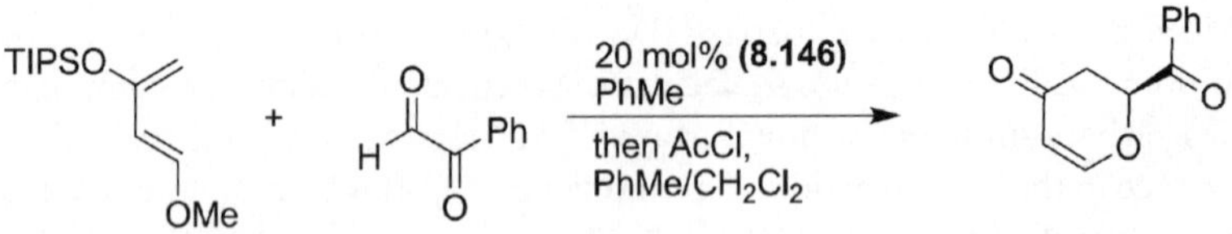

(8.147)   **(8.148)**   **(8.149)** 87% ee

have been observed to undergo asymmetric aza-Diels–Alder reaction with diene **(8.123)** in the presence of silver complexes of phosphine **(8.153)**. In this case ees of up to 95% can be achieved at catalysts loadings as low as 1 mol%.[91] Jørgensen and coworkers have studied the effects of a range of Lewis acid catalysts in the aza-Diels–Alder reaction of imino esters such as **(8.154)** and have shown that copper(I) complexes of phosphinooxazolines such as **(8.155)** are effective in the cycloaddition with activated dienes such as **(8.123)**, while copper complexes of BINAP **(8.156)** are broadly active over a range of dienes including relatively unactivated substrates such as **(8.138)**.[92]

Enantiomerically pure Brønsted acid catalysts are beginning to see application in the aza-Diels–Alder reaction.[93] For example, the aza-Diels–Alder reaction of 2-hydroxyaniline-derived aromatic imines, including (**8.151**), with diene (**8.123**) proceeds with up to 91% ee in the presence of 10 mol% of the phosphoric acid (**8.158**).[93a] While the strong acidity of this catalyst precludes its use in reactions of labile substrates, the pyridinium salt of the related acid (**8.159**) catalyses the aza-Diels–Alder reaction with less stable dienes such as Brassard's diene (**8.134**) with ees between 92 and 99%.[93b]

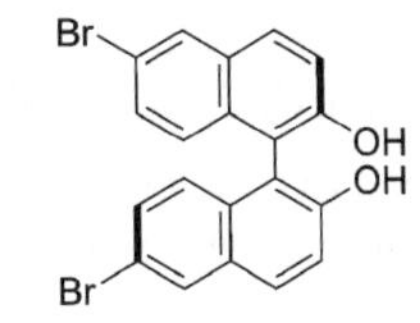

(8.150)

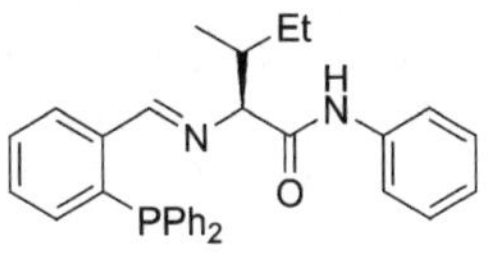

(8.153)

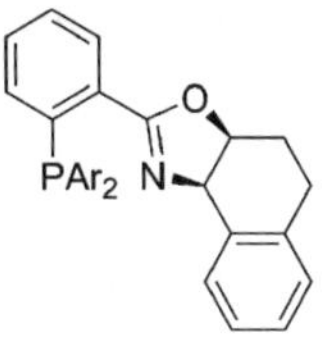

(8.155)
Ar = *p*-tol

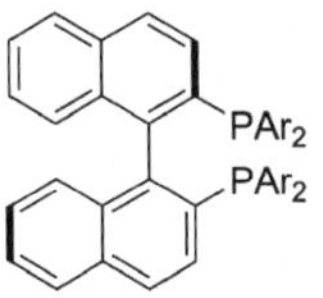

(8.156)
Ar = *p*-tol

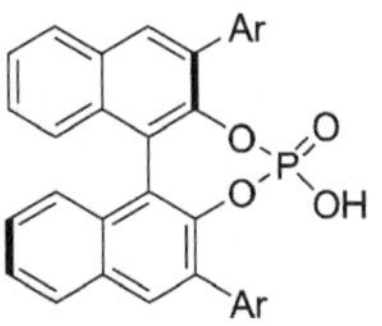

(8.158) Ar = 2,4,6-(*i*Pr)$_3$C$_6$H$_2$
(8.159) Ar = 9-anthryl

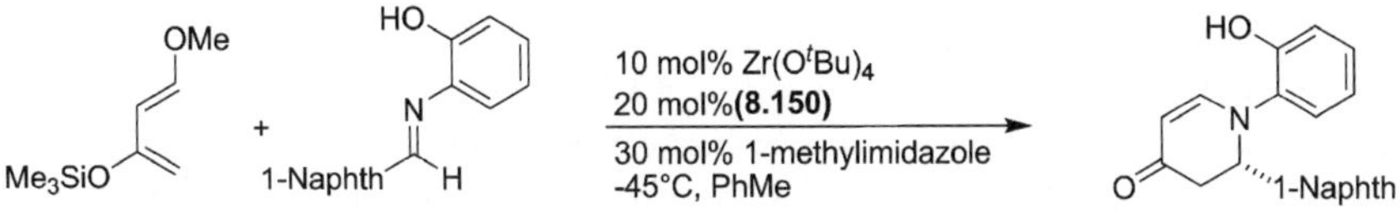

(8.123)   **(8.151)**

10 mol% Zr(O*t*Bu)$_4$
20 mol%(**8.150**)
———————————
30 mol% 1-methylimidazole
-45°C, PhMe

(8.152) 82% ee

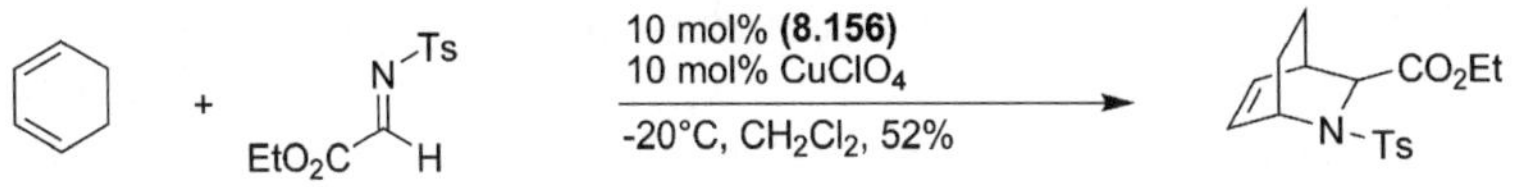

(8.138)   **(8.154)**

10 mol% (**8.156**)
10 mol% CuClO$_4$
———————————
-20°C, CH$_2$Cl$_2$, 52%

(8.157)
endo:exo 1:7.4
95% ee

Heterodienes such as oxadienes and azadienes can also be employed in the Diels–Alder reaction. In this process the diene is often the electron-deficient component and the Lewis acid catalyst binds to this substrate. Evans has demonstrated the use of copper(II) complexes of bis-oxazoline (**8.68**) in the reaction of oxadienes bearing extra binding functionality such as β,γ−unsaturated α-ketoesters, and phosphonates such as crotonyl phosphonate (**8.160**), with ethyl vinyl ether (**8.161**) and vinyl sulfides.[94] While two chelating functional groups are required when using bis-oxazoline catalysts, Jacobsen and coworkers have achieved high ees in the Diels–Alder reaction of unsubstituted oxadienes using the Schiff base chromium catalyst (**8.131**).[95] Alternatively, asymmetric oxadiene Diels–Alder reaction can be achieved using enamines or enolates as dienophiles, generated by reaction with organocatalysts. For instance the hetero-Diels–Alder reaction of β,γ-unsaturated α-ketoesters such as (**8.163**) and enamines, prepared *in situ* by reaction of aldehydes such as (**8.164**) with 10 mol% pyrrolidine (**8.165**), proceeds with good ee in the presence of silica gel.[96] The silica gel is added to regenerate the catalyst by hydrolysis of the cycloadduct (**8.166**). The resulting hemiacetals are isolated as the lactone (**8.168**). Bode and coworkers have developed a novel approach to the design of an enantioselective oxa-Diels–Alder process that uses enolates derived from *N*-heterocyclic carbenes.[97] In this reaction, α-chloroaldehydes such as (**8.169**) are converted into enolates such as (**8.170**) on reaction with the carbene generated from the triazolium precatalyst (**8.171**). These enolates participate as the dienophile in the asymmetric hetero-Diels–Alder reaction with β,γ−unsaturated α-ketoesters to give cycloadducts with ees ranging from 95 to 99%.

The catalytic asymmetric azadiene-Diels–Alder reaction has also received some recent attention.[89] Initial work in this area used hydroxyaniline-derived imine (**8.173**), which behaves as a diene in the reaction with cyclopentadiene (**8.06**) catalysed by the enantiomerically pure ytterbium complex (**8.174**).[98] The initial product rearomatises to give the 2-aryltetrahydroquinoline (**8.175**), with fairly good enantioselectivity. The enantioselective Lewis acid-catalysed Diels–Alder reaction of electron-rich azadienes such as (**8.176**) with oxazolidinones occurs with high ee in the presence of the copper bis-oxazoline complex formed with ligand (**8.68**),[99] and electron-deficient *N*-sulfonylazadienes react with ethyl vinyl ether using the nickel complex of ligand (**8.179**) to give cycloadducts with ees generally ranging from 77 to 92%.[100]

Some work has been directed towards the use of organocatalysts in the azadiene Diels–Alder reaction. For instance, the cycloaddition/aromatisation of imine (**8.173**) and derivatives with a range of cyclic and acyclic vinyl ethers occurs in the presence of the Brønsted acid catalyst (**8.159**) giving access to a variety of 2-aryltetrahydroquinolines.[101] Bode and coworkers have applied the triazolium precatalyst (**8.171**) used in the oxadiene Diels–Alder reaction to an *N*-heterocyclic carbene-catalysed cycloaddition with azadienes.[102] In this case, enolates are obtained by reaction of (**8.171**) with electron-poor α,β-enals such as (**8.180**) and undergo Diels–Alder reaction with electron-deficient *N*-sulfonylazadienes, including (**8.181**), to give the cycloadducts (**8.182**) with ees ranging from 97 to 99%.

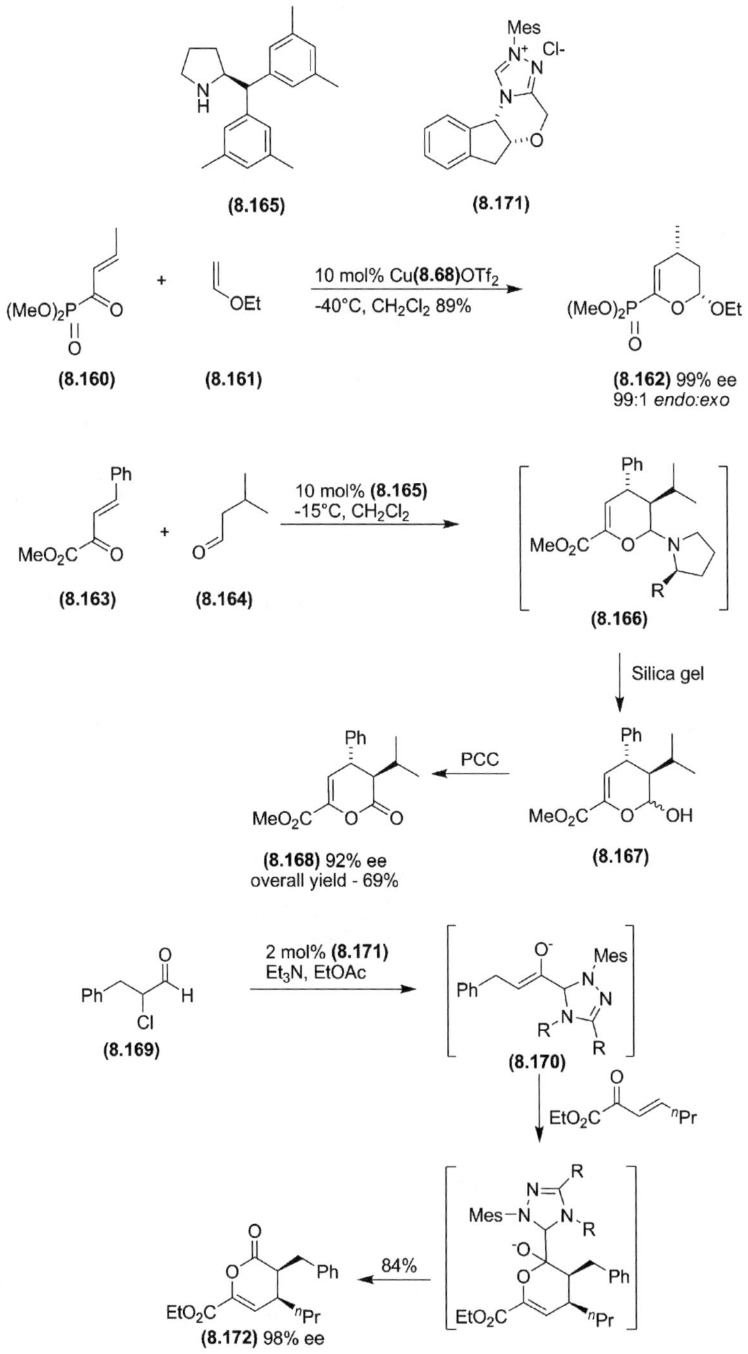

(8.165)

(8.171)

(MeO)$_2$P + ⫯OEt → (MeO)$_2$P⫯OEt

(8.160)   (8.161)   10 mol% Cu(8.68)OTf$_2$ -40°C, CH$_2$Cl$_2$ 89%   (8.162) 99% ee 99:1 *endo:exo*

(8.163) + (8.164) → (8.166)

10 mol% (8.165) -15°C, CH$_2$Cl$_2$

Silica gel

(8.168) 92% ee overall yield - 69%   ←PCC—   (8.167)

(8.169) → (8.170)

2 mol% (8.171) Et$_3$N, EtOAc

EtO$_2$C ⟍^{n}Pr

(8.172) 98% ee   ←84%—

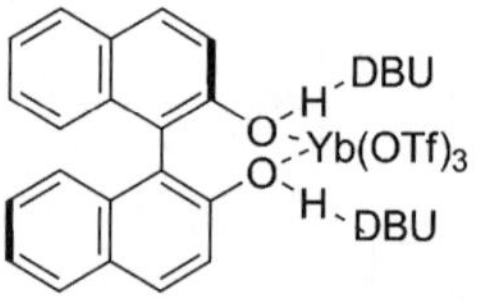

(8.174)

(8.179)

DBU = 1,8-diazabicyclo[5.4.0]undec-7-ene

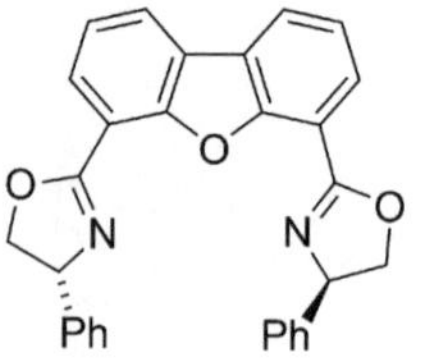

(8.173) + **(8.06)**

20 mol% **(8.174)**

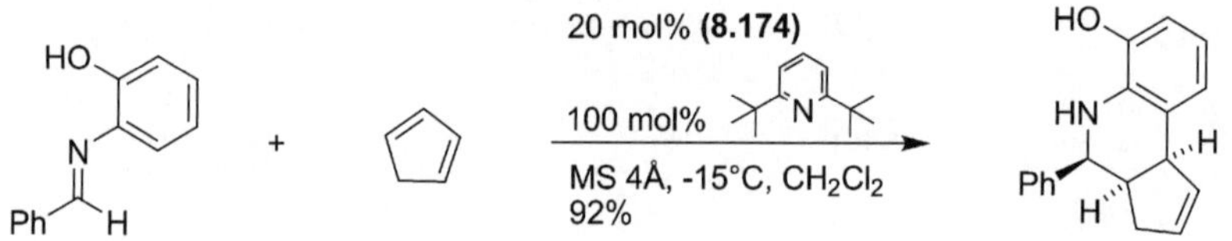

100 mol% [pyridine structure]

MS 4Å, -15°C, CH_2Cl_2
92%

(8.175) 71% ee
99:1 *cis/trans*

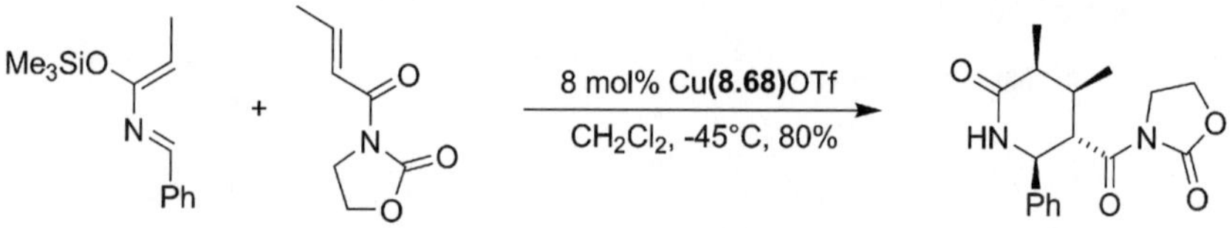

(8.176) + **(8.177)**

8 mol% Cu**(8.68)**OTf

CH_2Cl_2, -45°C, 80%

(8.178)
>99:1 *exo:endo*
95.1% ee

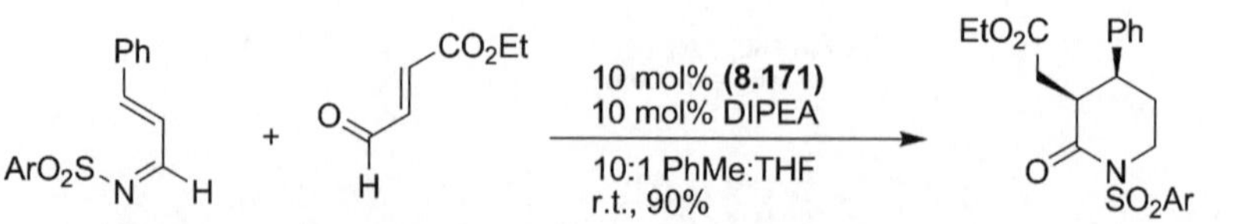

(8.181) + **(8.180)**

10 mol% **(8.171)**
10 mol% DIPEA

10:1 PhMe:THF
r.t., 90%

(8.182) 99% ee

8.4   1,3-Dipolar Cycloaddition Reactions

Whilst there are many cycloaddition reactions which could be subjected to asymmetric catalysis, the majority of work has been involved with the Diels–Alder and related reactions. Nevertheless, 1,3-dipolar cycloadditions have provided fairly good results as well. Asymmetric catalytic 1,3-dipolar cycloadditions have focussed on the cycloaddition reactions of nitrones[103] and *in situ*-generated azomethine ylides with alkenes.[103c, 104]

The asymmetric 1,3-dipolar cycloaddition of nitrones is generally catalysed by enantiomerically pure Lewis acids that coordinate to either the dipole or the dipolarophile. As the resulting isoxazolidines are readily cleaved by reduction of the N−O bond this strategy provides a route to enantioenriched γ-aminoalcohols. The first example of such a catalytic reaction was reported by the groups of Scheeren and coworkers in 1994, using the boron catalyst (**8.183**), which provided reasonable enantioselectivities in the reaction between nitrones and electron-rich alkenes such as ketene acetals and vinyl ethers.[105] Higher yields and enantioselectivities have been obtained by the group of Jørgensen using BINOL-aluminium complexes (**8.184**) and a range of *C,N*-diaryl nitrones including (**8.185**).[106] In this process the Lewis acid coordinates to the nitrone oxygen, lowering the energy of the dipole LUMO, and catalyst (**8.184**) is also effective in the asymmetric cycloaddition with cyclic aromatic nitrones such as tetrahydroisoquinoline *N*-oxides.[107]

Asymmetric catalysis of the 'normal' electron demand nitrone cycloaddition, i.e. using electron-poor alkenes has received most attention. In this process the Lewis acid must coordinate to functionality on the alkene and, in general, high ees have only been obtained using oxazolidones such as (**8.51**), which allow two-point binding to the catalyst and thus bind in preference to the dipole. Jørgensen's group used titanium-TADDOL complexes (**8.187**) to catalyse 1,3-dipolar cycloaddition of *C,N*-diaryl nitrones and oxazolidinone (**8.51**) in moderate ee. Higher ees have been obtained using the bis-titanium complex (**8.188**) developed by Maruoka and coworkers, which does not require bidentate dipolarophiles and effectively catalyses the dipolar cycloaddition of the simple α,β-enal acrolein (**8.63**) with aromatic and aliphatic *N*-benzyl nitrones such as (**8.189**).[108]

Kanemasa and coworkers have investigated the activity of a variety of metal complexes formed from DBFOX ligand (**8.179**) in the nitrone cycloaddition and have discovered that while the nickel complex does effect highly enantioselective reactions of *C,N*-diaryl nitrones with oxazolinone (**8.51**),[109] this complex, along with the magnesium and zinc species, also catalyses the cycloaddition with monodentate α-substituted acroleins. The zinc species are especially active and effect formation of the cycloadducts with ees between 97 and 99%.[110] In this case it is postulated that a nitrone catalyst complex is in fact the active catalyst that undergoes further binding to the dipolarophile. The iron half-sandwich (**8.191**) and the corresponding ruthenium derivative are examples of Lewis acids that bind preferentially to the dipolarophile and thus also function well in the cycloaddition with monodentate dipolarophiles.[111] While these catalysts perform poorly in the reaction of *C,N*-diaryl nitrones with methacrolein (**8.124**) the iron catalyst affords high ees in the cycloaddition with cyclic nitrones such as the pyrroline *N*-oxide (**8.192**). Related rhodium and iridium half-sandwich complexes also catalyse the nitrone cycloaddition with methacrolein and, in this instance, good ees are obtained with acyclic C,N-diaryl nitrones. [112] In addition, cobalt(III) ketaminato complexes have also been observed to catalyse the nitrone cycloaddition with monodentate acroleins with moderate to good ee. [113]

The magnesium and copper bis-oxazoline catalysts used in the asymmetric Diels–Alder reaction (see Section 8.1) display high levels of selectivity in the nitrone cycloaddition with bidentate, electron-deficient dipolarophiles. The magnesium bis-oxazoline catalysts effect moderate to good ee in the cycloaddition with oxazolidines such as (8.51), [114] while copper bis-oxazoline complexes show good selectivity in the addition to bidentate pyrazolidinones rather than oxazolidinones and also α-hydroxyenones such as (8.75) used as substrates in the copper bis-oxazoline-catalysed Diels–Alder reaction. [115]

Furukawa and coworkers have used palladium/BINAP complexes to catalyse 1,3-dipolar cycloaddition of nitrones to alkenes and whilst enantioselectivity is high (up to 91% ee), control of diastereoselectivity (*endo* versus *exo*) is poor. [116]

High ees in the nitrone cycloaddition with oxazolidinone (8.51) have been achieved by Kobayashi and Kawamura using the ytterbium catalyst (8.194). [117] The BINOL controls the sense of asymmetric induction with achiral amines (up to 78% ee). But the correct enantiomer of the amine (8.195) enhanced the enantioselectivity still further, such that the cycloadduct (8.197) is formed with up to 96% ee.

MacMillan's imidazolidinone salts, used successfully in the organocatalysed Diels–Alder reaction (see Section 8.1) also function as effective catalysts in the asymmetric nitrone cycloaddition with simple monodentate dipolarophiles. [118] Thus acrolein (8.63) and crotonaldehyde (8.99) both react with acyclic C-aryl, N-benzyl nitrones and C-aryl N-alkyl nitrones such as (8.198) with high ees ranging from 90 to 99% in the presence of the perchlorate salt of imidazolidinone (8.91).

The cycloaddition of *in situ*-generated azomethine ylides with electron-deficient alkenes is a useful method for the generation of stereodefined, substituted pyrrolidines, and there has been some recent interest in the development of a catalytic asymmetric variant. While a variety of methods for the generation of azomethine ylides have been developed, treatment of an α-iminoester (8.200) with an amine base in the presence of metal salts is the process most commonly employed in the asymmetric variant, which generally uses an enantiomerically pure metal complex of copper, silver or zinc to give an N-metallated ylide (8.201) (Figure 8.6). [119]

Zhang and coworkers screened a variety of phosphine ligands in the silver-catalysed cycloaddition of the azomethine ylide generated from aryl imines such as (8.202) with dimethyl maleate (8.203), and discovered that the highest ees are achieved with the bis-ferrocinyl amide phosphine (8.204). [120] During Zhang's studies it was shown that silver complexes with BINAP are poor catalysts in the addition with dimethyl maleate, however Sansano and coworkers have discovered

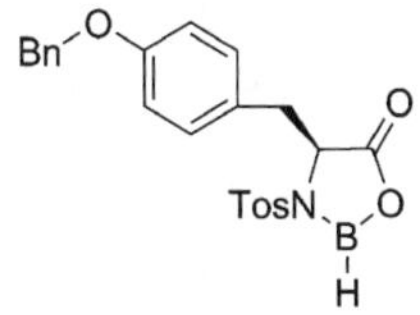

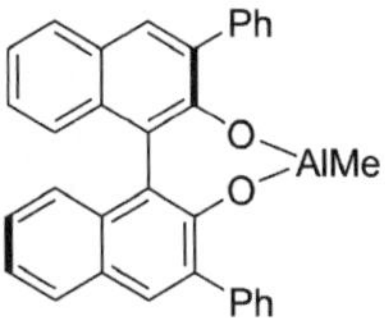

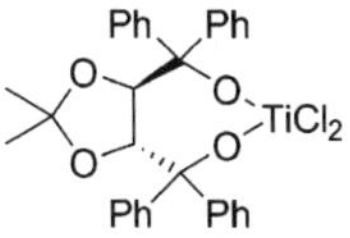

(8.183) (8.184) (8.187)

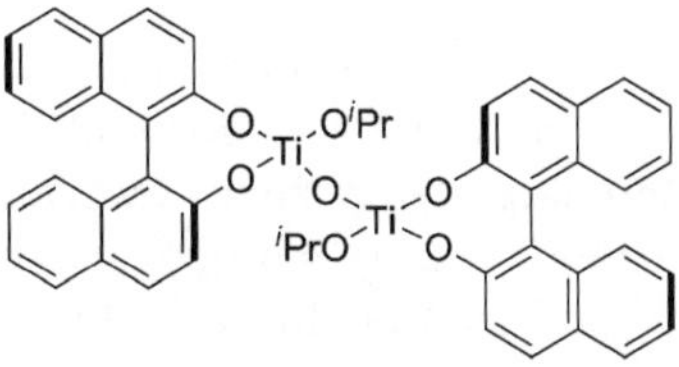

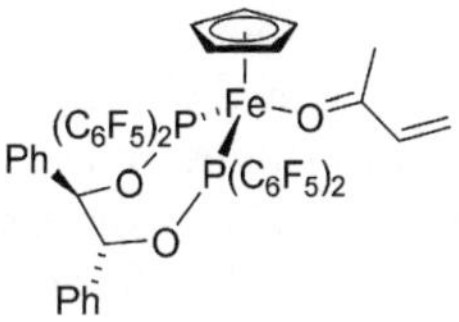

(8.188) (8.191)

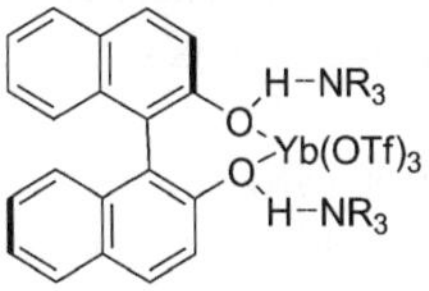

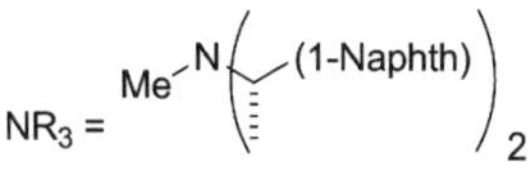

(8.194) (8.195)

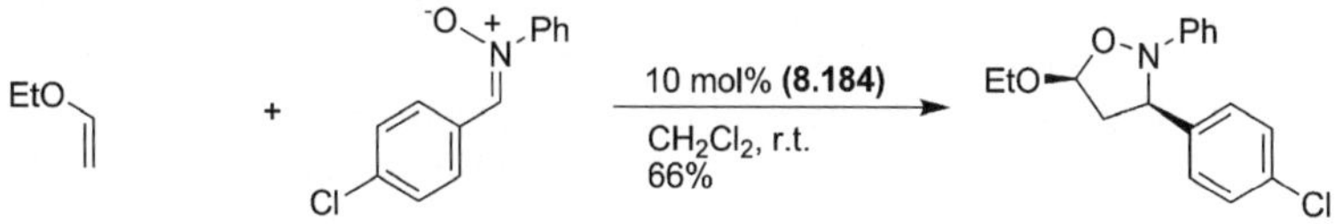

EtO (8.161) + (8.185) → 10 mol% (8.184), CH₂Cl₂, r.t., 66% → (8.186) 97% ee >95:5 *exo:endo*

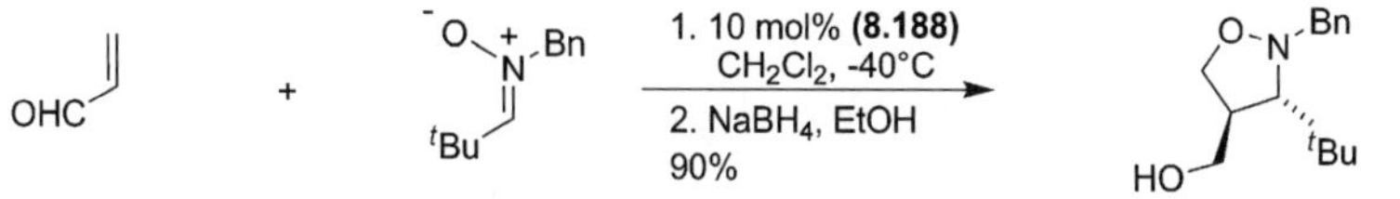

(8.63) + (8.189) → 1. 10 mol% (8.188) CH₂Cl₂, -40°C 2. NaBH₄, EtOH 90% → (8.190) 97% ee

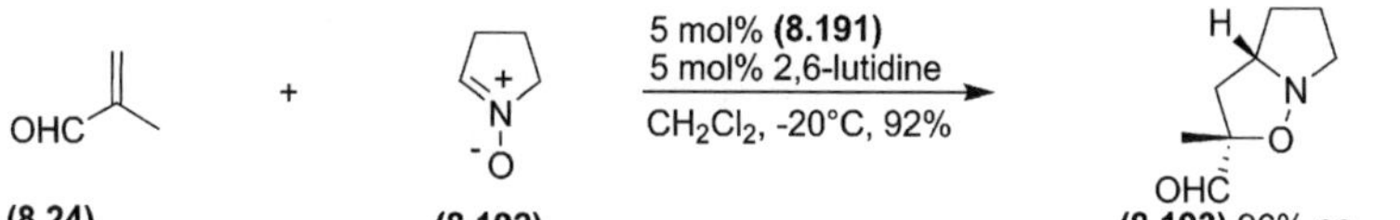

(8.24) + (8.192) → 5 mol% (8.191) 5 mol% 2,6-lutidine CH₂Cl₂, -20°C, 92% → (8.193) 96% ee

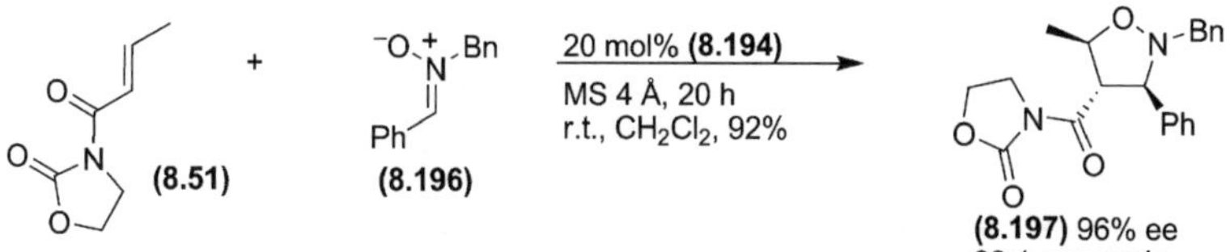

(8.51) + (8.196) → 20 mol% (8.194) MS 4 Å, 20 h r.t., CH₂Cl₂, 92% → (8.197) 96% ee 99:1 *exo:endo*

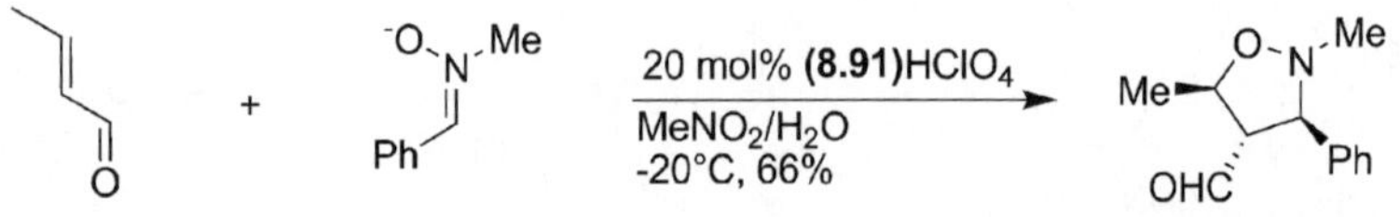

(8.99)	(8.198)	(8.199) 99% ee
		exo:endo 5:95

20 mol% (8.91)HClO$_4$
MeNO$_2$/H$_2$O
-20°C, 66%

that catalysts of this type do effect the cycloaddition of amino acid-derived aryl imines with N-methylmaleimide, with ees up to 90%.[121] Some of the highest ees in the azomethine ylide cycloaddition have been obtained by Schreiber and coworkers using silver complexes of QUINAP (8.206) and this catalyst is active with monoactivated olefins such as *tert*-butyl acrylate (8.207).[122] In an alternate approach silver salts have been used in combination with the base hydrocinchonine giving moderate to good ee in the cycloaddition with acrylates.[123] In this process, enantioselectivity arises through formation of an ion pair between the protonated base and ylide.

A variety of copper-based catalysts have been used with success in the asymmetric azomethine ylide cycloaddition. Kanemasa and coworkers have obtained good ees in the cycloaddition with N-phenylmaleimide, using complexes generated from copper(II) triflate and the bis-phosphines BINAP and SEGPHOS (8.210).[124] Higher ees have been obtained using copper catalysts formed with ferrocene-based ligands. For instance, the cycloaddition with N-methylmaleimide (8.211) proceeds with 99% ee using copper(I) complex formed with ligand (8.212)[125] while phosphino-oxazolines (8.214) have been used in the enantioselective cycloaddition with acrylates[126] and nitroalkenes.[127]

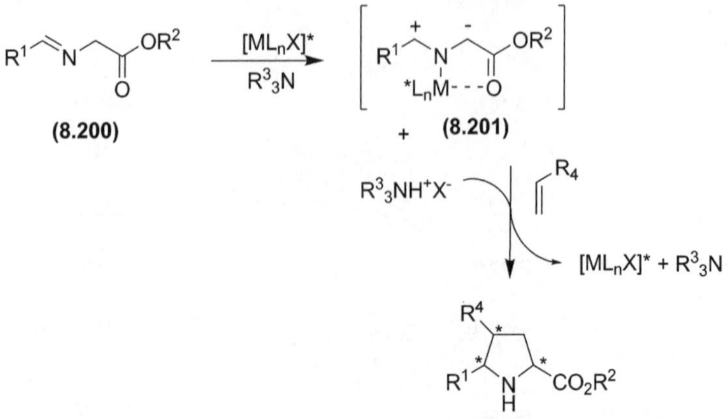

Figure 8.6 Metal complex/base catalysed azomethine ylide cycloaddition

Bis-oxazolines of much use in both the Diels–Alder and nitrone cycloadditions, have also been shown, by Jørgensen and coworkers, to be good ligands in the azomethine ylide cycloaddition. In particular, the zinc complex formed with ^{t}Bu-BOX ligand (**8.68**) and zinc(II) triflate has been used to effect good to high ees in the reaction of ylides generated from iminoesters such as (**8.202**) with acrylates.[128]

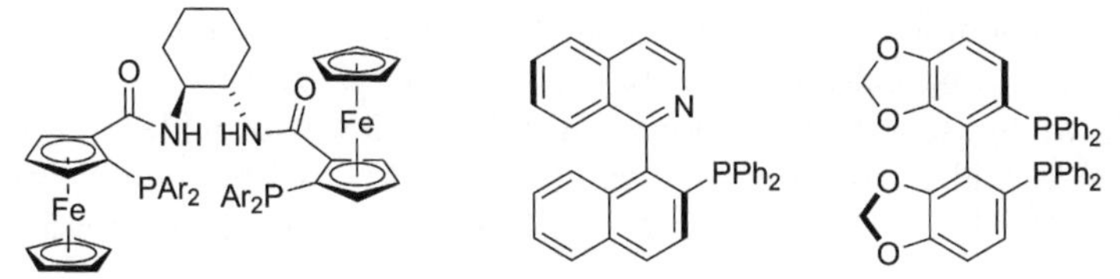

O
NH HN
O
Fe
PAr2 Ar2P
Fe
N
PPh2
O
O
PPh2
PPh2
O
O

(8.204) **(8.206)** **(8.210)**

Ar = 3,5-dimethylphenyl

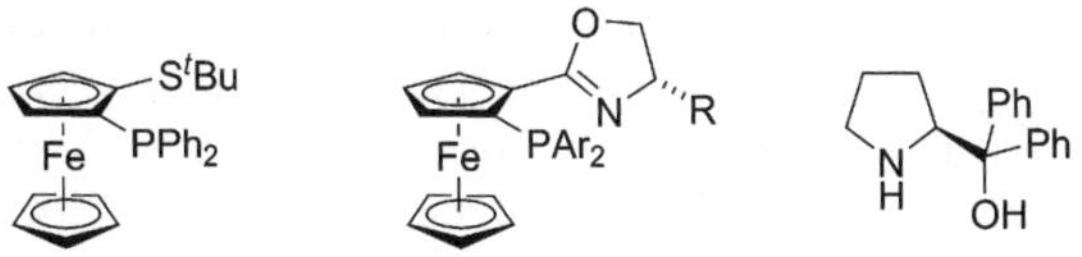

S^{t}Bu
Fe PPh2
O
N
R
Fe PAr2
N
H
Ph
Ph
OH

(8.212) **(8.214)** **(8.215)**

R = ^{t}Bu, ^{i}Pr

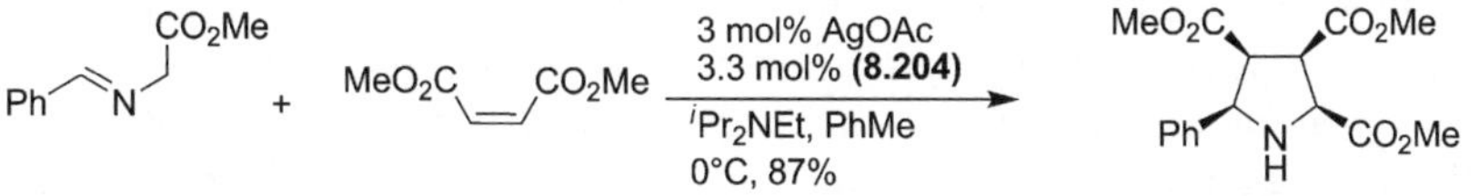

CO2Me
Ph N
+
MeO2C CO2Me
3 mol% AgOAc
3.3 mol% (8.204)
^{i}Pr2NEt, PhMe
0°C, 87%
MeO2C CO2Me
Ph N CO2Me
H

(8.202) **(8.203)** **(8.205)** 87% ee

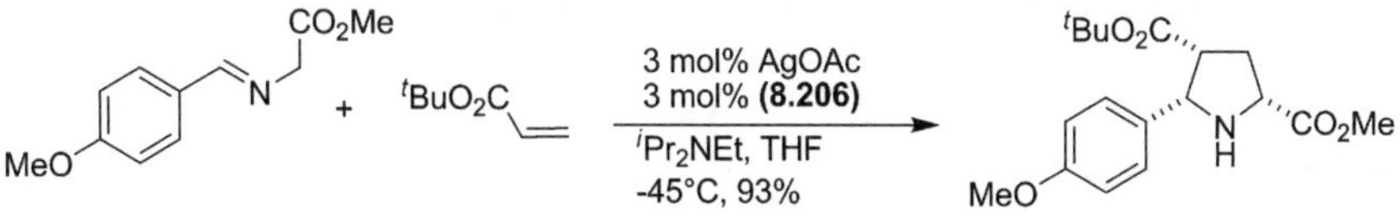

CO2Me
N
MeO
+
^{t}BuO2C
3 mol% AgOAc
3 mol% (8.206)
^{i}Pr2NEt, THF
-45°C, 93%
^{t}BuO2C
N
H
CO2Me
MeO

(8.208) **(8.207)** **(8.209)** 95% ee

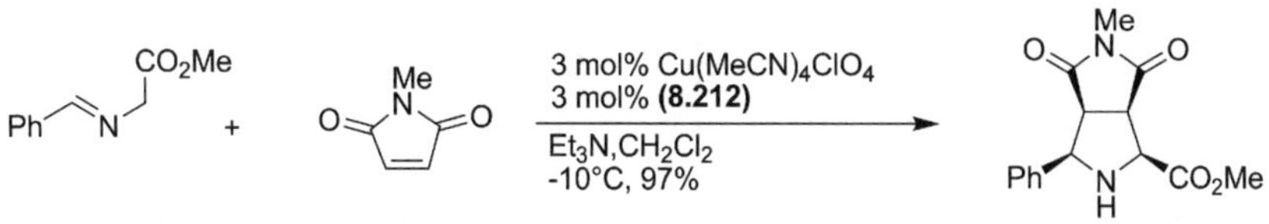

CO2Me
Ph N
+
Me
O N O
3 mol% Cu(MeCN)4ClO4
3 mol% (8.212)
Et3N,CH2Cl2
-10°C, 97%
Me
O N O
Ph N CO2Me
H

(8.202) **(8.211)** **(8.213)** >99% ee

One example of an organocatalysed azomethine ylide cycloaddition has been developed by Vicario and coworkers using pyrrolidine (**8.215**) as catalyst.[129] In this example, the α,β-enal dienophile is activated by conversion to the iminium ion *in situ*, leading to the formation of cycloadducts with ees ranging from 93 to 99%.

8.5 [2+2] Cycloadditions

The [2+2] cycloaddition is the main method for the synthesis of cyclobutanes and 4-membered ring heterocycles. The thermal reaction between two alkenes is not a synchronous, pericyclic process, which is symmetry forbidden, but is a two-step, Lewis acid-catalysed procedure involving a Michael reaction between an electron-rich alkene and an electron-poor partner followed by cyclisation (Figure 8.7).

A small number of enantiomerically pure Lewis acid catalysts have been investigated in an effort to develop a catalytic asymmetric process.[130] Initial work in this area was carried out by Narasaka and coworkers using the titanium complex derived from diol (**8.216**) in the cycloaddition of electron-deficient oxazolidinones such as (**8.217**) with ketene dithioacetal (**8.218**), alkenyl sulfides and alkynyl sulfides.[131] Cyclic alkenes can be used in this reaction and up to 73% ee has been obtained in the [2+2] cycloaddition of thioacetylene (**8.220**) and derivatives with 2-methoxycarbonyl-2-cyclopenten-1-one (**8.221**) using the copper catalyst generated with bis-pyridine (**8.222**).[132] Furthermore, up to 99% ee has been obtained in the [2+2] cycloaddition of norbornene with alkynyl esters using rhodium/H_8-BINAP catalysts.[133] This reaction is not restricted to the use of transition metal-based Lewis acids. Canales and Corey have shown that aluminium bromide-activated oxazaborrolidine (**8.224**) used as a catalyst in the Diels–Alder reaction[17] also functions effectively in the [2+2] cycloaddition of cyclic silyl enol ethers such as (**8.225**) with trifluoroethyl acrylate (**8.226**) to give the adducts with ees ranging from 92 to 99%.[134]

This asymmetric cycloaddition has also been achieved using the ammonium salt derived from amine (**8.228**) as catalyst.[135] In this approach the acceptor α,β-enal (**8.230**) is further activated by conversion to iminium ion *in situ* and even unactivated alkenes such as (**8.229**) can be used as the donor.

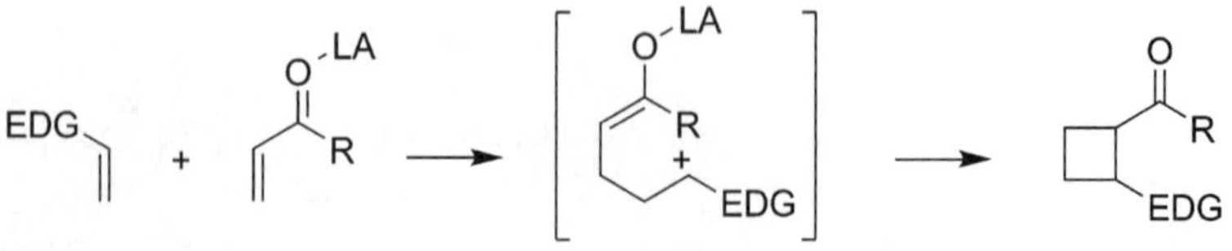

Figure 8.7 Lewis acid catalysed thermal [2+2] cycloaddition. EDG = Electron donating group. LA = Lewis acid

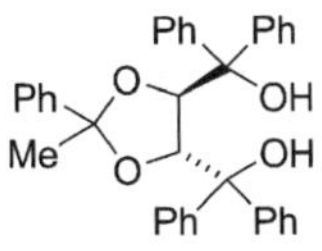

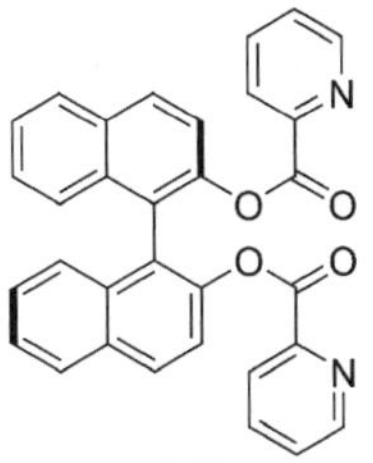

(8.216)

(8.222)

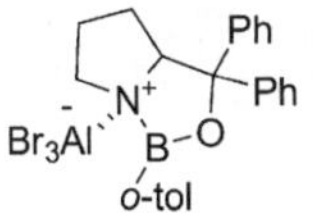

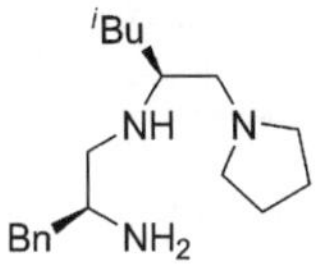

(8.224)

(8.228)

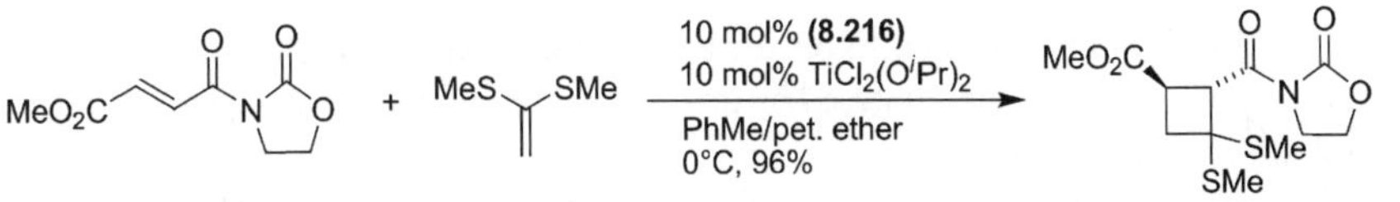

(8.217) + (8.218)

10 mol% **(8.216)**
10 mol% TiCl$_2$(O^{i}Pr)$_2$

PhMe/pet. ether
0°C, 96%

(8.219) 98% ee

(8.221) + (8.220)

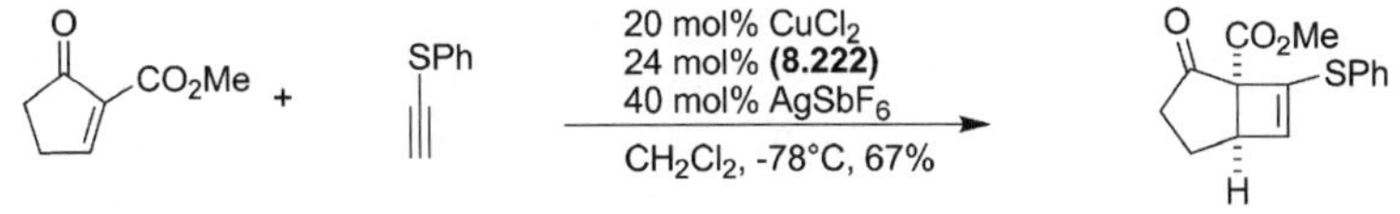

20 mol% CuCl$_2$
24 mol% **(8.222)**
40 mol% AgSbF$_6$

CH$_2$Cl$_2$, -78°C, 67%

(8.223)

(8.225) + (8.226)

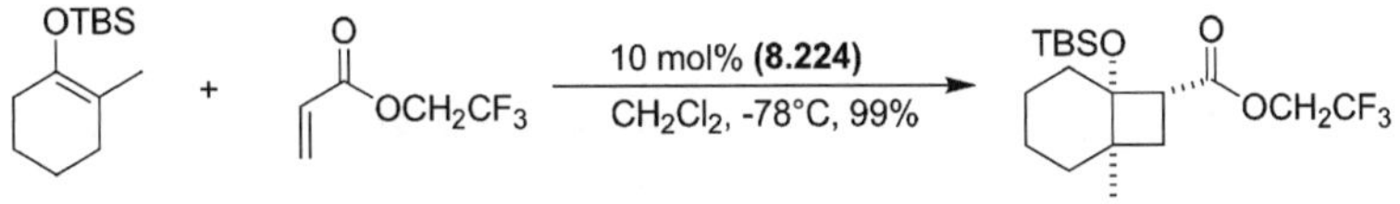

10 mol% **(8.224)**
CH$_2$Cl$_2$, -78°C, 99%

(8.227)

(8.229) + (8.230)

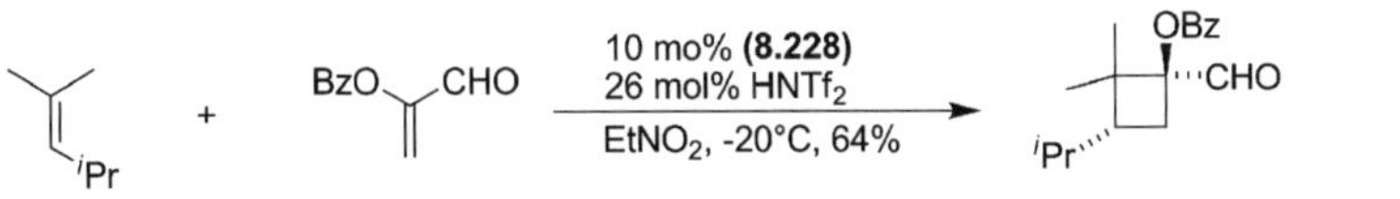

10 mo% **(8.228)**
26 mol% HNTf$_2$

EtNO$_2$, -20°C, 64%

(8.231) 85% ee

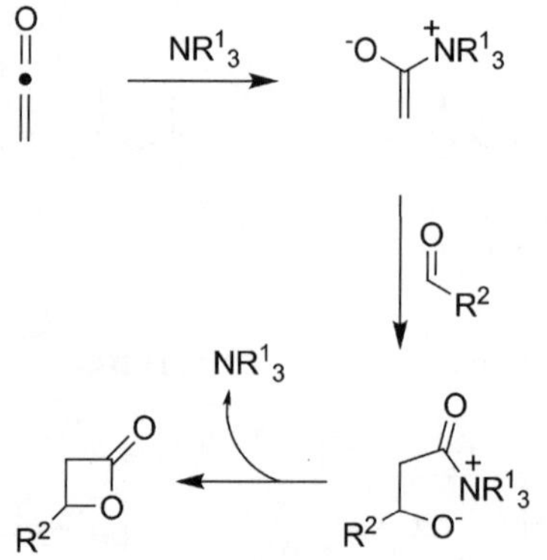

Figure 8.8 Amine catalysed [2+2] cycloaddition

The [2+2] cycloaddition also occurs between ketenes and aldehydes to give oxetanones. This reaction is catalysed by tertiary amines which form acylammonium enolates *in situ* (Figure 8.8).

Wynberg and coworkers developed an asymmetric variant utilising cinchona alkaloids as amine catalyst, and up to 98% ee is obtained using chloral and derivatives as the aldehyde component. [136] The group of Nelson has recently expanded the scope of this process by performing the reaction in the presence of the Lewis acid $LiClO_4$, which allows the use of less electrophilic aldehydes. [137] In this reaction the *O*-trimethylsilyl-protected quinine or quinidine (**8.232**) provides the highest enantioselectivities and the ketene is generated from an acid chloride by reaction with $^iPr_2NEt$.

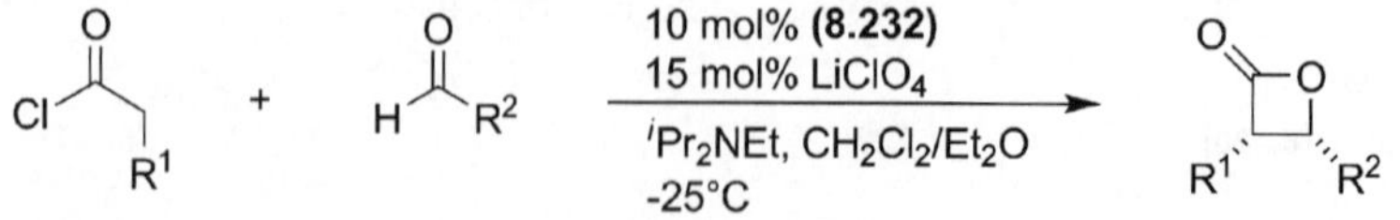

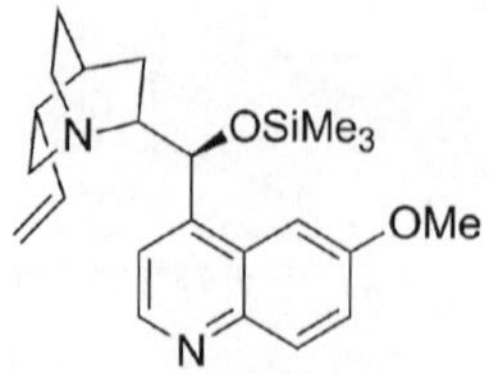

R[1]	R[2]	Yield	ee
H	CMe_3	71%	96%
H	$c\text{-}C_6H_{11}$	85%	94%
H	CH_2CH_2Ph	80%	92%
Me	$o\text{-}C_6H_4Cl$	80%	>99%

8.6 Pauson–Khand-Type Reactions

The Pauson–Khand reaction is a cobalt mediated [2+2+1] cycloaddition of an alkene (8.233), an alkyne (8.234) and carbon monoxide to give a cyclopentenone (8.235).

The mechanism for this process is depicted in Figure 8.9. Initial formation of the cobalt complex (8.236) is followed by alkene insertion and carbonyl insertion to gives the acyl cobalt complex (8.238), which undergoes reductive elimination to give the cyclopentenone.

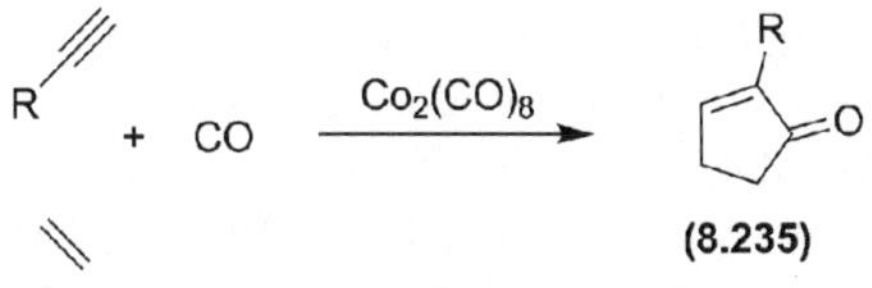

The Pauson–Khand reaction is also mediated by complexes of other transition metals including those of titanium, rhodium and iridium. A number of complexes of these metals have been used to catalyse the intramolecular asymmetric Pauson–Khand reaction of 1,6-enynes to give bicyclo[3.3.0]octenes.[138] The first catalytic asymmetric Pauson–Khand reaction was developed by Hicks and

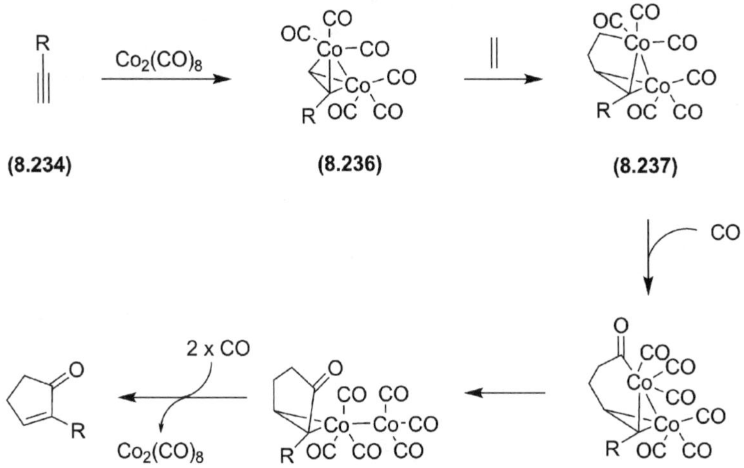

Figure 8.9 Mechanism of the Pauson-Khand reaction

Buchwald in 1999 using the enantiomerically pure titanocene (**8.240**).[139] A variety of all carbon-1,6-enynes, such as (**8.241**) and substrates incorporating oxygen or nitrogen atoms at the 4-position, undergo cyclisation with high ee using this catalyst under elevated pressures of CO.

Rhodium and iridium catalysts incorporating bisphosphine ligands such as BINAP and derivatives have also been shown to be effective in the asymmetric Pauson–Khand reaction of 1,6-enynes, and some high ees have been obtained with substrates incorporating heteroatoms.[140] As an example, the allylpropargylamine (**8.243**) is converted into the bicyclic product (**8.244**) with high ee in the presence of an iridium/TolBINAP catalyst.[140a] It has been shown that, in some cases, aldehydes can be used as a CO source, thus avoiding the use of high pressures of this gas.[140c,d] For instance, Kwong, Chan and coworkers have developed an operationally simple aqueous procedure for the enantioselective cyclisation of ethers such as (**8.245**)

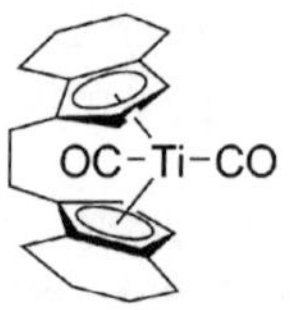

(**8.240**) 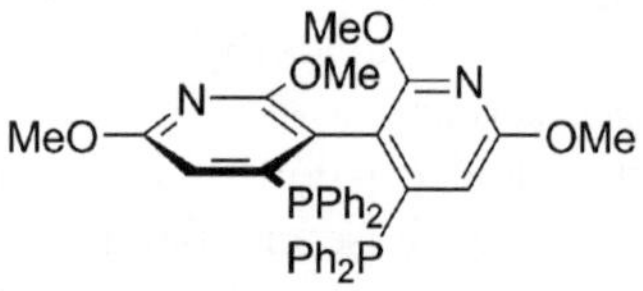   (**8.246**)

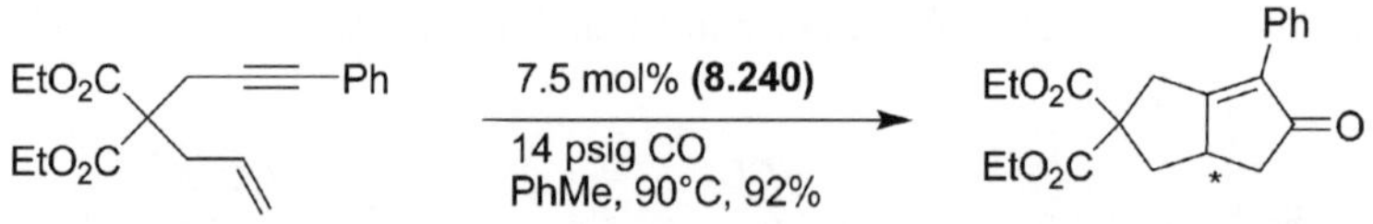

EtO₂C⟍ ≡—Ph	7.5 mol% (**8.240**)	Ph
EtO₂C⟋ ⟍	14 psig CO PhMe, 90°C, 92%	EtO₂C⟍ ⟍=O EtO₂C⟋ *
(**8.241**)		(**8.242**) 94% ee

$$EtO_2C,\ \equiv\!-Ph \quad \xrightarrow[\text{14 psig CO}]{\text{7.5 mol\% (8.240)}} \quad$$

(**8.241**)   (**8.242**) 94% ee

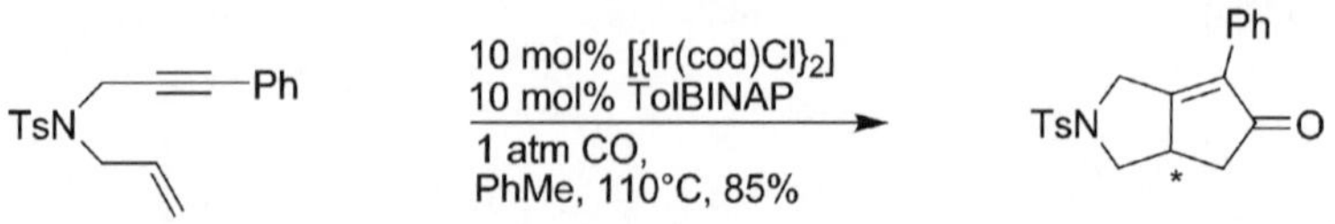

$$TsN,\ \equiv\!-Ph \quad \xrightarrow[\substack{\text{1 atm CO,}\\ \text{PhMe, 110°C, 85\%}}]{\substack{\text{10 mol\% [\{Ir(cod)Cl\}_2]}\\ \text{10 mol\% TolBINAP}}} \quad$$

(**8.243**)   (**8.244**) 95% ee

$$O,\ \equiv\!-Et \quad \xrightarrow[\substack{\text{1.5 equiv. cinnamaldehyde}\\ \text{H}_2\text{O, 100°C, 60\%}}]{\substack{\text{3 mol\% [\{Rh(cod)Cl\}_2]}\\ \text{6 mol\% (8.246)}}} \quad$$

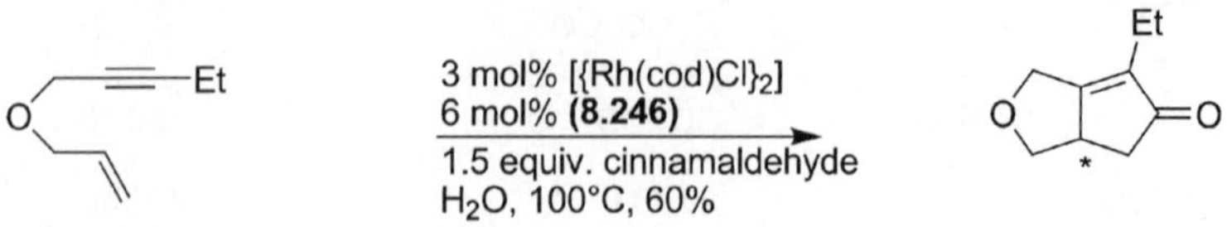

(**8.245**)   (**8.247**) 95% ee utilising the air and moisture stable rhodium catalyst incorporating bisphosphine (8.246) and 1.5 equivalents of cinnamaldehyde.[140d]

References

1. For general reviews on the catalytic asymmetric Diels–Alder reaction see: (a) H. B. Kagan and O. B. Riant, *Chem. Rev.*, **1992**, *92*, 1007. (b) D. A. Evans and J. S. Johnson, in *Comprehensive Asymmetric Catalysis*, Vol. 3, ed. E. N. Jacobsen, A. Pfaltz and H. Yamamoto, Springer-Verlag, Berlin, **1999**, 1178. (c) E. J. Corey, *Angew. Chem. Int. Ed.*, **2002**, *41*, 1650.

2. M. Takasu and H. Yamamoto, *Synlett*, **1990**, 194.

3. D. Sartor, J. Saffrich and G. Helmchen, *Synlett*, **1990**, 197.

4. E. J. Corey and T.-P. Loh, *J. Am. Chem. Soc.*, **1991**, *113*, 8966.

5. E. J. Corey, T.-P. Loh, T. D. Roper, M. D. Azimioara and M. C. Noe, *J. Am. Chem. Soc.*, **1992**, *114*, 8290.

6. E. J. Corey, A. Guzman-Perez and T.-P. Loh, *J. Am. Chem. Soc.*, **1994**, *116*, 3611.

7. E. J. Corey and T.-P. Loh, *Tetrahedron Lett.*, **1993**, *34*, 3979.

8. K. Furuta, Y. Miwa, K. Iwanaga and H. Yamamoto, *J. Am. Chem. Soc.*, **1988**, *110*, 6254.

9. K. Ishihara, Q. Gao and H. Yamamoto, *J. Org. Chem.*, **1993**, *58*, 6917.

10. K. Furuta, A. Kanematsu, H. Yamamoto and S. Takaoka, *Tetrahedron Lett.*, **1989**, *30*, 7231.

11. K. Ishihara, Q. Gao and H. Yamamoto, *J. Am. Chem. Soc.*, **1993**, *115*, 10412.

12. (a) K. Ishihara and H. Yamamoto, *J. Am. Chem. Soc.*, **1994**, *116*, 1561. (b) K. Ishihara, H. Kurihara, M. Matsumoto and H. Yamamoto, *J. Am. Chem. Soc*, **1998**, *120*, 6920.

13. Y. Hayashi, J. J. Rohde and E. J. Corey, *J. Am. Chem. Soc.*, **1996**, *118*, 5502.

14. (a) E. J. Corey, T. Shibata and T. W. Lee, *J. Am. Chem. Soc.*, **2002**, *124*, 3808. (b) D. H. Ryu and E. J. Corey, *J. Am .Chem. Soc.*, **2002**, *124*, 9992.

15. D. H. Ryu and E. J. Corey, *J. Am. Chem. Soc.*, **2003**, *125*, 6388.

16. J. N. Payette and H. Yamamoto, *J. Am. Chem. Soc.*, **2007**, *129*, 9536.

17. D. Liu, E. Canales and E. J. Corey, *J. Am. Chem. Soc.*, **2007**, *129*, 1498.

18. K. Futatsugi and H. Yamamoto, *Angew. Chem. Int. Ed.*, **2005**, *44*, 1484.

19. For example, see; (a) S. Kobayashi, M. Murakami, T. Harada and T. Mukaiyama, *Chem. Lett.*, **1991**, 1341. (b) D. Kaufmann, R. Boese, *Angew. Chem., Int. Ed. Engl.*, **1990**, *29*, 545. (c) K. Ishihara, H. Kurihara and H. Yamamoto, *J. Am. Chem. Soc.*, **1996**, *118*, 3049.

20. J. M. Hawkins and S. Loren, *J. Am. Chem. Soc.*, **1991**, *113*, 7794.

21. J. M. Hawkins, M. Nambu and S. Loren. *Org. Lett.*, **2003**, *5*, 4293.

22. S. Hashimoto, N. Komeshima and K. Koga, *J. Chem. Soc., Chem. Commun.*, **1979**, 437.

23. E. J. Corey, R. Imwinkelried, S. Pikul and Y. B. Xiang, *J. Am. Chem. Soc.*, **1989**, *111*, 5493.

24. J. Bao, W.D. Wulff and A.L. Rheingold, *J. Am. Chem. Soc.*, **1993**, *115*, 3814.

25. D. J. Ramón and M. Yus, *Chem. Rev.*, **2006**, *106*, 2126.

26. K. Narasaka, N. Iwasawa, M. Inoue, T. Yamada, M. Nakashima and J. Sugimori, *J. Am. Chem. Soc.*, **1989**, *111*, 5340.

27. (a) C. Haase, C. R. Sarko and M. DiMare, *J. Org. Chem.*, **1995**, *60*, 1777 . (b) D. Seebach, R. Dahinden, R. E. Marti, A. K. Beck, D. A. Plattner and F. N. M. Kühnle, *J. Org. Chem.*, **1995**, *60*, 1788. (c) K. V. Gothelf and K. A. Jørgensen, *J. Org. Chem.*, **1995**, *60*, 6847.

28. B. Altava, M. I. Burguete, J. I. García, S. V. Luis, J. A. Mayoral and M. J. Vicent, *Tetrahedron: Asymmetry*, **2001**, *12*, 1829.

29. K. Mikami, Y. Motoyama and M. Terada, *J. Am. Chem. Soc.*, **1994**, *116*, 2812.

30. M. Breuning and E. J. Corey, *Org. Lett.*, **2001**, *3*, 1559.

31. Y. Motoyama, M. Terada and K. Mikami, *Synlett*, **1995**, 967.

32. J. B. Jaquith, J. Guan, S. Wang and S. Collins, *Organometallics*, **1995**, *14*, 1079.

33. K. Maruoka, N. Murase and H. Yamamoto, *J. Org. Chem.*, **1993**, *58*, 2938.

34. E. J. Corey, T. D. Roper, K. Ishihara and G. Sarakinos, *Tetrahedron Lett.*, **1993**, *34*, 8399.

35. G. Desimoni, G. Faita and K. A. Jorgensen, *Chem. Rev.*, **2006**, *106*, 3561.

36. E. J. Corey, N. Imai and H.-Y. Zhang, *J. Am. Chem. Soc.*, **1991**, *113*, 728.

37. E. J. Corey and K. Ishihara, *Tetrahedron Lett.*, **1992**, *33*, 6807.

38. T. Fujisawa, T. Ichiyanagi, M. Shimizu, *Tetrahedron Lett.*, **1995**, *36*, 5031.

39. G. Desimoni, G. Faita and P. P. Righetti, *Tetrahedron Lett.*, **1996**, *37*, 3027.

40. (a) D. A. Evans, S. J. Miller and T. Lectka, *J. Am. Chem. Soc.*, **1993**, *115*, 6460. (b) D. A. Evans, S. J. Miller, T. Letcka and P. von Matt, *J. Am. Chem. Soc.*, **1999**, *121*, 7559. (c) D. A. Evans, D. M. Barnes, J. S. Johnson, T. Letcka, P. von Matt, S. J. Miller, J. A. Murry, R. D. Norcross, E. Shaughnessy and K. R. Campos, *J. Am. Chem. Soc.*, **1999**, *121*, 7582.

41. D. A. Evans and D. M. Barnes, *Tetrahedron Lett.*, **1997**, *38*, 57.

42. D. A. Evans and J. S. Johnson, *J. Org. Chem.*, **1997**, *62*, 786.

43. V. K. Aggarwal, E. S. Anderson, D. Elfyn Jones, K. B. Obierey and R. Giles, *Chem. Commun.*, **1998**, 1985. (b) V. K. Aggarwal, D. Elfyn Jones and A. M. Martin-Castro, *Eur. J. Org. Chem.*, **2000**, 2939.

44. C. Palomo, M. Oiarbide, J. M. García, A. González and E. Arceo, *J. Am. Chem. Soc.*, **2003**, *125*, 13942.

45. (a) D. A. Evans, J. A. Murry, P. von Matt, R. D. Norcross and S. J. Miller, *Angew. Chem., Int. Ed. Engl.*, **1995**, *34*, 798. (b) A. K. Ghosh, P. Mathivanan and J. Cappiello, *Tetrahedron Lett.*, **1996**, *37*, 3815. (c) I. W. Davies, L. Gerena, D. Cai, R. D. Larsen, T. R. Verhoeven and P. J. Reider, *Tetrahedron Lett.*, **1997**, *38*, 1145.

46. I. Sagasser and G. Helmchen, *Tetrahedron Lett.*, **1998**, *39*, 261.

47. A. J. Davenport, D. L. Davies, J. Fawcett, S. A. Garrett, L. Lad and D. R. Russell, *J. Chem. Soc., Chem. Commun.*, **1997**, 2347.

48. D. L. Davies, J. Fawcett, S. A. Garratt and D. R. Russell, *J. Chem. Soc., Chem. Commun.*, **1997**, 1351.

49. G. Desimoni, G. Faita, M. Guala, A. Laurenti and M. Mella, *Chem. Eur. J.*, **2005**, *11*, 3816.

50. D. A. Evans and J. Wu, *J. Am. Chem. Soc.*, **2003**, *125*, 10162.

51. S. Kanemasa, Y. Oderaotoshi, S.-I. Sakaguchi, H. Yamamoto, J. Tanaka, E. Wada and D. P. Curran, *J. Am. Chem. Soc.*, **1998**, *120*, 3074.

52. T. Morita, T. Arai, H. Sasai and M. Shibasaki, *Tetrahedron: Asymmetry*, **1998**, *9*, 1445.

53. Y. Yamashita and T. Katsuki, *Synlett*, **1995**, 829.
54. (a) Y. Huang, T. Iwama and V. K. Rawal, **2000**, *122*, 7843. (b) Y. Huang, T. Iwama and V. H. Rawal, *Org. Lett.*, **2002**, *4*, 1163. (c) Y. Huang, T. Iwama and V. H. Rawal, *J. Am. Chem. Soc.*, **2002**, *124*, 5950.
55. E. P. Kündig, B. Bourdin and G. Bernardinelli, *Angew. Chem., Int. Ed. Engl.*, **1994**, *33*, 1856.
56. J. W. Faller, B. J. Grimmond and D. G. D'Alliessi, *J. Am. Chem. Soc.*, **2001**, *123*, 2525.
57. S. Kobayashi, H. Ishitani, M. Araki and I. Hachiya, *Tetrahedron Lett.*, **1994**, *35*, 6325.
58. S. Kobayashi, M. Araki and I. Hachiya, *J. Org. Chem.*, **1994**, *59*, 3758.
59. S. Kobayashi, H. Ishitani, I. Hachiya and M. Araki, *Tetrahedron*, **1994**, *50*, 11623.
60. (a) H. Brunner, M. Muschiol and F. Prester, *Angew. Chem., Int. Ed. Engl.*, **1990**, *29*, 652. (b) M. Lautens, J. C. Lautens and A. C. Smith, *J. Am. Chem. Soc.*, **1990**, *112*, 5627.
61. For reviews on the organocatalytic Diels–Alder reaction see: (a) A. Berkessel and H. Gröger, *Asymmetric Organocatalysis*, Wiley-VCH, Weinheim, **2005**, 256.(b) A. Berkessel and H. Gröger, *Asymmetric Organocatalysis*, Wiley-VCH, Weinheim, **2005**, 256.(c) H. Pellisier, *Tetrahedron*, **2007**, *63*, 9267.
62. (a) O. Riant and H. B. Kagan, *Tetrahedron Lett.* **1989**, *30*, 7403. (b) O. Riant, H. B. Kagan and L. Ricard, *Tetrahedron*, **1994**, *50*, 4543.
63. K. A. Ahrendt, C. J. Borths and D. W. C. MacMillan, *J. Am. Chem. Soc.*, **2000**, *122*, 4243.
64. R. M. Wilson, W. S. Jen and D. W. C. MacMillan, *J. Am. Chem. Soc.*, **2005**, *127*, 11616.
65. A. B. Northrup and D. W. C. MacMillan, *J. Am. Chem. Soc.*, **2002**, *124*, 2458
66. (a) M. Lemay and W. W. Ogilvie, *Org. Lett.*, **2005**, *7*, 4141. (b) T. Kano, Y. Tanaka and K. Maruoka, *Org. Lett.*, **2006**, *8*, 2687. (c) K. Ishihara and K. Nakano, *J. Am. Chem. Soc.*, **2005**, *127*, 10504.
67. A. N. Thadani, A. R. Stankovic and V. K. Rawal, *Proc. Natl. Acad. Sci.*, **2004**, *101*, 5846.
68. D. Nakashima and H. Yamamoto, *J. Am. Chem. Soc.*, **2006**, *128*, 9626.
69. I. E. Markó and G. R. Evans, *Tetrahedron Lett.*, **1994**, *35*, 2771.
70. G. H. Posner, F. Eydoux, J.K. Lee and D. S. Bull, *Tetrahedron Lett.*, **1994**, *35*, 7541.
71. For reviews on the catalytic asymmetric hetero-Diels Alder reaction see: (a) T. Ooi and K. Maruoka, in *Comprehensive Asymmetric Catalysis*, Vol. 3, ed. E. N. Jacobsen, A. Pfaltz and H. Yamamoto, Springer-Verlag, Berlin, **1999**, 1237. (b) K. A. Jørgensen, *Angew. Chem. Int. Ed.*, **2000**, *39*, 3559. (c) K. A. Jørgensen, *Eur. J. Org. Chem.* **2004**, 2093.
72. M. Bednarski, C. Maring and S. Danishefsky, *Tetrahedron Lett.*, **1983**, *24*, 3451.
73. K. Maruoka, T. Itoh, T. Shirasaka and H. Yamamoto, *J. Am. Chem. Soc.*, **1988**, *110*, 310.
74. (a) Q. Gao, K. Ishihara, T. Maruyama, M. Mouri and H. Yamamoto, *Tetrahedron*, **1994**, *50*, 979. (b) Q. Gao, T. Maruyama, M. Mouri and H. Yamamoto, *J. Org. Chem.*, **1992**, *57*, 1951.
75. (a) S. E. Schaus, J. Brånalt and E. N. Jacobsen, *J. Org. Chem.* **1998**, *63*, 403. (b) A. Berkessel and N. Vogl, *Eur. J. Org. Chem.*, **2006**, 5029.
76. Q. Fan, L. Lin, J. Liu, Y. Huang, X. Feng and G. Zhang, *Org. Lett.*, **2004**, *6*, 2185.
77. J. Long, J. Hu, X. Shen, B. Ji and K. Ding, *J. Am. Chem. Soc.*, **2002**, *124*, 10.

78. M. Anada, T. Washio, N. Shimada, S. Kitagaki, M. Nakajima, M. Shiro and S. Hashimoto, *Angew. Chem. Int. Ed.*, **2004**, *43*, 2665.
79. M. Johannsen and K. A. Jørgensen, *J. Org. Chem.*, **1995**, *60*, 5757.
80. M. Johannsen and K. A. Jørgensen, *Tetrahedron*, **1996**, *52*, 7321.
81. (a) M. Johannsen, S. Yao and K.A. Jørgensen, *J. Chem. Soc., Chem. Commun.*, **1997**, 2169. (b) S. Yao, M. Johannsen, H. Audrain, R. G. Hazell and K. A. Jørgensen, *J. Am. Chem. Soc.*, **1998**, *120*, 8599.
82. B. Bolm and O. Simić, *J. Am. Chem. Soc.*, **2001**, *123*, 3830.
83. K. Seki, M. Ueno and S. Kobayashi, *J. Chem. Soc., Chem. Commun.*, **2007**, *5*, 1347.
84. Y. Huang, A. K. Unni, A. N. Thadani and V. K. Rawal, *Nature*, **2003**, *424*, 146.
85. (a) A. K. Unni, N. Takenaka, H. Yamamoto and V. K. Rawal, *J. Am. Chem. Soc.*, **2005**, *127*, 1336. (b) X. Zhang, H. Du, Z. Wang, Y.-D. Wu and K. Ding, *J. Org. Chem.*, **2006**, *71*, 2862.
86. S. Rajaram and M. S. Sigman, *Org. Lett.*, **2005**, *7*, 5473. (b) K. H. Jensen and M. S. Sigman, *Angew. Chem. Int. Ed.*, **2007**, *46*, 4748.
87. W. Zhuang, T. B. Poulsen and K. A. Jørgensen, *Org. Biomol. Chem.*, **2005**, *3*, 3284.
88. T. Tonoi and K. Mikami, *Tetrahedron Lett.*, **2005**, *46*, 6355.
89. G. B. Rowland, E. B. Rowland, Q. Zhang and J. C. Antilla, *Curr. Org. Chem.*, **2006**, *10*, 981.
90. (a) S. Kobayashi, S. Komiyama and H. Ishitani, *Angew. Chem., Int. Ed. Engl.*, **1998**, *37*, 979.(b) S. Kobayashi, K.-I. Kusakabe, S. Komiyama and H. Ishitani, *J. Org. Chem.* **1999**, *64*, 4220.(c) S. Kobayashi, K.-I. Kusakabe and H. Ishitani, *Org. Lett.*, **2000**, *2*, 1225.
91. N. S. Josephsohn, M. L. Snapper and A. H. Hoveyda, *J. Am. Chem. Soc.*, **2003**, *125*, 4018.
92. (a) S. Yao, M. Johannsen, R. G. Hazell and K. A. Jørgensen, *Angew. Chem. Int. Ed.*, **1998**, *37*, 3121. (b) S. Yao, S. Saaby, R. G. Hazel and K. A. Jørgensen, *Chem. Eur. J.*, **2000**, *6*, 2435.
93. (a) T. Akiyama, Y. Tamura, J. Itoh, H. Morita and K. Fuchibe, *Synlett*, **2006**, 141. (b) J. Itoh, K. Fuchibe and T. Akiyama, *Angew. Chem. Int. Ed.*, **2006**, *45*, 4796. (b) H. Liu, L.-F. Cun, A.-Q. Mi, Y.-Z. Jiang and L.-Z. Gong, *Org. Lett.*, **2006**, *8*, 6023.
94. D. A. Evans and J. S. Johnson, *J. Am. Chem. Soc.*, **1998**, *120*, 4895. (b) D. A. Evans, J. S. Johnson and E. J. Olhava, *J. Am. Chem. Soc.*, **2000**, *122*, 1635.
95. K. Gademann, D. E. Chavez and E. N. Jacobsen, *Angew. Chem. Int. Ed.*, **2002**, *41*, 3059.
96. K. Juhl and K. A. Jørgensen, *Angew. Chem. Int. Ed.*, **2003**, *42*, 1498.
97. M. He, G. J. Uc and J. W. Bode, *J. Am. Chem. Soc.*, **2006**, *128*, 15088.
98. H. Ishitani and S. Kobayashi, *Tetrahedron Lett.*, **1996**, *37*, 7357.
99. E. Jnoff and L. Ghosez, *J. Am. Chem. Soc.*, **1999**, *121*, 2617.
100. J. Esquivias, R. G. Arrayás and J. C. Carretero, *J. Am. Chem. Soc.*, **2007**, *129*, 1480.
101. T. Akiyama, H. Morita and K. Fuchibe, *J. Am. Chem. Soc.*, **2006**, *128*, 13070.
102. M. He, J. R. Struble and J. W. Bode, *J. Am. Chem. Soc.*, **2006**, *128*, 8418.
103. For reviews covering the catalytic asymmetric cycloaddition reactions of nitrones see; (a) M. Frederickson, *Tetrahedron*, **1997**, *53*, 403. (b) K. V. Gothelf and K. A. Jørgensen, *Chem. Commun.*, **2000**, 1449. (c) H. Pellissier, *Tetrahedron*, **2007**, *63*, 3235.
104. G. Pandey, P. Banerjee and S. R. Gadre, *Chem. Rev.*, **2006**, *106*, 4484.

105. J.-P. G. Seerden, A. W. A. Scholte op Reimer and H. W. Scheeren, *Tetrahedron Lett.*, 1994, 35, 4419. (b) J.-P. G. Seerden, M. M. M. Kuypers and H. W. Scheeren, *Tetrahedron: Asymmetry*, 1995, 6, 1441. (c) J.-P. G. Seerden, M. M. M. Boeren and W. Scheeren, *Tetrahedron*, 1997, 53, 11843.

106. K. B. Simonsen, P. Bayón, R. G. Hazell, K. V. Gothelf and K. A. Jørgensen, *J. Am. Chem. Soc.*, 1999, 121, 3845.

107. K. B. Jensen, M. Roberson and K. A. Jørgensen, *J. Org. Chem.* 2000, 65, 9080.

108. T. Kano, T. Hashimoto and K. Maruoka, *J. Am. Chem. Soc.*, 2005, 127, 11926.

109. S. Kanemasa, Y. Oeratoshi, J. Tanaka and E. Wada, *J. Am. Chem. Soc.*, 1998, 120, 12355.

110. M. Shirahase, S. Kanemasa and Y. Oderatoshi, *Org. Lett.*, 2004, 6, 675.

111. F. Viton, G. Bernardinelli and E. P. Kündig, *J. Am. Chem. Soc.*, 2002, 124, 4968.

112. (a) D. Carmona, M. P. Lamata, F. Viguri, R. Rodriguez, L. A. Oro, A. I. Balana, F. J. Lahoz, T. Tejero, P. Merino, S. Franco and I. Montesa, *J. Am. Chem. Soc.*, 2004, 126, 2716. (b) D. Carmona, M. P. Lamata, F. Viguri, R. Rodriguez, L. A. Oro, F. J. Lahoz, A. I. Balana, T. Tejero and P. Merino, *J. Am. Chem. Soc.*, 2005, 127, 13386.

113. (a) T. Mita, N. Ohtsuki, T. Ikeno and T. Yamada, *Org. Lett.*, 2002, 4, 2457. (b) N. Ohtsuki, S. Kezuka, Y. Kogami, T. Mita, T. Ashizawa, T. Ikeno and T. Yamada, *Synthesis*, 2003, 9, 1462.

114. (a) K. V. Gothelf, R. G. Hazell and Jørgensen, *J. Org. Chem.*, 1996, 61, 346. (b) K. V. Gothelf, R. G. Hazell and K. A. Jørgensen, *J. Org. Chem.* 1998, 63, 5483.

115. M. P. Sibi, Z. Ma and C. P. Jasperse, *J. Am. Chem. Soc.*, 2004, 126, 718. (b) C. Palomo, M. Oiarbide, E. Arceo, J. M. Garcia, R. López, R. González and A. Linden, *Angew. Chem. Int. Ed.*, 2005, 44, 6187.

116. K. Hori, H. Kodama, T. Ohta and I. Furukawa, *Tetrahedron Lett.*, 1996, 37, 5947.

117. S. Kobayashi and M. Kawamura, *J. Am. Chem. Soc.*, 1998, 120, 5840.

118. W. S. Jen, J. J. M. Wiener and D. W. C. MacMillan, *J. Am. Chem. Soc.*, 2000, 122, 9874.

119. C. Nájera and J. M. Sansano, *Angew. Chem. Int. Ed.*, 2005, 44, 6272.

120. J. M. Longmire, B. Wang and X. Zhang, *J. Am. Chem. Soc.*, 2002, 124, 12400.

121. D. Nájera, M. de Gracia Retamosa and J. M. Sansano, *Org. Lett.*, 2007, 9, 4025.

122. C. Chen, X. Li and S. L. Schreiber, *J. Am. Chem. Soc.*, 2003, 125, 10174.

123. C. Alemparte, G. Blay and K. A. Jørgensen, *Org. Lett.*, 2005, 7, 4569.

124. Y. Oderatoshi, W. Cheng, S. Fujitomi, Y. Kasano, S. Minikata and M. Komatsu, *Org. Lett.*, 2003, 5, 5043.

125. S. Cabrera, R. G. Arrayás and J. C. Carretero, *J. Am. Chem. Soc.*, 2005, 127, 16394.

126. W. Gao, X. Zhang and M. Raghunath, *Org. Lett.*, 2005, 7, 4241.

127. X.-X. Yan, Q. Peng, Y. Zhang, K. Zhang, W. Hong, X.-L. Hou and Y.-D. Wu, *Angew. Chem. Int. Ed.*, 2006, 45, 1979.

128. A. S. Gothelf, K. V. Gothelf, R. G. Hazell and K. A. Jørgensen, *Angew. Chem. Int. Ed.*, 2002, 41, 4236.

129. J. L. Vicario, S. Reboredo, D. Badía and L. Carillo, *Angew. Chem. Int. Ed.*, 2007, 46, 5168.

130. For reviews see: (a) Y. Hayashi and K. Narasaka, in *Comprehensive Asymmetric Catalysis*, Vol. 3, ed. E. N. Jacobsen, A. Pfaltz and H. Yamamoto, Springer-Verlag, Berlin, 1999, 1255. (b) E. Lee-Ruff and G. Mladenova, *Chem. Rev.*, 2003, 103, 1449. (c) R. M. Ortuño, A. G. Moglioni and G. Y. Moltrasio, *Curr. Org. Chem.*, 2005, 9, 237.

131. K. Narasaka, Y. Hayashi, H. Shimadzu and S. Niihata, *J. Am. Chem. Soc.*, **1992**, *114*, 8869.
132. (a) H. Ito, M. Hasegawa, Y. Takenaka, Y. Kobayashi and K. Iguchi, *J. Am. Chem. Soc.*, **2004**, *126*, 4520. (b) Y. Takenaka, H. Ito, M. Hasegawa and K. Iguchi, *Tetrahedron*, **2006**, *62*, 3380. (c) Y. Takenaka, H. Ito and K. Iguchi, *Tetrahedron*, **2007**, *63*, 510.
133. T. Shibata, K. Takami and A. Kawachi, *Org. Lett.*, **2006**, *8*, 1343.
134. E. Canales and E. J. Corey, *J. Am. Chem. Soc.*, **2007**, *129*, 12686.
135. K. Ishihara and K. Nakano, *J. Am. Chem. Soc.*, **2007**, *129*, 8930.
136. (a) H. Wynberg and E. G. J. Staring, *J. Am. Chem. Soc.*, **1982**, *104*, 166. (b) H. Wynberg and E. G. J. Staring, *J. Org. Chem.*, **1985**, *50*, 1977.
137. C. Zhu, X. Shen and S. G. Nelson, *J. Am. Chem. Soc.*, **2004**, *126*, 5352.
138. S. L. Buchwald and F. A. Hicks, in *Comprehensive Asymmetric Catalysis*, Vol. 2, ed. E. N. Jacobsen, A. Pfaltz and H. Yamamoto, Springer-Verlag, Berlin, **1999**, 491.
139. (a) F. A. Hicks and S. L. Buchwald, *J. Am. Chem. Soc.*, **1996**, *118*, 11688. (b) F. A. Hicks and S. L. Buchwald, *J. Am. Chem. Soc.*, **1999**, *121*, 7026.
140. (a) T. Shibata and K. Takagi, *J. Am. Chem. Soc.*, **2000**, *122*, 9852. (b) N. Jeong, B. K. Sung and Y. K. Choi, *J. Am. Chem. Soc.*, **2000**, *122*, 6771. (c) T. Shibata, N. Toshida, M. Yamasaki, S. Maekawa and K. Takagi, *Tetrahedron*, **2005**, *61*, 9974. (d) F. Y. Kwong, Y. M. Li, W. H. Lam, L. Qiu, H. W. Lee, C. H. Yeung, K. S. Chan and A. S. C. Chan, *Chem. Eur. J.*, **2005**, *11*, 3872.

Chapter 9
Catalytic Reactions Involving Carbenes and Ylides

Carbenes and their metal complexes are the reactive intermediates in the cyclo-propanation of alkenes, insertion reactions and in the formation of ylides. In general these species are generated by decomposition of diazo compounds in the presence of catalytic quantities of transition metal complexes – mainly those of rhodium and copper. A number of carbene-mediated transformations occur in an enantioselective sense using metal catalysts bearing enantiomerically pure ligands, and some high ees can be obtained in both the intra- and intermolecular cyclo-propanation and insertion processes. The scope of the latter reaction has expanded recently and high ees have now been achieved in the insertion into C−H, O−H and N−H bonds. Ylides can be accessed by reaction of heteroatoms with carbenes and the final subsection of this chapter discusses the asymmetric rearrangement and cycloadditions of these species in the presence of enantiopure metal catalysts.

9.1   Cyclopropanation

The cyclopropane ring exists in a number of biologically active natural products. Furthermore, this structural moiety is readily converted into other functionality and the development of enantioselective cyclopropanation strategies has received much attention. This can be achieved by reaction of diazo compounds with alkenes in the presence of enantiomerically pure transition metal catalysts.[1] In this process the diazo compound decomposes in the presence of the catalyst to give a metal carbene that then reacts with the alkene. While a number of transition metal complexes are known to catalyse this transformation, the use of copper[2] and rhodium[3] catalysts has been most widely investigated in the catalytic asymmetric variant. Alternately, cyclopropanation of alkenes can be achieved by the Simmons–Smith process utilising iodomethylzinc reagents formed by insertion of zinc into diiodomethane. Catalytic asymmetric variants of this reaction have been developed using enantiopure Lewis acids and also iodomethylzinc reagents generated *in situ*.

Catalysis in Asymmetric Synthesis 2e   © 2009 Vittorio Caprio and Jonathan M.J. Williams

Electron-deficient alkenes undergo cyclopropanation by reaction with a range of metal-free ylides and some of the most recent work in this area has focussed on the development of an organocatalysed cyclopropanation using these species.

9.1.1   Copper-Catalysed Cyclopropanation

In 1966, Nozaki and coworkers reported a cyclopropanation reaction catalysed by an enantiomerically pure copper complex albeit with low enantioselectivity (6% ee) by today's standard.[4] This was the first example of a homogeneous synthetic catalyst providing enantioselectivity in an organic reaction. The ligand design was elaborated by Aratani to provide complex (9.01), which is very selective for certain cyclopropanation reactions.[5] The Aratani catalyst set the standard for catalytic cyclopropanation reactions, and the next significant advance was made by Pfaltz, who reported the use of semi-corrin ligands (9.02)[6] and azasemicorrins (9.03).[7] Styrene (9.04) undergoes cyclopropanation with these catalysts with good enantioselectivity, and the diastereoselectivity is reasonably good using bulky diazoesters (9.06).

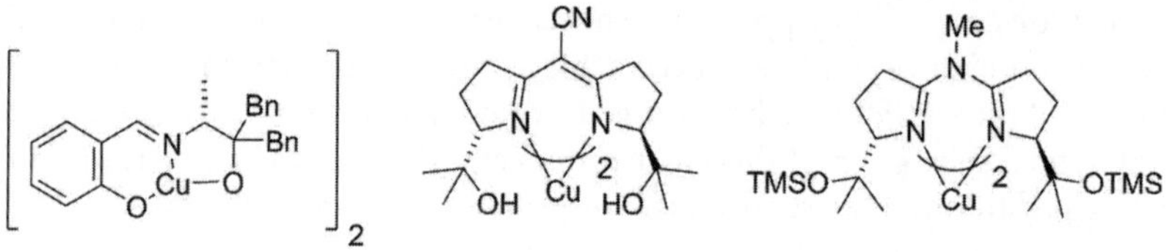

| (9.01) | (9.02) | (9.03) |

| catalyst (9.02) | 93% ee | 81:19 *trans:cis* |
| catalyst (9.03) | 96% ee | 86:14 *trans:cis* |

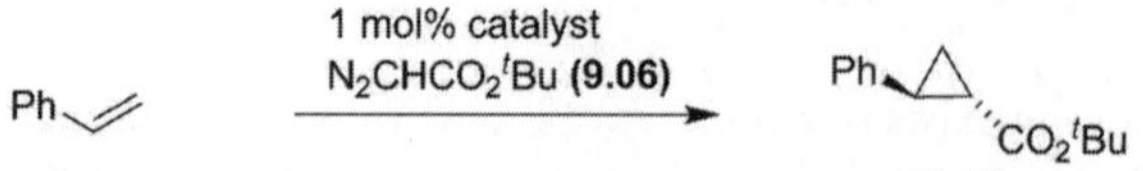

Bis-oxazolines provide a similar stereochemical environment around the copper, but are generally easier to prepare. At about the same time, the research groups of Masamune,[8] Evans[9] and Pfaltz[10] all reported the use of bis-oxazolines for copper-catalysed cyclopropanation. These ligands, including structures (9.07) and (9.08), provide excellent stereocontrol for many intermolecular examples of cyclopropanation. The use of the very hindered diazoester (9.09) enhances the *trans* selectivity of the cyclopropanation reaction of styrene (9.04). Monocyclopropanation of diene (9.11) is also achieved with very good stereocontrol.[11]

A number of alternative bis-oxazoline ligands have been reported, but with few benefits over the more straightforward bis-oxazolines already described.[12] Di-nitrogen ligands, including iminophosphoranes,[13] diamines[14] and bipyridines[15] and also enantiomerically pure Schiff bases related to Aratani's ligand (9.01)[16] have all been shown to provide good enantioselectivity in copper-catalysed cyclopropanation, with the best results often being achieved in the cyclopropanation of styrene (**9.04**) with hindered diazoesters.

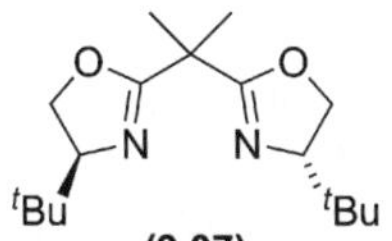
tBu
tBu

(9.07)

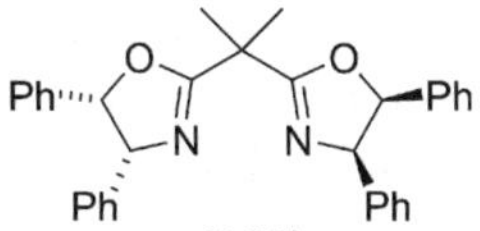
Ph
Ph
Ph
Ph

(9.08)

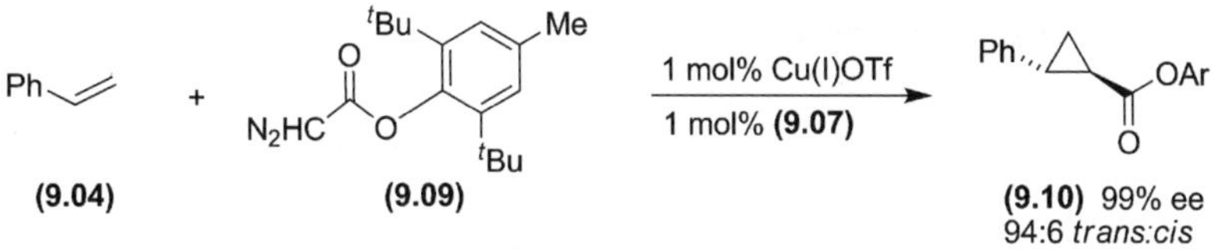
Ph
+
N2HC
tBu
Me
tBu
1 mol% Cu(I)OTf
1 mol% (9.07)
Ph
OAr
(9.04)
(9.09)
(9.10) 99% ee
94:6 trans:cis

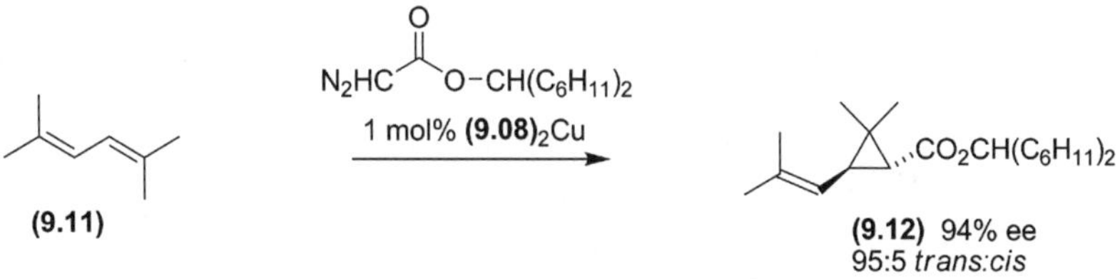
N2HC
O-CH(C6H11)2
1 mol% (9.08)2Cu
CO2CH(C6H11)2
(9.11)
(9.12) 94% ee
95:5 trans:cis

The copper-catalysed asymmetric cyclopropanation has been achieved using diazomethane gas, albeit with moderate yields and/or ee. Some of the best enantioselectivites (up to 75% ee) have been obtained by Charette and coworkers using copper bis-oxazoline ligands and cinnamate esters as substrates.[17]

While this cyclopropanation strategy is usually limited to the use of diazo compounds there has been one recent report of the use of phenyliodonium ylides as the carbene source. In this example the ylide generated from methyl nitroacetate (**9.13**) and iodosobenzene reacts with styrenes and also 1,3-butadiene in the presence of the copper complex derived from bis-oxazoline (**9.14**).[18]

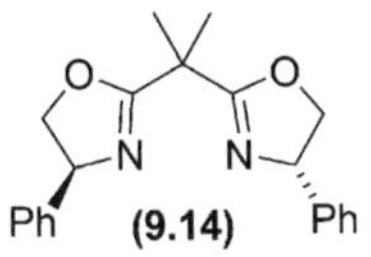
Ph
(9.14)
Ph

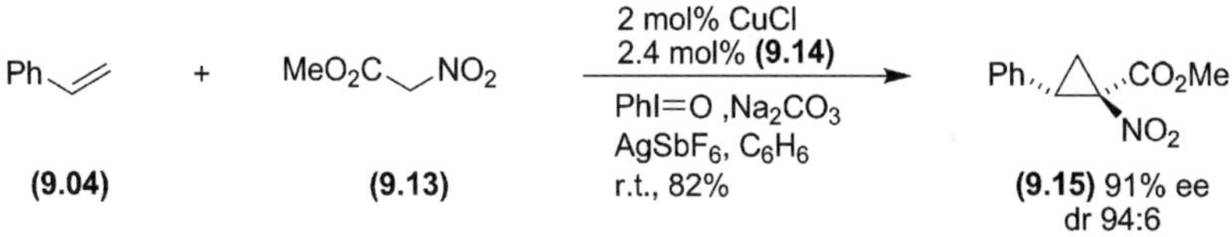
Ph
+
MeO2C
NO2
2 mol% CuCl
2.4 mol% (9.14)
PhI=O ,Na2CO3
AgSbF6, C6H6
r.t., 82%
Ph
CO2Me
NO2
(9.04)
(9.13)
(9.15) 91% ee
dr 94:6

Intramolecular cyclopropanation with copper bis-oxazoline catalysts have generally been less successful than with the Doyle rhodium catalyst (see Section 9.1.2).[19] However, Shibasaki has shown that using bis-oxazoline ligand (9.16), good enantioselectivity is obtained in the cyclisation of the silyl enol ether (9.17).[19b, 20] Bis-oxazolines such as (9.19) have also been used by Nakada and coworkers to effect cyclisation of α-diazo-β-keto sulfones such as (9.20).[19c-f] The product (9.21) is obtained with good ee and is a useful polyfunctional synthetic building block.

Pfaltz has shown that his original semi-corrin ligand complexes (9.02) are best suited to the cyclisation of the diazoketone (9.22). The corresponding reaction with a bis-oxazoline afforded only 77% ee.[19a]

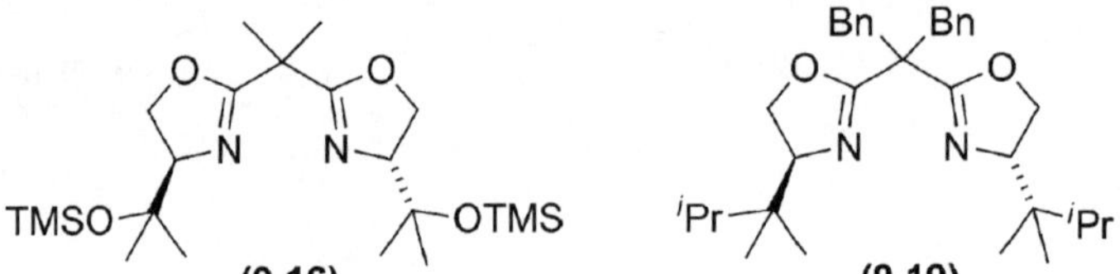

(9.16)

(9.19)

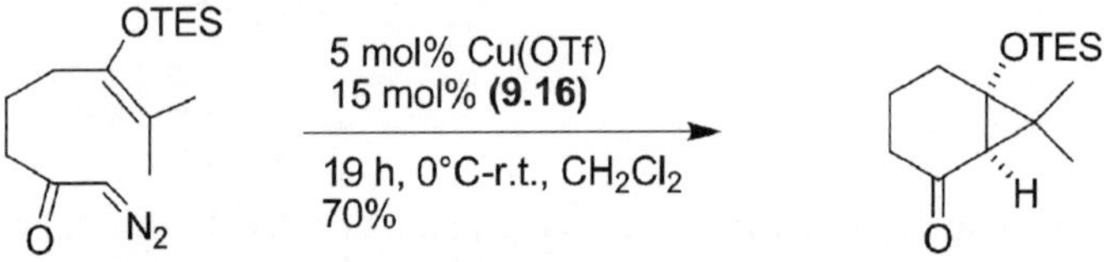

(9.17)

5 mol% Cu(OTf)
15 mol% (9.16)

19 h, 0°C-r.t., CH₂Cl₂
70%

(9.18) 92% ee

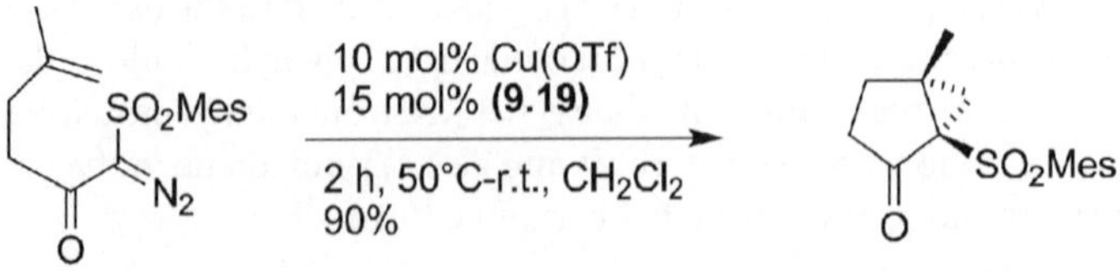

(9.20)

10 mol% Cu(OTf)
15 mol% (9.19)

2 h, 50°C-r.t., CH₂Cl₂
90%

(9.21) 98% ee

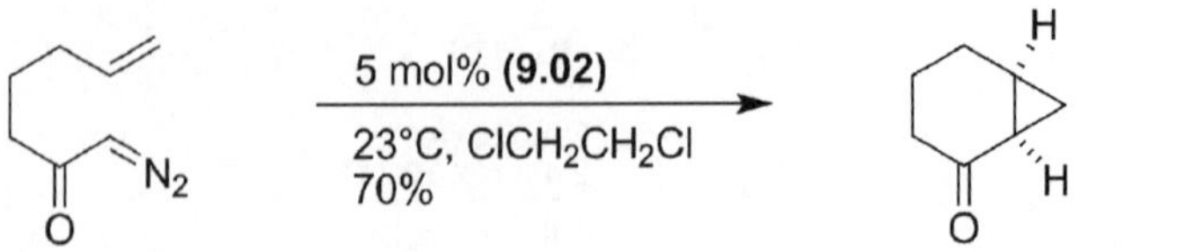

(9.22)

5 mol% (9.02)

23°C, ClCH₂CH₂Cl
70%

(9.23) 95% ee

9.1.2 Rhodium-Catalysed Cyclopropanation

Enantioselective rhodium-catalysed cyclopropanation reactions have enjoyed considerable success for intramolecular cases. However, by suitable choice of catalyst, intermolecular reactions have also been highly selective in some instances. [1,3]

The two main families of catalysts used are the Davies/McKervey dirhodium carboxylate catalysts (**9.24a–c**) [21] and Doyle's carboxamidates, including $Rh_2(MEPY)_4$ (**9.25**), $Rh_2(MPPIM)_4$ (**9.26**) and the azetidinones such as $Rh_2(IBAZ)_4$ (**9.27**). [22]

The best results using catalysts (**9.24a–c**) are achieved with carbenes generated from diazo compounds possessing both donor and acceptor groups such as phenyldiazoacetate (**9.28**). [21d, 23] The activity and stability of this type of catalyst is further enhanced by the presence of bridging ligands allowing the use of very low catalyst loadings. For instance, the rhodium catalyst formed from ligand (**9.30**) effects cyclopropanation of styrene with (**9.28**) with 85% ee at substrate/catalyst ratios of 100,000:1! [24]

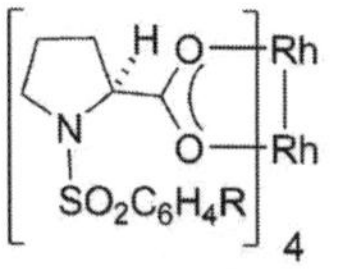

(**9.24a**) R = H
(**9.24b**) R = ^{t}Bu
(**9.24c**) R = $C_{12}H_{25}$

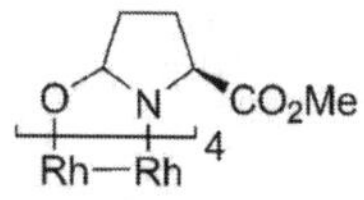

(**9.25**)

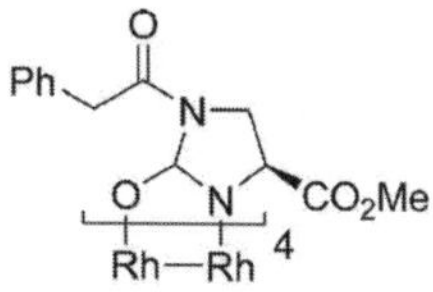

(**9.26**)

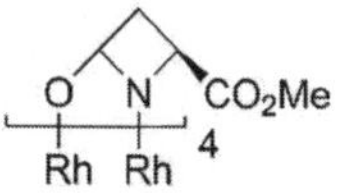

(**9.27**)

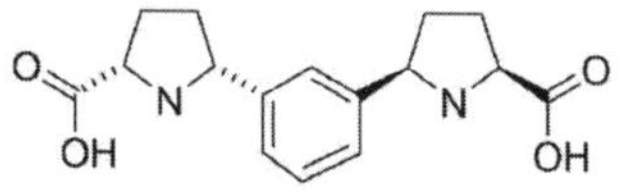

(**9.30**)

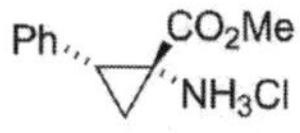

(**9.33**)

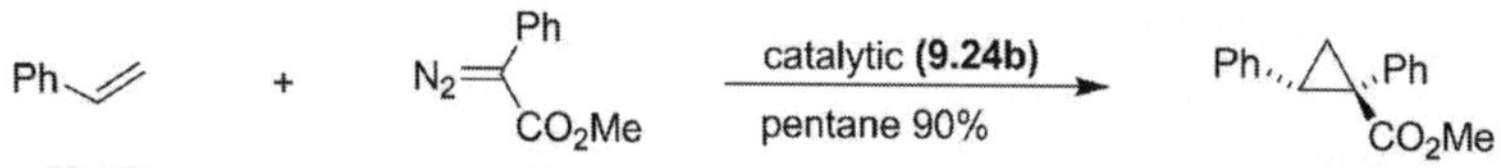

(**9.04**)

(**9.28**)

catalytic (**9.24b**)
pentane 90%

(**9.29**)
87% ee
98:2 *trans:cis*

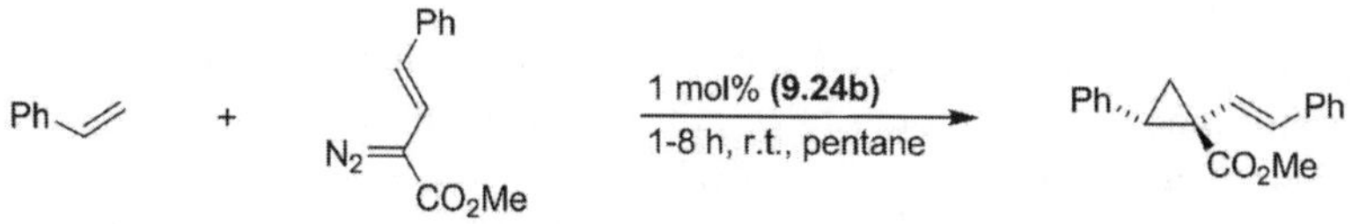

(**9.04**)

(**9.31**)
added over 10 min

1 mol% (**9.24b**)
1-8 h, r.t., pentane

(**9.32**)
90% ee
>40:1 *trans:cis*

Good ees are also obtained using cyclic[25] and acyclic vinyl diazoacetates, such as (9.31), and the product (9.32) is converted into the cyclopropyl amino acid (9.33).[21b] The same product (9.32) has also been converted into the antidepressant Sertraline by Corey and Gant.[16]

Diazomalonate esters are generally poor substrates for this reaction and only moderate ees have been obtained so far in the rhodium-catalysed addition to alkenes.[22b] Far better results have been obtained using the phenyliodonium ylide generated from dimethylmalonate and also Meldrum's acid (9.34) and iodosylbenzene in the presence of the rhodium catalyst bearing ligand (9.35).[27]

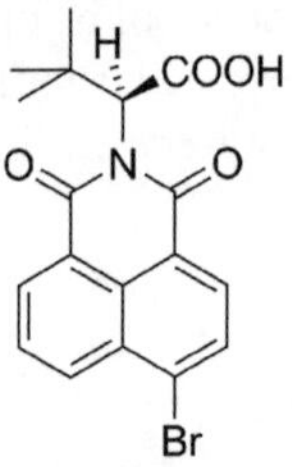

(9.35)

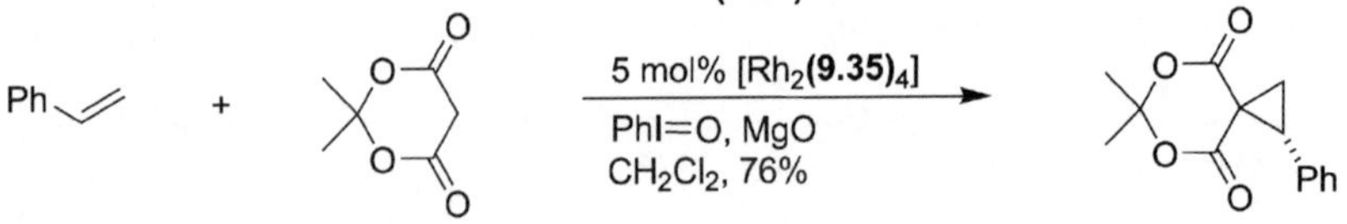

5 mol% [Rh$_2$(9.35)$_4$]

PhI=O, MgO
CH$_2$Cl$_2$, 76%

(9.04)   **(9.34)**   **(9.36)** 92% ee

Rhodium-catalysed intramolecular cyclopropanation has been achieved for a broad range of substrates,[28] and the enantioselectivity and diastereoselectivity are often reliably high. The best catalysts for this transformation are Doyle's complexes (9.25–9.27). Thus the Rh$_2$(MEPY)$_4$ catalyst (9.25) can be used for bicyclisation of the diazoesters (9.37). The selectivity is generally good, but gives poor results for *trans*-substituted alkenes, where much improved results are seen with the Rh$_2$(4S-MPPIM)$_4$ catalysts (9.26).[29,30] This is further illustrated by the reaction of the *trans*-alkene (9.38), which undergoes a highly enantioselective cyclopropanation with the Rh$_2$(MPPIM)$_4$ catalyst, but only provides 68% ee when the Rh$_2$(MEPY)$_4$ catalyst is employed.

The prochiral divinyl substrate (9.39) undergoes reaction selectively with one of the enantiotopic alkene groups to give the cyclisation product (9.43) with high enantioselectivity and diastereoselectivity.[31]

Cyclisation reactions have also been performed on diazoacetamides, where the intramolecular cyclopropanation of diazoacetamide (9.40) affords the bicyclic lactam (9.44).[32]

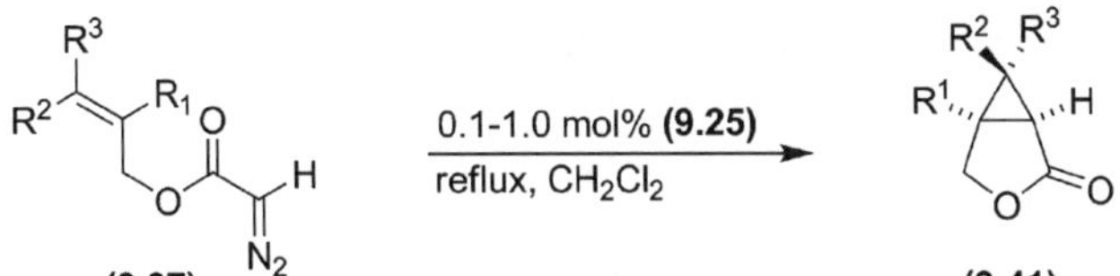

(9.37) → **(9.41)**

0.1-1.0 mol% **(9.25)**
reflux, CH$_2$Cl$_2$

R^1, R^2, R^3 = H, 75%, 95% ee
R^1= H, R^2, R^3= Me, 84%, 98% ee
R^1, R^2= H, R^3= Me, 72%, 7% ee
using Rh$_2$((4S)-MPPIM)$_4$ gives 75%, 89% ee

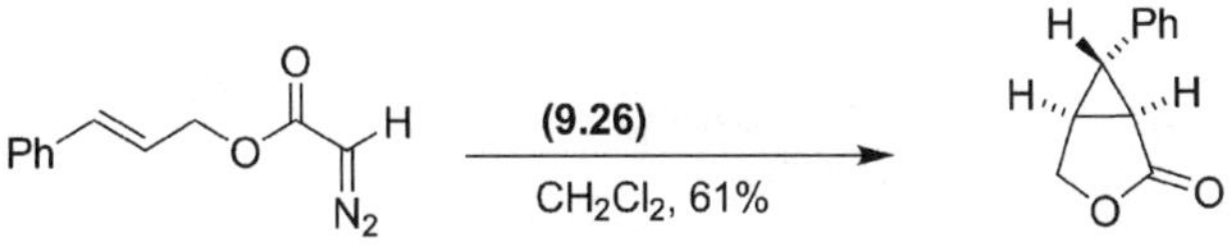

(9.38) → **(9.42)** 96% ee

(9.26)
CH$_2$Cl$_2$, 61%

68% ee with **(9.25)**

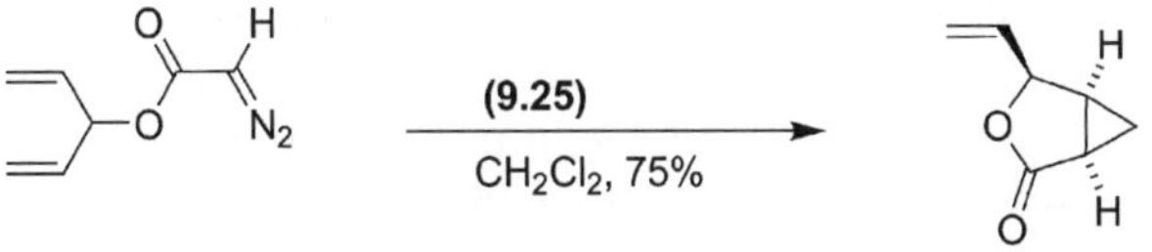

(9.39) → **(9.43)**

(9.25)
CH$_2$Cl$_2$, 75%

> 94% ee
>20:1 diastereoselectivity

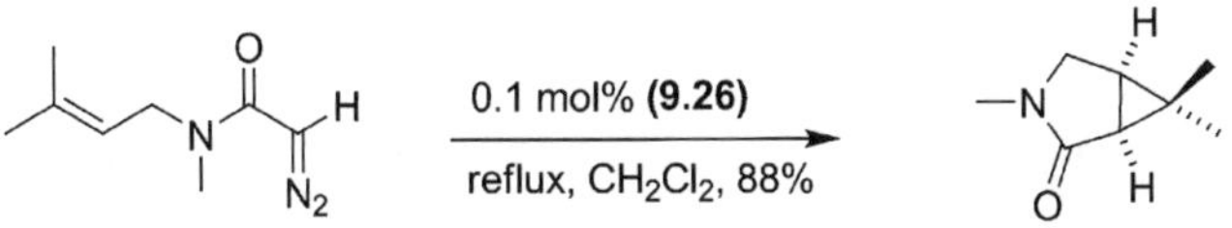

(9.40) → **(9.44)** 94% ee

0.1 mol% **(9.26)**
reflux, CH$_2$Cl$_2$, 88%

The Rh$_2$(MEPY)$_4$ catalysts are also able to catalyse enantioselective cyclopropenation reactions of a range of terminal alkynes such as **(9.45)**.[33] Corey and coworkers have designed a novel cyclopropenation catalyst **(9.46)** bearing an enantiomerically pure imidazolidinone ligand that exhibits some of the highest ees in the cyclopropenation of alkynes, including **(9.45)** and also **(9.47)**.[34]

Davies has extended his work with catalysts **(9.24)** to the reaction with dienes, where the initially formed products undergo a Cope rearrangement. For example, diene **(9.50)** undergoes cyclopropanation with diazoester **(9.31)** to give an intermediate cyclopropyl derivative **(9.51)**, which ring-expands by the Cope reaction, providing the seven-membered ring product **(9.52)** with good enantioselectivity.[35]

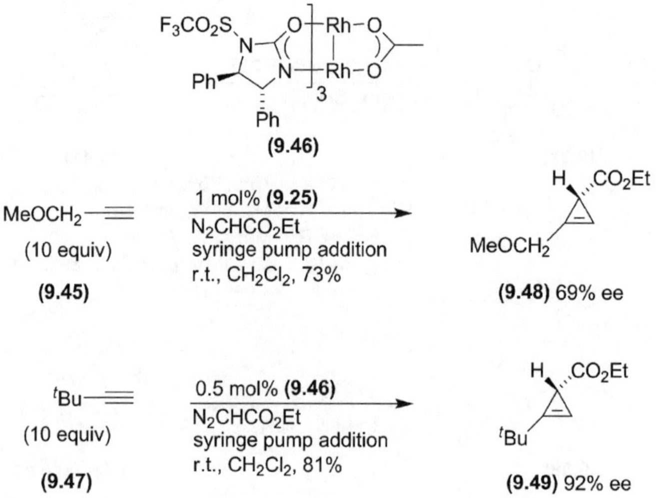

(9.46)

MeOCH$_2$—≡
(10 equiv)

(9.45)

1 mol% (9.25)
N$_2$CHCO$_2$Et
syringe pump addition
r.t., CH$_2$Cl$_2$, 73%

(9.48) 69% ee

^{t}Bu—≡
(10 equiv)

(9.47)

0.5 mol% (9.46)
N$_2$CHCO$_2$Et
syringe pump addition
r.t., CH$_2$Cl$_2$, 81%

(9.49) 92% ee

Reaction with the furan (9.53) affords the bicyclic product (9.54) after rearrangement. A similar reaction sequence has also been performed in an intramolecular fashion to provide the tricyclic tremulane skeleton.[36] This transformation can also be performed in an asymmetric sense with pyrroles using catalyst (9.56), giving access to functionalised tropanes such as (9.58).[37]

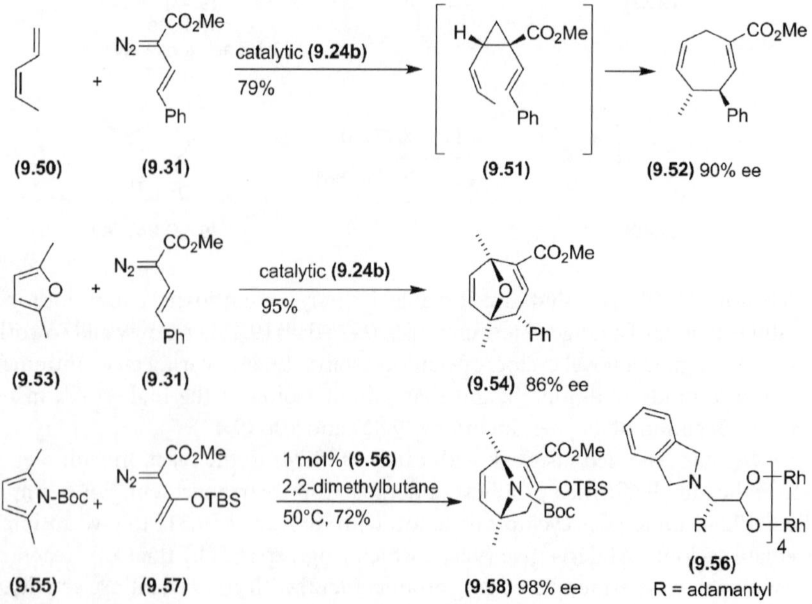

(9.50)   (9.31)   catalytic (9.24b) 79%   (9.51)   (9.52) 90% ee

(9.53)   (9.31)   catalytic (9.24b) 95%   (9.54) 86% ee

(9.55)   (9.57)   1 mol% (9.56) 2,2-dimethylbutane 50°C, 72%   (9.58) 98% ee   (9.56) R = adamantyl

The direct rhodium-catalysed cyclopropanation is limited to relatively electron-rich alkenes, as the metallocarbenoids formed *in situ* are electrophilic. Aggarwal and coworkers have developed a strategy for the catalytic enantioselective epoxidation of aldehydes utilising catalytic quantities of enantiomerically pure sulfides such as (9.59) in the presence of rhodium catalyst and diazo compounds (see Section 4.8) and this methodology can also be applied to the cyclopropanation of electron-deficient olefins.[38] In this process the key reactive species – a sulfonium ylide (9.60), formed *in situ*, reacts with a range of α,β-enones such as chalcone (9.61) to give the cyclopropane with high ee.

9.1.3 Simmons–Smith-Type Cyclopropanation

The Simmons–Smith reaction involves the cyclopropanation of alkenes using CH_2I_2 and a Zn/Cu couple.[39] The asymmetric Simmons–Smith cyclopropanation of allylic alcohols can be effected in the presence of catalytic amounts of ligand (9.63) (see also Section 6.1), although there is still some uncatalysed background reaction.[30] The reaction works well for allyl alcohols such as cinnamyl alcohol (9.64), and also has been extended to vinylsilanes and vinylstannanes.[41] Denmark has improved the yields (up to 99%) and enantioselectivity (up to 89% ee) of this reaction by employing a bis-methylsulfonamide variant of the ligand and a slightly modified procedure.[42,43] Higher ees in the cyclopropanation of cinnamyl alcohol (9.64) have been obtained by Charette and coworkers using the Lewis acidic titanium-TADDOLate complex albeit at relatively high catalyst loadings of 25 mol% (9.65).[44] This catalyst can also be used in the cyclopropanation of other 3-aryl-substituted allylic alcohols and also 3-alkyl derivatives with up to 74% ee. Shi and coworkers have discovered that the iodomethylzinc reagent formed *in situ* by reaction of diethylzinc, diiodomethane and catalytic amounts of N-Boc-L-Val-L-Pro-OMe is capable of effecting cyclopropanation of unfunctionalised olefins such as dihydronaphthalenes and indenes with ees ranging from 77 to 89%.[45] Particularly high ees are obtained in the cyclopropanation of aromatic silyl enol ethers such as (9.67) and alkynyl enol ethers using dipeptide (9.68) as catalyst.[46] An enantiomerically pure iodomethylzinc reagent can also be prepared *in situ* from phosphate (9.70) and up to 88% ee in the cyclopropanation of cinnamyl alcohol has been obtained using 10 mol% of this catalyst.[47]

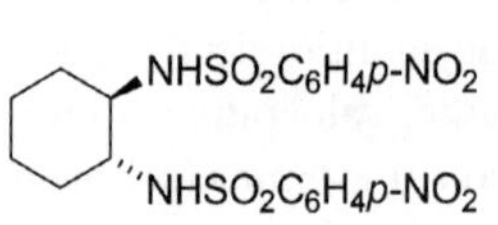

(9.63) 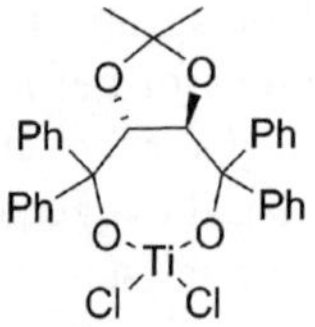   (9.65) 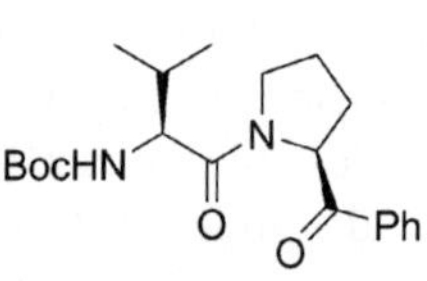   (9.68)

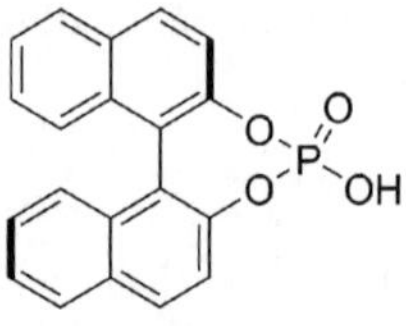

(9.70)

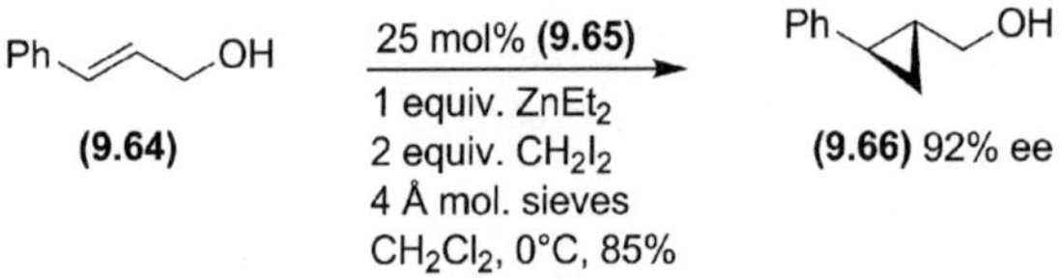

9.1.4 Other Metal-Based Cyclopropanation Catalysts

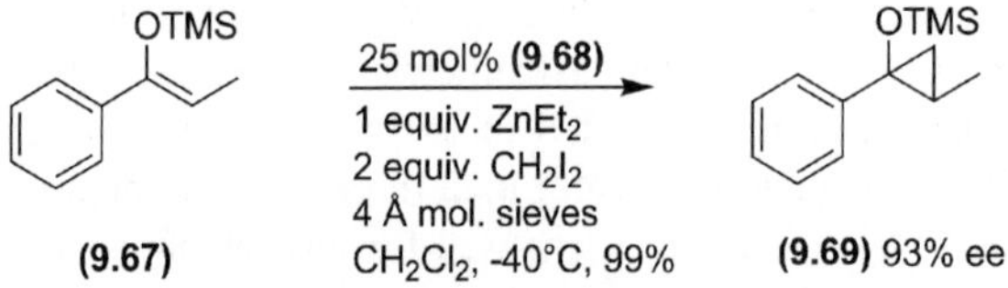

Whilst copper- and rhodium-catalysed cyclopropanation reactions have been the most widely investigated, many other metals also catalyse cyclopropanation reactions. In particular, ruthenium and cobalt complexes have been successful for enantioselective reactions and some of these catalysts display excellent scope and ee. Ruthenium PYBOX complexes, for example (9.71), provide good *trans* selectivity as well as high enantiocontrol in the cyclopropanation of styrene (9.04).[48] Ruthenium(salen) complexes have also been investigated as cyclopropanation catalysts.[49] Some of the best results have been obtained using salen complexes with pyridine ligands such as (9.72). These catalyse the cyclopropanation of styrene with up to 99% ee and also a range of electron-rich alkenes such as ethyl vinyl ether and electron-poor alkenes including methyl methacrylate (9.73), with high ee.[49b] In addition, ruthenium porphyrin complexes have also been used with enantioselectivity up to 90.8% ee and turnover numbers of up to 10 000.[40]

The cobalt complexes (**9.75**)[51] and (**9.76**)[52] have been used in cyclopropanation reactions, but have given poor control of diastereoselectivity, although high enantioselectivities, up to 97% ee with (**9.76**), have been recorded. In contrast the Co(III)(salen) complex (**9.77**) provides good diastereocontrol, with reasonable enantioselectivity.[53] Recently, cobalt porphyrins have been shown to exhibit

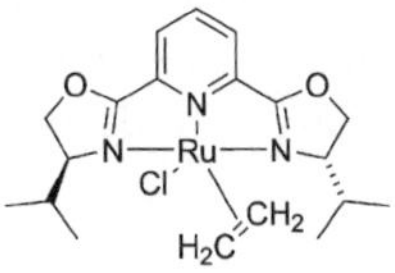

(**9.71**) 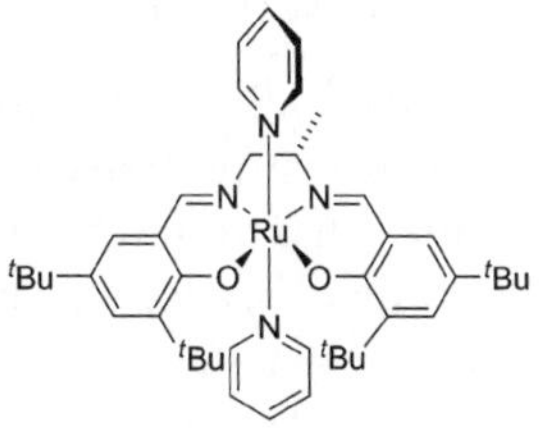   (**9.72**)

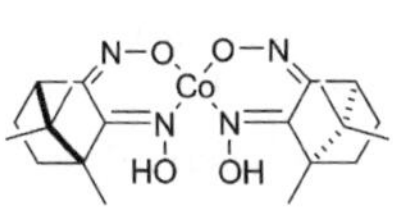

(**9.75**) 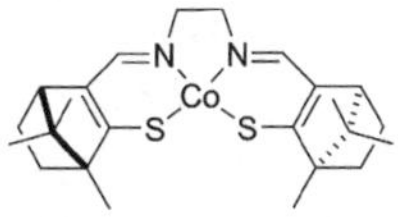   (**9.76**) 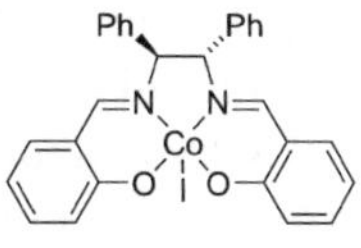   (**9.77**)

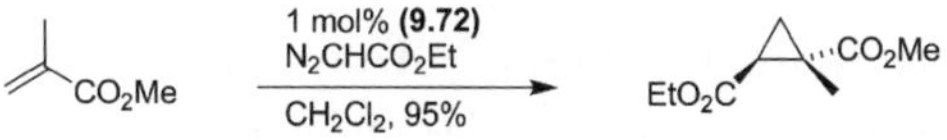

	1 mol% (**9.72**)	
	N₂CHCO₂Et	
(**9.73**)	CH₂Cl₂, 95%	(**9.74**) 95% ee
		100:1 *trans:cis*

$$\text{1 mol\% (9.72)}\quad N_2CHCO_2Et\quad CH_2Cl_2,\ 95\%$$

(**9.73**) → (**9.74**) 95% ee, 100:1 *trans:cis*

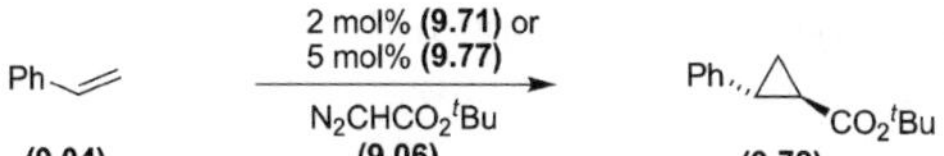

Ph (**9.04**) 5 equiv.

2 mol% (**9.71**) or 5 mol% (**9.77**)

N₂CHCO₂^{t}Bu (**9.06**)

4-12 h, 20-25°C CH₂Cl₂, 65 - 69%

Ph (**9.78**) CO₂^{t}Bu

using catalyst (**9.71**) 94% ee, 97:3 *trans:cis*

using catalyst (**9.77**) 74% ee, 96:4 *trans:cis*

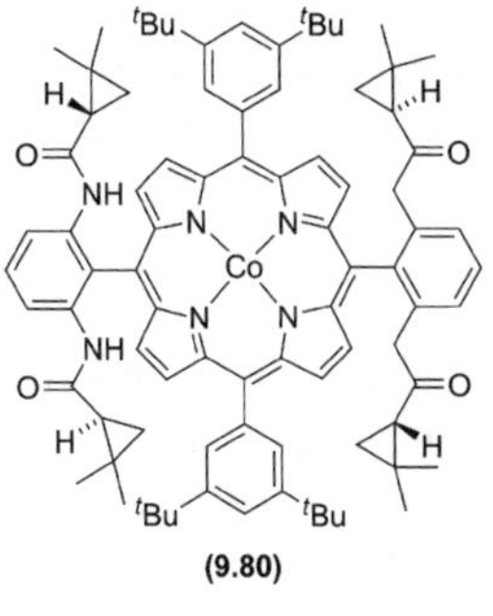

(**9.80**)

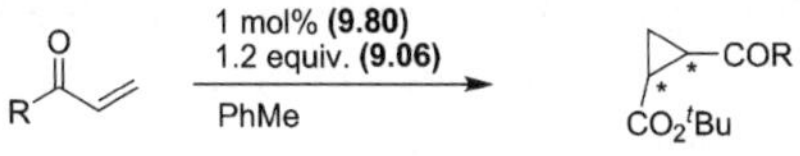

(**9.79**)

1 mol% (**9.80**), 1.2 equiv. (**9.06**), PhMe

(**9.81**) COR, CO₂^{t}Bu

R	Yield	*trans:cis*	ee
OEt	72%	99:1	90%
NH₂	66%	99:1	97%
Et	81%	99:1	94%
CN	83%	76:24	93%

high selectivity and exceptional substrate scope. For instance, high enantioselectivity with styrene and derivatives and also with a range of electron-deficient alkenes (**9.79**), including acrylates, acrylamides, acrylonitriles and alkyl vinyl ketones, have been obtained using 1 mol% cobalt porphyrin (**9.80**).[54]

9.1.5 Organocatalysed Cyclopropanation

The cyclopropanation of α,β-unsaturated carbonyls can be achieved using metal-free ylides. This is an attractive strategy avoiding the use of expensive metal-based catalysts and potentially hazardous diazo compounds. Two main approaches to the asymmetric ylide-mediated cyclopropanation have been developed, both utilising enantiopure amines as catalysts.

Gaunt, Ley and coworkers have devised an enantioselective route to cyclopropanes with high ee using ammonium ylides such as (**9.82**), generated *in situ* by reaction of α-halocarbonyls including α-bromoacrylates and acrylamides such as (**9.83**) with quinine derivative (**9.84**) followed by base. Reaction with enones such as (**9.85**) yields the cyclopropane (**9.86**) and regenerates the catalyst.[55] This reaction can be performed with quinidines to give the opposite enantiomer of the product and an intramolecular variant also proceeds with high ee.[56]

An alternate approach, pioneered by Kunz and MacMillan, proceeds by reaction of sulfur ylides with α,β-unsaturated iminium ions, formed by reaction of α,β-enals with enantiomerically pure amines.[57] In this approach the sulfur ylide (**9.87**) reacts with a range of β-aryl- and β-alkyl-substituted enals, including (**9.88**), in the presence of the dihydroindole (**9.89**) to give the cyclopropane with ees ranging from 89 to 96%. More recently, it has been shown that replacement of the carboxylate functional group in (**9.89**) with an isosteric tetrazolic acid gives an improved catalyst that effects cyclopropanation with 99% ees in all cases studied.[58]

9.2 Insertion Reactions

Enantioselective C–H insertion reactions have been successfully performed by various rhodium catalysts over a broad range of substrates in both an intramolecular and an intermolecular manner.[1,2,3,39b] Of the many examples reported, a few are highlighted here. McKervey and coworkers have obtained good diastereoselectivity and enantioselectivity in the C–H insertion reaction of compound (**9.91**) catalysed by rhodium complex (**9.92**).[59] Intramolecular cyclopropanation is not competitive in this case, since the alkene moiety is too remote.

Doyle and coworkers found that the $Rh_2(4S\text{-MACIM})_4$ catalyst (**9.94**) provides higher *cis*-selectivity than related 'Doyle' catalysts (**9.25–9.27**) in the reaction of substrate (**9.95**), where the insertion takes place highly selectively into one of four C–H bonds, affording essentially a single product (**9.96**).[60]

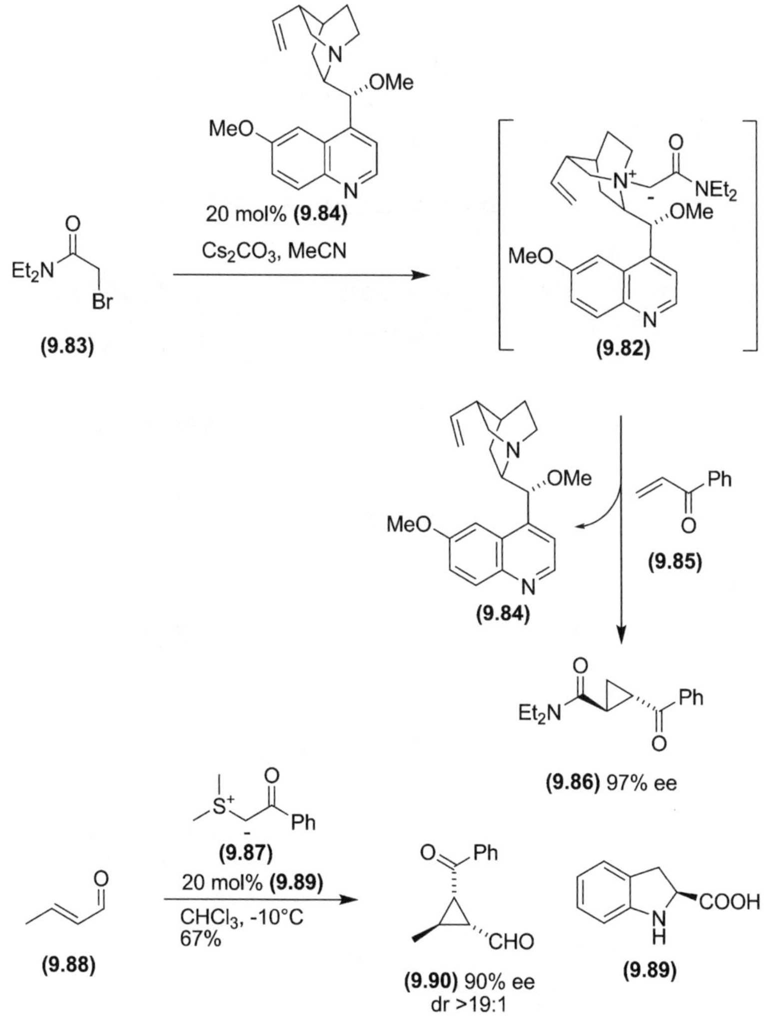

In an acyclic version of this reaction, substrate (**9.97**) was also found to undergo the C–H insertion with high selectivity using the $Rh_2(5R\text{-MEPY})_4$ catalyst (**9.25**).[61]

Bicyclic β-lactones have been prepared by enantioselective C–H insertion using the $Rh_2(5S\text{-MEPY})_4$ catalyst, including the formation of compound (**9.100**) from the diazoamide precursor (**9.99**).[62]

The Hashimoto catalyst (**9.101**) has been used to give very high stereocontrol in the selective insertion into the enantiotopic aryl C–H bonds of substrate (**9.102**), to give the product (**9.103**) with a quaternary chiral centre in up to 98% ee.[63]

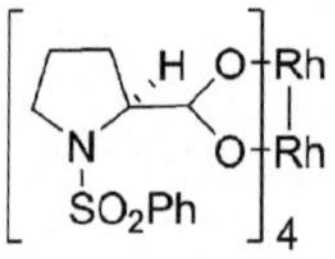

(9.92)

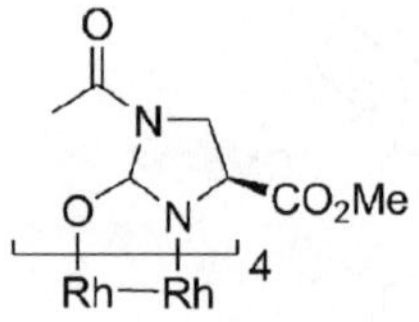

(9.94) Rh$_2$(4S-MACIM)$_4$

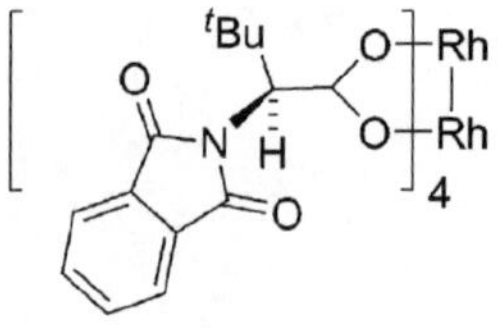

(9.101)

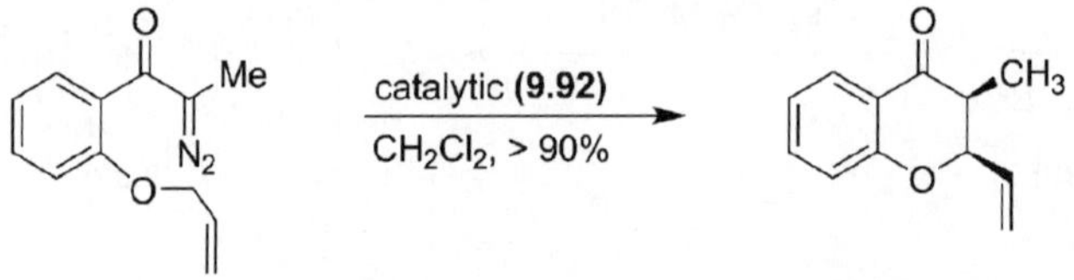

(9.91)

catalytic **(9.92)**
CH$_2$Cl$_2$, > 90%

(9.93) 79% ee
93:7 *cis:trans*

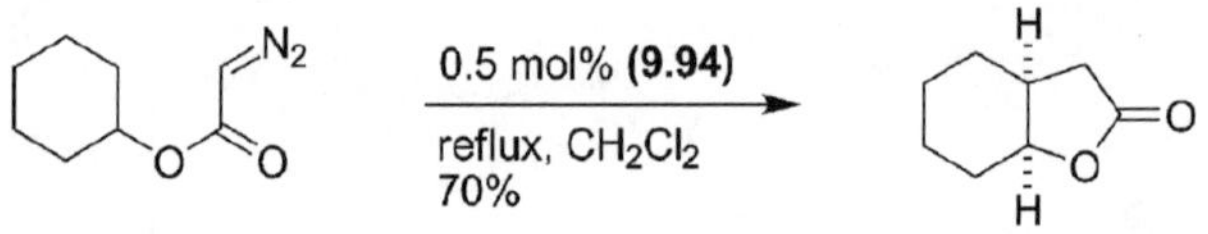

(9.95)

0.5 mol% **(9.94)**
reflux, CH$_2$Cl$_2$
70%

(9.96) 97% ee
99:1 *cis:trans*

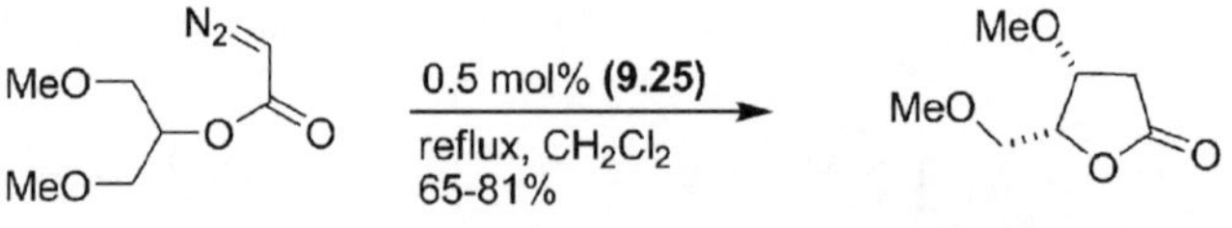

(9.97)

0.5 mol% **(9.25)**
reflux, CH$_2$Cl$_2$
65-81%

(9.98) 97% ee
93:7 *cis:trans*

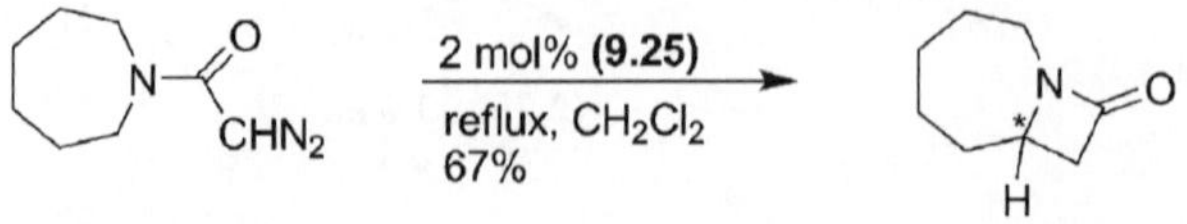

(9.99)

2 mol% **(9.25)**
reflux, CH$_2$Cl$_2$
67%

(9.100) 97% ee

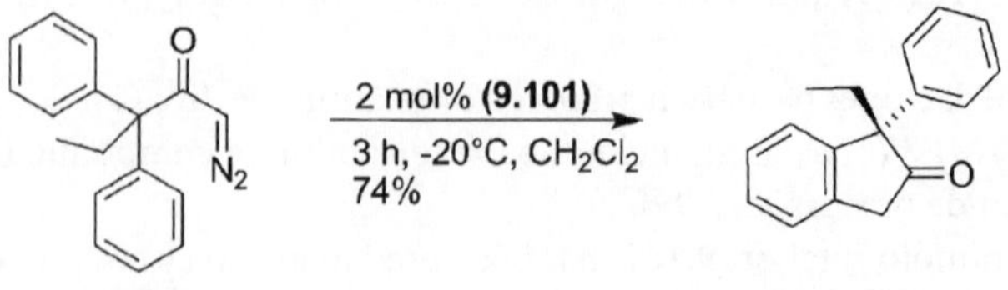

(9.102)

2 mol% **(9.101)**
3 h, -20°C, CH$_2$Cl$_2$
74%

(9.103) 98% ee

The catalytic asymmetric intermolecular C–H insertion has been achieved with a broad range of substrates by Davies and coworkers using catalyst (**9.24c**) in combination with donor acceptor carbenes generated from methylaryldiazoacetates such as (**9.28**).[64] These compounds undergo highly enantioselective insertion into cycloalkanes such as cyclopentane (**9.105**) and also cyclohexane,[1a,c] and into the C–H bond α to the heteroatom in THF (**9.106**)[1c] and cyclic amines such as *N*-Boc-pyrrolidine (**9.107**).[1b] Asymmetric C–H activation at allylic positions is also possible and this has been achieved with a wide range of substrates.[64e] The insertion into the allylic C–H bond of cyclic enol ethers such as (**9.108**) occurs regioselectively to provide products equivalent to those accessed by asymmetric Michael additions.[64d]

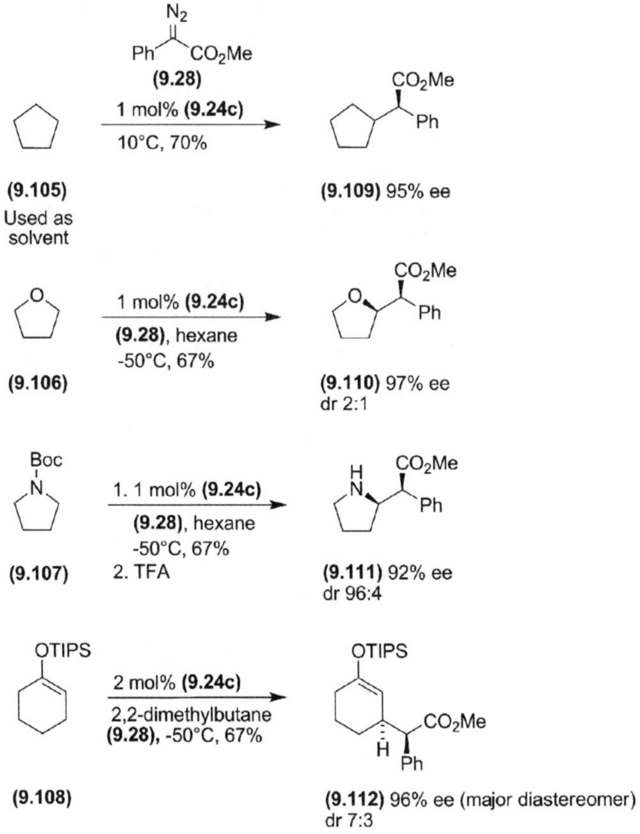

Asymmetric insertion into X–H bonds is also possible. Enantioselective Si–H insertion reactions have also been achieved using the $Rh_2(MEPY)_4$ catalyst (**9.25**).[65] The enantioselectivity of Si–H insertion into the silane (**9.113**) was improved up to 85% by the use of catalyst (**9.24c**).[66] While most success in the C–H insertion process has been achieved using enantiomerically pure rhodium catalysts, recent work by the groups of Fu and Zhou reveal that copper complexes are the most effective to date in the enantioselective insertion into O–H and N–H bonds. Zhou and coworkers have achieved high enantioselectivites (between 94 and 99% ee) in the carbenoid insertion into the N–H bond of anilines[67] and the O–H bond

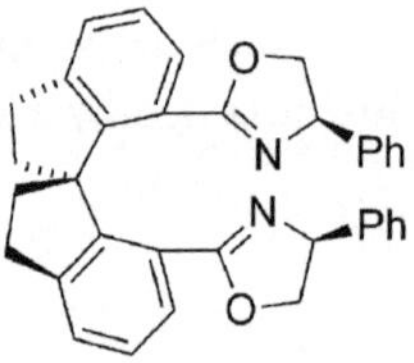
O
N
Ph
N
Ph
O

(9.114)

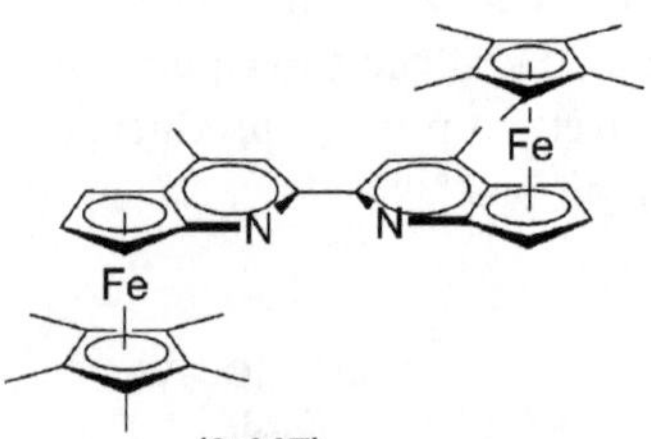
Fe
N
N
Fe

(9.117)

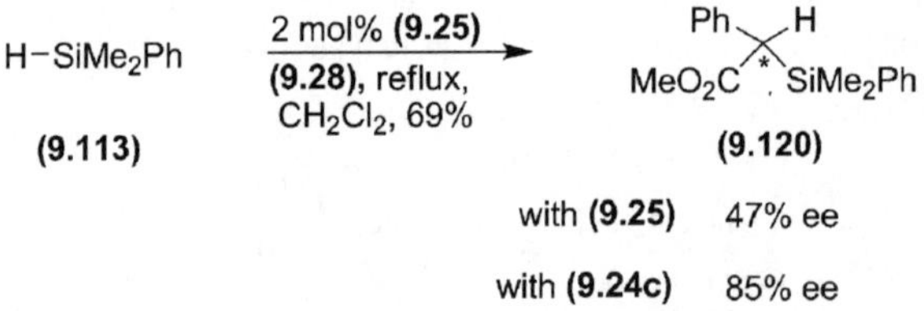
H–SiMe2Ph
(9.113)
2 mol% (9.25)
(9.28), reflux,
CH2Cl2, 69%
Ph
H
MeO2C
*
SiMe2Ph
(9.120)
with (9.25) 47% ee
with (9.24c) 85% ee

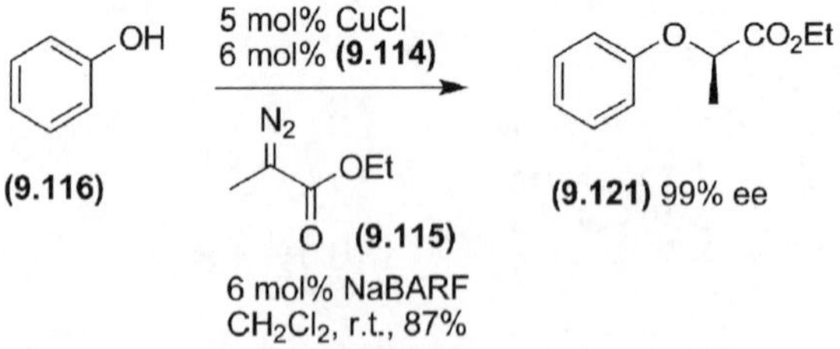
OH
(9.116)
5 mol% CuCl
6 mol% (9.114)
N2
OEt
O (9.115)
6 mol% NaBARF
CH2Cl2, r.t., 87%
O
CO2Et
(9.121) 99% ee

BARF = tetrakis[3,5-bis(trifluoromethyl)phenyl)]borate

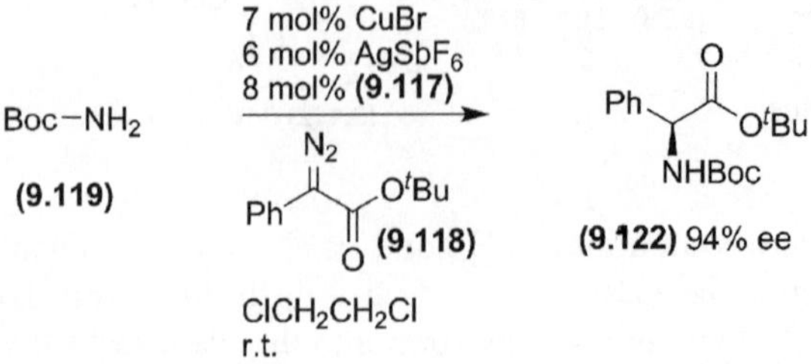
Boc–NH2
(9.119)
7 mol% CuBr
6 mol% AgSbF6
8 mol% (9.117)
N2
Ph
O^tBu
O (9.118)
ClCH2CH2Cl
r.t.
Ph
O
O^tBu
NHBoc
(9.122) 94% ee

of phenols[68] using the copper bis-oxazoline generated from spiro bis-oxazoline ligand (**9.114**) and diazoesters such as (**9.115**). Fu and coworkers have focussed their studies on the use of copper complexes generated with bipyridine such as (**9.117**). These complexes catalyse the insertion of *tert*-butylaryldiazoacetates such as (**9.118**) into the O–H bond of ethanol and derivatives with up to 98% ee,[69] and the N–H bond of Boc-NH_2 (**9.119**).[70]

9.3 Asymmetric Ylide Reactions

Metal carbenoid complexes will react with heteroatoms to provide the heteroatom ylides, which can then rearrange to products.[1] For example, on treatment with catalyst (**9.101**), the diazocompound (**9.123**) forms an intermediate oxonium ylide (**9.124**), which then undergoes a [2,3]-sigmatropic rearrangement to give the product (**9.125**) with good enantioselectivity.[61] Either the rhodium catalyst remains associated to the ylide to control the enantioselectivity of the rearrangement (more likely), or the free ylide rearranges more quickly than the rate of racemisation.

A similar transformation has been reported using enantiomerically pure copper catalysts, where the diazoketone (**9.126**) forms an initial oxonium ylide, which then rearranges to the cyclic product (**9.127**), with reasonable enantioselectivity, using a copper complex of ligand (**9.128**).[72]

The tandem ylide generation/[2,3]-sigmatropic rearrangement can also be performed with sulfur[73] and iodonium ylides and in these cases the reaction can proceed in intermolecular fashion.[74] For example, up to 78% ee has been obtained in the reaction of allyl-2-methylphenyl sulfide (**9.129**) with aryl diazoacetates such as (**9.130**) in the presence of the copper catalyst generated with bis-oxazoline (**9.07**),[73c] while the reaction of allyl iodide (**9.132**) with ethyl diazoacetate yields product (**9.133**) with 69% ee using this complex.[74]

Apparent C–O insertion reactions involving decomposition of diazocompounds mechanistically involve ylide formation and subsequent Stevens rearrangement.[75] An interesting example involves enantioselective ylide formation with one of the enantiotopic oxygens in the acetal (**9.134**).[76] The intermediate oxonium ylide (**9.135**) rearranges to the bicyclic product (**9.136**), which is isolated with fairly good enantiomeric excess using the $Rh_2(MPPIM)_4$ catalyst (**9.26**).

Decomposition of diazocompounds in the presence of a carbonyl group affords carbonyl ylides, which can be trapped to give enantiomerically enriched products.[1,77] Hodgson and coworkers have investigated the intramolecular tandem carbonyl ylide formation/cycloaddition of diazo precursors such as (**9.137**), proceeding through the intermediacy of ylide (**9.138**), and discovered that moderate enantioselectivities are obtained using catalyst (**9.24c**), but up to 90% ee is achieved with BINAP-based rhodium complexes such as (**9.140**) bearing substituents at the

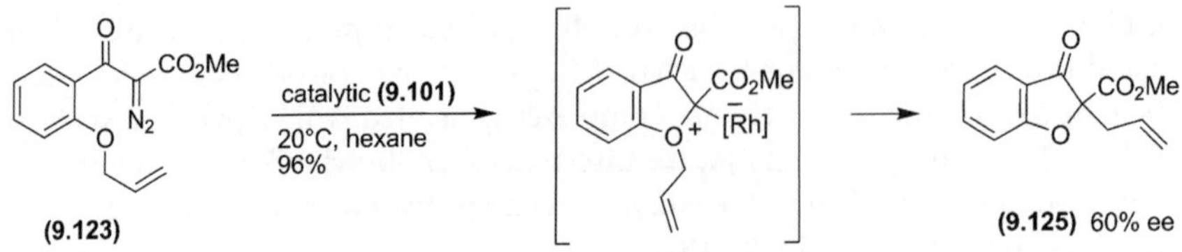

(9.123)   catalytic (9.101) 20°C, hexane 96%   (9.124)   (9.125) 60% ee

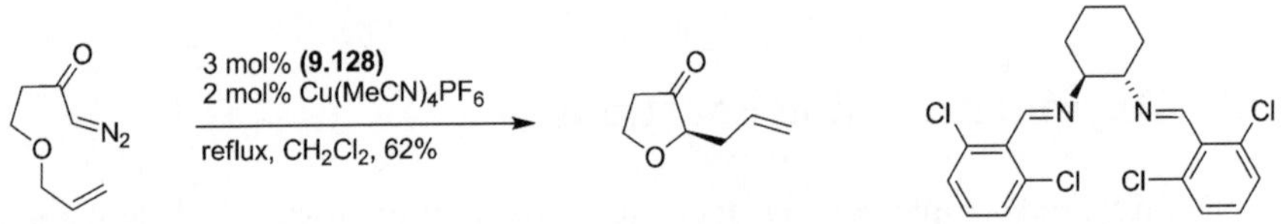

(9.126)   3 mol% (9.128) 2 mol% Cu(MeCN)₄PF₆ reflux, CH₂Cl₂, 62%   (9.127) 57% ee   (9.128)

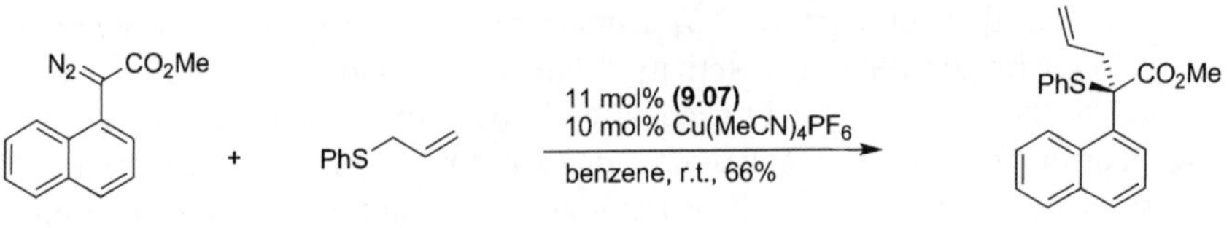

(9.130)   +   PhS   (9.129)   11 mol% (9.07) 10 mol% Cu(MeCN)₄PF₆ benzene, r.t., 66%   (9.131) 78% ee

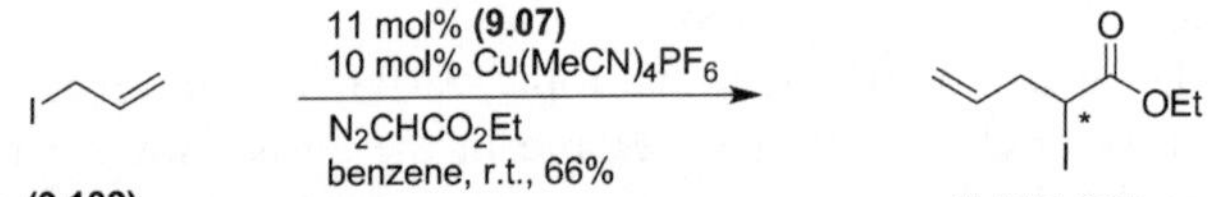

(9.132)   11 mol% (9.07) 10 mol% Cu(MeCN)₄PF₆ N₂CHCO₂Et benzene, r.t., 66%   (9.133) 69% ee

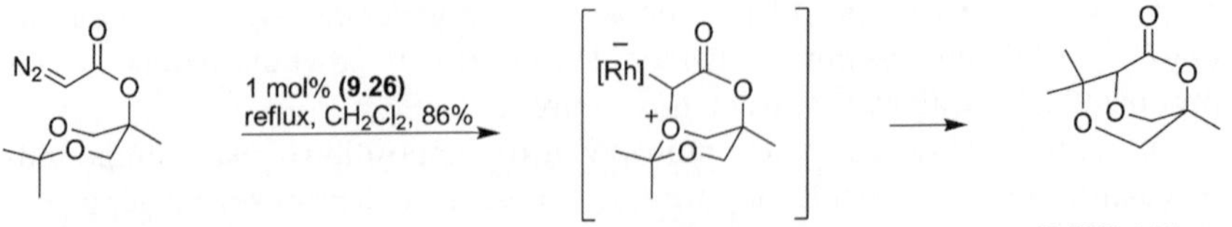

(9.134)   1 mol% (9.26) reflux, CH₂Cl₂, 86%   (9.135)   (9.136) 81% ee

6,6'-positions.[78] This process can also be performed in an intermolecular sense in the presence of alkynes and alkenes.[78e, 79] For instance, Hashimoto and coworkers have achieved high ee in the reaction of diazo ketone (9.141) with dimethyl acetylenedicarboxylate (9.142) using rhodium catalyst (9.143).[79a]

In the above examples, enantioselectivity is proposed to arise from formation of a enantiopure metal complex-associated ylide. Suga and coworkers have developed an alternate strategy utilising Lewis acids that bind to an electron-deficient bidentate dipolarophile. In this approach the 2-benzopyrilium-4-olate (9.146) formed by rhodium-catalysed decomposition of *o*-methoxycarbonyl-α-diazoacetophenone (**9.145**) undergoes cycloaddition with the carbonyl group of pyruvate esters and also the alkene moiety of oxazolidinone (**9.147**) with high ee in the presence of rare earth metal/bis(oxazolinyl)pyridine (PYBOX) complexes formed from ligand (**9.148**).[80] In this approach the Lewis acid can also exert stereocontrol by binding to the carbonyl oxygen of ylide (**9.146**) and enantioselective inverse electron demand cycloadditions have been achieved with this ylide and vinyl ethers.[81]

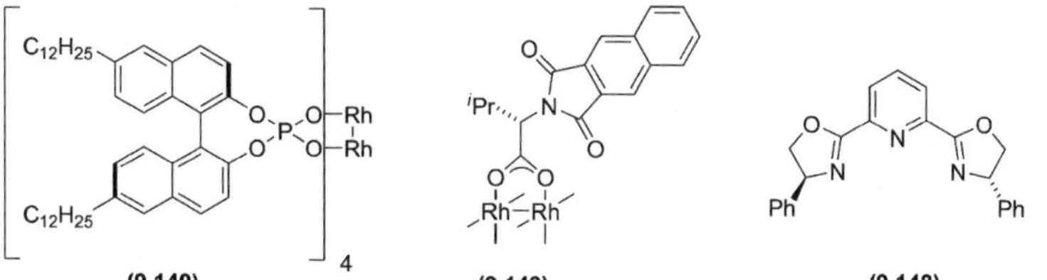

(**9.140**)   (**9.143**)   (**9.148**)

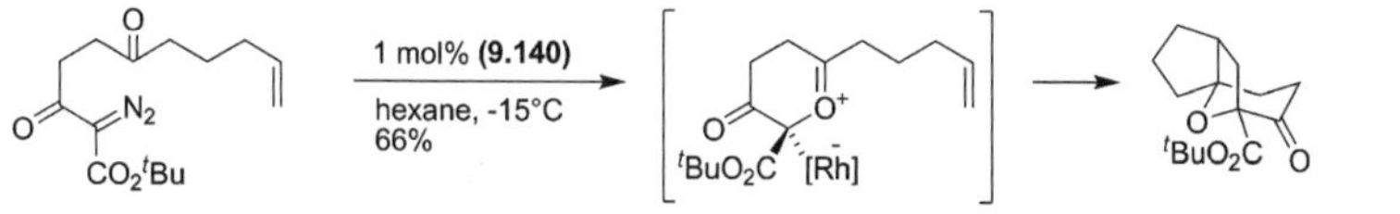

(**9.137**)   1 mol% (**9.140**)   (**9.138**)   (**9.139**) 90% ee
hexane, -15°C
66%

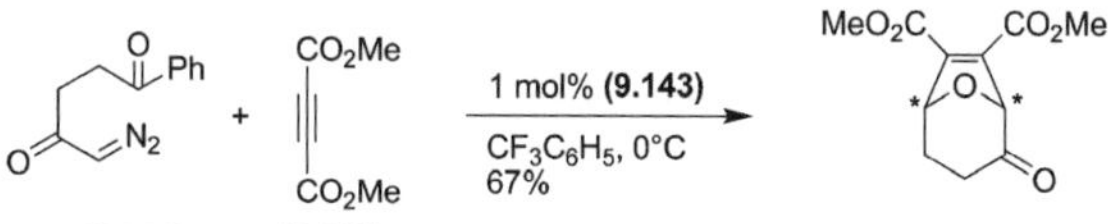

(**9.141**)   (**9.142**)   1 mol% (**9.143**)   (**9.144**) 92% ee
CF₃C₆H₅, 0°C
67%

(**9.145**)   2 mol% Rh(OAc)₄   (**9.146**)   (**9.147**)   (**9.149**) 98% ee
CH₂Cl₂, -25°C   10 mol% Yb(OTf)₃   88:12 *exo:endo*
10 mol% (**9.148**)

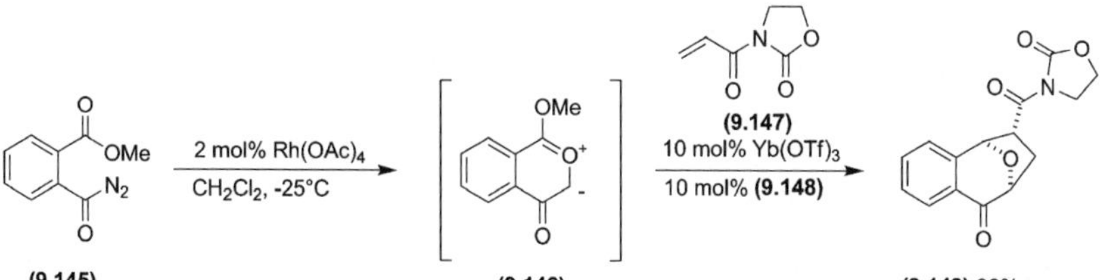

References

1. For general reviews covering catalytic asymmetric carbene reactions see: (a) M. P. Doyle and D. C. Forbes, *Chem. Rev.*, **1998**, *98*, 911. (b) D. C. Forbes and M. C. McMills, *Curr. Org. Chem.*, **2001**, *5*, 1091.
2. A. Pfaltz, in *Comprehensive Asymmetric Catalysis*, Vol. 2, ed. E. N. Jacobsen, A. Pfaltz and H. Yamamoto, Springer-Verlag, Berlin, **1999**, 513.

3. K. M. Lydon and M. A. McKervey, in *Comprehensive Asymmetric Catalysis*, Vol. 3, ed. E. N. Jacobsen, A. Pfaltz and H. Yamamoto, Springer-Verlag, Berlin, **1999**, 1178.
4. H. Nozaki, S. Moriuti, H. Takaya and R. Noyori, *Tetrahedron Lett.*, **1966**, 5239.
5. T. Aratani, *Pure Appl. Chem.*, **1985**, *57*, 1839.
6. H. Fritschi, U. Leutenegger and A. Pfaltz, *Helv. Chim. Acta*, **1988**, *71*, 1553.
7. U. Leutenegger, G. Umbricht, C. Fahrni, P. von Matt and A. Pfaltz, *Tetrahedron*, **1992**, *48*, 2143.
8. R. E. Lowenthal, A. Abiko and S. Masamune, *Tetrahedron Lett.*, **1990**, *31*, 6005.
9. D. A. Evans, K. A. Woerpel, M. M. Hinman and M. M. Faul, *J. Am. Chem. Soc.*, **1991**, *113*, 726.
10. D. Müller, G. Umbricht, B. Weber and A. Pfaltz, *Helv. Chim. Acta*, **1991**, *74*, 232.
11. R. E. Lowenthal and S. Masamune, *Tetrahedron Lett*, **1990**, *31*, 6005.
12. (a) A. V. Bedekar and P. G. Andersson, *Tetrahedron Lett.*, **1996**, *37*, 4073. (b) A. M. Harm, J. G. Knight and G. Stemp, *Tetrahedron Lett.*, **1996**, *37*, 6189. (c) Y. Uozumi, H. Kyota, E. Kishi, K. Kitayama and T. Hayashi, *Tetrahedron: Asymmetry*, **1996**, *7*, 1603. (d) H. Nishiyama, N. Soeda, T. Naito and Y. Motoyama, *Tetrahedron: Asymmetry*, **1998**, *9*, 2865. (e) K. Alexander, S. Cook and C. L. Gibson, *Tetrahedron Lett.*, **2000**, *41*, 7135. (f) D. M. Du, Z.-Y. Wang, D.-C. Xu and W.-T. Hua, *Synthesis*, **2002**, 2347. (g) M. Schinnerl, C. Bohm, M. Seitz and O. Reiser, *Tetrahedron: Asymmetry*, **2003**, *14*, 765. (h) D.-M. Du, B. Fu and W.-T. Hua, *Tetrahedron*, **2003**, *59*, 1933. (i) M. Itagaki, K. Masumoto and Y. Yamamoto, *J. Org. Chem.*, **2005**, *70*, 3292. (j) C. Mazet, V. Kohler and A. Pfaltz, *Angew. Chem. Int. Ed.*, **2005**, *44*, 4888. (k) B. Liu, S.-F. Zhu, L.-X. Wang and Q.-L. Zhou, *Tetrahedron: Asymmetry*, **2006**, *17*, 634. (l) X. Zhang, F. Xie, S. Matsuo, Y. Imahori, T. Kida, Y. Nakatsuji and I. Ikeda, *Tetrahedron: Asymmetry*, **2006**, *17*, 767.
13. (a) M. T. Reetz, E. Bohres and R. Goddard, *J. Chem. Soc., Chem. Commun.*, **1998**, 935. (b) J. M. Brunel, O. Legrand, S. Reymond and G. Buono, *J. Am. Chem. Soc.*, **1999**, *121*, 5807.
14. (a) S. Kanemasa, S. Hamura, E. Harada and H. Yamamoto, *Tetrahedron Lett.*, **1994**, *35*, 7985. (b) D. Tanner, F. Johansson, A. Harden and P. G. Andersson, *Tetrahedron*, **1998**, *54*, 15731. (c) C. Borriello, M. E. Cucciolito, A. Panunzi and F. Ruffo, *Tetrahedron: Asymmetry*, **2001**, *12*, 2467. (d) G. Lesma, C. Cattenati, T. Pilati, A. Sacchetti and A. Silvani, *Tetrahedron: Asymmetry*, **2007**, *18*, 659.
15. (a) K. Ito and T. Katsuki, *Tetrahedron Lett.*, **1993**, *34*, 2661. (b) H. L. Wong, Y. Tian and K. S. Chan, *Tetrahedron Lett.*, **2000**, *41*, 7723. (c) R. Ramon, J. Liang, M. M.-C. Lo and G. C. Fu, *J. Chem. Soc., Chem. Commun.*, **2000**, *5*, 377. (d) A. V. Malkov and P. Kocovsky, *Curr. Org. Chem.*, **2003**, *7*, 1737. (e) A. V. Malkov, D. Pernazza, M. Bell, M. Bella, A. Massa, F. Teply, P. Meghani and P. Kocovsky, *J. Org. Chem.*, **2003**, *68*, 4727. (f) M. P. A. Lyle and P. D. Wilson, *Org. Lett.*, **2004**, *6*, 855. (g) A. Bouet, B. Heller, C. Papamicael, G. Dupas, S. Oudeyer, F. Marsais and V. Levacher, *Org. Biomol. Chem.*, **2007**, *5*, 1397.
16. (a) L. Cai, H. Mamoud and Y. Han, *Tetrahedron: Asymmetry*, **1999**, *10*, 411. (b) Z. Li, G. Liu, Z. Zheng and H. Chen, *Tetrahedron*, **2000**, *56*, 7187. (c) Z. Li, Z. Zheng and H. Chen, *Tetrahedron: Asymmetry*, **2000**, *11*, 1157. (d) M. Itagaki, K. Hagiya, M. Kamitamari, K. Masumoto, K. Suenobu and Y. Y. Yamamoto, *Tetrahedron*, **2004**, *60*, 7835.
17. A. B. Charette, M. K. Janes and H. Lebel, *Tetrahedron: Asymmetry*, **2003**, *14*, 867.

18. B. Moreau and A. B. Charette, *J. Am. Chem. Soc.*, **2005**, *127*, 18014.
19. For examples of the Cu-catalysed asymmetric intramolecular cyclopropanation see: (a) C. Piqué, B. Fähndrich and A. Pfaltz, *Synlett*, **1995**, 491. (b) R. Tokunoh, H. Tomiyama, M. Sodeoka and M. Shibasaki, *Tetrahedron Lett.*, **1996**, *37*, 2449. (c) M. Honma, T. Sawada, Y. Fujisawa, M. Utsugi, H. Watanabe, A. Umino, T. Matsumura, T. Hagihara, M. Takano and M. Nakada, *J. Am. Chem. Soc.*, **2003**, *125*, 2860. (d) H. Takeda, H. Watanabe and M. Nakada, *Tetrahedron*, **2006**, *62*, 8054. (e) H. Takeda and M. Nakada, *Tetrahedron: Asymmetry*, **2006**, *17*, 2896. (f) R. Ida and M. Nakada, *Tetrahedron Lett.*, **2007**, *48*, 4855.
20. For more information on asymmetric cyclopropanation of silyl enol ethers, see: R. Schumacher, F. Dammast and H.-U. Reiβig, *Chem. Eur. J.*, **1997**, *3*, 614.
21. (a) M. Kennedy, M. A. McKervey, A. R. Maguire and G. H. P. Roos, *J. Chem. Soc., Chem. Commun.*, **1990**, 361. (b) H. M. L. Davies and D. K. Hutcheson, *Tetrahedron Lett.*, **1993**, *34*, 7243. (c) H. M. L. Davies, P. R. Bruzinski and M. J. Fall, *Tetrahedron Lett.*, **1996**, *37*, 4133. (d) H. M. L. Daves, P. R. Bruzinski, D. H. Lake, N. Kong and M. J. Fall, *J. Am. Chem. Soc.*, **1996**, *118*, 6897.
22. (a) M. P. Doyle, W. R. Winchester, J. A. A. Hoorn, V. Lynch, S. H. Simonsen and R. Ghosh, *J. Am. Chem. Soc.*, **1993**, *115*, 9968. (b) M. P. Doyle, S. B. Davies and W. Hu, *Org. Lett.*, **2000**, *2*, 1145. (c) M. P. Doyle, S. B. Davies and W. Hu, *Chem. Commun.*, **2000**, 867. (d) M. P. Doyle, *J. Org. Chem.*, **2006**, *71*, 9253.
23. M. P. Doyle, Q.-L. Zhou, C. Charnsangavej and M. A. Longoria, *Tetrahedron Lett.*, **1996**, *37*, 4129.
24. H. M. L. Davies and C. Venkataramani, *Org. Lett.*, **2003**, *5*, 1403.
25. D. Bykowski, K.-H. Wu and M. P. Doyle, *J. Am. Chem. Soc.*, **2006**, *128*, 16038.
26. E. J. Corey and T. G. Gant, *Tetrahedron Lett.*, **1994**, *35*, 5373.
27. P. Müller and A. Ghanem, *Org. Lett.*, **2004**, *6*, 4347.
28. M. P. Doyle, R. E. Austin, A. Scott Bailey, M. P. Dwyer, A. B. Dyatkin, A. V. Kalinin, M. M. Y. Kwan, S. Liras, C. J. Oalmann, R. J. Pieters, M. N. Protopopova, C. E. Raab, G. H. P. Roos, Q. -L. Zhou and S. F. Martin, *J. Am. Chem. Soc.*, **1995**, *117*, 5763.
29. M. P. Doyle, Q. -L. Zhou, A. B. Dyatkin and D. A. Ruppar, *Tetrahedron Lett.*, **1995**, *36*, 7569.
30. M. P. Doyle, C. S. Peterson, Q. -L. Zhou and H. Nishiyama, *J. Chem. Soc., Chem. Commun.*, **1997**, 211.
31. S. F. Martin, M. R. Spallar, S. Liras and B. Hartmann, *J. Am. Chem. Soc.*, **1994**, *116*, 4493.
32. M. P. Doyle and A. V. Kalinin, *J. Org. Chem.*, **1996**, *61*, 2179.
33. M. P. Doyle, M. Protopopova, P. Müller, D. Ene and E. A. Shapiro, *J. Am. Chem. Soc.*, **1994**, *116*, 8492.
34. Y. Lou, M. Horikawa, R. A. Kloster, N. A. Hawryluk and E. J. Corey, **2004**, *126*, 8916.
35. (a) H. M. L. Davies, Z.-Q. Peng and J. H. Houser, *Tetrahedron Lett.*, **1994**, *35*, 8939. (b) H. M. L. Davies, D. G. Stafford, B. D. Doan and J. H. Houser, *J. Am. Chem. Soc.*, **1998**, *120*, 3326.
36. H. M. L. Davies and B. D. Doan, *Tetrahedron Lett.*, **1996**, *37*, 3967.
37. R. P. Reddy and H. M. L. Davies, *J. Am. Chem. Soc.*, **2007**, *129*, 10312.
38. (a) V. K. Aggarwal, H. W. Smith, R. V. H. Jones and R. Fieldhouse, *Chem. Commun.*, **1997**, 1785. (b) V. K. Aggarwal, H. W. Smith, G. Hynd, R. V. H. Jones, R. Fieldhouse

and S. E. Spey, *J. Chem. Soc., Perkin Trans. 1*, **2000**, 3267.   (c) V. K. Aggarwal, E. Alsono, G. Fang, M. Ferrara, G. Hynd and M. Porcelloni, *Angew. Chem. Int. Ed.*, **2001**, *40*, 1433.

39. For reviews on asymmetric cyclopropanation with iodomethylzinc reagents see; (a) A. B. Charette and J.-F. Marcoux, *Synlett*, **1995**, 1197.   (b) A. B. Charrette and H. Lebel, in *Comprehensive Asymmetric Catalysis*, Vol. 2, ed. E. N. Jacobsen, A. Pfaltz and H. Yamamoto, Springer-Verlag, Berlin, **1999**, 581.

40. H. Takahashi, M. Yoshioka, M. Ohno and S. Kobayashi, *Tetrahedron Lett.*, **1992**, *33*, 2575.

41. N. Imai, K. Sakamoto, H. Takahashi and S. Kobayashi, *Tetrahedron Lett.*, **1994**, *35*, 7045.

42. S. E. Denmark and S. P. O'Connor, *J. Org. Chem.*, **1997**, *62*, 584.

43. S. E. Denmark, S. P. O'Connor and S. R. Wilson, *Angew. Chem., Int. Ed. Engl.*, **1998**, *37*, 1149.

44. (a) A. B. Charette and C. Brochu, *J. Am. Chem. Soc.*, **1995**, *117*, 11367.   (b) A. B. Charette, C. Molinaro and C. Brochu, *J. Am. Chem. Soc.*, **2001**, *123*, 12160.   (c) A. B. Charette, C. Molinaro and C. Brochu, *J. Am. Chem. Soc.*, **2001**, *123*, 12168.

45. J. Long, H. Du, K. Li and Y. Shi, *Tetrahedron Lett.*, **2005**, *46*, 2737.

46. H. Du, J. Long and Y. Shi, *Org. Lett.*, **2006**, *8*, 2827.

47. M.-C. Lacasse, C. Poulard and A. B. Charette, *J. Am. Chem. Soc.*, **2005**, *127*, 12440.

48. (a) H. Nishiyama, Y. Itoh, H. Matsumoto, S.-B. Park and K. Itoh, *J. Am. Chem. Soc.*, **1994**, *116*, 2223.   (b) S. Iwasa, F. Takezawa, Y. Tuchiya and H. Nishiyama, *J. Chem. Soc., Chem. Commun.*, **2001**, 59.

49. (a) T. Uchida, R. Irie and T. Katsuki, *Tetrahedron*, **2000**, *56*, 3501.   (b) J. A. Miller, W. Jin and S. T. Nguyen, *Angew. Chem. Int. Ed.*, **2002**, *41*, 2953.

50. (a) W. -C. Lo, C. -M. Che, K. -F. Cheng and T. C. W. Mak, *J. Chem. Soc., Chem. Commun.*, **1997**, 1205.   (b) C. -M. Che, J. -S. Huang, F. -W. Lee, Y. Li, T. -S. Lai, H. -L. Kwong, P. -F. Teng, W. -S. Lee, W. -C. Lo, S. -M. Peng and Z. Y. Zhou, *J. Am. Chem. Soc.*, **2001**, *123*, 4119.

51. A. Nakamura, *Pure Appl. Chem.*, **1978**, *50*, 37.

52. G. Jommi, R. Pagliarin, G. Rizzi and M. Sisti, Synlett, **1993**, 833.

53. T. Fukuda and T. Katsuki, Synlett, **1995**, 825.

54. (a) L. Huang, Y. Chen, G. -Y. Gao and X. P. Zhang, *J. Org. Chem.*, **2003**, *68*, 8179.   (b) Y. Chen, K. B. Fields and X. P. Zhang, *J. Am. Chem. Soc.*, **2004**, *126*, 14718.   (c) Y. Chen and X. P. Zhang, *J. Org. Chem.*, **2007**, *72*, 5931.   (d) Y. Chen, J. V. Ruppel and X. P. Zhang, *J. Am. Chem. Soc.*, **2007**, *129*, 12074.

55. C. D. Papageorgiou, M. A. Cubillo de Dios, S. V. Ley and M. J. Gaunt, *Angew. Chem. Int. Ed.*, **2004**, *43*, 4641.

56. (a) N. Bremeyer, S. C. Smith, S. V. Ley and M. J. Gaunt, *Angew. Chem. Int. Ed.*, **2004**, *43*, 2681.   (b) C. C. C. Johansson, N. Bremeyer, S. V. Ley, D. R. Owen, S. C. Smith and M. J. Gaunt, *Angew. Chem. Int. Ed.*, **2006**, *45*, 6024.

57. R. K. Kunz and D. W. C. MacMillan, *J. Am. Chem. Soc.*, **2005**, *127*, 3240.

58. A. Hartikka and P. I. Anderson, *J. Org. Chem.*, **2007**, *72*, 5874.

59. T. Ye, C. F. Garcia and M. A. McKervey, *J. Chem. Soc., Perkin Trans. 1*, **1995**, 1373.

60. M. P. Doyle, A. B. Dyatkin, G. H. P. Roos, F. Cañas, D. A. Pierson, A. van Basten, P. Müller and P. Polleux, *J. Am. Chem. Soc.*, **1994**, *116*, 4507.

61. M. P. Doyle, A. B. Dyatkin and J. S. Tedrow, *Tetrahedron Lett.*, **1994**, *35*, 3853.

62. M. P. Doyle and A. V. Kalinin, Synlett, **1995**, 1075.

63. (a) N. Watanbe, T. Ogawa, Y. Ohtake, S. Ikegami and S.-I. Hashimoto, *Synlett*, **1996**, 85. (b) N. Watanabe, Y. Ohtake, S.-I. Hashimoto, M. Shiro and S. Ikegami, *Tetrahedron Lett.*, **1995**, *36*, 1491.

64. (a) H. M. L. Davies and T. Hansen, *J. Am. Chem. Soc.*, **1997**, *119*, 9075. (b) H. M. L. Davies, T. Hansen, D. W. Hopper and S. A. Panaro, *J. Am. Chem. Soc.*, **1999**, *121*, 6509. (c) H. M. L. Davies, T. Hansen and M. R. Churchill, *J. Am. Chem. Soc.*, **2000**, *122*, 3063. (d) H. M. L. Davies and P. Ren, *J. Am. Chem. Soc.*, **2001**, *123*, 2070. (e) H. M. L. Davies and J. Nikolai, *Org. Biomol. Chem.*, **2005**, *3*, 4176.

65. R. T. Buck, M. P. Doyle, M. J. Drysdale, L. Ferris, D. C. Forbes, D. Haigh, C. J. Moody, N. D. Pearson and Q.-L. Zhou, *Tetrahedron Lett.*, **1996**, *37*, 7631.

66. H. M. L. Davies, T. Hansen, J. Rutberg and P. R. Bruzinski, *Tetrahedron Lett.*, **1997**, *38*, 1741.

67. B. Liu, S. -F. Zhu, W. Zhang, C. Chen and Q.-L. Zhou, *J. Am. Chem. Soc.*, **2007**, *129*, 5834.

68. C. Chen, S. -F. Zhu, B. Liu, L. -X. Wang and Q.-L. Zhou, *J. Am. Chem. Soc.*, **2007**, *129*, 12616.

69. T. C. Maier and G. C. Fu, *J. Am. Chem. Soc.*, **2006**, *128*, 4594.

70. E. C. Lee and G. C. Fu, *J. Am. Chem. Soc.*, **2007**, *129*, 12066.

71. N. Pierson, C. Fernández-García and M. A. McKervey, *Tetrahedron Lett.*, **1997**, *38*, 4705.

72. J. S. Clark, M. Fretwell, G. A. Whitlock, C. J. Burns and D. N. A. Fox, *Tetrahedron Lett.*, **1998**, *39*, 97.

73. (a) Y. Nishibayashi, K. Ohe and S. Uemura, *J. Chem. Soc., Chem. Commun.*, **1995**, 1245. (b) D. W. McMillen, N. Varga, B. A. Reed and C. King, *J. Org. Chem.*, **2000**, *65*, 2532. (c) X. Zhang, Z. Qu, Z. Ma, W. Shi, X. Jin and J. Wang, *J. Org. Chem.*, **2002**, *67*, 5621.

74. M. P. Doyle, D. C. Forbes, M. M. Vasbinder and C. S. Peterson, *J. Am. Chem. Soc.*, **1998**, *120*, 7653.

75. K. Ito and T. Katsuki, Chem. Lett., **1994**, 1857.

76. M. P. Doyle, D. G. Ene, D. C. Forbes and J. S. Tedrow, *Tetrahedron Lett.*, **1997**, *38*, 4367.

77. M. P. Doyle, D. C. Forbes, M. N. Protopopova, S. A. Stanley, M. M. Vasbinder and K. R. Xavier, *J. Org. Chem.*, **1997**, *62*, 7210.

78. (a) D. M. Hodgson, P. A. Stupple and C. Johnstone, *Tetrahedron Lett.*, **1997**, *138*, 6471. (b) D. M. Hodgson, P. A. Stuple, F. Y. T. M. Pierard, A. H. Labande and C. Johnstone, *Chem. Eur. J.*, **2001**, *7*, 4465. (c) D. M. Hodgson, P. A. Stupple and C. Johnstone, *J. Chem. Soc., Chem. Commun.*, **1999**, 2185. (d) D. M. Hodgson, D. A. Seldon and A. G. Dossetter, *Tetrahedron: Asymmetry*, **2003**, *14*, 3841. (e) D. M. Hodgson, R. Glen, G. H. Grant and A. J. Redgrave, *J. Org. Chem.*, **2003**, *68*, 581. (f) D. M. Hodgson, A. H. Labande, F. Y. T. M. Pierard and M. A. Expósito Castro, *J. Org. Chem.*, **2003**, *68*, 6153.

79. (a) S. Kitagaki, M. Ananda, O. Kataoka, K. Matsuno, C. Umeda, N. Watanabe and S. Hashimoto, *J. Am. Chem. Soc.*, **1999**, *121*, 1417. (b) D. M. Hodgson, A. H. Labande, R. Glen and A. J. Redgrave, *Tetrahedron: Asymmetry*, **2003**, *14*, 921.

80. (a) H. Suga, K. Inoue and A. Kakehi, *J. Am. Chem. Soc.*, **2002**, *124*, 14836. (b) H. Suga, K. Inoue, A. Kakehi and M. Shiro, *J. Org. Chem.*, **2005**, *70*, 47. (c) H. Suga, T. Suzuki, K. Inoue and A. Kakehi, *Tetrahedron*, **2006**, *62*, 9218.

81. H. Suga, D. Ishimoto, S. Higuchi, M. Ohtsuka, T. Arikawa, T. Tsuchida, A. Kakehi and T. Baba, *Org. Lett.*, **2007**, *9*, 4359.

Chapter 10
Catalytic Carbon–Carbon Bond-Forming Reactions

This chapter describes carbon–carbon bond-forming reactions that are metal catalysed. The reactions discussed include simple cross-coupling reactions and variations on this theme. The palladium-catalysed cross coupling of secondary alkyl Grignards and organozincs occurs with high ee in the presence of a variety of enantiomerically pure P,N-ligands. However, the substrate scope is still largely limited to coupling of secondary phenylethyl Grignard and zinc species. The enantioselective synthesis of axially chiral biaryls has also been achieved, either by the cross coupling of achiral aryl-metal precursors or by desymmetrisation of achiral biaryl substrates using enantiopure palladium or nickel catalysts.

The enantioselective palladium-catalysed allylic alkylation occurs with high ee using a range of P,N-, P,P- and P,S-based ligands. The two major limitations with the use of palladium catalysts are the restriction to the use of soft nucleophiles and the propensity for the formation of achiral linear products when utilising nonsymmetrical substrates with a terminal methylene groups. Both of these drawbacks have been overcome by the use of other transition metal catalysts. Complexes of iridium, molybdenum, tungsten and nickel all favour the formation of chiral branched products, and those of copper have been used to good effect in the substitution with hard organometallic nucleophiles such as Grignard reagents. Related C–X and C–H bond-forming reactions are also considered in this subchapter and the allylic substitution with amines and oxygen and sulfur nucleophiles has been achieved with high ee using palladium and iridium catalysts.

The asymmetric Heck reaction is catalysed by enantiomerically pure palladium catalysts formed with chelating biphosphines, especially BINAP, and this has proved an effective method for the synthesis of sterically constrained carbon centres, including quaternary centres. This chapter concludes with a brief discussion of enantioselective alkylmetallations using Grignard and organoaluminium species, which have proved useful in the diastereo- and enantioselective synthesis of polyene systems.

Catalysis in Asymmetric Synthesis 2e   © 2009 Vittorio Caprio and Jonathan M.J. Williams

10.1 Cross-Coupling Reactions

The cross-coupling of organometallic reagents with organohalides using a transition metal catalyst is a very powerful method for C—C bond formation.[1] While these reactions do not typically involve the union of alkyl groups, the Kumada coupling, involving racemic sp^3-hybridised Grignard reagents, or the Negishi coupling, using organozinc species, to sp^2-hybridised organohalides has been achieved with asymmetric induction.[2] The most commonly studied reaction is the coupling of vinyl bromide (**10.01**) with racemic 1-phenylethyl Grignard reagent (**10.02**), using either a palladium- or a nickel-based catalyst.[3,4] Enantioselectivity in this process arises by selective interaction of one enantiomer of the Grignard reagent with the catalyst during the transmetallation step. High yields can be obtained in this kinetic resolution, as racemisation of the Grignard reagent occurs at a similar rate to cross coupling.

The ligands (**10.03–10.07**) are amongst the best for catalysing these reactions with good enantioselectivity. Addition of zinc chloride causes transmetallation of the Grignard reagent prior to the coupling reaction and leads to enhanced enantioselectivity in some cases. Moderate to good levels of enantioselectivity have also been obtained using (E)-β-bromostyrene (**10.08**) as coupling partner,[5] with some of the highest ees achieved using C_2-symmetric ligand (**10.06**).[5b]

The coupling of silylated Grignard reagents such as (**10.09**) also occurs with high ee.[6] Fu and coworkers have expanded the scope of the catalytic asymmetric cross coupling by the development of catalysts capable of effecting the Negishi reaction of primary alkyl zinc reagents with secondary alkyl halides. To date, good results have been obtained in the cross coupling of primary alkylzinc bromides such as (**10.12**) with both α-bromoamides[7] and a range of 1-bromindanes, including (**10.11**), in the presence of the nickel catalyst derived from PYBOX (**10.07**).[8]

Asymmetric carbon—carbon bond-forming catalysis has been used to generate selectivity in the axial chirality of biaryls. In one impressive example, coupling of achiral naphthylmagnesium bromide (**10.17**) and bromonaphthalene (**10.18**) has been achieved with high yield and enantioselectivity using the nickel catalyst formed with ligand (**10.19**).[9] Chiral biaryls can also be accessed with high enantioselectivites by Suzuki coupling of aryl boronic acids with aryl halides.[10] For instance, Buchwald and coworkers have achieved couplings of phenyl and naphthyl boronic acids with sterically hindered bromonaphthalenes such as (**10.22**), with up to 92% ee in the presence of ligand (**10.23**).[10c] Chiral binaphthalenes have also recently been accessed with high ee by asymmetric Negishi coupling.[11]

An alternate approach to the asymmetric synthesis of axially chiral biaryls has been developed based on the nickel-catalysed Kumada cross coupling of dibenzothiophenes.[12] Cleavage of the carbon—sulfur bond of achiral dibenzothiophene (**10.25**) with the nickel catalyst generated from ligand (**10.26**) leads to the

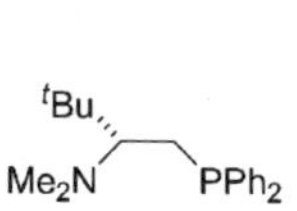

ᵗBu
Me2N
PPh2

(10.03)

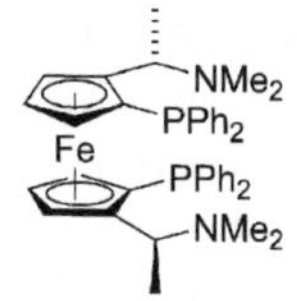

NMe2
PPh2
Fe
PPh2
NMe2

(10.04)

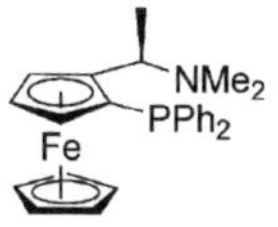

NMe2
PPh2
Fe

(10.05)

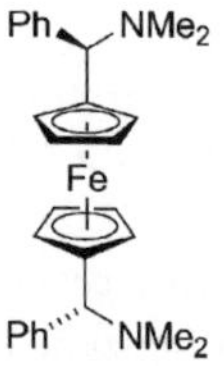

Ph
NMe2
Fe
Ph
NMe2

(10.06)

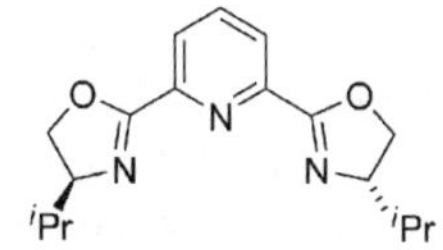

O
N
N
O
N
N
ⁱPr
ⁱPr

(10.07)

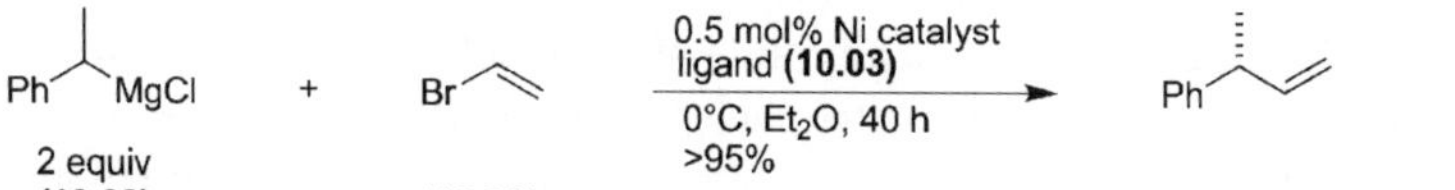

Ph
MgCl
+
Br
0.5 mol% Ni catalyst
ligand (10.03)
0°C, Et2O, 40 h
>95%
Ph
2 equiv

(10.02) (10.01) (10.13) 83% ee

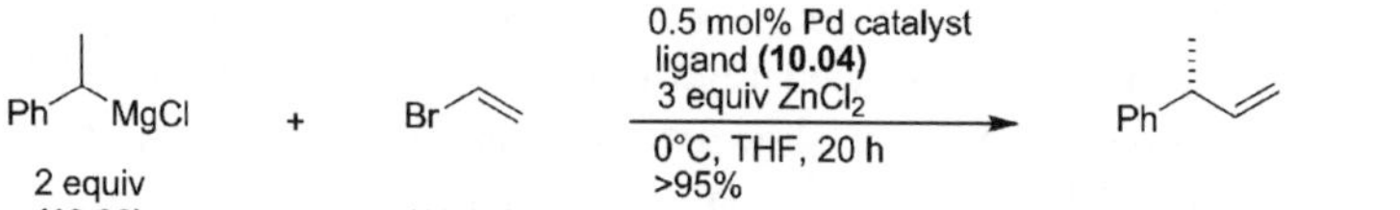

Ph
MgCl
+
Br
0.5 mol% Pd catalyst
ligand (10.04)
3 equiv ZnCl2
0°C, THF, 20 h
>95%
Ph
2 equiv

(10.02) (10.01) (10.13) 93% ee

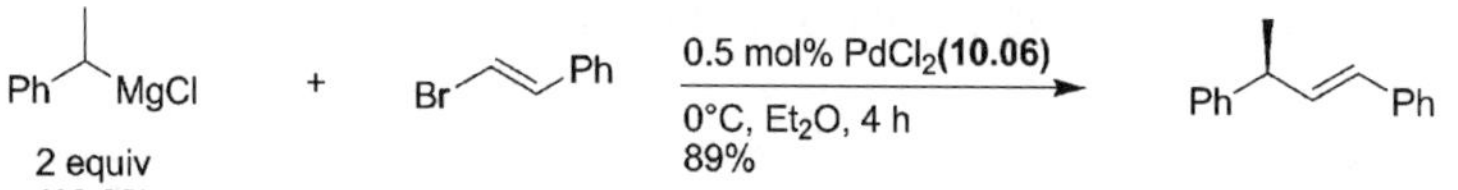

Ph
MgCl
+
Br
Ph
0.5 mol% PdCl2(10.06)
0°C, Et2O, 4 h
89%
Ph
Ph
2 equiv

(10.02) (10.08) (10.14) 93% ee

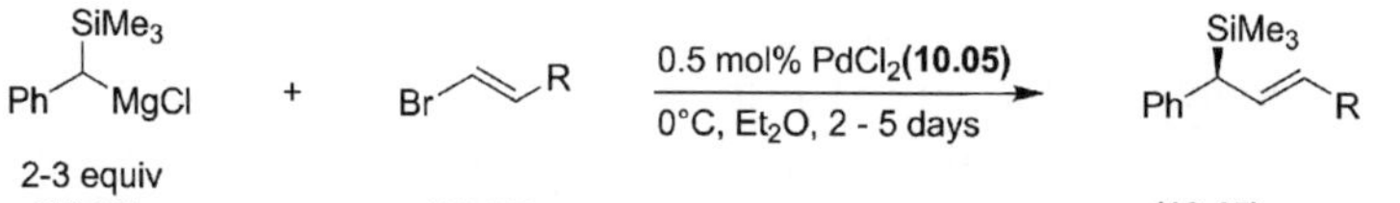

SiMe3
Ph
MgCl
+
Br
R
0.5 mol% PdCl2(10.05)
0°C, Et2O, 2 - 5 days
SiMe3
Ph
R
2-3 equiv

(10.09) (10.10) (10.15)

H	42%, 95% ee
Me	77%, 85% ee
Ph	93%, 95% ee

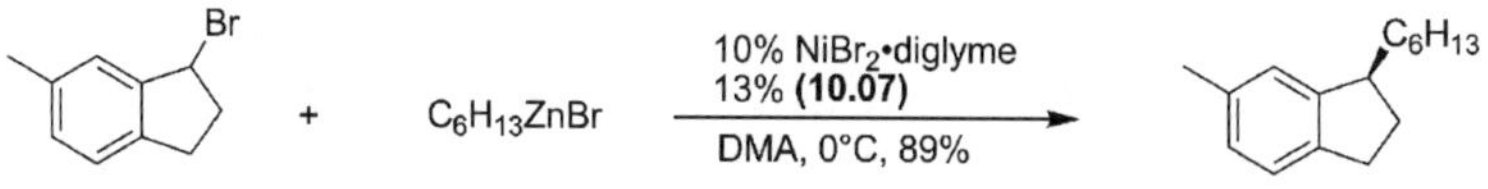

Br
+
C6H13ZnBr
10% NiBr2•diglyme
13% (10.07)
DMA, 0°C, 89%
C6H13

(10.11) (10.12) (10.16) 96% ee

formation of a nickelacycle (**10.27**) which undergoes transmetallation with a range of Grignard reagents to give a 2-mercapto-2′-binaphthyl with up to 95% ee.

Hayashi's group has induced axial chirality by the use of enantioposition-selective cross-coupling reactions. Selective replacement of one of the triflate groups in substrate (**10.29**) leads to the axially chiral product (**10.31**) with good enantioselectivity.[13] The selectivity is particularly high using alkynyl Grignard reagents such as (**10.30**) as coupling partner.[14] To some extent the reaction is self-correcting. When the initial coupling reaction affords the wrong enantiomer of product, this is more likely to undergo a second cross-coupling step to afford the di-substituted achiral product (**10.32**). Hence the enantioselectivity of the product (**10.31**) increases with longer reaction times.

This desymmetrisation strategy can be applied to access compounds possessing planar chirality. Suzuki reaction of *ortho*-dichlorobenzene chromium tricarbonyl complex with aryl and ethynylboronic acids using enantiomerically pure palladium complexes gives the chiral monosubstitution product with moderate ee as the major product with formation of small amounts of the disubstituted compound.[15] The desymmetrisation of *meso*-ditriflates such as (**10.34**) is also possible by selective Suzuki coupling with arylboronic acids, giving chiral compounds with quaternary centres with high ee.[16]

Buchwald and coworkers have recently described the cross coupling of ketone enolates with aryl bromides using enantiomerically pure palladium/BINAP complexes.[17] The ketone (**10.38**) was shown to be a particularly suitable substrate, affording products (**10.39**) with high enantioselectivity.

10.2 Metal-Catalysed Allylic Substitution

Enantioselective metal-catalysed allylic substitution reactions have attracted considerable attention, especially over recent years.[18] The metal that has been most widely investigated for allylic substitution reactions is palladium. The mechanism of palladium-catalysed allylic substitution typically involves a double inversion, resulting in overall retention of relative stereochemistry.[19] So, if the stereochemistry of the product is simply based on the stereochemistry of the starting material, how can an asymmetric synthesis be possible? The answer lies in the choice of substrate for the enantioselective version of the palladium-catalysed allylic substitution reaction. For example, the substrate (**10.40**) proceeds via a *meso* intermediate complex (**10.41**). Which end of the allyl group the nucleophile adds to dictates which enantiomer of product will be formed, (**10.42**) or *ent*-(**10.42**).

The most commonly employed test reaction for enantioselective palladium-catalysed allylic substitution reactions is the 1,3-diphenylpropenyl acetate (**10.43**), using dimethylmalonate anion as the incoming nucleophile, to provide the substitution product (**10.44**).

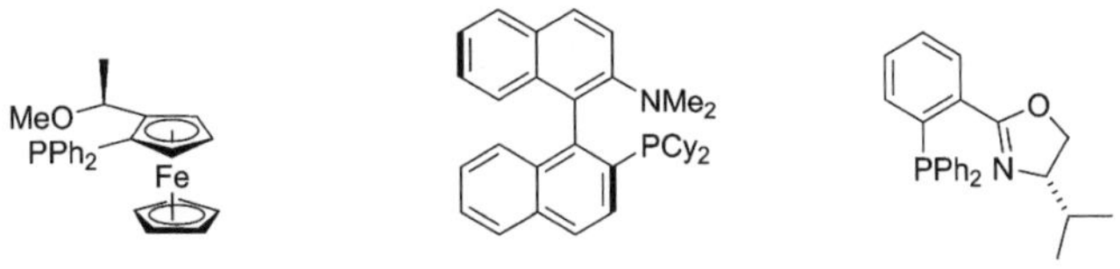

(10.19)   **(10.23)**   **(10.26)**

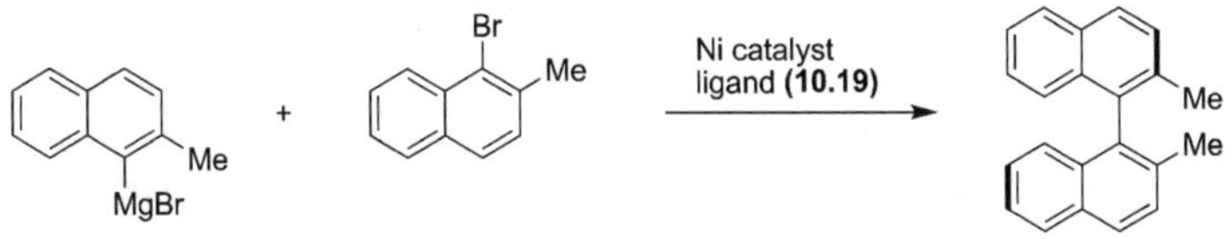

(10.17)   **(10.18)**   **(10.20)** 95% ee

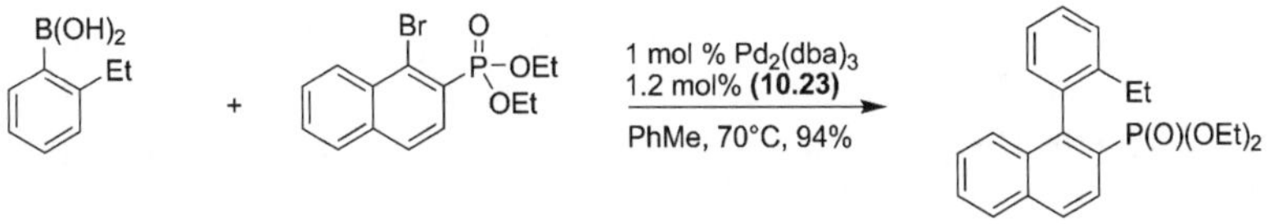

(10.21)   **(10.22)**   **(10.24)** 92% ee

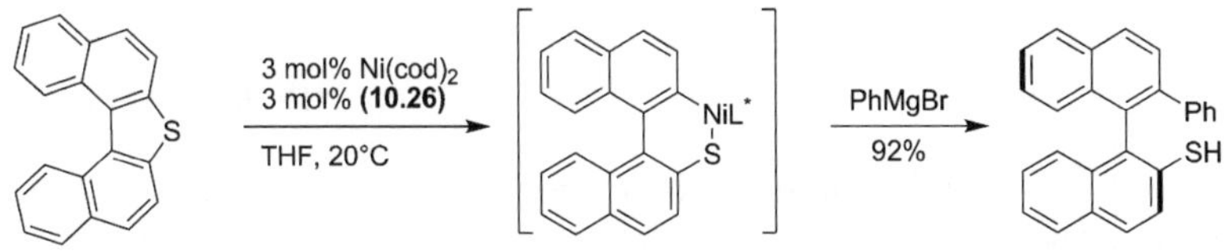

(10.25)   **(10.27)**   **(10.28)** 95% ee

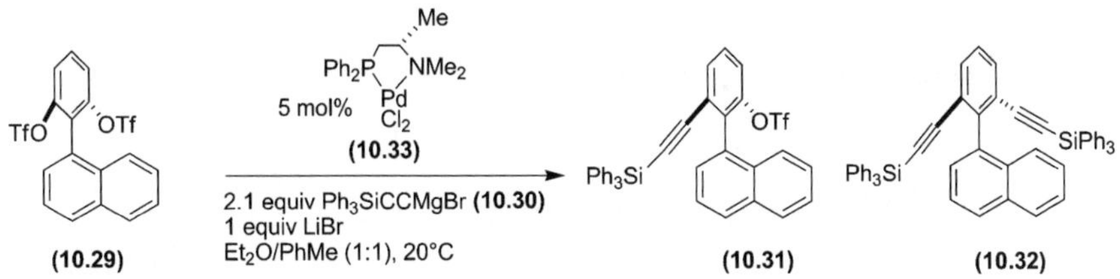

(10.29)   **(10.31)**   **(10.32)**

4 h	91% (88%ee)	0%
17 h	53% (>99% ee)	43%

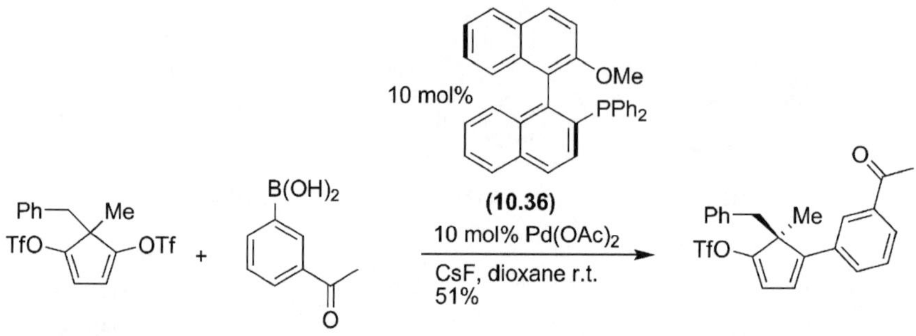
10 mol%
OMe
PPh2
B(OH)2
Ph
Me
TfO
OTf
+
O
(10.36)
10 mol% Pd(OAc)2
CsF, dioxane r.t.
51%
Ph
Me
TfO
O

(10.34) **(10.35)** **(10.37)** 86% ee

There have been many ligands used for this reaction, and often enantioselectivities in excess of 90% ee have been achieved. The first ligands to work with this level of selectivity were designed by Hayashi, where the ligand (**10.45**) is anticipated to bind to the palladium and the 'arm' on the ligand is able to reach round to the other side of the allyl group to direct the approach of the incoming nucleophile, as indicated by structure (**10.46**).[20,21]

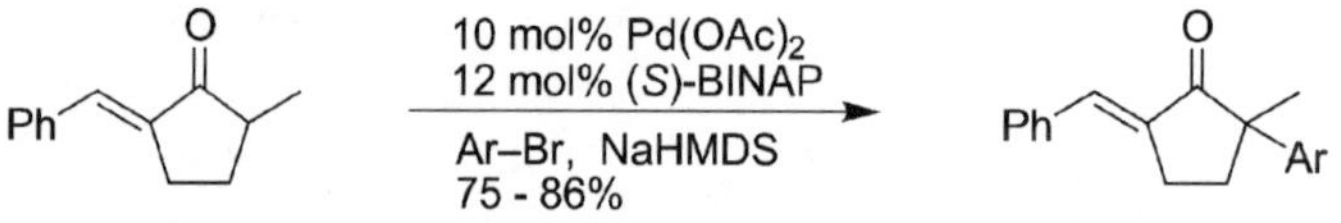
O
Ph
10 mol% Pd(OAc)2
12 mol% (S)-BINAP
Ar–Br, NaHMDS
75 - 86%
O
Ph
Ar

(10.38) **(10.39)** 94 -98% ee

Indeed, directing the approach of the nucleophile to one end of the allyl group appears to be a formidable task, since the ligand is on the wrong side. However, the ligand can perturb the symmetry of the allyl group by more direct

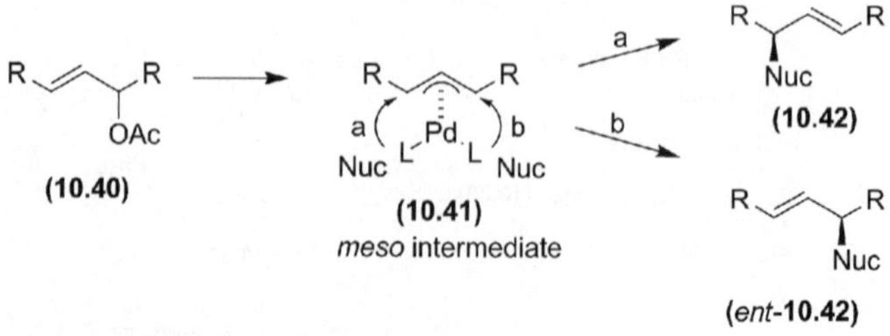
R
R
OAc
(10.40)
R
R
a
Pd
L
L
b
Nuc
Nuc
(10.41)
meso intermediate
a
b
R
R
Nuc
(10.42)
R
R
Nuc
(ent-10.42)

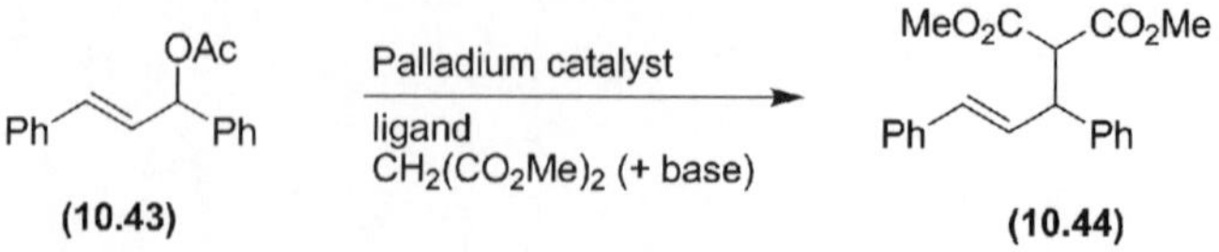
OAc
Ph
Ph
Palladium catalyst
ligand
CH2(CO2Me)2 (+ base)
MeO2C
CO2Me
Ph
Ph

(10.43) **(10.44)**

steric interactions. In the case of C_2-symmetric bidentate ligands, repulsive substrate–ligand interactions can result in one end of the allyl group being pushed further away from the metal. The incoming nucleophile is expected to add to this terminus, as shown in structure (**10.48**). Bidentate nitrogen ligands including sparteine,[22] bis-oxazoline (**10.47**)[23] and bis-aziridines,[24] as well as bidentate phosphines[25] and phosphates,[26] have all been used in allylic substitution reactions.

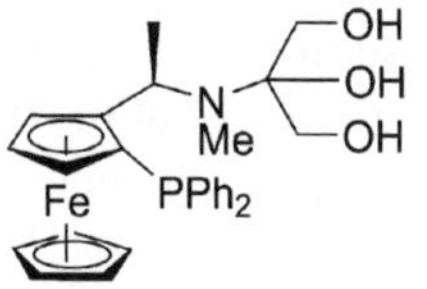

(10.45)

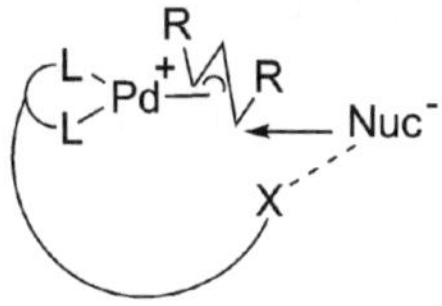

(10.46)

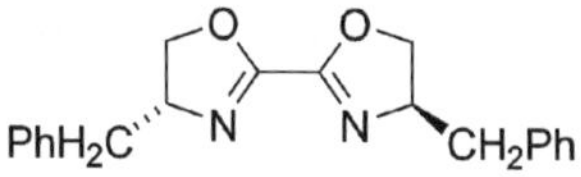

(10.47)

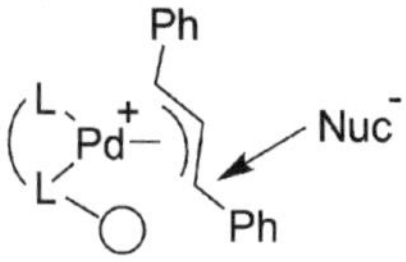

(10.48)

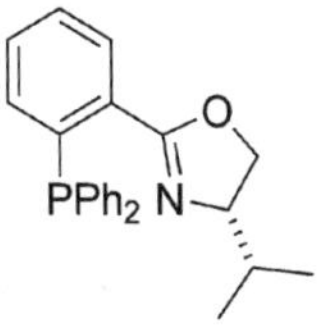

(10.49)

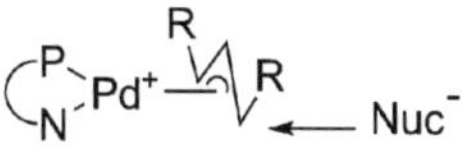

(10.53)

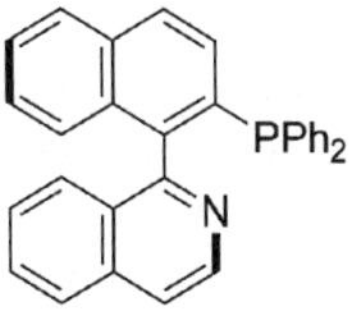

(10.50)

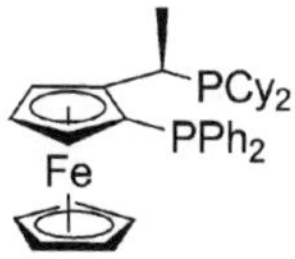

(10.51)

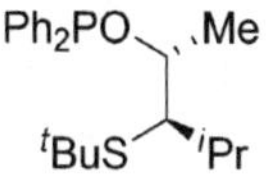

(10.52)

An alternative concept for disrupting the geometry of the allyl palladium intermediate involves using ligands with two different donor atoms. If one of the donor atoms is a better π-acceptor, then it is expected that the nucleophile will approach *trans* to that atom, as shown in structure (**10.53**). The use of ligands with an oxazoline moiety and a phosphorus atom, such as phosphinooxazoline (**10.49**), demonstrates this principle well.[27] Other ligands with various combinations of different donor atoms have also been used successfully. These ligands include Brown's QUINAP ligand (**10.50**),[28] JOSIPHOS (**10.51**), which has two electronically different phosphorus donor atoms,[29] the mixed sulfur/phosphorus ligands, such as (**10.52**) designed by Evans and coworkers,[30] and other ligand designs.[31]

The substrate (**10.43**) does not give a good indication of how useful particular ligands will be across a broad range of different substrates. In particular, many of these ligands give lower enantioselectivities with cyclic substrates. However, ligands such as (**10.54**), developed by Trost seem to be generally applicable to cyclic substrates to provide excellent asymmetric induction.[32] Thus, cyclohexenyl acetate (**10.55**) is converted into the substitution product (**10.56**) with excellent selectivity. Helmchen has used enantiomerically pure ligands containing a phosphorus donor atom tethered to a carboxylic acid, which provide excellent enantioselectivity in the same reaction[33] and some of the P,P-,[26b] the mixed P,N-[27e, 34] and S,P-ligands[30] developed by Dieguez and Evans, respectively, are also highly effective with cyclic substrates.

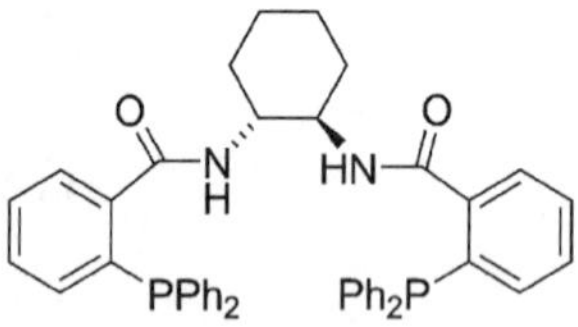

(**10.54**)

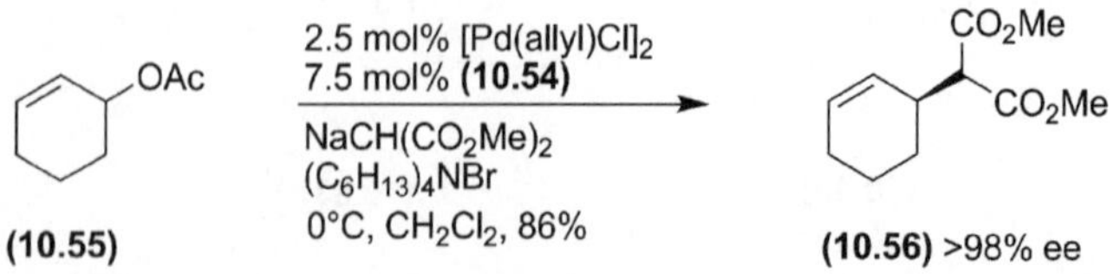

Substrates which do not proceed via symmetrical allyl intermediates such as (**10.41**) can also be subject to enantioselective allylic substitution reactions in some circumstances. Substrates which can equilibrate through a π−σ−π mechanism provide one option, however, the intermediate π-allyl group must contain two identical groups at one terminus for racemisation/epimerisation to occur. Thus the racemic compound (**10.57**) has been used with bidentate phosphines[35] and

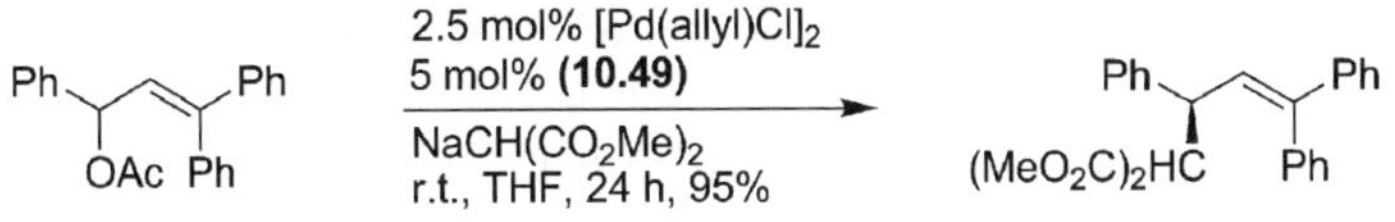
Ph
Ph
OAc Ph
2.5 mol% [Pd(allyl)Cl]2
5 mol% (10.49)
NaCH(CO2Me)2
r.t., THF, 24 h, 95%
Ph
Ph
(MeO2C)2HC Ph

(10.57) (10.58) 97% ee

with the phosphine/oxazoline ligand (**10.49**) to give very high enantioselectivity.[36] The product (**10.58**) could be converted into a range of substituted β-amino acids, succinic acids and other products.[37]

The asymmetric alkylation of substrates with a CH_2 group at one of the termini, such as linear alkene (**10.59**) or branched alkene (**10.61**), is also challenging, as addition of a nucleophile in the presence of palladium catalysts generally occurs to the least hindered end resulting in the formation of a nonchiral product. Nevertheless high regio- and enantiocontrol has been achieved using some ligands.[27e, 31e, 34,38] For example, high ees in the reaction of linear substrate (**10.59**) with dimethylmalonate has been achieved using ferrocenyl P,N-ligand (**10.60**),[31e] while the branched substrate (**10.61**) undergoes enantioselective and regioselective substitution in the presence of the phosphite-oxazoline (**10.62**).[27e]

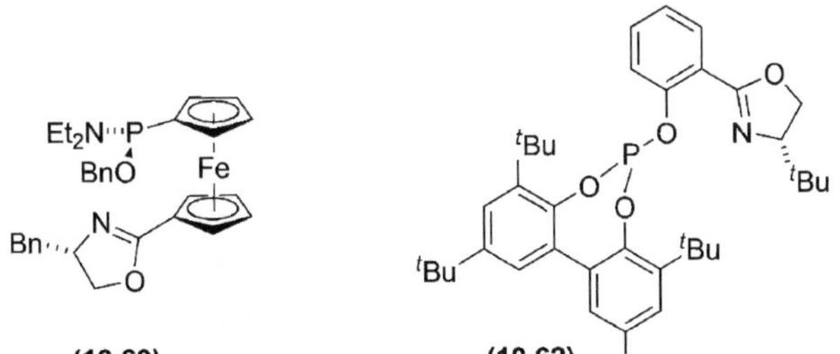
Et2N P
BnO
Fe
Bn
N
O
tBu
O
P
O
N
O
O
O
tBu
tBu
tBu
tBu
tBu

(10.60) (10.62)

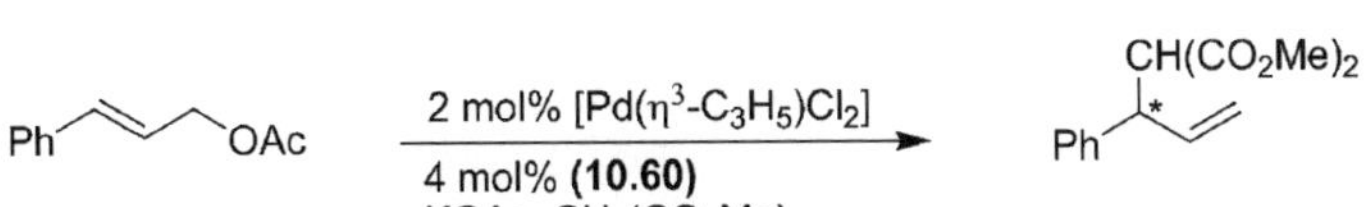
Ph
OAc
2 mol% [Pd(η3-C3H5)Cl2]
4 mol% (10.60)
KOAc, CH2(CO2Me)2
toluene, 98%
CH(CO2Me)2
Ph
*

(10.59) (10.63) 95% ee

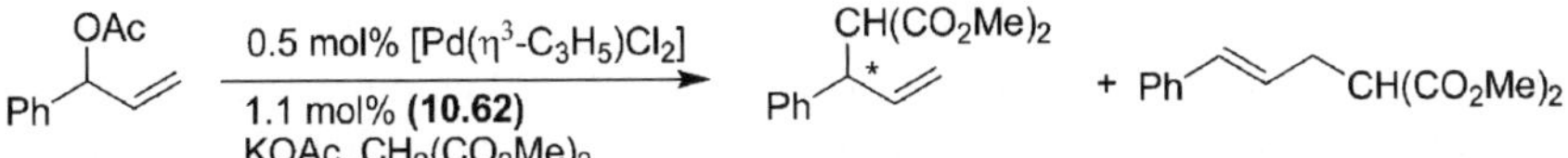
OAc
Ph
0.5 mol% [Pd(η3-C3H5)Cl2]
1.1 mol% (10.62)
KOAc, CH2(CO2Me)2
benzene
CH(CO2Me)2
Ph
*
+ Ph
CH(CO2Me)2

(10.61) (10.63) 86% ee 68% (10.64) 32%

Substrates which possess enantiotopic leaving groups provide another opportunity for an asymmetric reaction.[18c] The dibenzoate (**10.65**) undergoes selective substitution of one of the enantiotopic benzoate groups using the Trost ligand (**10.54**), with diketone (**10.66**) as the incoming nucleophile.[39] Alkylation of the corresponding diacetate with dimethyl malonate has been achieved with 96% ee using the P,S-ligands developed by Evans and coworkers.[30] Ligand (**10.54**) has also been used to control enantioselectivity in the displacement of one of the enantiotopic acetates of the gem-diacetate (**10.67**).[40] Interestingly, the same substrate (**10.67**) has been reacted with the prochiral nucleophile (**10.68**) to give excellent control of diastereoselectivity and enantioselectivity.[41] In fact, the control of stereochemistry in the prochiral nucleophile is not achieved easily, and generally proceeds with low enantioselectivity, with only a few exceptions.[42]

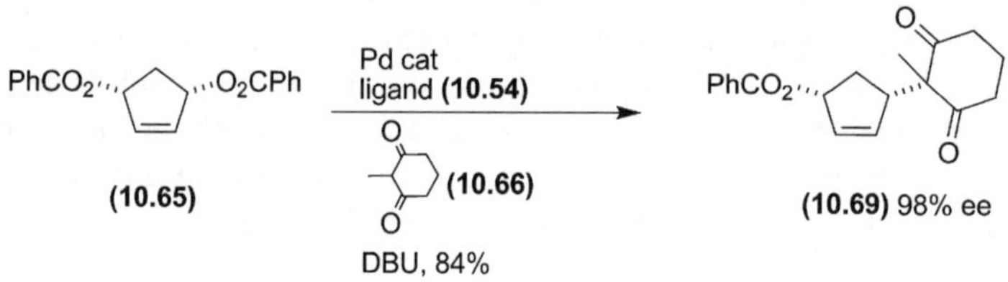

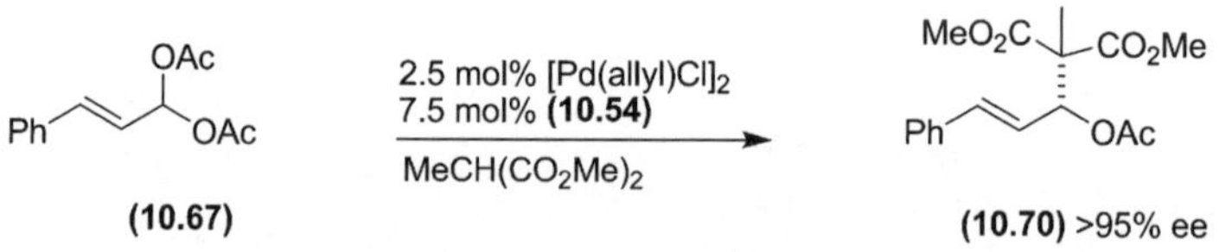

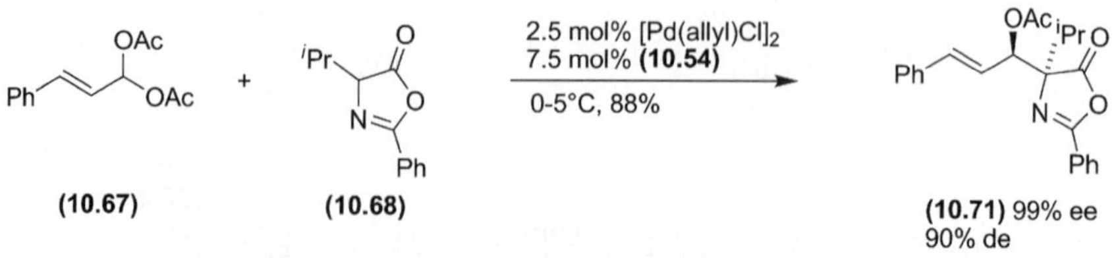

The most common nucleophiles employed in the allylic alkylation are soft species such as malonate esters. However, Trost and Schroeder have discovered that high ees can be achieved in the addition of cyclic lithium enolates using ligand (**10.54**) with one equivalent of trimethyltin chloride, which may act to soften the nucleophilic species by transmetallation to the tin enolate.[43] Enantioselective additions of nonstabilised ketone enolates can also be achieved using an alternate palladium-catalysed decarboxylation protocol. In this approach an allyl 3-ketoester (**10.72**) or allyl vinyl carbonate (**10.73**) undergoes decarboxylation in the presence of Pd(0) to

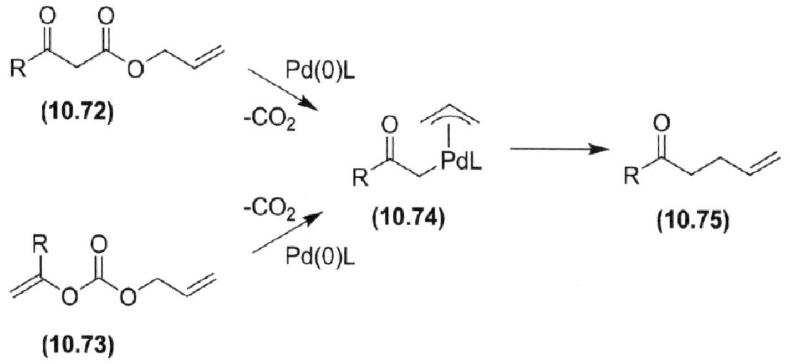

Figure 10.1 Palladium-catalysed decarboxylative allylation

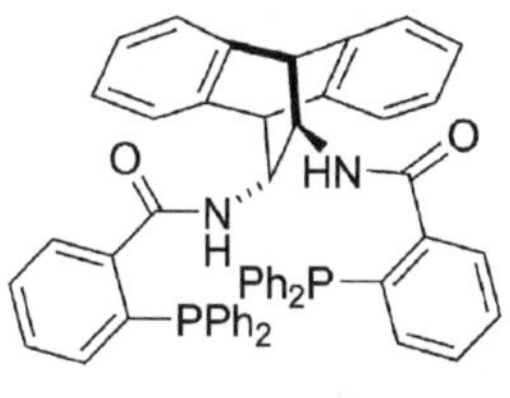

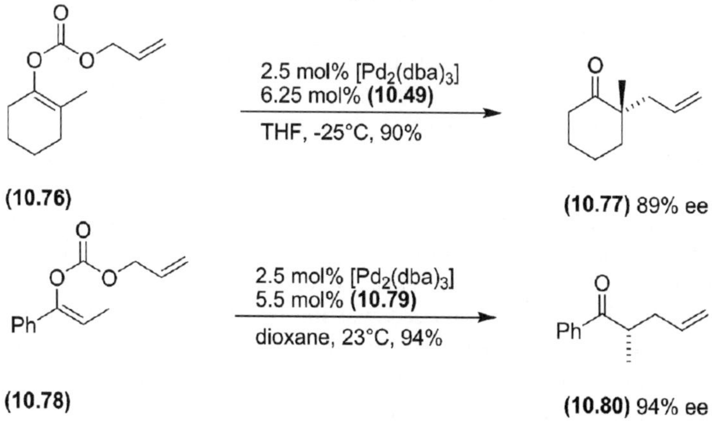

give a π-allyl palladium species (**10.74**), which undergoes reductive elimination to give the product (**10.75**) (Figure 10.1). [44] High levels of enantioselectivity in both the acyclic and cyclic variants of this reaction have been achieved in the presence of enantiomerically pure ligands. For example, allyl enol carbonate (**10.76**) is converted into ketone (**10.77**) with high ee in the presence of phosphino-oxazoline (**10.49**), [45] while high ees in the decarboxylation of acyclic substrates such as (**10.78**) have been achieved using ligand (**10.79**). [46]

Various cyclisation reactions have been achieved using enantiomerically pure palladium catalyst through an allylic substitution process. One of the first reported examples of such a cyclisation, with substrate (10.81), is also one of the most selective reactions, with control of both enantioselectivity and diastereoselectivity.[47] Enantioselective cyclisation reactions using phosphino-oxazoline ligands have been reported by Pfaltz, although the enantioselectivities (up to 87% ee) are lower than for most intermolecular reactions.[48]

(10.81)   (10.82) 95% ee, 95% de

Whilst palladium has been examined more than any other metal for the allylic substitution reaction, other transition metals have also been used successfully. One of the advantages of using an alternative metal is that the regiochemistry of the substitution process with challenging substrates such as (10.59) or (10.61) is often biased towards attack at the more substituted terminus. For palladium-catalysed allylic substitution, the regiochemical outcome is dependent upon the electronic nature of the ligand.[38] It has been shown that rhodium-catalysed allylic substitution occurs at the carbon bearing the leaving group, via formation of a σ-Rh complex, and proceeds with retention at that site owing to very slow σ–π–σ isomerisation leading to formation of a π-allyl-complex. Thus, the potential for a catalytic asymmetric alkylation of a racemic substrate seems limited. Nevertheless, the rhodium-catalysed substitution of branched substrate (10.61) with dimethyl malonate has been achieved with high ee in the presence of phosphino-oxazolines under conditions which maximise the lifetime of the π-allyl-Rh complex.[49] The iridium-catalysed procedure is believed to proceed via σ–π–σ rearrangement, especially in the presence of electron-deficient ligands[50] and, in contrast to the Pd-catalysed procedure, favours formation of the chiral branched product, even when using linear monosubstituted allylic substrates such as (10.59).[51] Some of the highest regio- and enantioselectivites in the iridium-catalysed alkylation of a variety of linear allylic acetates and carbonates, including (10.83), have been obtained using phosphoramidite ligands such as (10.84).[52] Various nucleophiles are compatible with the iridium-catalysed allylic alkylation and high ees have also been obtained using nitro compounds[53] and ketone enolates.[54]

Enantiopure copper complexes have also been used successfully in the asymmetric allylic alkylation of linear substrates, and these catalysts are most effective in the substitution of cinnamyl halides with hard organometallic nucleophiles, especially organozinc and Grignard reagents.[55] Many enantiomerically pure ligands have been explored in this process, with some of the highest ees obtained using phosphoramidites[56] and ferrocenyl amines and aminophosphines.[57] For instance, Dübner and Knochel obtained up to 98% ee in the addition of bulky dineopentylzinc to cinnamyl chloride (**10.85**) using the copper catalyst generated from ligand (**10.86**),[57b] while some of the best enantioselectivites in the addition of alkyl Grignards have been achieved by Feringa and coworkers using the ferrocenyl aminophosphine Taniaphos (**10.87**).[57c] High ees in the substitution with Grignard reagents have also been obtained by Alexakis and coworkers using phosphoramidite ligands.[56c]

Good levels of enantioselectivity in the allylic alkylation have been obtained using molybdenum catalysts.[58] The branched product is again favoured and the allylic carbonate (**10.89**) reacts with dimethyl malonate with high ee using the pyridine-containing ligand (**10.90**).[59] Pfaltz and coworkers have shown that replacement of the pyridines in (**10.90**) with oxazoline moieties result in ligands that also display good selectivity.[60]

The tungsten complex (**10.95**)[61] and the nickel complex (**10.96**)[62] have been used in allylic substitution reactions, providing over 90% ee in the formation of some products.

The use of heteroatom nucleophiles has provided the opportunity to extend the synthetic repertoire of allylic substitution reactions. Nitrogen nucleophiles have been particularly popular and the allylic amination has been achieved with high ee using a variety of both palladium[31d, 27e, 30,63] and iridium catalysts.[50,64] The phosphino-oxazoline ligand (**10.97**)[63d] has given slightly better results than other phosphino-oxazoline ligands in the amination of substrate (**10.43**) with potassium phthalimide (**10.98**).[63a, 63b] Similiar levels of enantioselectivity in the amination with benzylamine have been observed using the phosphite-oxazoline (**10.62**).[27e] Togni's ferrocenyl pyrazole ligand (**10.100**) also provides excellent enantioselectivity in the amination of the allylic carbonate (**10.101**).[63c] Trost has reported high selectivities for allylic amination of cyclic substrates using his ligands.32 For example, cycloheptenyl acetate (**10.103**) is converted into the allylic amine derivative (**10.104**) with excellent control of enantioselectivity.

The use of enantiomerically pure iridium catalysts allows the enantioselective allylic amination of linear substrates and this has been achieved with high ee using the iridium complex of phosphoramidite (**10.84**) and both acyclic and cyclic amines, including pyrrolidine and piperidine.[64]

The allylic amination also proceeds in intramolecular fashion. The *meso*-diol (**10.106**) forms an intermediate (**10.107**), which undergoes enantioselective cyclisation with high enantioselectivity to give the product (**10.108**). The use of

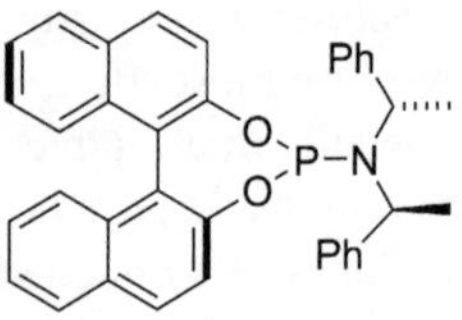

(10.84)

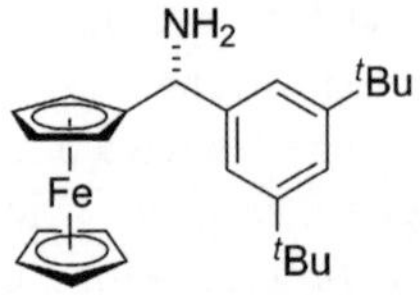

(10.86)

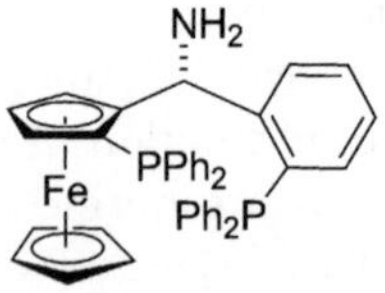

(10.87)

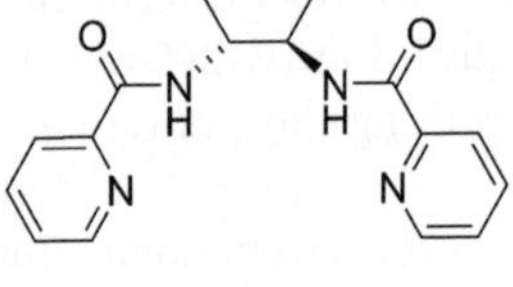

(10.90)

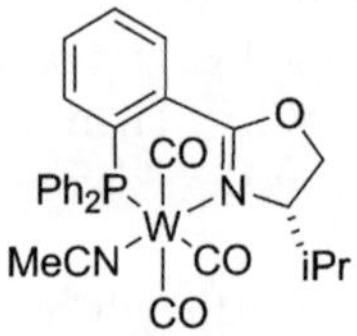

(10.95)

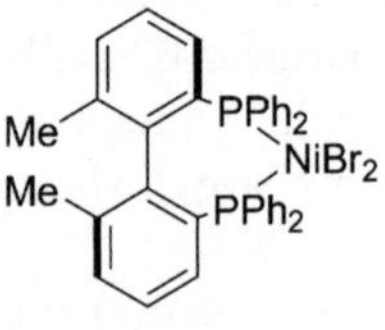

(10.96)

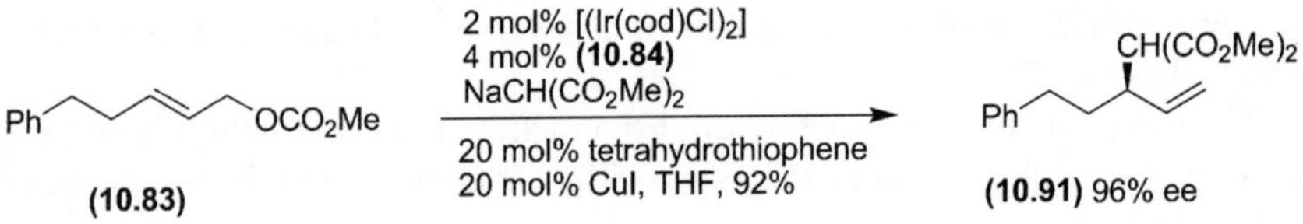

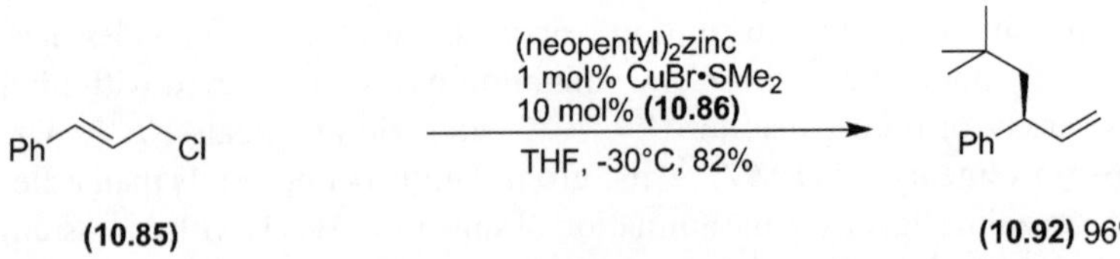

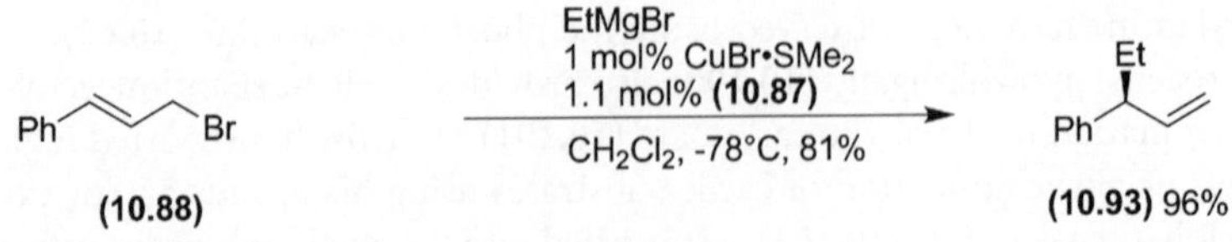

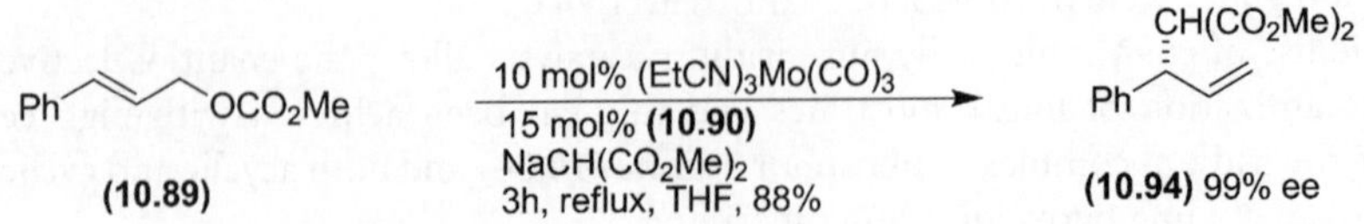

Ph ⌒⌒⌒ OCO₂Me

(10.83)

2 mol% [(Ir(cod)Cl)₂]
4 mol% (10.84)
NaCH(CO₂Me)₂
───────────────────
20 mol% tetrahydrothiophene
20 mol% CuI, THF, 92%

CH(CO₂Me)₂

(10.91) 96% ee

Ph ⌒⌒ Cl

(10.85)

(neopentyl)₂zinc
1 mol% CuBr·SMe₂
10 mol% (10.86)
───────────────────
THF, -30°C, 82%

(10.92) 96% ee

Ph ⌒⌒ Br

(10.88)

EtMgBr
1 mol% CuBr·SMe₂
1.1 mol% (10.87)
───────────────────
CH₂Cl₂, -78°C, 81%

Et

(10.93) 96% ee

Ph ⌒⌒ OCO₂Me

(10.89)

10 mol% (EtCN)₃Mo(CO)₃
───────────────────
15 mol% (10.90)
NaCH(CO₂Me)₂
3h, reflux, THF, 88%

CH(CO₂Me)₂

(10.94) 99% ee

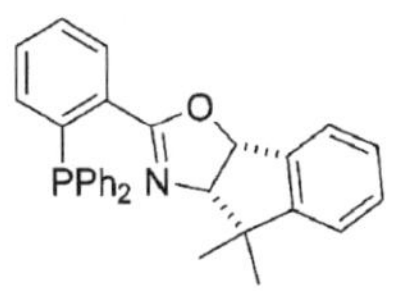
PPh2
N
O

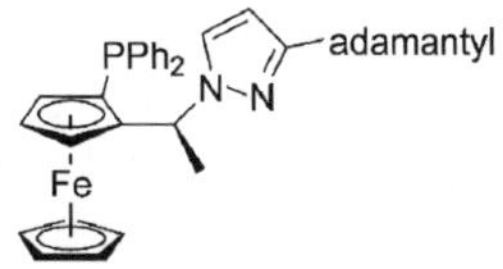
PPh2
adamantyl
N
N
Fe

(10.97) **(10.100)**

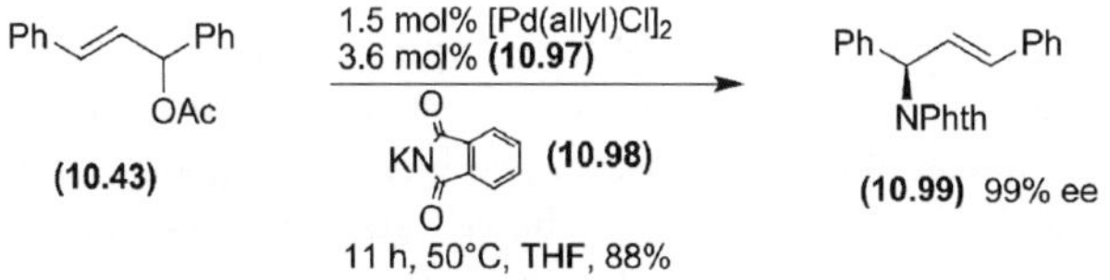
Ph
Ph
OAc
(10.43)
1.5 mol% [Pd(allyl)Cl]2
3.6 mol% (10.97)
KN
(10.98)
11 h, 50°C, THF, 88%
Ph
Ph
NPhth
(10.99) 99% ee

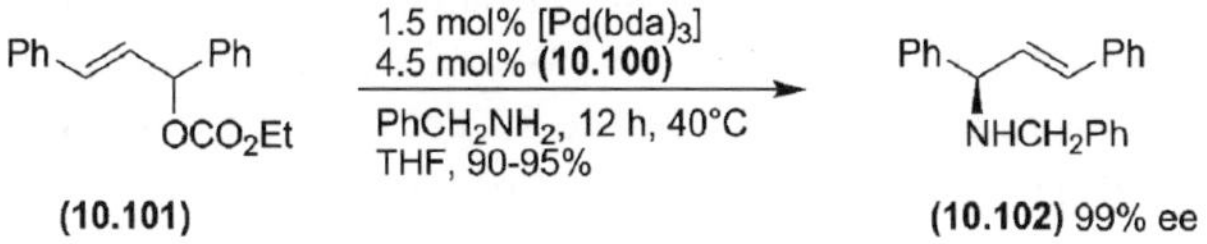
Ph
Ph
OCO2Et
(10.101)
1.5 mol% [Pd(bda)3]
4.5 mol% (10.100)
PhCH2NH2, 12 h, 40°C
THF, 90-95%
Ph
Ph
NHCH2Ph
(10.102) 99% ee

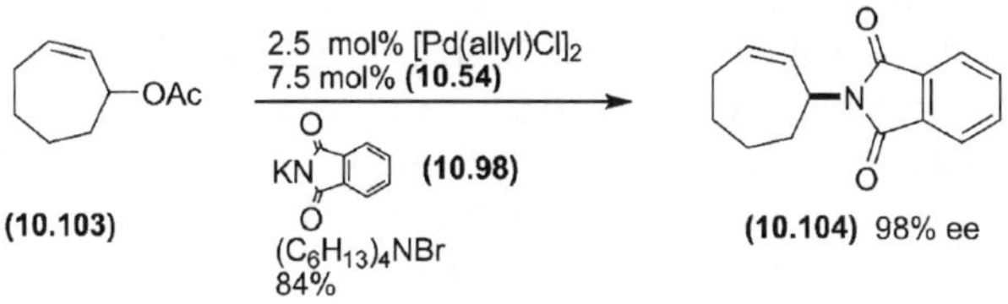
OAc
(10.103)
2.5 mol% [Pd(allyl)Cl]2
7.5 mol% (10.54)
KN
(10.98)
(C6H13)4NBr
84%
N
(10.104) 98% ee

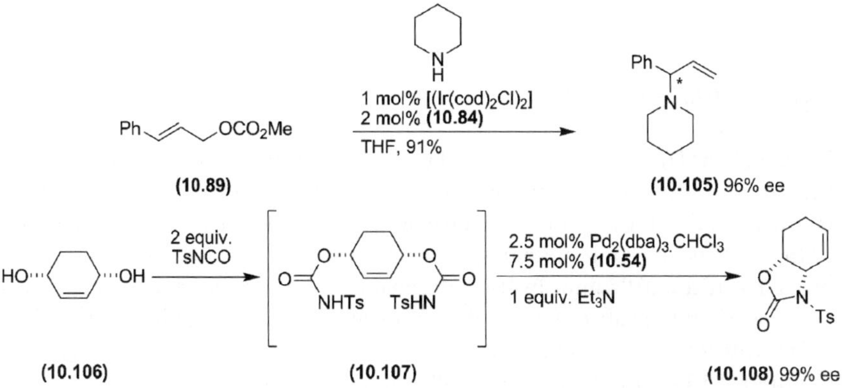
Ph
OCO2Me
(10.89)
N
H
1 mol% [(Ir(cod)2Cl)2]
2 mol% (10.84)
THF, 91%
Ph
*
N
(10.105) 96% ee
HO
OH
(10.106)
2 equiv.
TsNCO
NHTs
TsHN
(10.107)
2.5 mol% Pd2(dba)3.CHCl3
7.5 mol% (10.54)
1 equiv. Et3N
N
Ts
(10.108) 99% ee

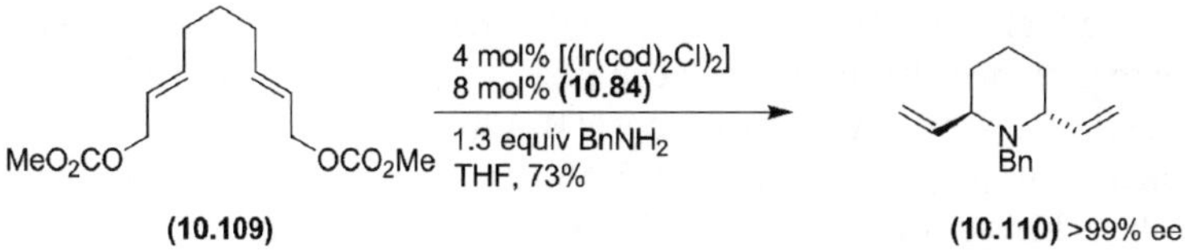
MeO2CO
OCO2Me
(10.109)
4 mol% [(Ir(cod)2Cl)2]
8 mol% (10.84)
1.3 equiv BnNH2
THF, 73%
N
Bn
(10.110) >99% ee

triethylamine was crucial to this selectivity.[65] A similar desymmetrisation performed on the five-membered ring analogue has been recently used in an enantioselective synthesis of the alkaloid (−)-swainsonine.[66] Iridium-phosphoramidite
catalysts have been used in the intramolecular amination. In an impressive example, bis-allylic carbonate (**10.109**) undergoes a sequential inter- followed by
intramolecular amination using phosphoramidites such as (**10.84**) to give the
piperidine (**10.110**) with very high ee.[67]

Cyclisation reactions involving a Wacker-type oxidative cyclisation have been
reported to take place with very high enantioselectivity in some cases.[68]

Even oxygen nucleophiles have been introduced with good enantioselectivity
using both palladium-[69,70] and iridium-based catalysts.[51] The conditions of the
reaction need to be sufficiently mild that the product does not become a substrate
for the allylic substitution, since this will ultimately lead to racemisation. Pivalate
(^{t}BuCO$_2$$^-$)[69] and phenols have been used as nucleophiles, in the presence of palladium catalysts, with good results,[70] while linear allylic carbonates are converted
into chiral branched products with high ee using phenolates,[71] alkoxides[72] and
also hydroxylamines[73] with iridium complexes. Sulfur nucleophiles have also been
used in enantioselective allylic substitution reactions.[74]

The reduction of allylic esters has been achieved using formate as the hydride
source. Although not widely investigated with a variety of ligands, Hayashi has
demonstrated that the monodentate MOP ligands (**10.111**) and (**10.112**) are certainly effective.[75] The reduction of substrates (**10.113**) and (**10.115**) has been
achieved with good control of enantioselectivity.

10.3   Heck Reactions

In general, the Heck reaction involves the alkylation or arylation of an alkene to give
the more substituted alkene product. In order for an asymmetric Heck reaction to
take place, either the reaction must take place on one of two enantiotopic alkenes,
or the standard reaction mechanism must be diverted. The reaction pathway can
be altered if the palladium hydride elimination step cannot be achieved, which
happens when the palladium and β-hydride are unable to align in a *syn* fashion.
This is true for cyclic substrates and also in certain other cases. For example, dihydrofuran (**10.117**) forms the initial palladium alkyl adduct (**10.118**). The hydride
next to the R group is unable to undergo β-hydride elimination. However, elimination to give the palladium alkene complex (**10.119**) is possible. This species can
either decomplex directly to give product (**10.120**), or undergoes further β-hydride
rearrangements to give the regioisomer (**10.121**). The asymmetric Heck reaction is
generally achieved using enantiomerically pure bidentate ligands, typically phosphines such as BINAP.[2d, 76] The association and insertion of an alkene proceeds via

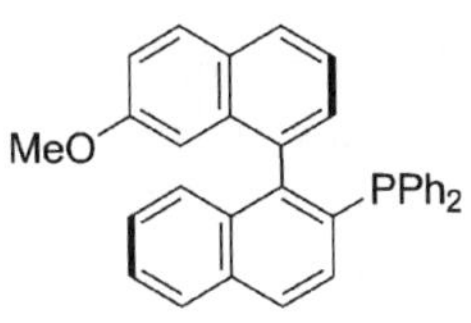

(10.111) 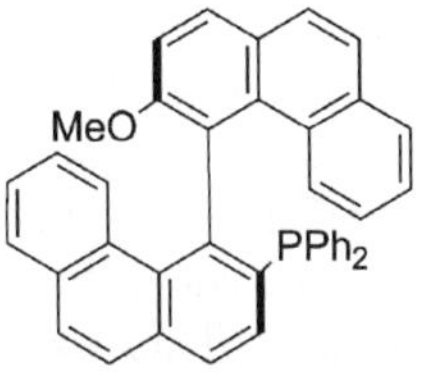   (10.112)

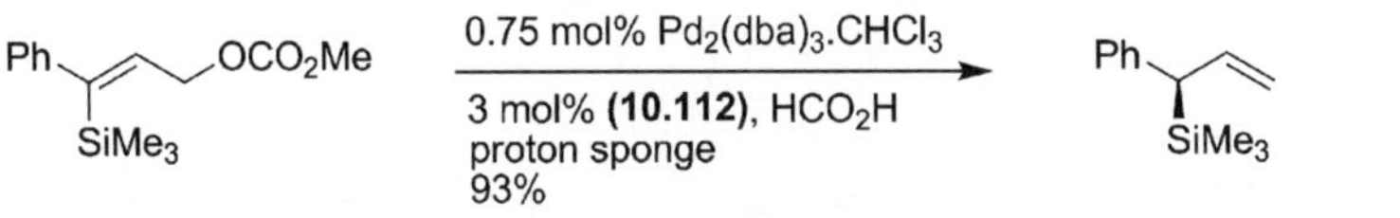

(10.113)   (10.114) 91%ee

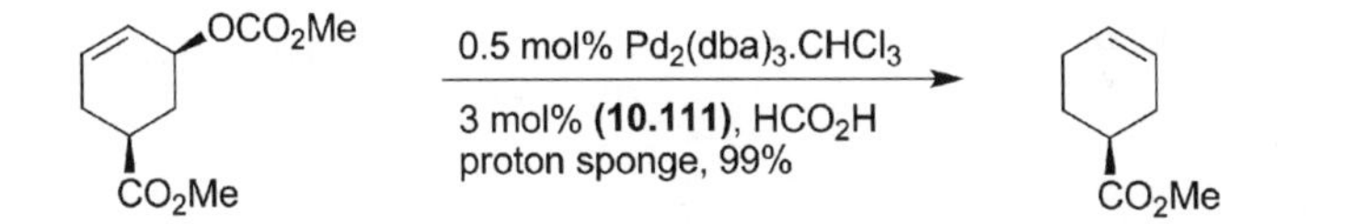

(±)-(10.115)   (10.116) 87% ee

proton sponge = 1,8-bis(dimethylamino)naphthalene

dissociation of a halide or one phosphine of the ligand from the palladium centre of the catalyst. The former, cationic process usually results in higher enantioselectivities and thus asymmetric Heck reactions employing bidentate phosphines are often performed in the presence of silver salts that act as halide scavengers.

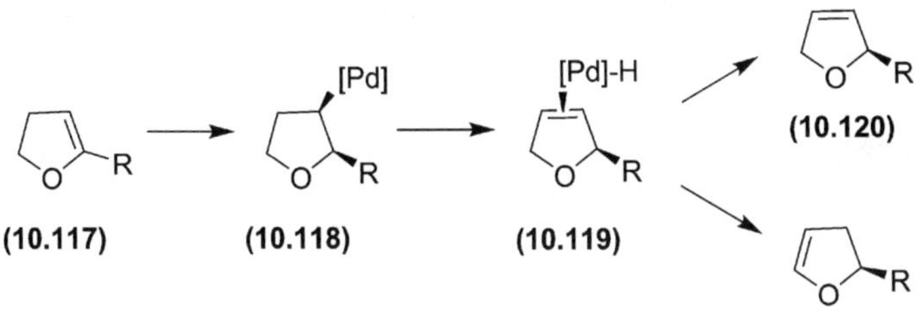

(10.117)   (10.118)   (10.119)   (10.120)   (10.121)

The first asymmetric Heck reactions were reported by Shibasaki in 1989.[77] Typical examples from this group include the desymmetrising cyclisation of vinyl iodide (**10.122**). In some cases, pinacol has been found to have a beneficial effect on yield and enantiomeric excess of the product, for example in the formation of the decalin (**10.125**). In the absence of pinacol, only 6% yield was obtained after 106 h, in 92% ee. When triflates are used as starting materials, silver salts are not required, since the triflate is more readily dissociated from the palladium to give cationic intermediates. This desymmetrisation can also be performed on other cyclic precursors.[78] For instance, Feringa and coworkers have achieved the cyclisation of quinone monoacetal (**10.126**) with high ee using phosphoramidite (**10.127**).[78a] In this transformation, the formation of a Pd-enolate intermediate

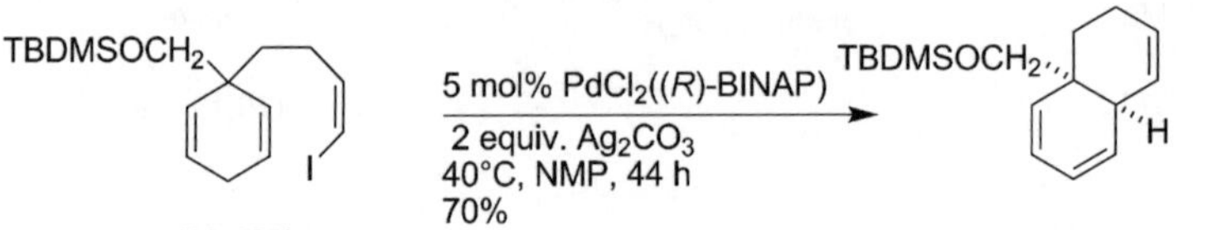
TBDMSOCH2
I
5 mol% PdCl2((R)-BINAP)
2 equiv. Ag2CO3
40°C, NMP, 44 h
70%
TBDMSOCH2
H

(10.122) **(10.123)** 44% ee

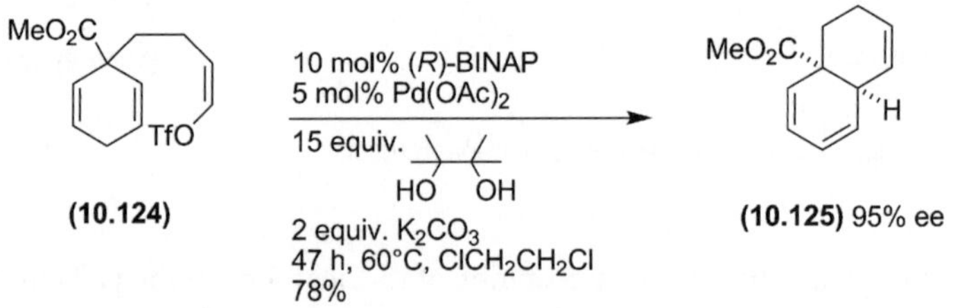
MeO2C
TfO
10 mol% (R)-BINAP
5 mol% Pd(OAc)2
15 equiv.
HO OH
2 equiv. K2CO3
47 h, 60°C, ClCH2CH2Cl
78%
MeO2C
H

(10.124) **(10.125)** 95% ee

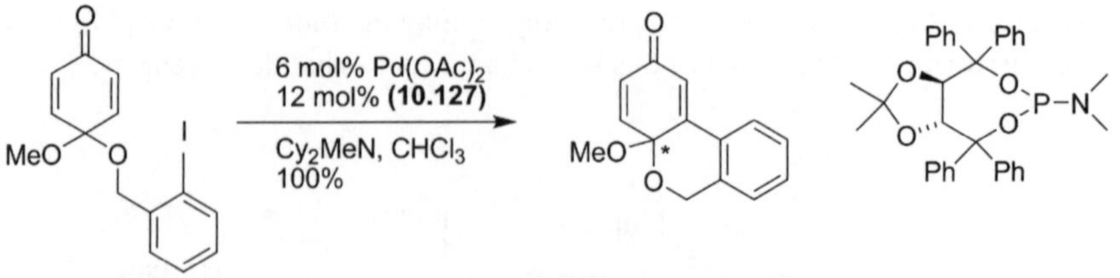
O
MeO O
I
6 mol% Pd(OAc)2
12 mol% (10.127)
Cy2MeN, CHCl3
100%
O
MeO
*
O
Ph Ph
O
O
O
P–N
O
Ph Ph

(10.126) **(10.128)** 96% ee **(10.127)**

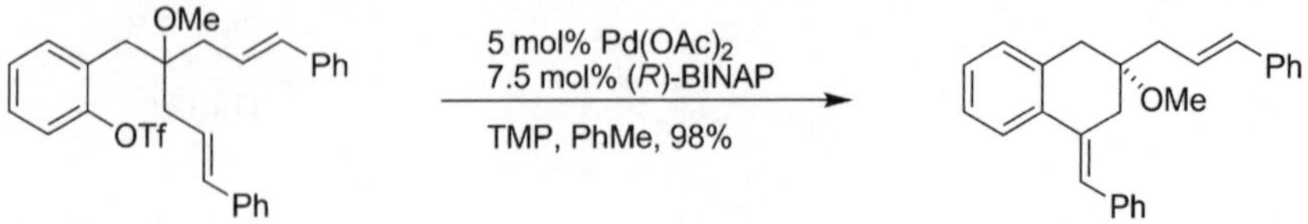
OMe
Ph
OTf
Ph
5 mol% Pd(OAc)2
7.5 mol% (R)-BINAP
TMP, PhMe, 98%
Ph
OMe
Ph

(10.129) **(10.130)** 97% ee

allows β-hydride elimination to occur and is one of the only examples where a monodentate ligand leads to high enantioselectivity and proceeds without additives. The desymmetrising Heck reaction of a small number of acyclic precursors has been achieved. In this process the tertiary methoxy group in substrates such as (**10.129**) is postulated to facilitate fast equilibration of the diastereomeric pair of palladium-alkene complexes, required for high enantioselectivity, by temporary binding to the Pd centre, and Heck cyclisation of the deoxygenated substrate proceeds with only poor ee.[79]

The asymmetric Heck reaction can be used to synthesise quaternary carbon centres. During studies towards the synthesis of 3,3-disubstituted oxindoles, Overman and coworkers have shown how the use of silver salts can change the sense of asymmetric induction of the cyclised product.[80] Thus, the iodide (**10.131**) can be converted into the product (**10.132**) with the (*S*)-enantiomer predominating, when the reaction is run in the presence of silver salts. In the absence of silver salts, the (*R*)-enantiomer is the major product.

Overman and Poon have shown that the mechanism without silver salts, indeed even with deliberately added halide salt (e.g. Bu$_4$NBr) can give high enantioselectivities, as demonstrated by the conversion of triflate (**10.133**) into the cyclised product (**10.134**). Furthermore high ees are also obtained using the iodide precursor in the absence of silver salts. This shows that the cationic mechanism is not a requirement for a successful outcome, and the authors suggest an intermediate in which halide, bidentate phosphine, alkene and aryl group are all coordinated to the palladium simultaneously (i.e. pentacoordinate).[81] In the absence of Bn$_4$NBr, a 72% yield was achieved, but with only 43% ee. Overman and coworkers have also achieved the formation of products such as (**10.136**) with sterically congested quaternary centres bearing two aryl groups using this chemistry.[82]

Dihydrofuran (**10.137**) has also proved to be a popular substrate for the asymmetric Heck reaction. Hayashi has reported that using a Pd/BINAP catalyst not only is the initial addition enantioselective, but that the diastereomeric intermediates, i.e. of structure (**10.119**) preferentially give different regioisomeric products (**10.138**) and (**10.139**). This effect is similar to that of a kinetic resolution (see Section 4.1).[83]

Other ligands which work well in this reaction include the bidentate phosphine (**10.140**), where the 3,5-dialkyl substituents are important to enantioselectivity (the 3,5-dialkyl *meta* effect).[84] The phosphino-oxazoline (**10.141**) has been used to great effect in this transformation.[85] Interestingly, these last two examples show how the regiochemistry of the reaction is controlled by the ligand and reaction conditions. The same phosphino-oxazoline (**10.141**) has also been applied to other enantioselective Heck reactions, including the coupling of triflates to give the products (**10.143**) and (**10.146**) with good to excellent stereocontrol. This ligand has also been employed in the intramolecular Heck arylation of cyclic enamides.[86] The success obtained with (**10.141**) has inspired the design of a variety of other

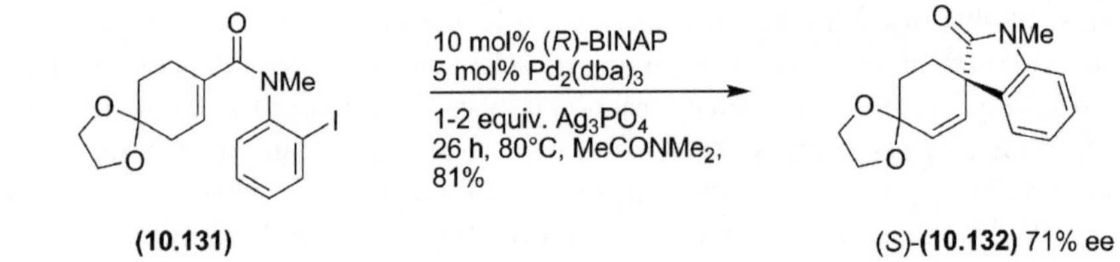

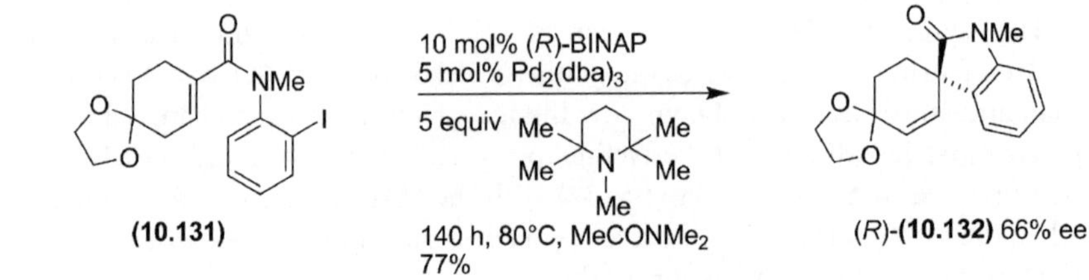

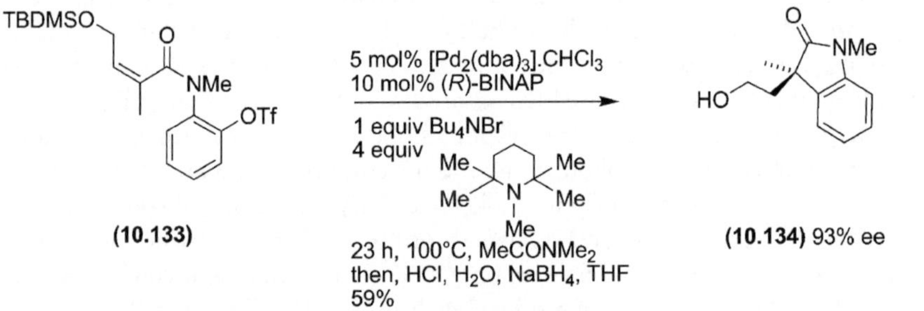

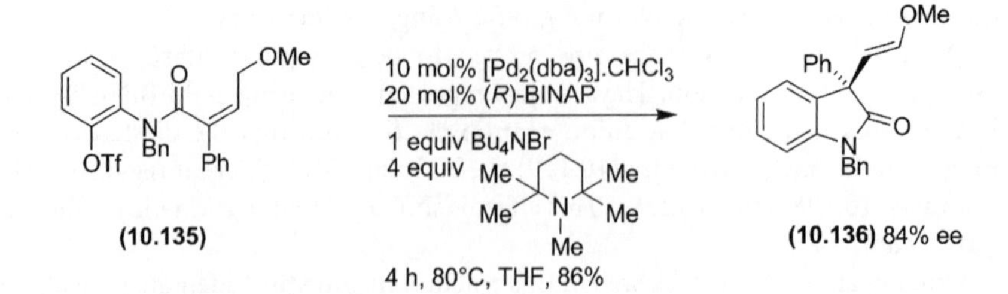

(10.131)

10 mol% (*R*)-BINAP
5 mol% Pd$_2$(dba)$_3$

1-2 equiv. Ag$_3$PO$_4$
26 h, 80°C, MeCONMe$_2$,
81%

(*S*)-(10.132) 71% ee

(10.131)

10 mol% (*R*)-BINAP
5 mol% Pd$_2$(dba)$_3$

5 equiv

140 h, 80°C, MeCONMe$_2$
77%

(*R*)-(10.132) 66% ee

(10.133)

5 mol% [Pd$_2$(dba)$_3$].CHCl$_3$
10 mol% (*R*)-BINAP

1 equiv Bu$_4$NBr
4 equiv

23 h, 100°C, MeCONMe$_2$
then, HCl, H$_2$O, NaBH$_4$, THF
59%

(10.134) 93% ee

(10.135)

10 mol% [Pd$_2$(dba)$_3$].CHCl$_3$
20 mol% (*R*)-BINAP

1 equiv Bu$_4$NBr
4 equiv

4 h, 80°C, THF, 86%

(10.136) 84% ee

phosphino-oxazolines as ligands in the Heck reaction. While these ligands have also shown good selectivity in the intermolecular Heck reactions of (10.137) and (10.144), the ees obtained rarely match those achieved with the parent phosphino-oxazoline (10.141).[87] Readily accessible carbohydrate-derived phosphite-oxazolines such as (10.147), developed by Diéguez and coworkers, do show selectivities comparable to (10.141) in the intermolecular Heck coupling and are more reactive, providing up to 99% ee in the arylation of (10.137) within

10 minutes under microwave irradiation.[88] The tunability of these ligands is another advantage, allowing the convenient preparation of ligand libraries.

In the presence of nucleophiles, Shibasaki and coworkers have extended their methodology to a Heck reaction/carbanion capture sequence, which gives good enantiomeric excess. For example, using nucleophile (**10.148**) provides the highest enantioselectivity in the cyclisation/nucleophilic capture of triflate (**10.149**).[89]

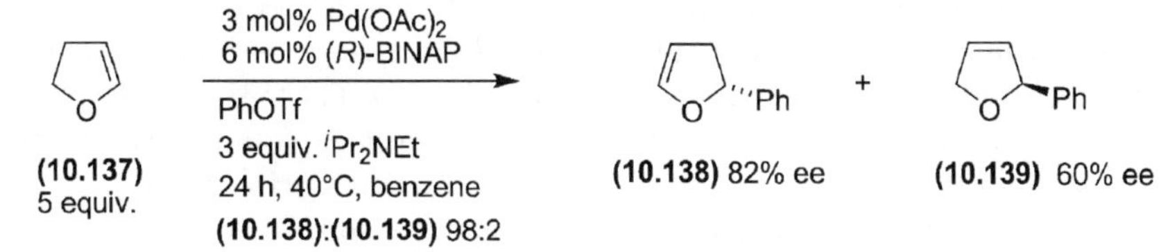

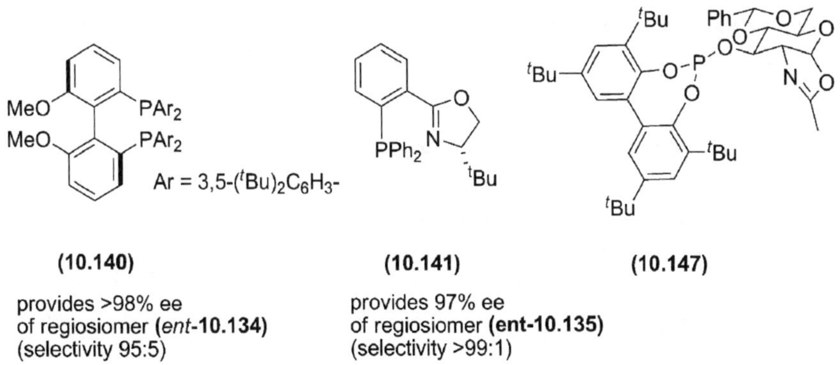

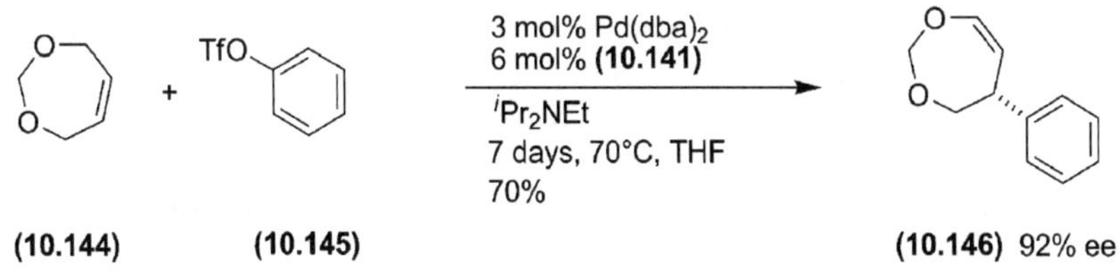

(10.137)
5 equiv.

3 mol% Pd(OAc)$_2$
6 mol% (*R*)-BINAP

PhOTf
3 equiv. ^{i}Pr$_2$NEt
24 h, 40°C, benzene
(10.138):(10.139) 98:2

(10.138) 82% ee   +   **(10.139)** 60% ee

MeO
MeO
PAr$_2$
PAr$_2$
Ar = 3,5-(^{t}Bu)$_2$C$_6$H$_3$-

(10.140)

provides >98% ee
of regiosiomer (*ent*-**10.134**)
(selectivity 95:5)

PPh$_2$
N
^{t}Bu

(10.141)

provides 97% ee
of regiosiomer (**ent-10.135**)
(selectivity >99:1)

(10.147)

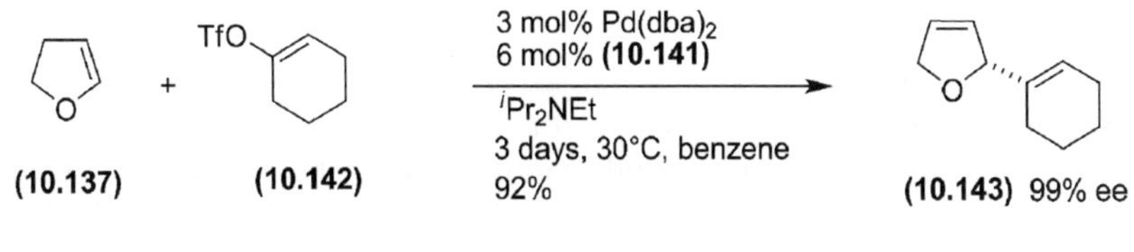

(10.137)   +   TfO   **(10.142)**

3 mol% Pd(dba)$_2$
6 mol% (**10.141**)

^{i}Pr$_2$NEt
3 days, 30°C, benzene
92%

(10.143) 99% ee

(10.144)   TfO   **(10.145)**

3 mol% Pd(dba)$_2$
6 mol% (**10.141**)

^{i}Pr$_2$NEt
7 days, 70°C, THF
70%

(10.146) 92% ee

This transformation can also be achieved using malonates and β-keto ester nucleophiles.[90] Tietze and Raschke have used a 'silane-terminated' Heck reaction, in which the substrate (10.151) loses the silyl group to give the cyclised product (10.152).[91] Incorporation of a nitrogen atom at the requisite position in the side chain allows the enantioselective synthesis of tetrahydroisoquinolines and benzazepines using this methodology.[92]

An asymmetric polyene cyclisation (asymmetric tandem Heck) has been reported in the synthesis of the pentacyclic polyketide (+)-xestoquinone.[93]

The use of a hydride source affords a hydroarylation of alkenes, which takes place when the initially formed alkene adduct is unable to undergo *cis*-β-hydride elimination.[94] For example, the Heck reaction of norbornene (10.153) with phenyltriflate fails because of the geometry of the first formed palladium alkyl intermediate. However, in the presence of a hydride source (Et_3N/HCO_2H), reductive elimination to give the product (10.154) takes place enantioselectively using ligand (10.155). In a more complex example, asymmetric reductive Heck cyclisation of aryl nonaflate (10.156) in the presence of ligand (10.158) and proton sponge as hydride source gives the indanone (10.157) with high ee.[95]

10.4 Alkylmetalation of Alkenes

The asymmetric alkylmagnesiation of alkenes has been achieved with enantiomerically pure zirconocene catalysts.[96–98] The reaction with allyl ethers is a useful procedure, typified by the reaction of the cyclic allylic ether (10.159), catalysed by the Brintzinger complex (10.160). The reaction is believed to proceed via the alkylmagnesium intermediate (10.161), which undergoes elimination to afford the product (10.162) in high enantioselectivity. The reaction has been employed for the ethylmagnesiation of larger ring systems,[99] as well as in an efficient kinetic resolution, as demonstrated by the reaction of racemic dihydropyran (10.163).[100] In some cases the starting material could be prepared *in situ* by ruthenium-catalysed ring-closing metathesis.

The enantioselective zirconium-catalysed alkylalumination of alkenes has been reported by Negishi.[1d, 101] The intermediate organoaluminium species can be quenched in a number of ways, including dioxygen quench, to provide an alcohol. For example, hex-l-ene (10.165) undergoes ethylalumination/oxidation with good enantioselectivity in the product alcohol (10.166), using the catalyst (10.167).[102] The intermediate organoalane derived from allylic alcohol (10.168) can also be quenched with iodine followed by *in situ* protection to give iodide (10.169). Further elaboration by Negishi cross coupling with a variety of organohalides occurs with no stereoisomerisation.[103] Furthermore, the organoalane derived by methylalumination of styrene (10.171) readily undergoes transmetallation to the zinc

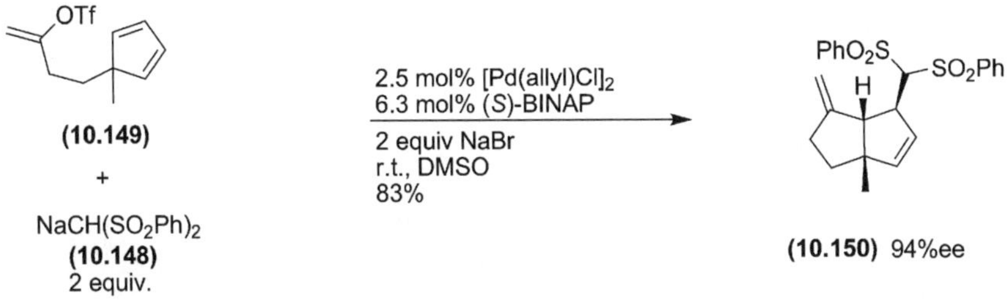
OTf
(10.149)
+
NaCH(SO2Ph)2
(10.148)
2 equiv.
2.5 mol% [Pd(allyl)Cl]2
6.3 mol% (S)-BINAP
2 equiv NaBr
r.t., DMSO
83%
PhO2S
SO2Ph
H
(10.150) 94%ee

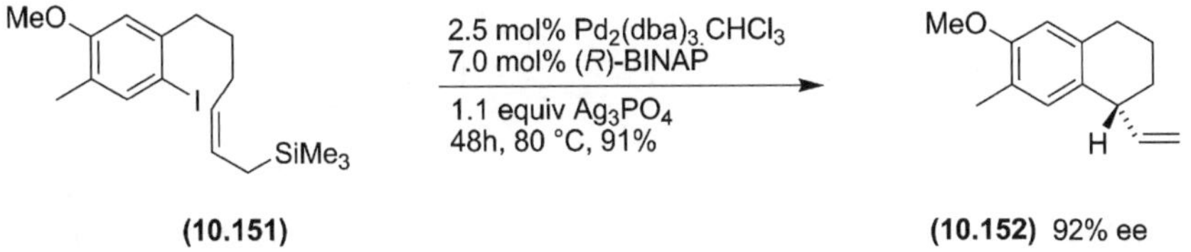
MeO
I
SiMe3
(10.151)
2.5 mol% Pd2(dba)3.CHCl3
7.0 mol% (R)-BINAP
1.1 equiv Ag3PO4
48h, 80 °C, 91%
MeO
H
(10.152) 92% ee

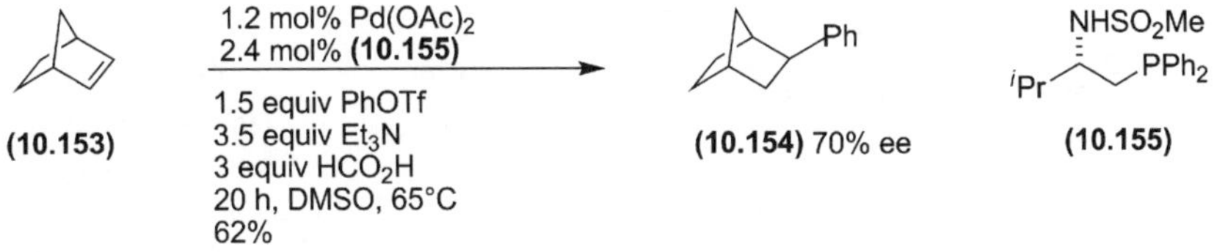
(10.153)
1.2 mol% Pd(OAc)2
2.4 mol% (10.155)
1.5 equiv PhOTf
3.5 equiv Et3N
3 equiv HCO2H
20 h, DMSO, 65°C
62%
Ph
(10.154) 70% ee
NHSO2Me
iPr
PPh2
(10.155)

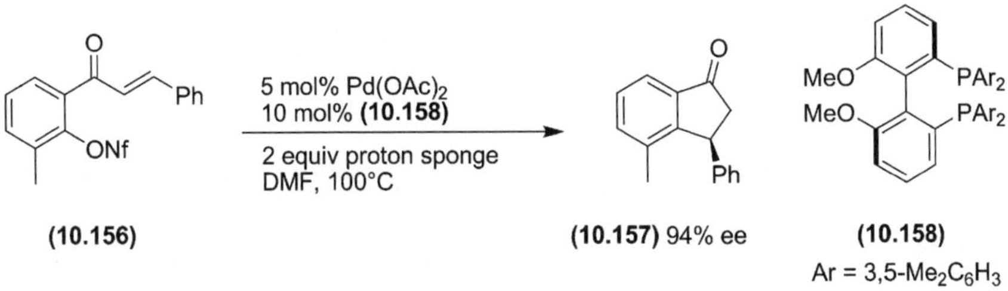
O
Ph
ONf
(10.156)
5 mol% Pd(OAc)2
10 mol% (10.158)
2 equiv proton sponge
DMF, 100°C
O
Ph
(10.157) 94% ee
MeO
MeO
PAr2
PAr2
(10.158)
Ar = 3,5-Me2C6H3

derivative followed by Pd-catalysed vinylation to give products such as (**10.172**) and an iterative carbometallation/vinylation has been applied to the stereoselective synthesis of polypropionates (**10.173**).[104]

Lautens has shown that the nickel-catalysed hydroalumination of certain alkenes can also be achieved with high enantioselectivity.[105] The mechanism of these reactions has been found to be dependent upon a number of factors, and does not necessarily involve an organoalane intermediate.[106]

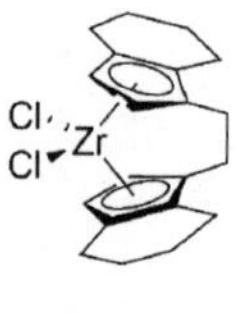

Cl
Cl
Zr

(10.160)

ⁱPr
ZrCl₂
2

(-)-$(NMI)_2ZrCl_2$ **(10.167)**

NMI = 1-neomenthylindenyl

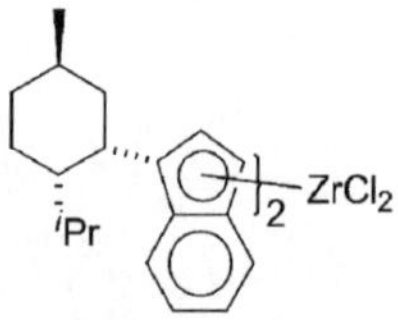

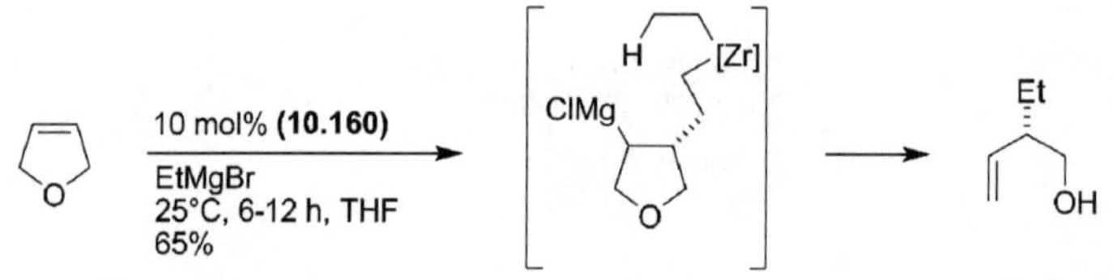

10 mol% (10.160)
EtMgBr
25°C, 6-12 h, THF
65%
H
[Zr]
ClMg
O
Et
OH

(10.159) **(10.161)** **(10.162)** >97% ee

10 mol% (10.160)
EtMgCl
70°C, 6-12 h, THF
60% conversion
O
C₃H₇
+
Et
OH
C₃H₇

(10.163) **(10.163)** >99% ee **(10.164)** 94% ee

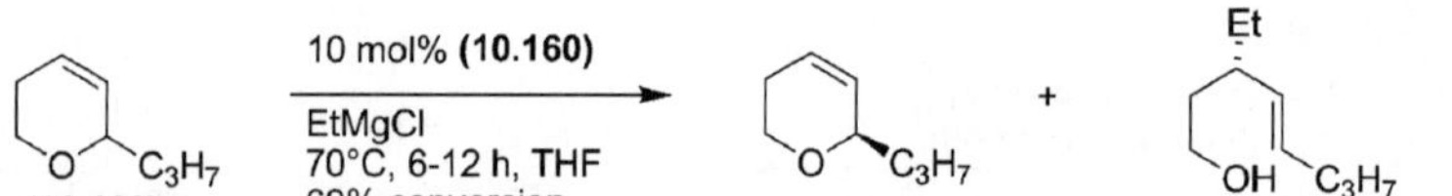

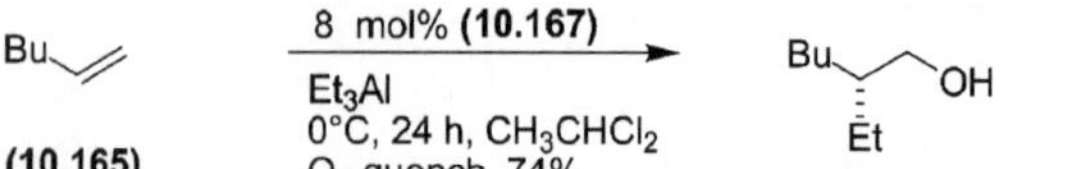

Bu
8 mol% (10.167)
Et₃Al
0°C, 24 h, CH₃CHCl₂
O₂ quench, 74%
Bu
OH
Et

(10.165) **(10.166)** 93% ee

HO
1. 5 mol% (+)-(NMI)₂ZrCl₂
Me₃Al, CH₂Cl₂, MAO,
2. I₂, THF,
3. TBSCl, DMA, 80%
TBSO
I
Zn/RX
TBSO
R

(10.168) **(10.169)** 82%ee **(10.170)**

R = alkenyl, alkynyl, benzyl acyl

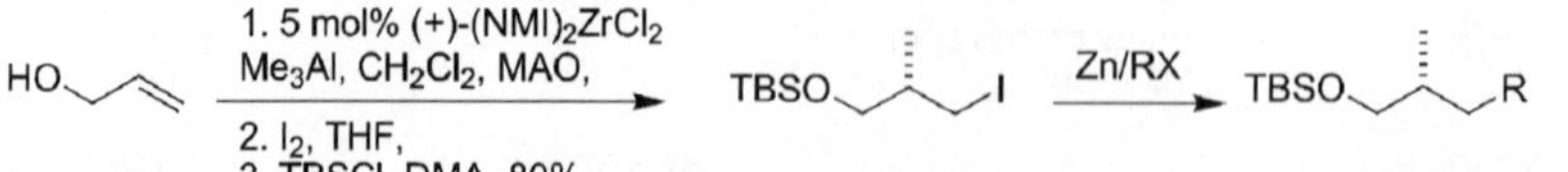

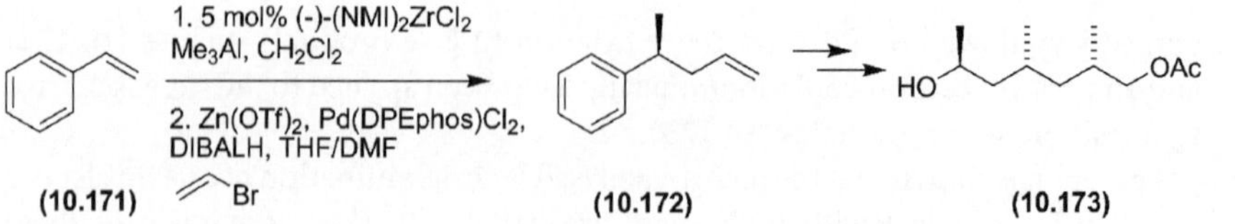

1. 5 mol% (-)-(NMI)₂ZrCl₂
Me₃Al, CH₂Cl₂
2. Zn(OTf)₂, Pd(DPEphos)Cl₂,
DIBALH, THF/DMF
Br
HO
OAc

(10.171) **(10.172)** **(10.173)**

References

1. (a) J. Tsuji, *Palladium Reagents and Catalysts*, John Wiley & Sons, Ltd, Chichester, **1995**. (b) D. J. Cardenas, *Angew. Chem. Int. Ed.*, **1999**, *38*, 3018. (c) K. C. Nicolaou, P. G. Bulger and D. Sarlah, *Angew. Chem. Int. Ed.*, **2005**, *44*, 4442. (d) E. Negishi, *Bull. Chem. Soc. Jpn.*, **2007**, *80*, 233.

2. (a) T. Hayashi in *Catalytic Asymmetric Synthesis*, ed. I. Ojima, VCH, New York, **1993**, 325. (b) T. Hayashi in *Comprehensive Asymmetric Catalysis*, Vol. *2*, ed. E. N. Jacobsen, A. Pfaltz and H. Yamamoto, Springer-Verlag, Berlin, **1999**, 887. (c) T. Hayashi, *J. Organomet. Chem.*, **2002**, *653*, 41. (d) L. F. Tietze, H. Ila and H. P. Bell, *Chem. Rev.*, **2004**, *104*, 3453.

3. T. Hayashi, M. Konishi, M. Fukushima, K. Kanehira, T. Hioki and M. Kumada, *J. Org. Chem.*, **1983**, *48*, 2195.

4. T. Hayashi, A. Yamamoto, M. Hojo and Y. Ito, *J. Chem. Soc., Chem. Commun.*, **1989**, 495.

5. (a) G. C. Lloyd-Jones and C.P. Butts, *Tetrahedron*, **1998**, *54*, 901. (b) L. Schwink and P. Knochel, *Chem. Eur. J.*, **1998**, *4*, 950. (B) H. Horibe, K. Kazuta, M. Kotoku, K. Kondo, H. Okuno, Y. Murakami and T. Aoyama, *Synlett*, **2003**, 2047. (c) H. Horibe, Y. Fukuda, K. Kondo, H. Okuno, Y. Murakami and T. Aoyama, *Tetrahedron*, **2004**, *60*, 10701.

6. T. Hayashi, M. Konishi, Y. Okamoto, K. Kabeta and M. Kumada, *J. Org. Chem.*, **1986**, *51*, 3772.

7. C. Fischer and G. C. Fu, *J. Am. Chem. Soc.*, **2005**, *127*, 4594.

8. F. O. Arp and G. C. Fu, *J. Am. Chem. Soc.*, **2005**, *127*, 10482.

9. T. Hayashi, K. Hayashizaki, T. Kiyoi and Y. Ito, *J. Am. Chem. Soc.*, **1988**, *110*, 8153.

10. (a) K. C. Nicolaou, H. Li, C. N. C. Boddy, J. M. Ramanjulu, T.-Y. Yue, S. Natarajan, X.-J. Chu, S. Bräse and F. Rübsam, *Chem. Eur. J.*, **1999**, *5*, 2584. (b) A. N. Cammidge and K. V. L. Crépy, *Chem. Commun.*, **2000**, 1723. (c) J. Yin and S. L. Buchwald, *J. Am. Chem. Soc.*, **2000**, *122*, 12051.

11. M. Genov, B. Fuentes, P. Espinet and B. Pelaz, *Tetrahedron: Asymmetry*, **2006**, *17*, 2593.

12. (a) T. Shimada, Y.-H. Cho and T. Hayashi, *J. Am. Chem. Soc.*, **2002**, *124*, 13396. (b) Y.-H. Cho, A. Kina, T. Shimada and T. Hayashi, *J. Org. Chem.* **2004**, *69*, 3811.

13. T. Hayashi, S. Niizuma, T. Kamikawa, N. Suzuki and Y. Uozumi, *J. Am. Chem. Soc.*, **1995**, *117*, 9101.

14. T. Kamikawa, Y. Uozumi and T. Hayashi, *Tetrahedron Lett.*, **1996**, *37*, 3161.

15. M. Uemura and H. Nishimura, *J. Organomet. Chem.*, **1994**, *473*, 129.

16. M. C. Willis, L. H. Powell, C. K. Claverie and S. J. Watson, *Angew. Chem. Int. Ed.*, **2004**, *43*, 1249.

17. (a) J. Åhman, J. P. Wolfe, M.V. Troutman, M. Palucki and S. L. Buchwald, *J. Am. Chem. Soc.*, **1998**, *120*, 1918. (b) T. Hamada, A. Chieffi, J. Åhamn and S. L. Buchwald, *J. Am. Chem. Soc.*, **2001**, *124*, 1261.

18. (a) B.M. Trost, *Chem. Rev.*, **1996**, *96*, 395. (b) A. Pfaltz and M. Lautens, in *Comprehensive Asymmetric Catalysis*, Vol. *2*, ed. E. N. Jacobsen, A. Pfaltz and H. Yamamoto, Springer-Verlag, Berlin, **1999**, 834. (c) T. Graening and H.-G. Schmalz, *Angew. Chem. Int. Ed.*, **2003**, *42*, 2580. B. M. Trost and M. L. Crawley, *Chem. Rev.*, **2003**, *103*, 2921.

19. C. G. Frost, J. Howarth and J. M. J. Williams, *Tetrahedron: Asymmetry*, **1992**, *3*, 1089.
20. T. Hayashi, A. Yamamoto, T. Hagihara and Y. Ito, *Tetrahedron Lett.*, **1986**, *27*, 191.
21. T. Hayashi, *Pure Appl. Chem.*, **1998**, *60*, 7.
22. A. Togni, *Tetrahedron: Asymmetry*, **1991**, *2*, 683.
23. A. Pfaltz, *Acc. Chem. Res.*, **1993**, *26*, 339.
24. D. Tanner, P. G. Andersson, A. Harden and P. Somfai, *Tetrahedron Lett.*, **1994**, *35*, 4631.
25. (a) C. Bolm, D. Kaufman, S. Gessler and K. Harms, *J. Organomet. Chem.*, **1995**, *502*, 47. (b) G. Zhu, M. Terry and X. Zhang, *Tetrahedron Lett.*, **1996**, *37*, 4475. (c) T. Imamoto, M. Nishimura, A. Koide and K. Yoshida, *J. Org. Chem.*, **2007**, *72*, 7413.
26. (a) O. Pàmies, G. P. F. van Strijdonck, M. Diéguez, S. Deerenberg, G. Net, A. Ruiz, C. Claver, P. C. J. Kamer and P. W. N. M. van Leeuwen, *J. Org. Chem.*, **2001**, *66*, 8867. (b) M. Diéguez, O. Pàmies and C. Claver, *J. Org. Chem.*, **2005**, *70*, 3363.
27. (a) J. Sprinz and G. Helmchen, *Tetrahedron Lett.*, **1993**, *34*, 1769. (b) G. J. Dawson, C. G. Frost, J. M. J. Williams and S. J. Coote, *Tetrahedron Lett.*, **1993**, *34*, 3149. (c) P. von Matt and A. Pfaltz, *Angew. Chem., Int. Ed. Engl.*, **1993**, *32*, 566. (d) G. Helmchen and A. Pfaltz, *Acc. Chem. Res.*, **2000**, *33*, 336. (e) O. Pàmies, M. Diéguez and C. Claver, *J. Am. Chem. Soc.*, **2005**, *127*, 3636.
28. J. M. Brown, D. I. Hulmes and P. J. Guiry, *Tetrahedron*, **1994**, *50*, 4493.
29. H. C. L. Abbenhuis, U. Burckhardt, V. Gramlich, C. Kollner, P. S. Pregosin, R. Salzman and A. Togni *Organometallics*, **1995**, *14*, 759.
30. D. A. Evans, K. R. Campos, J. S. Tedrow, F. E. Michael and M. R. Gagné, *J. Am. Chem. Soc.*, **2000**, *122*, 7905.
31. (a) J. V. Allen, S. J. Coote, G.J. Dawson, C. G. Frost, C. J. Martin and J. M. J. Williams, *J. Chem. Soc, Perkin Trans. 1*, **1994**, 2065. (c) H. Kubota and K. Koga, *Tetrahedron Lett.*, **1994**, *35*, 6689. (d) W.-P. Deng, S.-L. You, X.-L. Hou, L.-X. Dai, Y.-H. Yu, W. Xia and J. Sun, *J. Am. Chem. Soc.*, **2001**, *123*, 6508. (e) S.-L. You, X.-Z. Zhu, Y.-M. Luo, X.-L. Hou and L.-X. Dai, *J. Am. Chem. Soc.*, **2001**, *123*, 7471.
32. B. M. Trost and R. C. Bunt, *J. Am. Chem. Soc.*, **1994**, *116*, 4089.
33. G. Knühl, P. Sennhenn and G. Helmchen, *J. Chem. Soc., Chem. Commun.*, **1995**, 1845.
34. E. Raluy, C. Claver, O. Pàmies and M. Diéguez, *Org. Lett.*, **2007**, *9*, 49.
35. P. R. Auburn, P. B. Mackenzie and B. Bosnich, *J. Am. Chem. Soc.*, **1985**, *107*, 2033.
36. G. J. Dawson, J. M. J. Williams and S. J. Coote, *Tetrahedron Lett.*, **1995**, *36*, 461.
37. J. M. J. Williams, *Synlett*, **1996**, 705.
38. R. Prétot and A. Pfaltz, *Angew. Chem. Int. Ed.*, **1998**, *37*, 323.
39. B. M. Trost, D.L. van Vranken and C. Bingel, *J. Am. Chem. Soc.*, **1992**, *114*, 9327.
40. B. M. Trost, C. B. Lee and J. M. Weiss, *J. Am. Chem. Soc.*, **1995**, *117*, 7247.
41. B. M. Trost and X. Ariza, *Angew. Chem., Int. Ed. Engl*, **1997**, *36*, 2635.
42. (a) B. M. Trost, R. Radinov and E. M. Grenzer, *J. Am. Chem. Soc.*, **1997**, *119*, 7879. (b) B. M. Trost and X. Ariza, *J. Am. Chem. Soc.*, **1999**, *121*, 10727.
43. (a) B. M. Trost and G. M. Schroeder, *J. Am. Chem. Soc.*, **1999**, *121*, 6759. (b) B. M. Trost and G. M. Schroeder, *Chem. Eur. J.*, **2005**, *11*, 174.
44. S.-L. You and L.-X. Dai, *Angew. Chem., Int. Ed.*, **2006**, *45*, 5246.
45. D. C. Behenna and B. N. Stolz, *J. Am. Chem. Soc.*, **2004**, *126*, 15044.
46. B. M. Trost and J. Xu, *J. Am. Chem. Soc.*, **2005**, *127*, 17180.

47. N. Kardos and J.-P. Genêt, *Tetrahedron: Asymmetry*, **1994**, *5*, 1525.
48. G. Koch and A. Pfaltz, *Tetrahedron: Asymmetry*, **1994**, *5*, 1525.
49. T. Hayashi, A. Okada, T. Suzuka and M. Kawatsura, *Org. Lett.*, **2003**, *5*, 1713.
50. B. Bartels, C. García-Yebra, F. Rominger and G. Helmchen, *Eur. J. Inorg. Chem.*, **2002**, 2569.
51. G. Helmchen, A. Dahnz, M. Schelwies and R. Weihofen, *J. Chem. Soc., Chem. Commun.*, **2007**, 675.
52. G. Lipowsky, N. Miller and G. Helmchen, *Angew. Chem. Int. Ed.*, **2004**, 4595.
53. A. Dahnz and G. Helmchen, *Synlett*, **2006**, 697.
54. T. Graening and J. F. Hartwig, *J. Am. Chem. Soc.*, **2005**, *127*, 17192.
55. (a) A. Sofia, E. Karlström and J.-E. Bäckvall, in *Modern Organocopper Chemistry*, ed. N. Krause, Wiley-VCH, Weinheim, **2002**, 259. (b) H. Yoremitsu and K. Oshima, *Angew. Chem. Int. Ed.*, **2005**, *44*, 4435.
56. (a) B. L. Feringa, *Acc. Chem. Res.*, **2000**, *33*, 346. (b) H. Malda, A. W. van Zijl, L. A. Arnold and B. L. Feringa, *Org. Lett.*, **2001**, *3*, 1169. (c) K. Tissot-Crosset, D. Polet and A. Alexakis, *Angew. Chem. Int. Ed.*, **2004**, *43*, 2426.
57. (a) F. Dübner and P. Knochel, *Angew. Chem. Int. Ed.*, **1999**, *38*, 379. (b) F. Dübner and P. Knochel, *Tetrahedron Lett.*, **2000**, *41*, 9233. (c) F. López, A. W. van Zijl, A. J. Minnaard and B. L. Feringa, *J. Chem. Soc., Chem. Commun.*, **2006**, 409.
58. O. Belda and C. Moberg, *Acc. Chem. Res.*, **2004**, *37*, 159.
59. B.M. Trost and I. Hachiya, *J. Am. Chem. Soc.*, **1998**, *120*, 1104.
60. F. Glorius and A. Pfaltz, *Org. Lett.*, **1999**, *1*, 141.
61. G.C. Lloyd Jones and A. Pfaltz, *Angew. Chem., Int. Ed. Engl.*, **1995**, *34*, 462.
62. G. Consiglio and A. Indolese, *Organometallics*, **1991**, *10*, 3425.
63. (a) P. von Matt, O. Loiseleur, G. Koch and A. Pfaltz, *Tetrahedron: Asymmetry*, **1994**, *5*, 573. (b) R. Jumnah, A.C. Williams and J.M.J. Williams, *Synlett*, **1995**, 821. (c) A. Togni, U. Burckhardt, V. Gramlich, P.S. Pregosin and R. Salzmann, *J. Am. Chem. Soc.*, **1996**, *118*, 1031. (d) A. Sudo and K. Saigo, *J. Org. Chem.*, **1997**, *62*, 5508.
64. T. Ohmura and J. F. Hartwig, *J. Am. Chem. Soc.*, **2002**, *124*, 15164.
65. B.M. Trost and D.E. Patterson, *J. Org. Chem.*, **1998**, *63*, 1339.
66. N. Buschmann, A. Rückert and S. Blechert, *J. Org. Chem.*, **2002**, *67*, 4325.
67. C. Welter, A. Dahnz, B. Brunner, S. Sreiff, P. Dübon and G. Helmchen, *Org. Lett.*, **2005**, *7*, 1239.
68. Y. Uozumi, K. Kato and T. Hayashi, *J. Am. Chem. Soc.*, **1997**, *119*, 5063.
69. B.M. Trost and M. G. Organ, *J. Am. Chem. Soc.*, **1994**, *116*, 10320.
70. B.M. Trost and F.D. Toste, *J. Am. Chem. Soc.*, **1998**, *120*, 815.
71. (a) F. Lopez, T. Ohmura and J. F. Hartwig, *J. Am. Chem. Soc.*, **2003**, *125*, 3426. (b) A. Leitner, C. Shu, and J. F. Hartwig, *Org. Lett.*, **2005**, *7*, 1093.
72. C. Shu and J. F. Hartwig, *Angew. Chem. Int. Ed.*, **2004**, *43*, 4794.
73. H. Miyabe, K. Yoshida, M. Yamauchi and Y. Takemoto, *J. Org. Chem*, **2005**, *70*, 2148.
74. H. Eichelmann and H.-J. Gais, *Tetrahedron Asymmetry*, **1995**, *6*, 643.
75. T. Hayashi, H. Iwamura, Y. Uozumi, Y. Matsumoto and F. Ozawa, *Synthesis*, **1994**, 526.
76. For reviews on the catalytic asymmetric Heck reaction see: (a) M. Shibasaki, C. D. J. Boden and A. Kojima, *Tetrahedron*, **1997**, *53*, 7371. (b) M. Shibasaki and E. M. Vogl, in *Comprehensive Asymmetric Catalysis*, Vol. *1*, ed. E. N. Jacobsen, A. Pfaltz and H.

Yamamoto, Springer-Verlag, Berlin, **1999**, 458. (c) A. B. Dounay and L. E. Overman, *Chem. Rev.*, **2003**, *103*, 2945. (d) P. J. Guiry and D. Kiely, *Curr. Org. Chem.*, **2004**, *8*, 781.

77. (a) Y. Sato, M. Sodeoka and M. Shibasaki, *J. Org. Chem.*, **1989**, *54*, 4738. (b) K. Ohrai, K. Kondo, M. Sodeoka and M. Shibasaki, *J. Am. Chem. Soc.*, **1994**, *116*, 11737.

78. (a) R. Imbos, A. J. Minaard and B. L. Feringa, *J. Am. Chem. Soc.*, **2002**, *124*, 184. (b) M. Lautens and V. Zunic, *Can. J. Chem.*, **2004**, *82*, 399. (c) M. E. P. Lormann, M. Nieger and S. J. Bräse, *J. Organomet. Chem.*, **2006**, *691*, 2159.

79. (a) M. Oestreich, F. Sempere-Culler and A. B. Macotta, *Angew. Chem. Int. Ed.*, **2005**, *44*, 149. (b) A. B. Machotta, B. F. Straub and M. Oestreich, *J. Am. Chem. Soc.*, **2007**, *129*, 13455.

80. A. Ashimori and L. E. Overman, *J. Org. Chem.*, **1992**, *57*, 4571.

81. L. E. Overman and D. J. Poon, *Angew. Chem., Int. Ed. Engl.*, **1997**, *36*, 518.

82. A. B. Dounay, K. Hatanaka, J. J. Kodanko, M. Oestriech, L. E. Overman, L. A. Pfeifer and M. M. Weiss, *J. Am. Chem. Soc.*, **2003**, *125*, 6261.

83. F. Ozawa, A. Kubo, Y. Matsumoto and T. Hayashi, *Organometallics*, **1993**, *12*, 4188.

84. G. Trabesinger, A. Albinati, N. Feiken, R. W. Kunz, P. S. Pregosin and M. Tschoerner, *J. Am. Chem. Soc.*, **1997**, *119*, 6315.

85. (a) O. Loiseleur, M. Hayashi, N. Schmees and A. Pfaltz, *Synthesis*, **1997**, 1338. (b) O. Loiseleur, P. Meier and A. Pfaltz, *Angew. Chem. Int. Ed. Engl.*, **1996**, *35*, 200.

86. L. Ripa and A. Hallberg, *J. Org. Chem.*, **1997**, *62*, 595.

87. (a) Y. Hashimoto, Y. Horie, M. Hayashi and K. Saigo, *Tetrahedron: Asymmetry*, **2000**, *11*, 2205. (b) S. R. Gilbertson, D. G. Genov and A. L. Rheingold, *Org. Lett.*, **2000**, *2*, 2885. (c) S. R. Gilbertson and Z. Fu, *Org. Lett.*, **2001**, *3*, 161. (d) S. R. Gilbertson, D. Xie and Z. Fu, *J. Org. Chem.*, **2001**, *66*, 7240. (e) X.-L. Hou, D. X. Dong and K. Yuan, *Tetrahedron: Asymmetry*, **2004**, *15*, 2189. (f) R. A. Gossage, H. A. Jenkins and N. P. Yadav, *Tetrahedron Lett.*, **2004**, *45*, 7689. (g) T. G. Kilroy, P. G. Cozzi, N. End and P. J. McGuiry, *Synthesis*, **2004**, 1879. (h) D. Liu, Q. Dai and X. Zhang, *Tetrahedron*, **2005**, *61*, 6460.

88. (a) Y. Mata, M. Diéguez, O. Pámies and C. Claver, *Org. Lett.*, **2005**, *7*, 5597. (b) Y. Mata, O. Pámies and M. Diéguez, *Chem. Eur. J.*, **2007**, *13*, 3296.

89. T. Ohshima, K. Kagechika, M. Adachi, M. Sodeoka and M. Shibasaki, *J. Am. Chem. Soc.*, **1996**, *118*, 7108.

90. W. Itano, T. Ohshima and M. Shibasaki, *Synlett*, **2006**, 3053.

91. L.F. Tietze and T. Raschke, *Synlett*, **1995**, 597.

92. L. F. Tietze, K. Thede, R. Schimpf and F. Sannicoló, *J. Chem. Soc., Chem. Commun.*, **2000**, 583.

93. S.P. Maddaford, N.G. Andersen, W.A. Cristofoli and B.A. Keay, *J. Am. Chem. Soc.*, **1996**, *118*, 10766.

94. S. Sakuraba, K. Awano and K. Achiwa, *Synlett*, **1994**, 291.

95. A. Minatti, X. Zheng and S. L. Buchwald, *J. Org. Chem.*, **2007**, *72*, 9253.

96. . A. H. Hoveyda and N. M. Heron, in *Comprehensive Asymmetric Catalysis*, Vol. *1*, ed. E. N. Jacobsen, A. Pfaltz and H. Yamamoto, Springer-Verlag, Berlin, **1999**, 431.

97. L. Bell, R.J. Whitby, R.V.H. Jones and M.C.H. Standen, *Tetrahedron Lett.*, **1998**, *39*, 7139.

98. J.P. Morken, M.T. Didiuk and A.H. Hoveyda, *J. Am. Chem. Soc.*, **1993**, *115*, 6997.
99. M.S. Visser, N.M. Heron, M.T. Didiuk, J.F. Sagal and A.H. Hoveyda, *J. Am. Chem. Soc.*, **1996**, *118*, 4291.
100. M.S. Visser and A.H. Hoveyda, *Tetrahedron*, **1995**, *51*, 4383.
101. (a) D.Y. Kondakov and E. Negishi, *J. Am. Chem. Soc.*, **1995**, *117*, 10771. (b) E. Negishi and S. Huo, *Pure Appl. Chem.*, **2002**, *74*, 151.
102. D.Y. Kondakov and E.-I. Negishi, *J. Am. Chem. Soc.*, **1996**, *118*, 1577.
103. B. Liang, T. Novak, Z. Tan and E. Negishi, *J. Am. Chem. Soc.*, **2006**, *128*, 2770.
104. (a) T. Novak, Z. Tan, B. Liang and E. Negishi, *J. Am. Chem. Soc.*, **2005**, *127*, 2838. (b) Z. Tan and E. Negishi, *Angew. Chem. Int. Ed.*, **2004**, *43*, 2911.
105. M. Lautens, P. Chiu, S. Ma and T. Rovis, *J. Am. Chem. Soc.*, **1995**, *117*, 532.
106. M. Lautens and T. Rovis, *J. Am. Chem. Soc.*, **1997**, *119*, 11090.

Chapter 11
Conjugate Addition Reactions

The 1,4-addition of nucleophiles to alkenes attached to electron-withdrawing groups (ketones, aldehydes, esters, nitriles, etc.) is often referred to as conjugate addition. Nucleophiles which undergo conjugate addition rather than a direct nucleophilic attack on the electron-withdrawing group include enolates, thiolates and cuprates (or copper-catalysed addition of other organometallic reagents). A variety of transition metal complexes, which function as enantiomerically pure Lewis acids to activate the acceptor or via the formation of enolates from the donor, have been used in the asymmetric Michael reaction. In general, good levels of selectivity have only been achieved using stabilised enolates such as malonates and β–ketoesters. Bifunctional metal catalysts that activate both the donor and acceptor have also been applied successfully to the asymmetric Michael addition of stabilised enolates and also some nonstabilised ketone enolates. The major recent advance in this area is the application of organocatalysts. Enantiopure amine catalysts and hydrogen bonding catalysts have been used with much success in the addition of stabilised and nonstabilised enolates to α,β-enals and enones.

The asymmetric conjugate addition of nonstabilised organometallic species has also been achieved with high ee. A range of simple alkyl Grignard reagents and diethyl zinc add to cyclic and acyclic enones and enoates in the presence of copper catalysts and enantiomerically pure ligands, while the asymmetric conjugate addition of sp²-based organometallics can be achieved using rhodium catalysts. The catalyst systems used in the addition of carbon-based nucleophiles to Michael acceptors have also proved successful in the addition of heteroatom-based nucleophiles. Thus, high ees in the sulfur Michael and aza-Michael reaction have been obtained using enantiopure Lewis acids, bifunctional metal catalysts and organocatalysts.

11.1 Conjugate Addition of Enolates

The Michael addition reaction involves the conjugate addition of enolates to α,β-unsaturated carbonyl compounds. The reaction can be catalysed either by activation of the nucleophile or activation of the Michael acceptor. In some cases, the activation of both partners probably occurs, and this transformation can be catalysed with high ee using a variety of enantiomerically pure metal complexes, amine salts and crown ethers.[1]

Rhodium complexes of the *trans*-chelating phosphine ligand, TRAP (**11.01**) probably coordinate to the nitrile to facilitate enolisation of the nucleophile (**11.02**), and control the subsequent stereochemistry with good enantiocontrol in additions to Michael acceptors such as enal (**11.03**) and also α,β-enones.[2] Higher ees, in the addition with electron-deficent nitriles such as (**11.04**) onto α,β-unsaturated imides, has been achieved using the dimeric μ–oxo (salen) complex (**11.05**).[3] α-Cyano Weinreb amides also work well as nucleophiles in this reaction, and the products are readily converted into the corresponding aldehydes and ketones.[4] For example, the Weinreb amide (**11.07**) undergoes Michael addition with the unsaturated ketone (**11.08**) to give the product (**11.11**) in excellent yield and with good enantioselectivity.

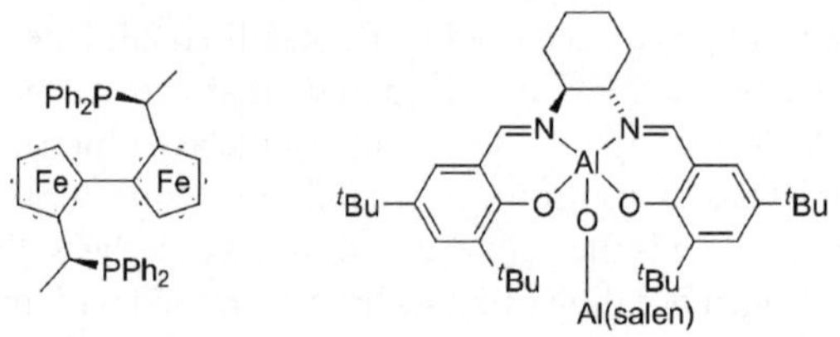
Ph2P
Fe
Fe
PPh2
N
N
Al
O
O
O
tBu
tBu
tBu
tBu
Al(salen)

(11.01) **(11.05)**

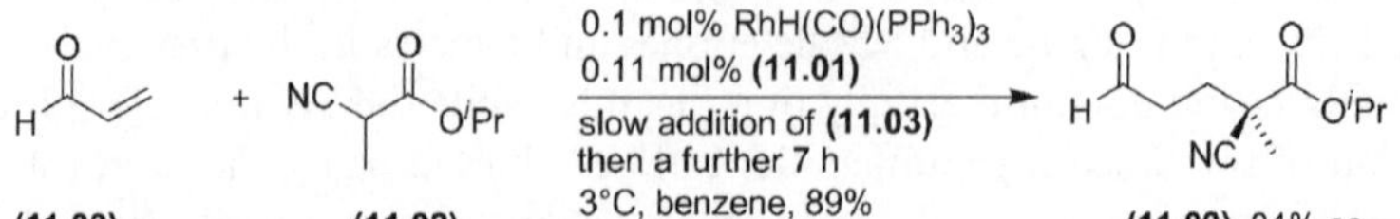
O
H
+ NC
O
OiPr
0.1 mol% RhH(CO)(PPh3)3
0.11 mol% (11.01)
slow addition of (11.03)
then a further 7 h
3°C, benzene, 89%
O
H
NC
O
OiPr

(11.03) **(11.02)** **(11.09)** 84% ee

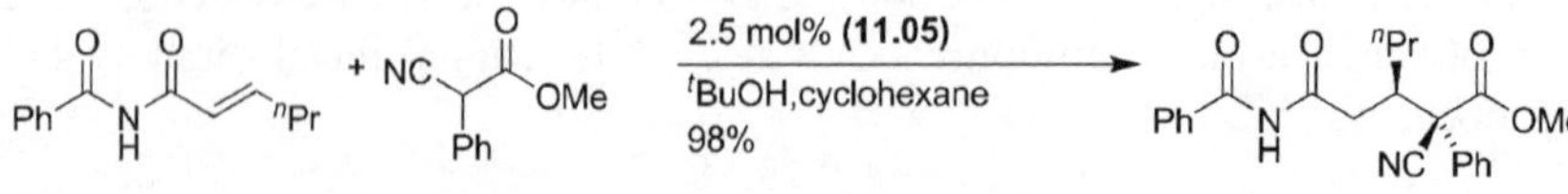
O
O
Ph
N
H
nPr
+ NC
O
OMe
Ph
2.5 mol% (11.05)
tBuOH,cyclohexane
98%
O
O
nPr
O
Ph
N
H
NC
Ph
OMe

(11.06) **(11.04)** **(11.10)** 97% ee

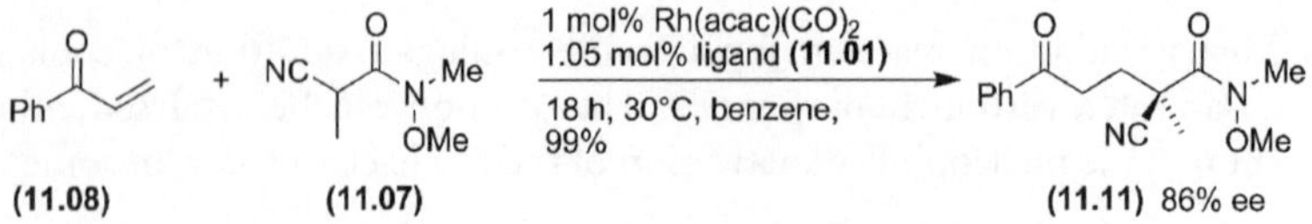
O
Ph
+ NC
O
N
Me
OMe
1 mol% Rh(acac)(CO)2
1.05 mol% ligand (11.01)
18 h, 30°C, benzene,
99%
O
Ph
NC
O
N
Me
OMe

(11.08) **(11.07)** **(11.11)** 86% ee

Enantioselective Michael addition reactions are certainly not restricted to nitrile-containing nucleophiles, and the conjugate addition of 1,3-dicarbonyls to cyclic and acyclic Michael acceptors occurs with high ee using a range of transition metal-based catalysts. The nickel-catalysed conjugate additions of malonates and β-keto esters have been achieved onto several Michael acceptors.[5] For instance the Ni(II)-TolBINAP complex (**11.12**) acts as a Lewis acid to effect high enantioselectivity in the addition of malonates and β-ketoesters such as (**11.13**) to α,β-thiazolidinethiones including (**11.14**). The initial products cyclise in the presence of DBU to give dihydropyrones.[5b]

Copper catalysts are also effective in the asymmetric Michael reaction. The addition of cyclic and acyclic stabilised enolates onto enones occurs with good ee in the presence of Lewis acidic copper bis-oxazoline complexes,[6] while the Michael reaction of enamides with alkylidene malonates proceeds with up to 94% ee using the copper diamine complex prepared from ligand (**11.15**), which also functions via binding to the acceptor.[7] The cyclic β-ketoester (**11.16**) undergoes an enantioselective Michael addition catalysed by the copper(II) complex (**11.17**).[8] In this instance the catalyst forms an enolate with the donor. An enolate also forms *in situ* on interaction of β-ketoesters and β-diketones such as (**11.19**) with catalytic quantities of the palladium aqua complex (**11.18**), and high ees are obtained in the addition of these species to enones. Ruthenium diamine complexes such as (**11.20**) developed by Noyori and coworkers, active in the asymmetric hydrogenation of ketones (see Section 3.1), have also been applied, with great effect, to the Michael addition of malonates onto cyclic enones.[9] The active species in this process is postulated to be a ruthenium bound donor formed by deprotonation of malonate by the catalyst.

Shibasaki's heterobimetallic complexes, active in the asymmetric aldol reaction (see Section 7.1) provide the opportunity to activate both the nucleophile and Michael acceptor. Whilst the aluminium lithium bis-BINOL complex (ALB) (**11.28**) does not catalyse conjugate addition of α-phosphonate ester (**11.29**) with cyclopentenone (**11.30**) by itself, addition of sodium *tert*-butoxide allows a highly enantioselective reaction to take place.[10] A postulated model for enantioinduction in this process involves simultaneous binding of the metallated nucleophile and acceptor to the catalyst by interaction with a BINAP oxygen and the aluminium centre respectively. Heterobimetallic complexes have also been used to catalyse the addition of α-nitroesters[11] and malonates[12] to Michael acceptors.

The asymmetric Michael addition of nonstabilised ketone enolates has proved difficult, with most success achieved using 1,3-dicarbonyls as donors. However, Shibasaki and coworkers have achieved high ees in the addition of α-hydroxyketones with both aromatic Michael acceptors such as (**11.32**) and also cyclic enones and alkyl vinyl ketones, using bifunctional zinc catalysts prepared from linked BINOL (**11.33**).[13] These catalysts are also effective in the asymmetric aldol reaction (see Section 7.1) and incorporate two zinc atoms, one of which activates the acceptor carbonyl group and the other forms a zinc enolate with the donor. In addition, catalysts of this type have been used to good effect in the addition of β-ketoesters to cyclic enones.[14]

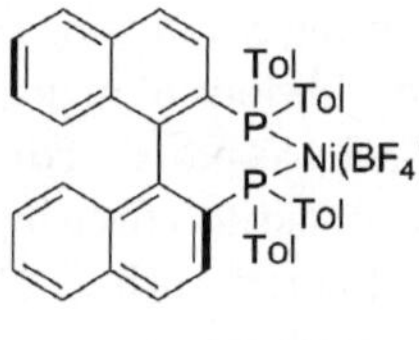

(11.12)

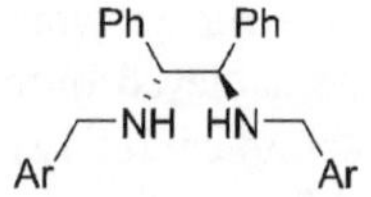

(11.15)

Ar = 9-anthranyl

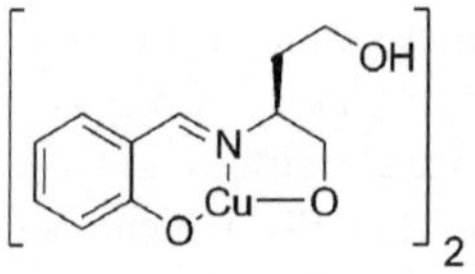

(11.17)

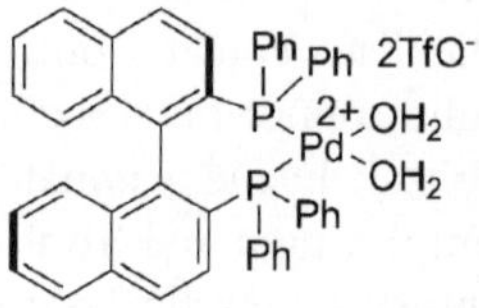

(11.18)

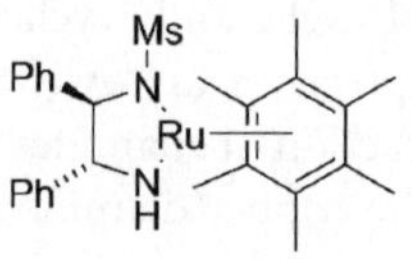

(11.20)

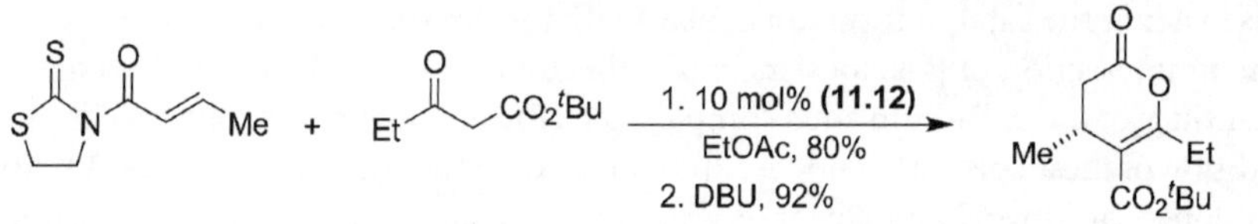

(11.14)   **(11.13)**

1. 10 mol% **(11.12)**
EtOAc, 80%
2. DBU, 92%

(11.24) 95%ee

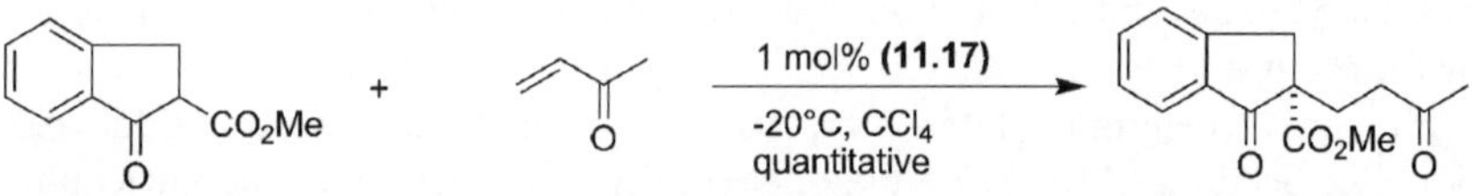

(11.16)   +   **(11.22)**

1 mol% **(11.17)**
-20°C, CCl₄
quantitative

(11.25) 75% ee

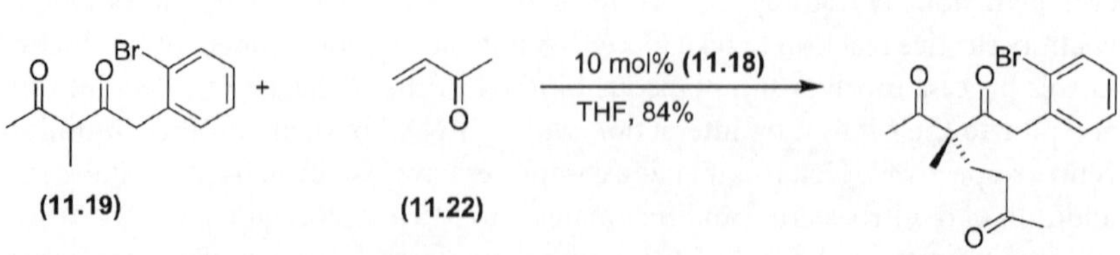

(11.19)   +   **(11.22)**

10 mol% **(11.18)**
THF, 84%

(11.26) 90% ee

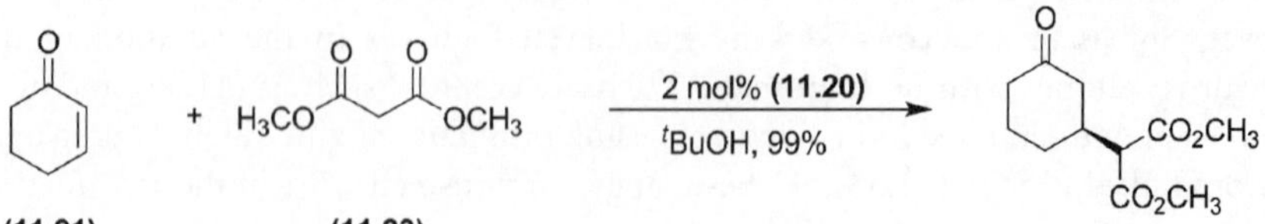

(11.21)   +   **(11.23)**

2 mol% **(11.20)**
^{t}BuOH, 99%

(11.27) 98%ee

The catalytic asymmetric Michael reaction using silyl enol esters (Mukaiyama–Michael reaction) as the pronucleophiles has been reported using a titanium/BINOL catalyst (in up to 90% ee). Considering furan (**11.36**) as a silyl enol ether, this has been shown to undergo nucleophilic addition to the Michael acceptor (**11.37**). The product (**11.38**) can be obtained with excellent diastereocontrol with the scandium complex of ligand (**11.39**), or with excellent enantiocontrol

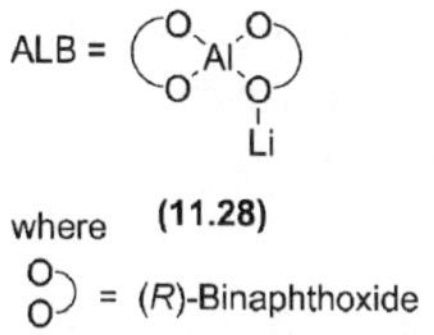

ALB =

where **(11.28)**

$\overset{O}{\underset{O}{\Big)}}$ = (*R*)-Binaphthoxide

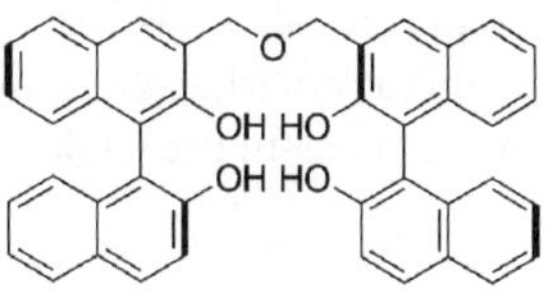

(11.33)

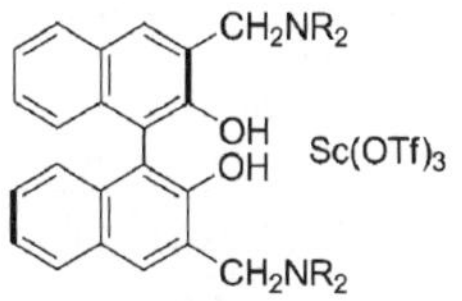

(11.39)

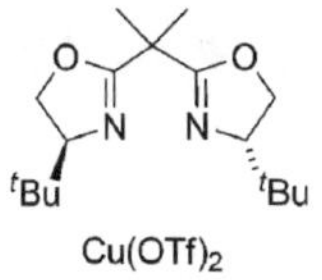

Sc(OTf)$_3$

(11.40)

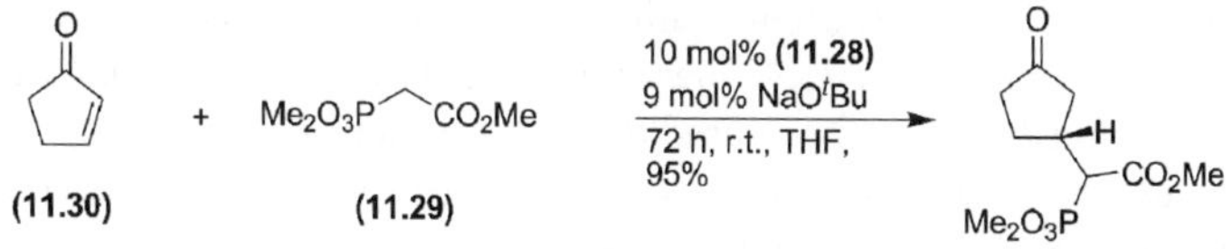

Cu(OTf)$_2$

(11.30) + Me$_2$O$_3$P$\diagdown$CO$_2$Me **(11.29)**

10 mol% **(11.28)**
9 mol% NaO^{t}Bu
72 h, r.t., THF,
95%

Me$_2$O$_3$P

(11.31) 95% ee

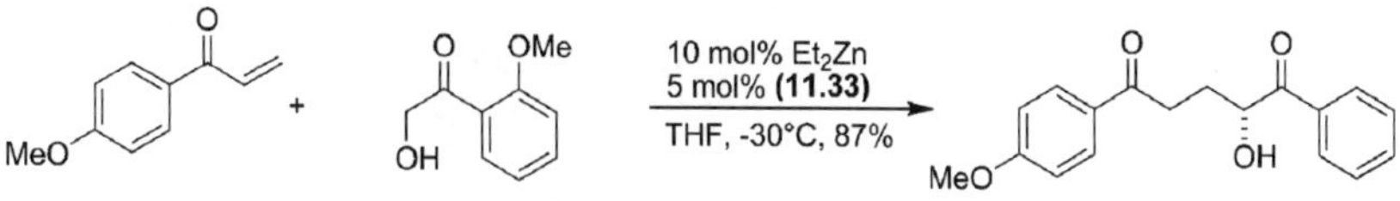

(11.32) + **(11.34)**

10 mol% Et$_2$Zn
5 mol% **(11.33)**
THF, -30°C, 87%

(11.35) 98% ee

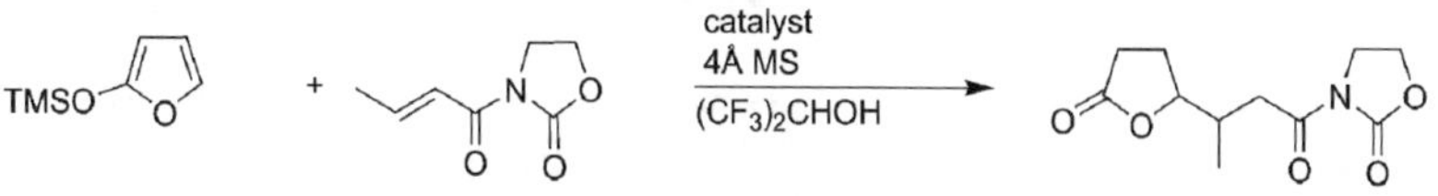

TMSO **(11.36)** + **(11.37)**

catalyst
4Å MS
(CF$_3$)$_2$CHOH

(11.38)

(11.39) provides;
68% ee (*anti:syn* > 50:1)

(11.40) provides;
95% ee (*anti:syn* > 8.5:1)

using the copper complex of bis-oxazoline (**11.40**).[15, 16, 17] The copper bis-oxazoline complex (**11.40**) also provides up to 99% ee in the enantioselective addition of silyl ketene acetals to alkylidene malonates.[18]

The intermediate enolate formed in a Michael reaction normally undergoes protonation to give the ketone product. However, in the presence of an aldehyde, a Michael/Aldol cascade can occur. Using the racemic Michael acceptor (**11.41**), Shibasaki has demonstrated an enantioselective and diastereoselective Michael/Aldol cascade involving the malonate nucleophile (**11.42**) and the aldehyde (**11.43**). The process also involves a kinetic resolution of the starting material.[19] Not only is this a remarkable example of several aspects of stereoselectivity, the product (**11.44**) is a useful prostaglandin precursor.

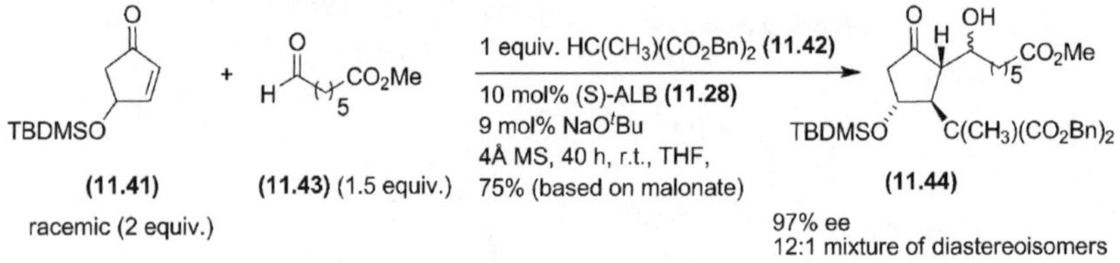

The asymmetric Michael reaction can be catalysed by enantiomerically pure crown ethers in the presence of base. For instance the addition of cyclic donor (**11.16**) with acceptor (**11.22**) occurs with up to 99% ee using an enantiomerically pure crown ether in the presence of potassium *tert*-butoxide.[20] In fact, enantiomerically pure crown ethers have been used to catalyse other Michael reactions,[21] including the use of crown ether (**11.45**) in the conjugate addition reaction between the ester (**11.46**) and Michael acceptor (**11.47**).[22] The reaction is remarkably rapid (one minute at $-78°C$).

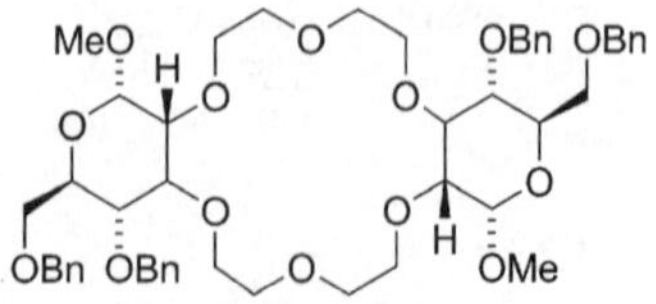

(11.45)

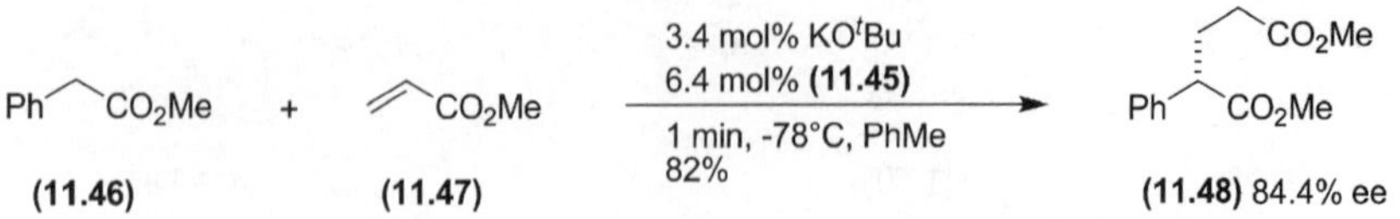

A catalytic enantioselective Michael addition of diisopropyl malonate to α,β-unsaturated ketones and aldehydes has been achieved using 5 mol% of the rubidium salt of L-proline.[23] While this process is postulated to proceed via the formation of an iminium ion precursor, as in the organocatalysed Diels–Alder reaction (see Section 8.1) and transfer hydrogenation (see Section 2.5), use of free proline as an organocatalyst leads to poor ees.[24] Nevertheless, the organocatalysed Michael addition of enolisable carbonyls with α,β-enones, enals and conjugated nitroolefins and vinyl sulfones occurs with high ee in the presence of a wide range of substituted proline derivates and cyclic diamines. The mechanism of addition onto the latter two types of acceptor probably involves the formation of enamines in a similar fashion to the organocatalysed aldol reaction (see Section 7.1).[25] A few examples are illustrated here. The first highly enantioselective organocatalysed addition of malonates to β-substituted enones such as benzylidene acetone (**11.49**) was developed by Jørgensen and coworkers using imidazolidine (**11.50**).[26] Simple aldehydes such as (**11.52**) can also be used as donors and some of the highest ees obtained with these substrates have been achieved using the diphenylprolinol (**11.53**) as catalyst. The Mukaiyama–Michael reaction is also susceptible to asymmetric organocatalysis and MacMillan and coworkers have synthesised γ-butenolides with up to 99% ee by conjugate addition of silyloxyfurans, including (**11.55**), with acyclic α,β-enals in the presence of the imidazolidinone salt (**11.56**).[27]

A number of enantiomerically pure amines catalyse the asymmetric Michael reaction of ketones with nitroolefins.[25c] Amongst these some of the most successful are the pyrrolidine-pyridine (**11.58**)[28] and the bipyrrolidine (**11.59**).[29]

Cyclic amines are not the only effective organocatalysts in the asymmetric Michael reaction. High ees in the addition to conjugated carbonyls and nitroolefins have been achieved using catalytic amounts of thioureas such as (**11.63**), which function via hydrogen bonding to the acceptor.[30] Cinchona alkaloids such as (**11.64**) act as bifunctional catalysts in the Michael addition interacting with both donor and acceptor and have been used to effect high ees in the addition to α,β-enals.[31]

11.2 Conjugate Addition of Sulfur Nucleophiles

The asymmetric conjugate addition of sulfur-based nucleophiles to Michael acceptors has been achieved using enantiomerically pure metal-based catalysts and organocatalysts.[32] Sulfur nucleophiles are soft, and preferentially react by conjugate addition with α,β-unsaturated carbonyl compounds. Only catalytic amounts of the lithium thiolate (**11.65**) are required, since addition to the enone (**11.66**) generates the enolate (**11.67**). The enolate is then able to deprotonate thiol (**11.68**), regenerating thiolate (**11.65**) with formation of the product (**11.69**).

Since the lithium will coordinate to an appropriate enantiomerically pure ligand, a asymmetric catalytic reaction can be achieved. Tomioka and coworkers

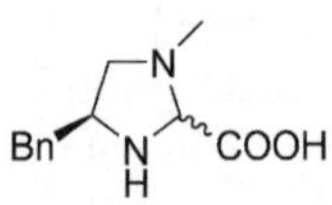

(11.50) 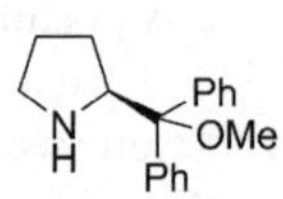   **(11.53)**   •2,4-(NO₂)₂C₆H₃COO⁻ **(11.56)** 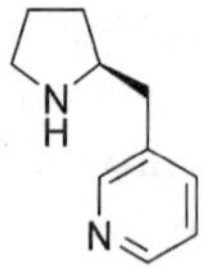   **(11.58)**

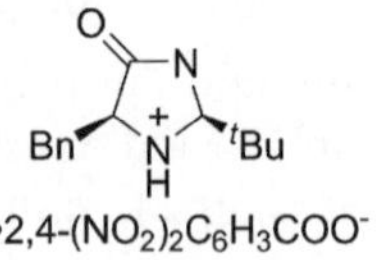

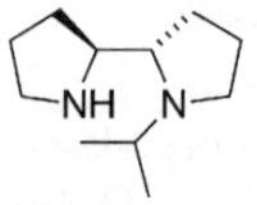

(11.59) 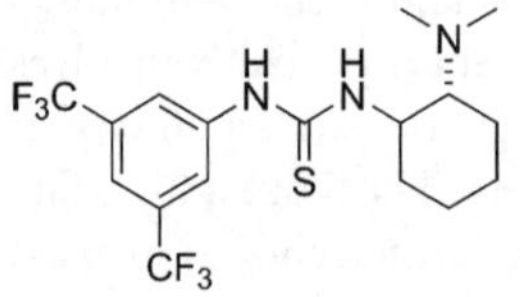   **(11.63)** 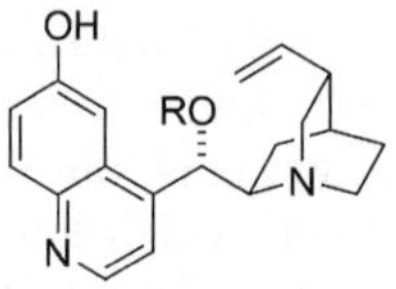   **(11.64)**

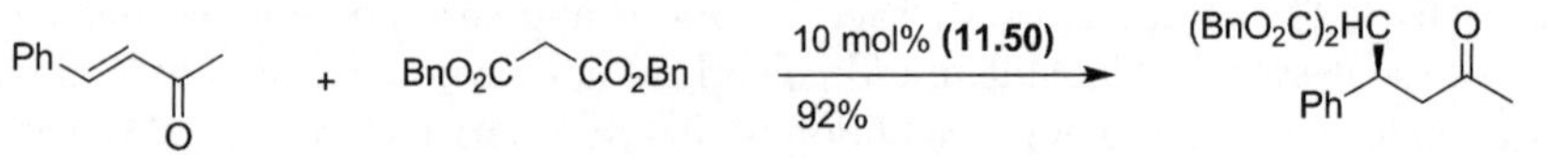

Ph — (11.49) + BnO₂C–CO₂Bn (11.51) → 10 mol% (11.50), 92% → (BnO₂C)₂HC ... Ph ... (11.60) >99% ee

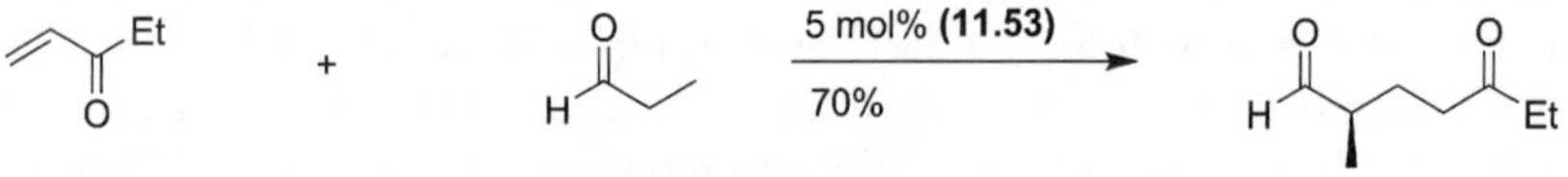

(11.54) + (11.52) → 5 mol% (11.53), 70% → (11.61) 99% ee

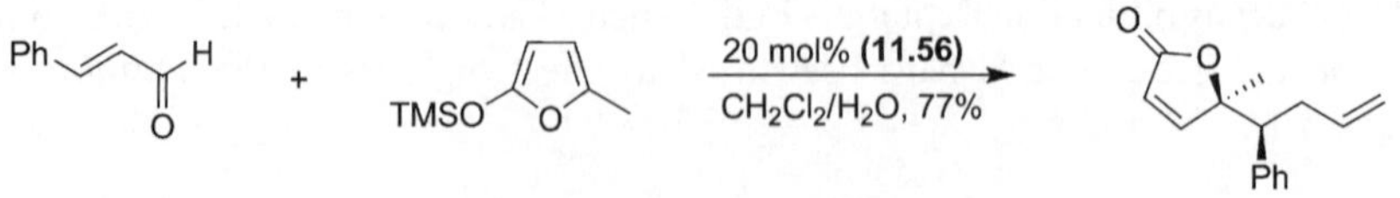

Ph (11.57) + TMSO (11.55) → 20 mol% (11.56), CH₂Cl₂/H₂O, 77% → Ph (11.62) 99% ee

have established this principle using ligand (**11.70**).[33] Methyl crotonate (**11.71**) and thiophenol (**11.72**) afford the conjugate addition product (**11.73**) with fairly good enantioselectivity. Improved enantioselectivities were obtained using 2-(trimethylsilyl)thiophenol as the incoming nucleophile (up to 97% ee)[34] and this methodology has also been applied to the addition to conjugated enones.[35] Shibasaki and coworkers have applied their bifunctional heterobimetallic catalysts, active in the asymmetric conjugate addition of enolates (see Section 11.1), to this transformation and have achieved good enantiocontrol in the conjugate addition of benzylthiol (**11.74**) to cyclohexenone (**11.21**) using LSB (**11.75**) (lanthanum sodium BINOL).[36]

Amongst the other metal catalysts employed in this reaction, high levels of enantioselectivity in the addition of aromatic thiols to conjugated oxazolidinones such as (**11.37**) have been achieved using the nickel complex of DBFOX ligand (**11.77**), which functions as a Lewis acid activator.[37]

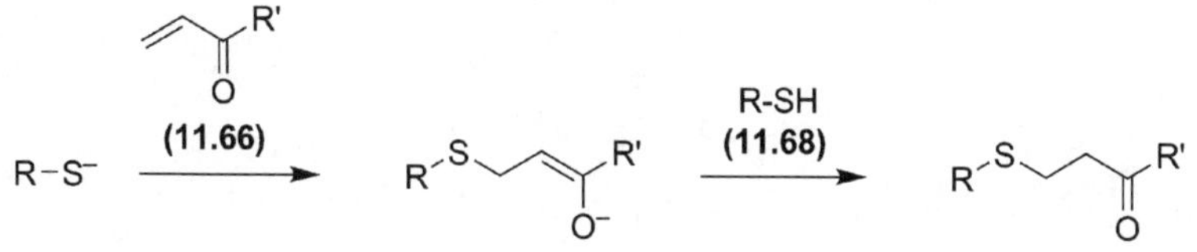

R'
O
(11.66)
R-S⁻
R-SH
(11.68)
O⁻

(11.65) **(11.67)** **(11.69)**

Ph
Me₂N
Ph
MeO
Na
La
Na
Na
where = (R)-Binaphthoxide
Ph
Ph

(11.70) **(11.75)** **(11.77)**

CO₂Me
8 mol% (11.70)/PhSLi
3 equiv. PhSH (11.72)
3 h, - 20°C, PhMe
95%
CO₂Me
SPh

(11.71) **(11.73)** 75% ee

O
10 mol% (11.75)
PhCH₂SH (11.74)
14 h, - 40°C, PhMe:
THF 60:1, 86%
O
SCH₂Ph
H

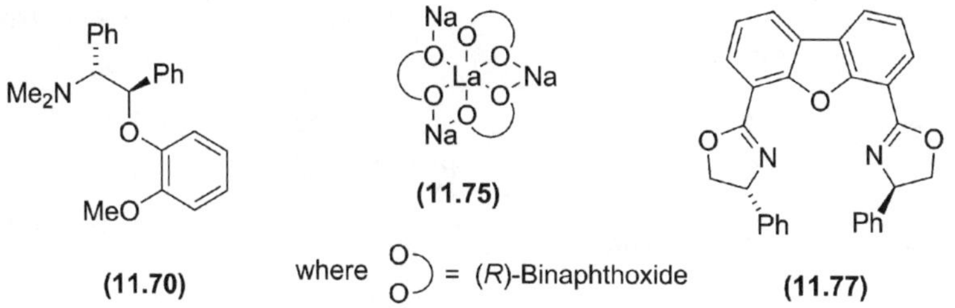

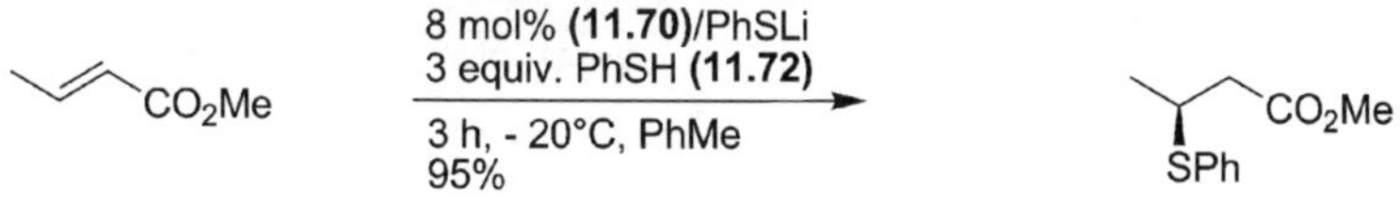

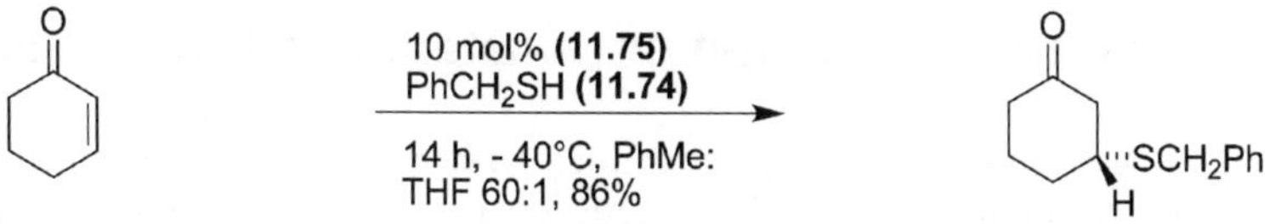

(11.21) **(11.76)** 90% ee

Organocatalysts have been used in the enantioselective conjugate addition of sulfur nucleophiles.[32] Aromatic, benzyl and alkyl thiols such as (**11.78**) give the conjugate addition product with ees ranging between 89 and 97% in the presence of the enantiomerically pure pyrrolidine (**11.79**).[38] High ees in the addition of aromatic thiols to cyclic enones have been achieved using the cinchona alkaloid catalyst $(DHQD)_2PYR$ (**11.81**),[39] and Wang and coworkers have developed a bifunctional catalyst (**11.82**) incorporating both a cinchona moiety and a thiourea, capable of hydrogen bonding to an oxazolidinone acceptor and donor. Use of this catalyst in the addition of 2-mercaptobenzaldehyde (**11.83**) results in a Michael–aldol cascade, giving rise to the formation of enantioenriched benzothiopyrans.[40]

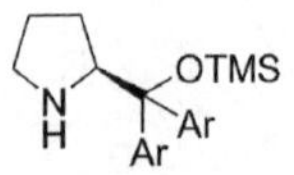

(**11.79**)

Ar = 3,5-(CF$_3$)$_2$C$_6$H$_3$

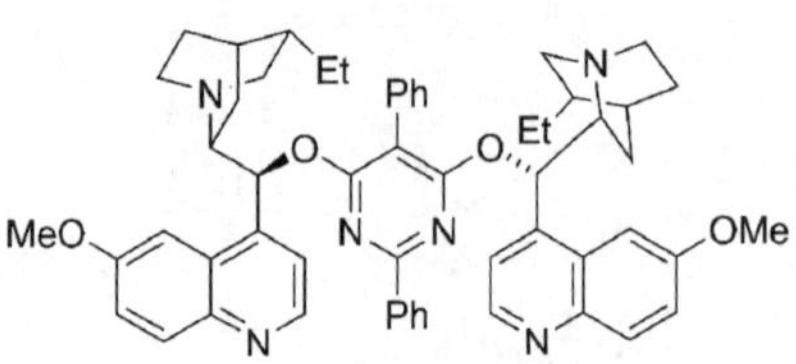

(**11.81**)

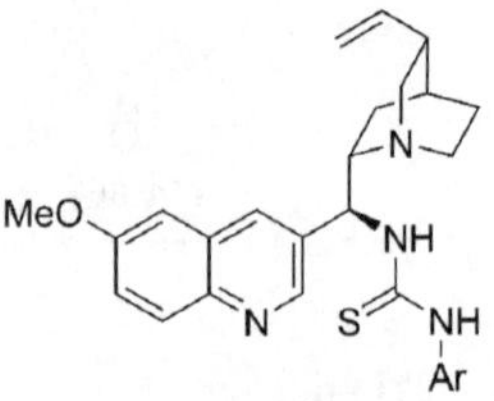

(**11.82**)

Ar = 3,5-(CF$_3$)$_2$C$_6$H$_3$

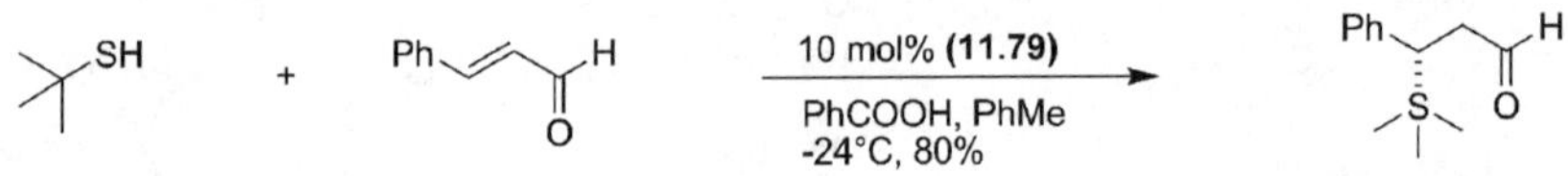

(**11.78**)   (**11.57**)

10 mol% (**11.79**)

PhCOOH, PhMe
-24°C, 80%

(**11.80**) 97% ee

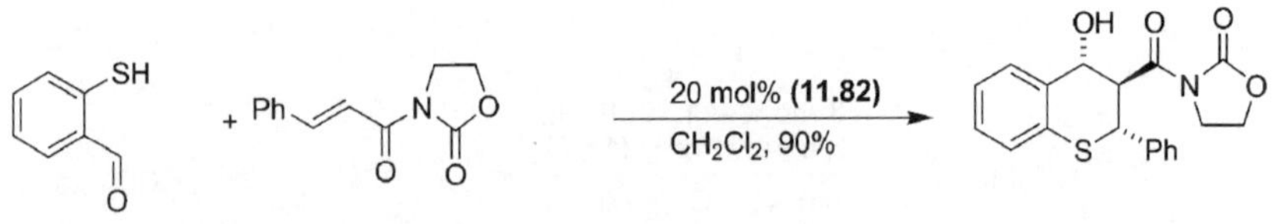

(**11.83**)   (**11.84**)

20 mol% (**11.82**)

CH$_2$Cl$_2$, 90%

(**11.85**) 99% ee

11.3   Conjugate Addition of Nonstabilised Nucleophiles

The conjugate addition of organocuprates with α,β-unsaturated enones and enoates is an important synthetic procedure. Whilst there are several examples of enantiomerically pure ligands which work well when used stoichiometrically,[41] it is only fairly recently that significant advances have been made for catalytic reactions. In order to achieve a catalytic reaction, the ligand design must be such that there will be preferential binding to the copper salt rather than the metal counterion of the carbanion.

The first developments in this area were made in the early 1990s. However, while Tanaka and coworkers, in 1993, reported high ees in the addition of methyllithium to enone (**11.86**) using substoichiometric quantities of ligand (**11.87**),[42] the copper-catalysed conjugate addition of Grignard reagents using catalytic amounts of external enantiomerically pure ligands was rarely achieved with ees above 70–80%

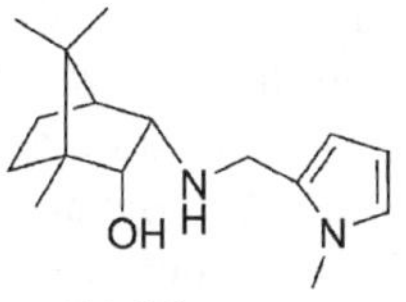

OH
H
N
N

(11.87)

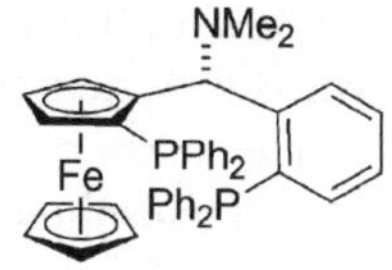

NMe2
PPh2
Fe
Ph2P

(11.89)

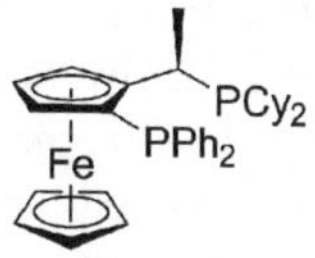

PCy2
PPh2
Fe

(11.91)

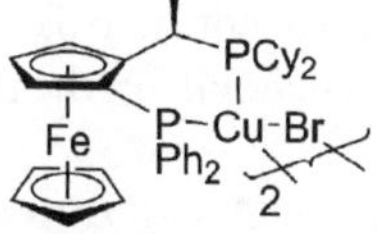

PCy2
P–Cu-Br
Fe
Ph2
2

(11.94)

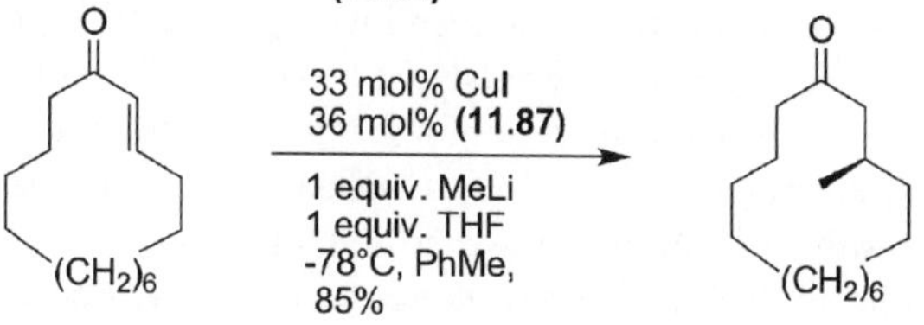

O
(CH2)6
33 mol% CuI
36 mol% (11.87)
1 equiv. MeLi
1 equiv. THF
-78°C, PhMe,
85%
O
(CH2)6

(11.86)

11.88 99% ee

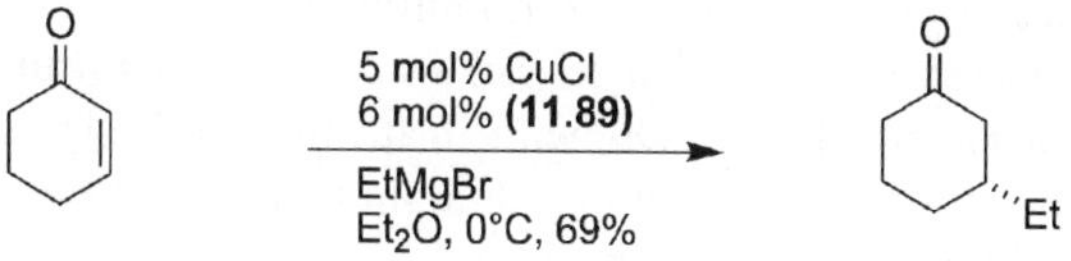

O
5 mol% CuCl
6 mol% (11.89)
EtMgBr
Et2O, 0°C, 69%
O
Et

(11.21)

(11.90) 96%ee

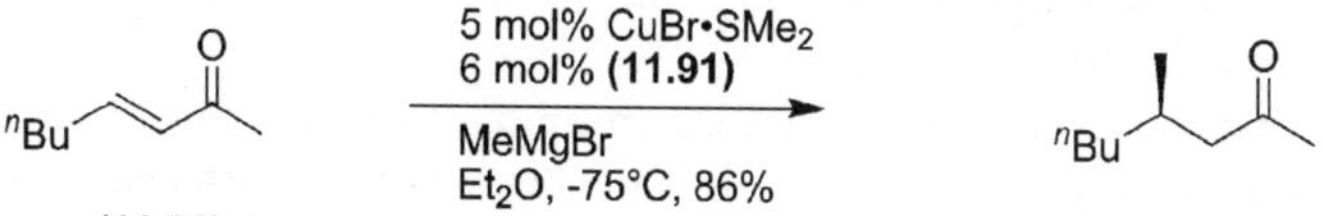

O
nBu
5 mol% CuBr•SMe2
6 mol% (11.91)
MeMgBr
Et2O, -75°C, 86%
O
nBu

(11.92)

(11.93) 98% ee

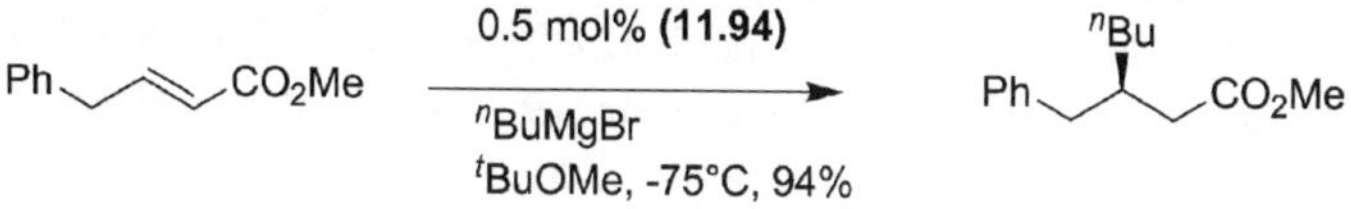

Ph
CO2Me
0.5 mol% (11.94)
nBuMgBr
tBuOMe, -75°C, 94%
nBu
Ph
CO2Me

(11.95)

(11.96) 92% ee

during this decade.[1c, 43] The major breakthrough in this area came in 2004 with the demonstration, by Feringa and coworkers, that high ees in the conjugate addition of a range of Grignard reagents to a variety of enones and enoates are achieved in the presence of bidentate ferrocenyl phosphines.[1d, 44] Simple Grignard reagents undergo highly enantioselective 1,4-addition with both cyclic enones, such as cyclohexenone **(11.21)**, in the presence of TANIAPHOS ligand **(11.89)**[44a] while the best ees in the addition to acyclic enones occur using JOSIPHOS **(11.91)**.[45] The conjugate addition of Grignard reagents to α,β-unsaturated esters such as **(11.95)** also proceeds with high enantioselectivity in the presence of the air stable dinuclear copper complex **(11.94)**.[46]

The enantioselective copper-catalysed conjugate addition of diethylzinc to Michael acceptors has received a great deal of attention and many enantiomerically pure phosphorus-bearing ligands have been investigated in this proceess.[1c,d, 41,47] Interestingly, this reaction is usually catalysed by copper(II) salts rather than the copper(I) salts previously mentioned, although it is not really clear which oxidation state is required in the catalytic cycle. In fact, nickel(II) salts can also be effective.[48] Substrate scope is still fairly limited, however, and high ees have only been obtained to date in the addition to simple cyclic enones such as 2-cyclohexenone, chalcone and some nitroolefins.

Some of the highest ees in the addition of diethylzinc to cyclic enones have been achieved using monodenate phosphoramidites such as **(11.97)**.[49] Alexakis and coworkers have shown that high ees can also be obtained using more simple, easily accessible ligands such as **(11.99)** with an atropisomerically flexible biphenol moiety. In this case the diamine subunit is thought to induce atropisomerisation in the biphenol, avoiding the need for the preparation of enantiopure biaryls during ligand synthesis.[50] Up to 99% ee in the addition to cyclohexenone is achieved with these ligands under the optimal conditions using copper(I) thiophenecarboxylate in diethyl ether and high ees are also obtained in the addition to nitrocyclohexene **(11.100)** and also nitrostyrene.[51]

Somewhat lower enantioselectivities are obtained using phosphoramidites in the conjugate additions to acyclic substrates. The best ees with chalcone **(11.102)** as substrate have been achieved using the monodentate phosphonite **(11.103)**[52] and several bidentate ligands, including BINAP-based P,O-ligand **(11.104)** have also proved successful.[53] In addition, functionalised zinc reagents, such as $Zn[(CH_2)_3CH(OEt)_2]_2$ and longer chain analogues have also been used by the Feringa group to give high enantioselectivity (97% ee) in the catalysed conjugate addition to cyclohexenone **(11.21)**[54] and cyclopentenones.[55]

The copper-catalysed conjugate addition process is not limited to the use of Grignard reagents and dialkylzincs. Some success has been achieved using trimethylaluminium as nucleophile and up to 97% ee has been obtained in the addition to 2-cyclohexenones using copper-phosphoramidite catalysts.[56]

The 1,4-addition of aryl groups can be achieved by rhodium-catalysed addition of arylboronic acids with α,β-unsaturated carbonyls. In this process a

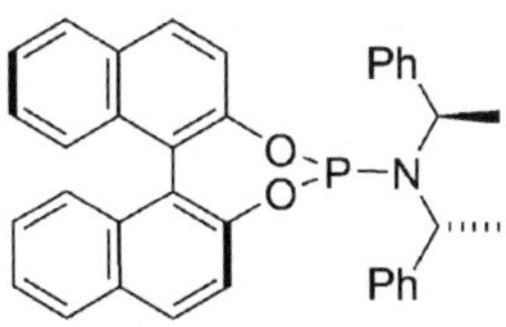

(11.97)

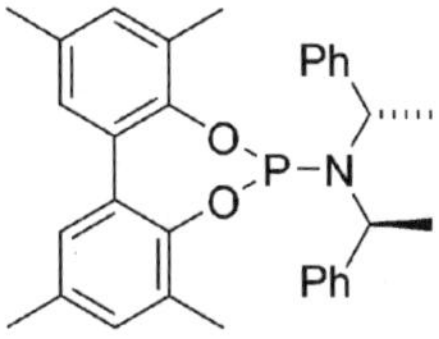

(11.99)

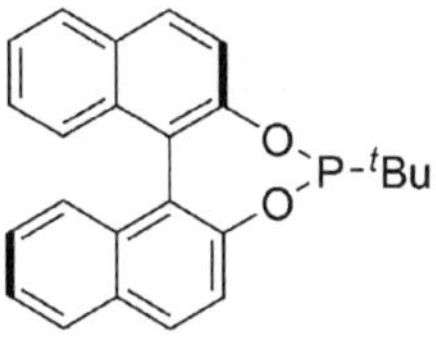

(11.103)

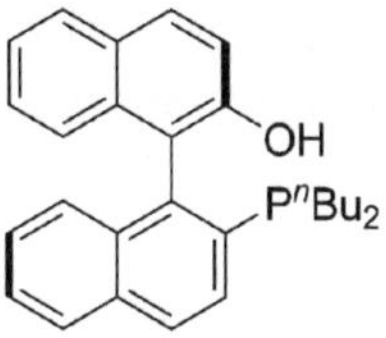

(11.104)

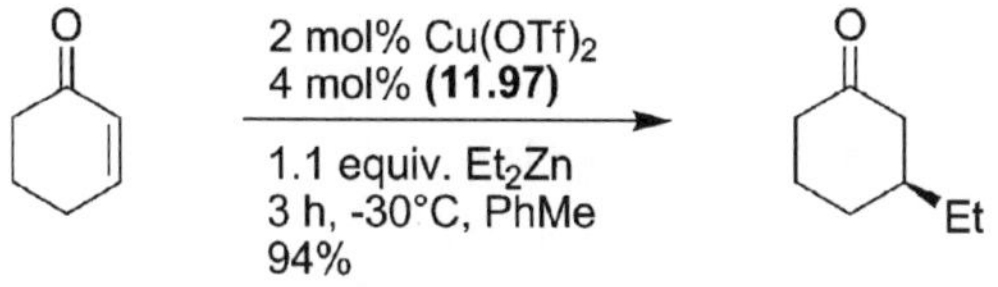

(11.21)

2 mol% Cu(OTf)$_2$
4 mol% (11.97)

1.1 equiv. Et$_2$Zn
3 h, -30°C, PhMe
94%

(11.98) >98%ee

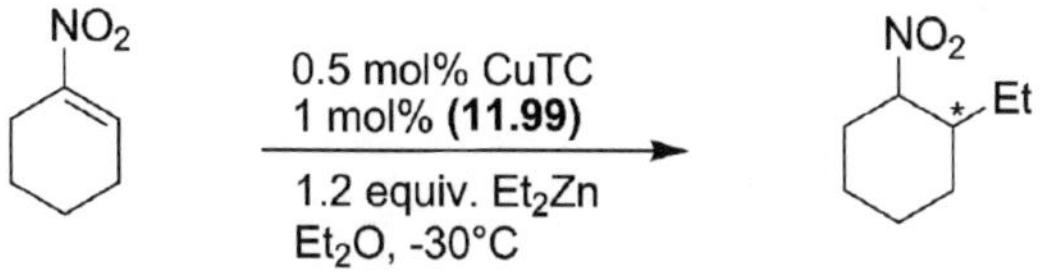

(11.100)

0.5 mol% CuTC
1 mol% (11.99)

1.2 equiv. Et$_2$Zn
Et$_2$O, -30°C

(11.101) 93% ee

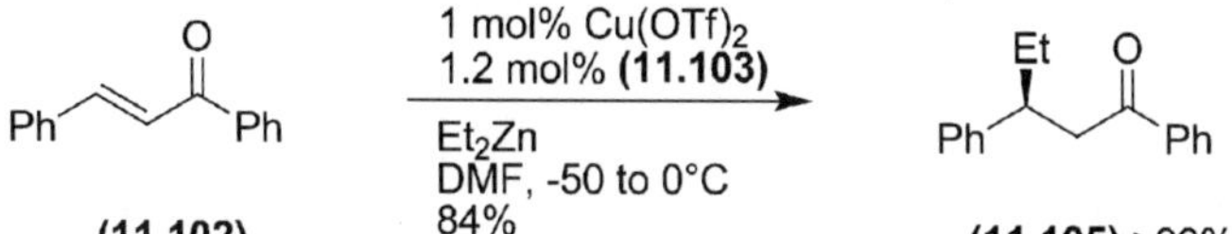

Ph

Ph

(11.102)

1 mol% Cu(OTf)$_2$
1.2 mol% (11.103)

Et$_2$Zn
DMF, -50 to 0°C
84%

(11.105) >99%ee

phenylrhodium species, formed by transmetallation, undergoes phenylrhodation across the double bond, giving an intermediate rhodium enolate with C–C bond formation at the β-position. In 1998 Hayashi and Miyaura reported that high ees could be obtained in this transformation using BINAP as a ligand.[57] For instance, cyclohexenone (**11.21**) and *trans*-acyclic enones such as (**11.106**) react with phenylboronic acid (**11.107**) to afford the products of conjugate addition (**11.108**) and (**11.109**) with high ee. Since then a variety of enantiomerically pure phosphorus-bearing ligands have been investigated for their efficacy in the asymmetric variant of this reaction.[1d, 58,59]

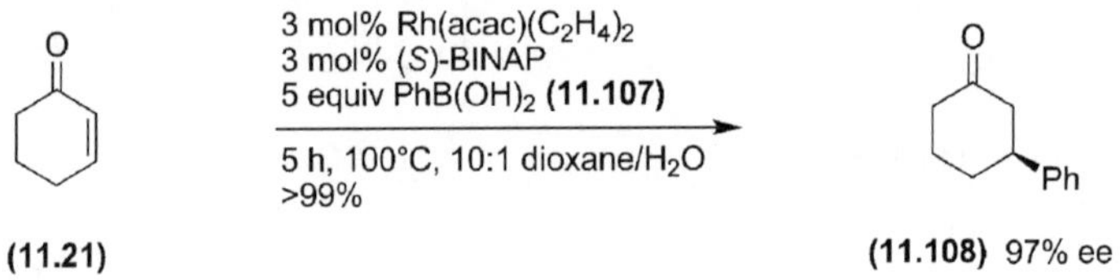

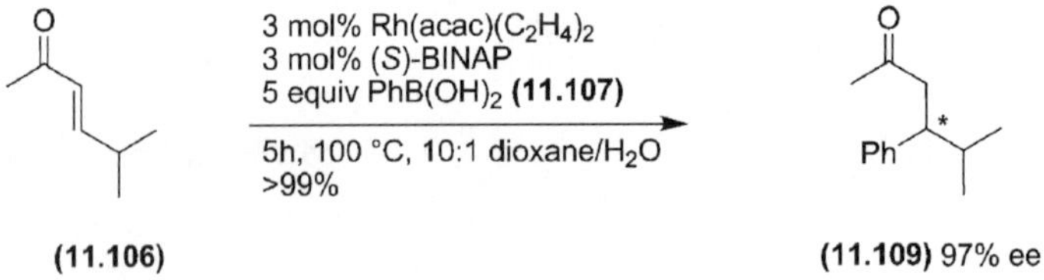

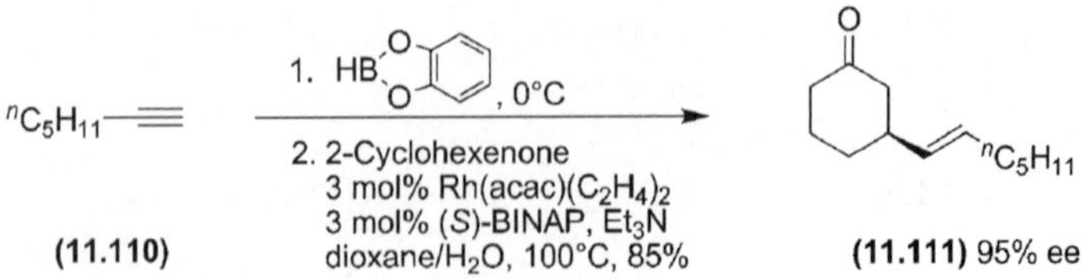

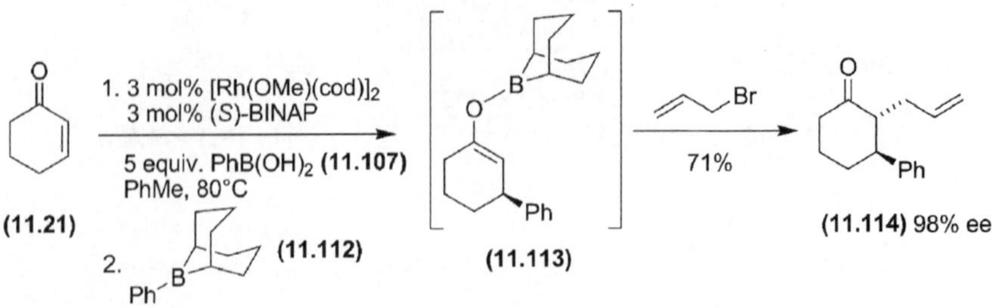

Alkenyl boranes including heptenylcatecholborane, generated by hydroboration of the corresponding alkyne (**11.110**) also add to cyclic and acyclic enones with high ee.[60] The presence of water is required in the reaction mixture to regenerate the catalyst from the rhodium enolate. Performing the addition in dry toluene in the presence of 9-borabicyclo[3.3.1]nonane (**11.112**) results in transmetallation to a boron enolate (**11.113**) which may be quenched with electrophiles such as allyl bromide and also aldehydes, to give the *trans*-disubstituted product with high ee. [61]

Success has been obtained using other organometallic reagents and some high enantioselectivities in the rhodium-catalysed addition of aryl and alkenylsilanes [62] and also aryltitanium species [63] to cyclic and acyclic enones has been achieved.

The asymmetric conjugate arylation can also be mediated by organocatalysts. MacMillan and coworkers have shown that the iminium ions, formed in the conjugate addition of silyloxyfurans in the presence of imidazolidinone (**11.56**), are also sufficiently activated to undergo 1,4-addition with electron-rich arenes. [64] Thus pyrroles undergo highly enantioselective Friedal–Crafts alkylation with α,β-enals such as (**11.57**) using the imidazolidinone salt (**11.115**), [64a] while anilines are alkylated with high ee in the presence of the related amine catalyst (**11.118**). [64c]

(**11.115**)   (**11.118**)

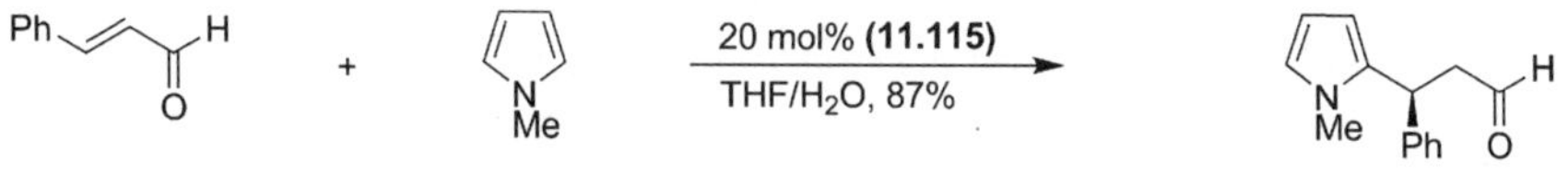

(**11.57**)   (**11.116**)   20 mol% (**11.115**)   (**11.117**) 93% ee
                               THF/H₂O, 87%

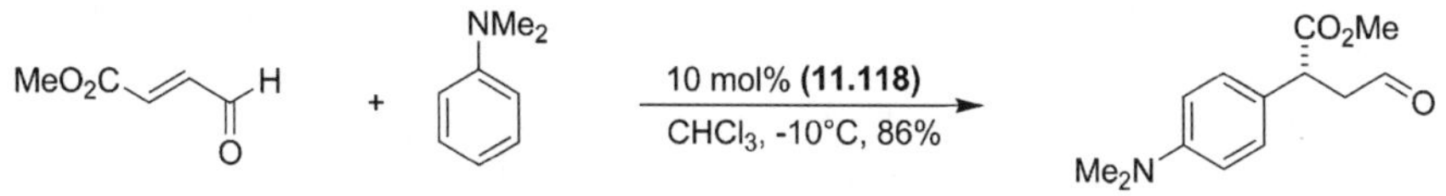

(**11.119**)   (**11.120**)   10 mol% (**11.118**)   (**11.121**) 96% ee
                                CHCl₃, -10°C, 86%

11.4 Conjugate Addition with Nitrogen-Based Nucleophiles and Electrophiles

The enantioselective conjugate addition of nitrogen-based nucleophiles provides products useful in the synthesis of β-amino acids, and this reaction proceeds with high ee using both enantiomerically pure Lewis acidic metal-based catalysts and organocatalysts. [65]

Jørgensen and coworkers have achieved up to 90% ee in the addition of aromatic amines to conjugated oxazolidinones such as (**11.37**) using Lewis acidic nickel complexes of DBFOX ligand (**11.77**).[66] Oxazolidinones of this type are also effectively activated with palladium catalysts and high ees in the addition of both aromatic and aliphatic amines have been obtained in the presence of Pd-BINAP complexes.[67] The use of a basic palladium aqua complex (**11.18**) in combination with amine salts results in slow release of active nucleophile thus suppressing any competing uncatalysed addition and catalyst deactivation by the nucleophile. In the presence of this catalyst high ees are obtained using salts of reactive electron-rich aniline and also benzylamine (**11.122**).[67b]

The enantioselective conjugate addition of hydroxylamines to a range of α,β-carbonyls occurs with moderate to good ee in the presence of Lewis acidic magnesium- and zinc-based bis-oxazolines,[68] and with high ee using the scandium-phosphate catalyst (**11.124**).[69]

The bifunctional heterobimetallic complexes developed by Shibasaki and coworkers, which have been used to effect high ees in a variety of conjugate additions discussed in this chapter, are also effective in the conjugate additions of hydroxylamines to enones and up to 99% ee has been obtained using the ytterbium lithium binaphthoxide YLi$_3$tris(binaphthoxide) (YLB).[70]

In addition to amines and hydroxylamines, very high levels of enantioselection have been obtained during the conjugate addition of carbamates and hydrazoic acid using copper bis-oxazolines[71] and aluminium(salen) complexes, respectively.[72]

The asymmetric amine-catalysed conjugate addition of nitrogen-based nucleophiles is challenging, as any secondary amine catalyst used may undergo competing 1,4-addition, while the nucleophile chosen may participate in iminium ion formation. Thus, most success in this arena has been obtained in the conjugate addition of *N*-heterocycles. Nevertheless, MacMillan and coworkers have achieved high ees in the 1,4-addition of amine moieties with acyclic α,β-enals using the more nucleophilic *N*-silyloxycarbamates such as (**11.126**) in the presence of the imidazolidone salt (**11.127**).[73]

Aromatic nitrogen heterocycles, especially triazoles and tetrazoles are the most common nucleophiles used in the organocatalysed aza-Michael reaction. 5-Phenyltetrazole undergoes conjugate addition with acyclic enals using the trifluoracetate salt of imidazolidinone (**11.127**),[74] while the 1,4-addition of phenyltetrazole and 1,2,4-triazole (**11.130**) occurs with high ee using the enantiopure pyrrolidine catalyst (**11.79**).[75]

High enantioselectivites in the aza-Michael reaction have been achieved using alternate organocatalysts, and the addition of benzotriazole to nitroolefins occurs with up to 94% ee using bifunctional catalysts such as (**11.64**).[76] Guerin and Miller have developed an alternate approach to the enantioselective introduction of triazoles based on the asymmetric conjugate addition of azide followed by a 1,3-dipolar cycloaddition of the product with an alkyne. In this approach, the addition of hydrazoic acid to Michael acceptors such as (**11.133**) proceeds with good ee in the presence of the dipeptide (**11.134**).[77]

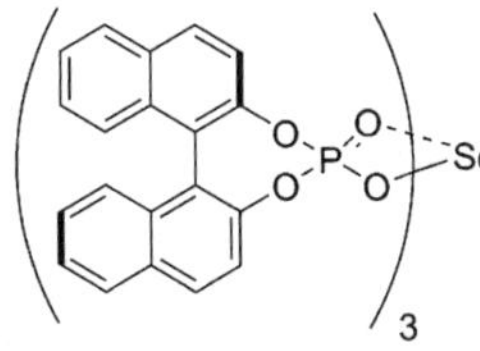

(11.124) 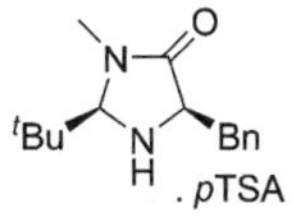   **(11.127)**

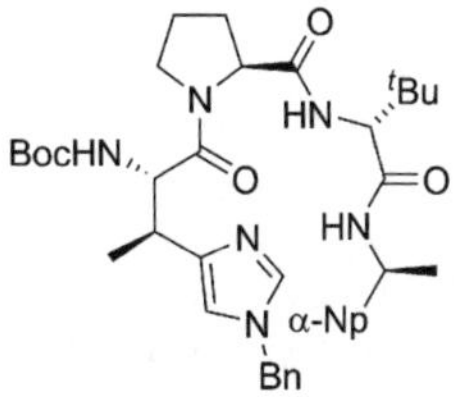

(11.134)

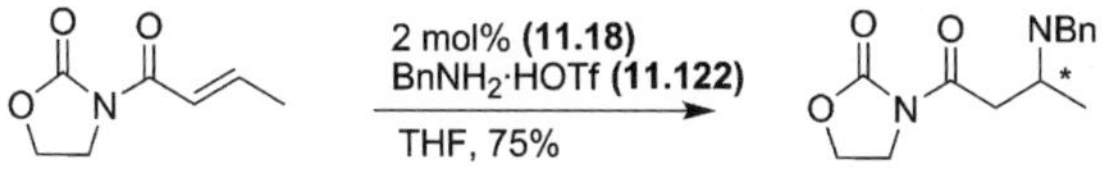

(11.37)

2 mol% **(11.18)**
BnNH$_2$·HOTf **(11.122)**
THF, 75%

(11. 123)

(11.102)

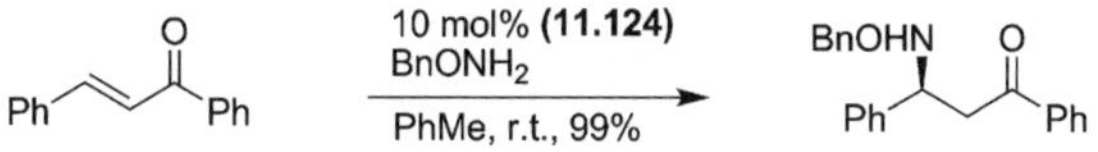

10 mol% **(11.124)**
BnONH$_2$
PhMe, r.t., 99%

(11.125) 99% ee

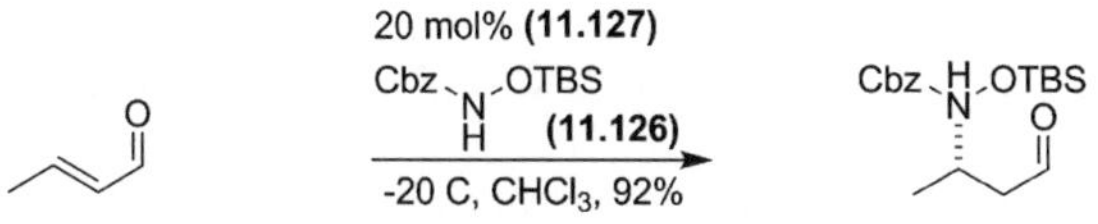

(11.128)

20 mol% **(11.127)**
Cbz$_{\text{N}}$OTBS
     H   **(11.126)**
-20 C, CHCl$_3$, 92%

(11.129) 92% ee

(11.130)
10 mol% **(11.79)**
PhMe, PhCOOH
r. t., 76%

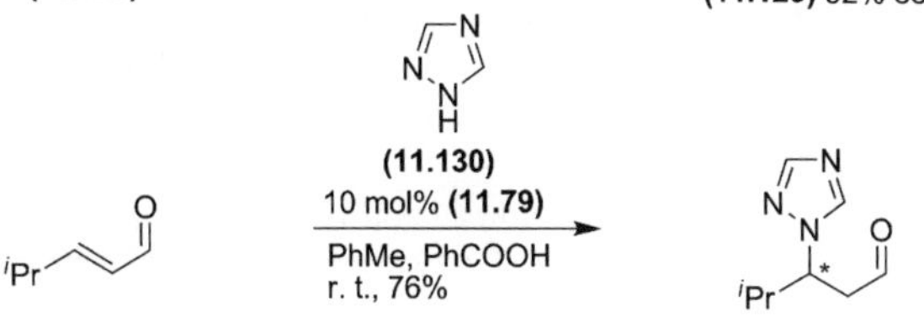

(11.131)

(11.132) 94% ee

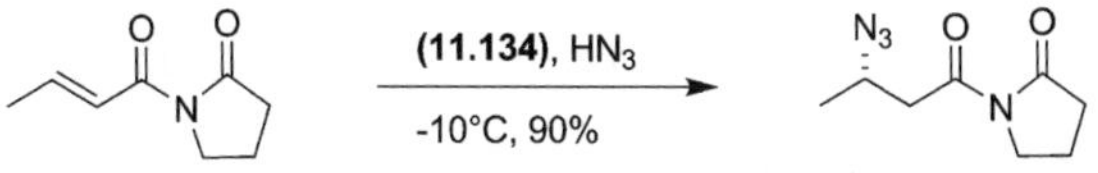

(11.133)

(11.134), HN$_3$
-10°C, 90%

(11.135) 86% ee

References

1. For reviews covering the asymmetric Michael reaction see: (a) M. Yamaguchi in, *Comprehensive Asymmetric Catalysis*, Vol. 3, ed. E. Jacobsen, A. Pfaltz and H. Yamamoto, Springer-Verlag, Berlin, **1999**, 1121. (b) M. P. Sibi and S. Manyem, *Tetrahedron*, **2000**, *56*, 8033. (c) N. Krause and A. Hoffman-Röder, *Synthesis*, **2001**, 171. (d) J. Christoffers, G. Koripelly, A. Rosiak and M. Rössle, *Synthesis*, **2007**, 1279.

2. M. Sawamura, H. Hamashima and Y. Ito, *J. Am. Chem. Soc.*, **1992**, *114*, 8295.

3. M. S. Taylor and E. N. Jacobsen, *J. Am. Chem. Soc.*, **2003**, *125*, 11204.

4. M. Sawamura, H. Hamashima, H. Shinoto and Y. Ito, *Tetrahedron Lett.*, **1995**, *36*, 6479.

5. (a) K. Itoh, M. Hasegasawa, J. Tanaka and S. Kanemasa, *Org. Lett.*, **2005**, *7*, 979. (b) D. A. Evans, R. J. Thomson and F. Franco, *J. Am. Chem. Soc.*, **2005**, *127*, 10816. (c) D. A. Evans and D. Seidel, *J. Am. Chem. Soc.*, **2005**, *127*, 9958.

6. (a) J. Comelles, M. Moreno-Mañas, E. Perez, A. Roglans, R. M. Sebastian and A. Vallribera, *J. Org. Chem.*, **2004**, *69*, 6834. (b) N. Halland, T. Velgaard and K. A. Jorgensen, *J. Org. Chem.*, **2003**, *68*, 5067.

7. F. Berthiol, R. Matsubara, N. Kawai and S. Kobayashi, *Angew. Chem., Int. Ed.*, **2007**, *46*, 7803.

8. G. Desimoni, G. Dusi, G. Faita, P. Quadrelli and P. P. Righetti, *Tetrahedron*, **1995**, *51*, 4131.

9. M. Watanabe, K. Murata and T. Ikariya, *J. Am. Chem. Soc.*, **2003**, *125*, 7508.

10. T. Arai, H. Sasai, K. K. Yamaguchi and M. Shibasaki, *J. Am. Chem. Soc.*, **1998**, *120*, 441.

11. E. Keller, N. Veldman, A. L. Spek and B. L. Feringa, *Tetrahedron: Asymmetry*, **1997**, *8*, 3403.

12. G. Manickam and G. Sundararajan, *Tetrahedron: Asymmetry*, **1997**, *8*, 2271.

13. S. Harada, N. Kumagai, T. Kinoshita, S. Matsunaga and M. Shibasaki, *J. Am. Chem. Soc.*, **2003**, *125*, 2582.

14. K. Majima, R. Takita, A. Okada, T. Ohshima and M. Shibasaki, *J. Am. Chem. Soc.*, **2003**, *125*, 15837.

15. S. Kobayashi, S. Suda, M. Yamada and T. Mukaiyama, *Chem. Lett.*, **1994**, 97.

16. (a) H. Nishikori, K. Ito and T. Katsuki, *Tetrahedron: Asymmetry*, **1998**, *9*, 1165. (b) H. Kitajima, K. Ito and T. Katsuki, *Tetrahedron*, **1997**, *53*, 17015.

17. For an earlier example of the use of Cu(II) bisoxazoline complexes in conjugate addition reactions see; A. Bernardi, G. Colombo and C. Scolastico, *Tetrahedron Lett.*, **1996**, *37*, 8921.

18. (a) D. A. Evans, T. Rovis, M. C. Kozlowski and J. S. Tedrow, *J. Am. Chem. Soc.*, **1999**, *121*, 1994. (b) D. A. Evans, T. Rovis, M. C. Kozlowski, C. W. Downey and J. S. Tedrow, *J. Am. Chem. Soc.*, **2000**, *122*, 9134.

19. K.-I. Yamada, T. Arai, H. Sasai and M. Shibasaki, *J. Org. Chem.*, **1998**, *63*, 3666.

20. D. J. Cram and G. D. Y. Sogah, *J. Chem. Soc., Chem. Commun.*, **1981**, 625.

21. S. Aoki, S. Sasaki and K. Koga, *Tetrahedron Lett.*, **1989**, *30*, 7229.

22. L. Tõke, P. Bakó, G. M. Keserü, M. Albert and L. Fenichel, *Tetrahedron*, **1998**, *54*, 213.

23. (a) M. Yamaguchi, T. Shiraishi and M. Hirama, *Angew. Chem. Int. Ed.*, **1993**, *32*, 1176. (b) M. Yamaguchi, T. Shiraishi and M. Hirama, *J. Org. Chem.*, **1996**, *61*, 3520.

24. K. Sakthivel, W. Notz, T. Bui and C. F. Barbas III, *J. Am. Chem. Soc.*, **2001**, *123*, 5260.
25. For reviews covering the organocatalysed Michael addition see: (a) P. I. Dalko and L. Moisan, *Angew. Chem. Int. Ed.*, **2004**, *43*, 5138. (b) H. Pellissier, *Tetrahedron*, **2007**, *63*, 9267. (c) S. Sulzer-Mossé and A. Alexakis, *J. Chem. Soc., Chem. Commun.*, **2007**, 3123. (d) S. B. Tsogoeva, *Eur. J. Org. Chem.*, **2007**, *11*, 1701.
26. N. Halland, P. S. Aburel and K. A. Jørgensen, *Angew. Chem., Int. Ed.*, **2003**, *42*, 661.
27. S. P. Brown, N. C. Goodwin and D. W. C. MacMillan, *J. Am. Chem. Soc.*, **2003**, *125*, 1192.
28. O. Andrey, A. Alexakis and G. Bernardinelli, *Org. Lett.*, **2003**, *5*, 2559.
29. T. Ishii, S. Fujioka, Y. Sekiguchi and H. Kotsuki, *J. Am. Chem. Soc.*, **2004**, *126*, 9558.
30. (a) T. Okino, Y. Hoashi and Y. Takemoto, *J. Am. Chem. Soc.*, **2003**, *125*, 12672. (b) J. Wang, H. Li, W. Duan, L. Zu and W. Wang, *Org. Lett.*, **2005**, *7*, 4713. (c) Y. Hoashi, T. Okino and Y. Takemoto, *Angew. Chem. Int. Ed.*, **2005**, *44*, 4032. (d) K. Liu, H.-F. Cui, J. Nie, K.-Y. Dong, X.-J. Li and J.-A. Ma, *Org. Lett.*, **2007**, *9*, 923.
31. F. Wu, R. Hong, J. Khan, X. Liu and L. Deng, *Angew. Chem. Int. Ed.*, **2006**, *45*, 4301.
32. For a recent review covering the catalytic asymmetric conjugate addition of sulfur nucleophiles see: D. Enders, K. Lütgen and A. A. Narine, *Synthesis*, **2007**, 959.
33. (a) K. Tomioka, M. Okuda, K. Nishimura, S. Manabe, M. Kanai, Y. Nagaoka and K. Koga, *Tetrahedron Lett.*, **1998**, *39*, 2141.
34. K. Nishimura, M. Ono, Y. Nagaoka and K. Tomioka, *J. Am. Chem. Soc.*, **1997**, *119*, 12974.
35. K. Nishimura and K. Tomioka, *J. Org. Chem.*, **2002**, *67*, 431.
36. E. Emori, T. Arai, H. Sasai and M. Shibasaki, *J. Am. Chem. Soc.*, **1998**, *120*, 4043.
37. (a) S. Kanemasa, Y. Oderaotoshi and E. Wada, *J. Am. Chem. Soc.*, **1999**, *121*, 8675. (b) S. Kanemasa and K. Itoh, *Eur. J. Org. Chem.*, **2004**, 4741.
38. M. Marigo, T. Schulte, J. Franzén and K. A. Jørgensen, *J. Am. Chem. Soc.*, **2005**, *127*, 15710.
39. P. McDaid, Y. Chen and L. Deng, *Angew. Chem. Int. Ed.*, **2002**, *41*, 338.
40. L. Zu, J. Wang, H. Li, H. Xie, W. Jiang and W. Wang, *J. Am. Chem. Soc.*, **2007**, *129*, 1036.
41. (a) B. E. Rossiter and N. M. Swingle, *Chem. Rev.*, **1992**, *92*, 771. (b) K. Tomioka and Y. Nagaoka in, *Comprehensive Asymmetric Catalysis*, Vol. 3, ed. E. Jacobsen, A. Pfaltz and H. Yamamoto, Springer-Verlag, Berlin, **1999**, 1105.
42. K. Tanaka, J. Matsui and H. Suzuki, *J. Chem. Soc., Perkin Trans. 1*, **1993**, 153.
43. (a) Q.-L. Zhou and A. Pfaltz, *Tetrahedron Lett.*, **1993**, *34*, 7725. (b) M. van Klaveren, F. Lambert, D. J. F. M. Eijkelkamp, D. M. Grove and G. van Koten, *Tetrahedron Lett.*, **1994**, *35*, 6135. (c) M. Kanai and K. Tomioka, *Tetrahedron Lett.*, **1995**, *36*, 4275.
44. (a) B. L. Feringa, R. Badorrey, D. Peña, S. R. Harutyunyan and A. J. Minnaard, *Proc. Natl. Acad. Sci. U.S.A*, **2004**, *101*, 5834. (b) S. Woodward, *Angew. Chem., Int. Ed.*, **2005**, *44*, 5560. (c) S. R. Harutyunyan, F. López, W. R. Browne, A. Correa, D. Peña, R. Badorrey, A. Meetsma, A. J. Minnaard and B. L. Feringa, *J. Am. Chem. Soc.*, **2006**, *128*, 9103.
45. F. Lopez, S. R. Harutyunyan, A. J. Minnaard and B. L. Feringa, *J. Am. Chem. Soc.*, **2004**, *126*, 12784.

46. F. Lopez, S. R. Harutyunyan, A. Meetsma, A. J. Minnaard and B. L. Feringa, *Angew. Chem. Int. Ed.*, **2005**, *44*, 2752.

47. (a) A. Alexakis, J. Frutos and P. Mangeney, *Tetrahedron: Asymmetry*, **1993**, *4*, 2427. (b) A. Alexakis, *Pure Appl. Chem.*, **2002**, *74*, 37. (b) A. Alexakis and C. Benham, *Eur. J. Org. Chem.*, **2002**, 3221. (c) B. L. Feringa, R. Naasz, R. Imbos and L. A. Arnold in *Modern Organocopper Chemistry*, ed. N. Krause, Wiley-VCH, Weinheim, **2002**, 224.

48. M. Asami, K. Usui, S. Higuchi and S. Inoue, *Chem. Lett.*, **1994**, 297.

49. (a) B. L. Feringa, M. Pineschi, L. A. Arnold, R. Imbos and A. H. M. de Vries, *Angew. Chem. Int., Ed. Engl.*, **1997**, *36*, 2620. (b) B. L. Feringa, *Acc. Chem. Res.*, **2000**, *33*, 346.

50. A. Alexakis, S. Rosset, J. Allamand, S. March, F. Guillen and C. Benham, *Synlett*, **2001**, 1375.

51. (a) A. Alexakis, C. Benham, S. Rosset and M. Humam, *J. Am. Chem. Soc.*, **2002**, *124*, 5262. (b) A. Alexakis, D. Polet, S. Rosset and S. March, *J. Org. Chem.*, **2004**, *69*, 5660. (c) H. Choi, Z. Hua and I. Ojima, *Org. Lett.*, **2004**, *6*, 2689. For a review covering the catalytic asymmetric conjugate addition of dialkylzincs to nitroolefins see: N. Sewald and A. Rimkus, *Synthesis*, **2004**, 135.

52. A. Martorell, R. Naasz, B. L. Feringa and P. G. Pringle, *Tetrahedron: Asymmetry*, **2003**, *4*, 3699.

53. K. Ito, S. Eno, B. Saito and T. Katsuki, *Tetrahedron Lett.*, **2005**, *46*, 3981.

54. B. L. Feringa, M. Pinesci, L. A. Arnold, R. Imbos and A. H. M. de Vries, *Angew. Chem. Int. Ed., Engl.*, **1997**, *36*, 2620.

55. (a) L. A. Arnold, R. Naasz, A. J. Minnaard and B. L. Feringa, *J. Am. Chem. Soc.*, **2001**, *123*, 5841. (b) L. A. Arnold, R. Naasz, A. J. Minnaard and B. L. Feringa, *J. Org. Chem.*, **2002**, *67*, 7244.

56. (a) A. Alexakis, V. Albrow, K. Biswas, M. d'Augustin, O. Prieto and S. Woodward, *J. Chem. Soc., Chem. Commun.*, **2005**, 2843. (b) M. d'Augustin, L. Palais and A. Alexakis, *Angew. Chem., Int. Ed.*, **2005**, *44*, 1376.

57. Y. Takaya, M. Ogasawara, T. Hayashi, M. Sakai and N. Miyaura, *J. Am. Chem. Soc.*, **1998**, *120*, 5579.

58. For reviews covering the rhodium-catalysed asymmetric 1,4-addition see: (a) K. Fagnou and M. Lautens, *Chem. Rev.*, **2003**, *103*, 169. (b) T. Hayashi and K. Yamasaki, *Chem Rev.*, **2003**, *103*, 2829. (c) T. Hayashi, *Pure Appl. Chem.*, **2004**, *76*, 465.

59. (a) T. Imamoto, K. Sugita and K. Yoshida, *J. Am. Chem. Soc.*, **2005**, *127*, 11934. (b) W.-L. Duan, H. Iwamura, R. Shintani and T. Hayashi, *J. Am. Chem. Soc.*, **2007**, *129*, 2130.

60. (a) Y. Takaya, M. Ogasawara and T. Hayashi, *Tetrahedron Lett.*, **1998**, *39*, 8479. (b) S. Darses and J.-P. Genet, *Eur. J. Org. Chem.*, **2002**, 3552.

61. K. Yoshida, M. Ogaswara and T. Hayashi, *J. Org. Chem.*, **2003**, *68*, 1901.

62. (a) S. Oi, A. Taira, Y. Honma and Y. Inoue, *Org. Lett.*, **2003**, *5*, 97 (b) Y. Nakao, J. Chen, H. Imanaka, T. Hiyama, Y. Ichikawa, W.-L. Duan, R. Shintani and T. Hatashi, *J. Am. Chem. Soc.*, **2007**, *129*, 9137.

63. T. Hayashi, N. Tokunaga, K. Yoshida and J. W. Han, *J. Am. Chem. Soc.*, **2002**, *124*, 12102.

64. (a) N. A. Paras and D. W. C. MacMillan, *J. Am. Chem. Soc.*, **2001**, *123*, 4370. (b) J. F. Austin and D. W. C. MacMillan, *J. Am. Chem. Soc.*, **2002**, *124*, 1172. (c) N. A. Paras and D. W. C. MacMillan, *J. Am. Chem. Soc.*, **2002**, *124*, 7894.

65. For a review on the catalytic asymmetric aza-Michael reaction see: L. W. Xu and C.-G. Xia, *Eur. J. Org. Chem.*, **2005**, 633.
66. W. Zhuang, R. G. Hazell and K. A. Jørgensen, *J. Chem. Soc., Chem. Commun.*, **2001**, 1240.
67. (a) K. Li and K. K. Hii, *J. Chem. Soc., Chem. Commun.*, **2003**, 1132. (b) Y. Hamashima, H. Somei, Y. Shimura, T. Tamura and M. Sodeoka, *Org. Lett.*, **2004**, 6, 1861.
68. (a) M. P. Sibi, J. J. Shay, M. Liu and C. P. Jasperse, *J. Am. Chem. Soc.*, **1998**, 120, 6615. (b) M. P. Sibi and M. Liu, *Org. Lett.*, **2001**, 3, 4181. (c) M. P. Sibi, U. Gorikunti and M. Liu, *Tetrahedron*, **2002**, 58, 8357.
69. X. L. Jin, H. Sugihara, K. Daikai, H. Tateishi, Y. Z. Jin, H. Furuno and J. Inanaga, *Tetrahedron*, **2002**, 58, 8321.
70. N. Yamagiwa, S. Matsunaga and M. Shibasaki, *J. Am. Chem. Soc.*, **2003**, 125, 16178.
71. A. Palomo, M. Oiarbide, R. Halder, M. Kelso, E. Gómez-Bengoa and J. M. Garcia, *J. Am. Chem. Soc.*, **2004**, 126, 9188.
72. J. K. Meyers and E. N. Jacobsen, *J. Am. Chem. Soc.*, **1999**, 121, 8959.
73. Y. K. Chen, M. Yoshida and D. W. C. MacMillan, *J. Am. Chem. Soc.*, **2006**, 128, 9328.
74. U. Uria, J. L. Vicario, D. Badia and L. Carillo, *J. Chem. Soc., Chem. Commun.*, **2007**, 2509.
75. P. Diner, M. Nielsen, M. Marigo and K. A. Jørgensen, *Angew. Chem. Int. Ed.*, **2007**, 46, 1983.
76. J. Wang, H. Li, L. Zu and W. Wang, *Org. Lett.*, **2006**, 8, 1391.
77. D. J. Guerin and S. J. Miller, *J. Am. Chem. Soc.*, **2002**, 124, 2134.

Chapter 12
Further Catalytic Reactions

The final chapter contains a collection of important reactions which haven't found a comfortable home elsewhere in the book. Some of the following sections describe reactions that are topical and have been successfully achieved with relatively high ee, such as deprotonations, ester formation, alkylations of enolates and epoxide opening.

12.1   Isomerisations and Rearrangements

The isomerisation of allylic amines into the corresponding enamines is an excellent example of asymmetric catalysis, which has been exploited on a commercial basis. The isomerisation of the allylamine (**12.01**) with a rhodium/BINAP complex occurs with excellent yield and enantioselectivity to give the enamine (**12.02**) as the initial product.[1]

The story of the development of the lab-scale reaction into a 1000 ton per year process has been told in several reviews,[2] and detailed information on the mechanism of the reaction has been published previously.[3] The initially formed enamine (**12.02**) is converted into citronellal (**12.03**) by hydrolysis, with subsequent cyclisation to isopulegol (**12.04**) and reduction to menthol (**12.05**). The whole process is performed by Takasago International Corporation, and represents the biggest application (so far) of an enantioselective reaction catalysed by a transition metal complex. Interestingly, the alternative geometry of starting material (the (*Z*)-isomer) affords the opposite enantiomer of product.

Similar isomerisation reactions have been applied to other substrates with very high enantioselectivities for many trisubstituted allylic amines. In general, the rearrangement of allylic alcohols[4] and ethers[5] provides lower enantioselectivity. However, higher ees have been obtained in the isomerisation of cyclic acetals and the desymmetrisation of 4,7-dihydro-1,3-dioxepins such as (**12.06**) occurs with up to 92% ee in the presence of the nickel-DUPHOS catalyst (**12.07**).[6,7] Unfunctionalised alkenes have been isomerised enantioselectively using a titanocene catalyst.[8]

Both the Claisen and aza-Claisen rearrangement can be catalysed with high ee using enantiomerically pure metal complexes. The group of Hiersmann have used

Catalysis in Asymmetric Synthesis 2e   © 2009 Vittorio Caprio and Jonathan M.J. Williams

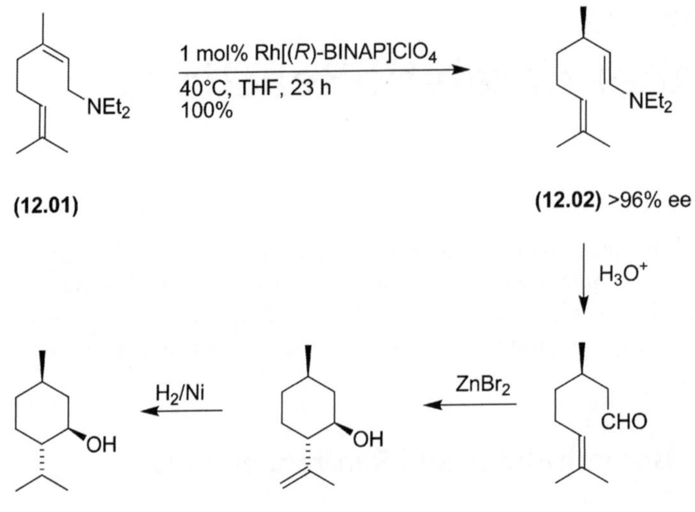

(12.01)

1 mol% Rh[(*R*)-BINAP]ClO$_4$
40°C, THF, 23 h
100%

(12.02) >96% ee

H$_3$O$^+$

menthol
(12.05)

H$_2$/Ni

(12.04)

ZnBr$_2$

(12.03)

the copper bis-oxazoline complex **(12.09)** as a Lewis acid in the highly enantiose-lective Claisen rearrangement of alkoxycarbonyl-substituted allyl vinyl ethers such as **(12.10)**[9] and applied this methodology in the enantioselective synthesis of a number of natural product targets.[10, 11]

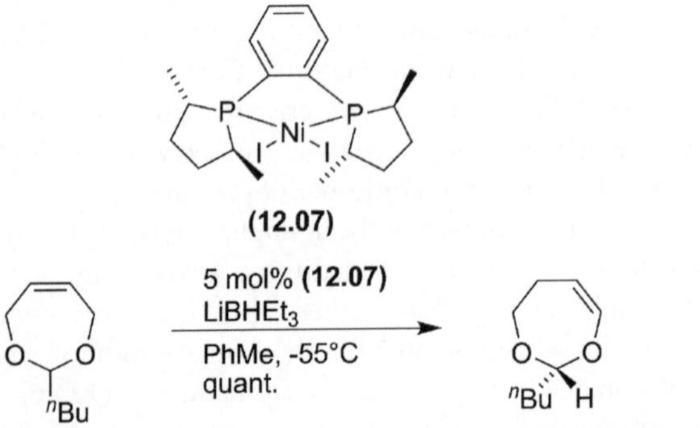

(12.07)

(12.06)

5 mol% **(12.07)**
LiBHEt$_3$

PhMe, -55°C
quant.

(12.08) 90% ee

The catalysed rearrangement of allylic imidates including allylic trichloroace-timidates such as **(12.12)** has been reported by Overman using palladium catalysts with some of the highest ees obtained using the cobalt oxazoline palladacycle **(12.13)**.[12]

Some enantioselective 1,2-carbon–carbon bond migrations have also been catalysed by enantiopure metal complexes. The rearrangement of achiral α,α-disubstituted silyloxy aldehydes to enantioenriched acyloins has been achieved with up to 92% ee using aluminium complexes,[13] while vinylcyclopropanols,

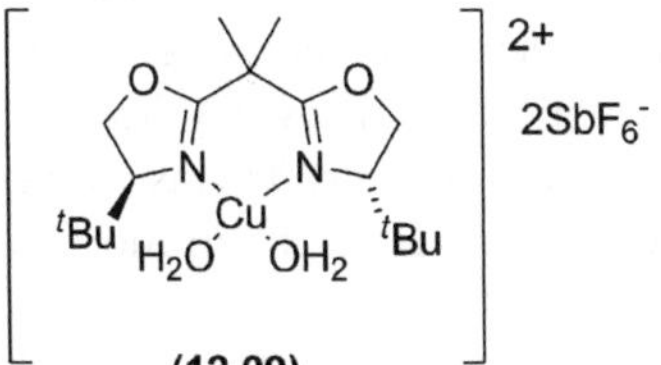

(12.09)

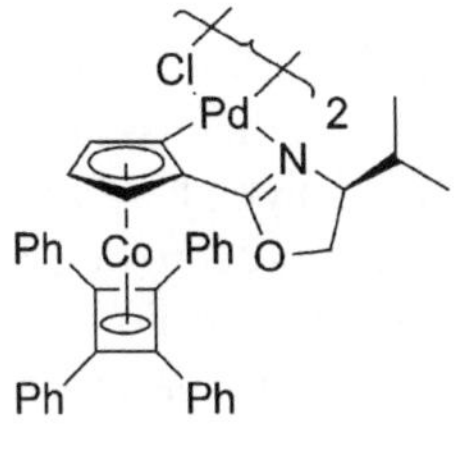

(12.13)

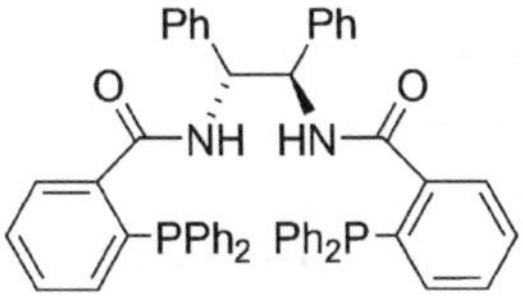

(12.16)

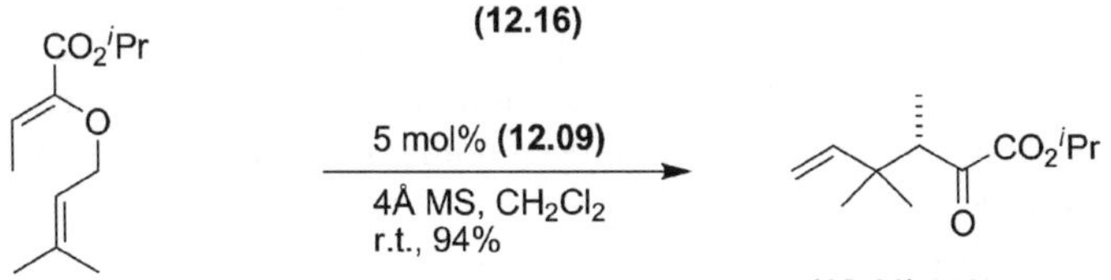

(12.10)

5 mol% (12.09)
4Å MS, CH₂Cl₂
r.t., 94%

(12.11) 99% ee

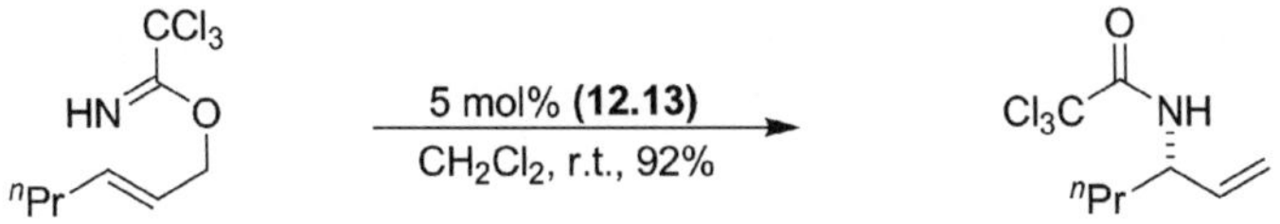

(12.12)

5 mol% (12.13)
CH₂Cl₂, r.t., 92%

(12.14) 98% ee

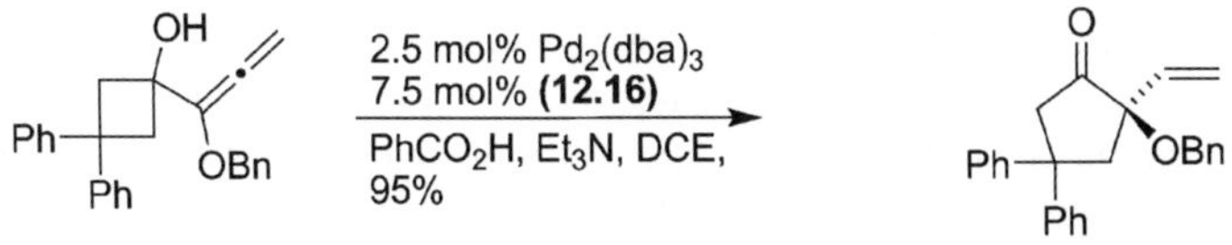

(12.15)

2.5 mol% Pd₂(dba)₃
7.5 mol% (12.16)
PhCO₂H, Et₃N, DCE,
95%

(12.17) 95% ee

vinylcyclobutanols and allenylcyclobutanols such as (**12.15**) undergo a catalytic Wagner–Meerwin shift via the formation of π-allylpalladium complexes in the presence of the catalyst generated from ligand (**12.16**).[14]

12.2   Deprotonation Reactions

The use of enantiomerically pure bases to catalyse asymmetric deprotonations is an exciting idea that has been shown to be technically feasible. The major difficulty is that the catalytic base must be continuously deprotonated under the reaction conditions. In order to be effective, whatever achiral base provides the continuous deprotonation must not directly deprotonate the substrate. This is conceptually similar to catalytic protonation reactions, which are described in more detail in the next section.

Using the catalytic base (**12.18**), Duhamel and coworkers have demonstrated that elimination of HBr from substrate (**12.19**) can be achieved with remarkable enantioselectivity in the formation of the alkene (**12.20**),[15] while diamine (**12.21**) has been used as a highly enantioselective base catalyst in the rearrangement of *meso*-epoxides such as (**12.22**).[16]

Sparteine and related compounds have been widely used in conjunction with butyllithium as an enantiomerically pure base. Deprotonation of the bicyclic epoxide (**12.24**) in the presence of catalytic amounts of isosparteine (**12.26**) results in formation of the product (**12.25**) by a cyclisation/rearrangement.[17] The catalytic asymmetric deprotonation of *N*-Boc pyrrolidine or *O*-alkylcarbamates has generally proceeded with poor yield and ee and this is attributed to chelation of the enantiomerically pure base with the product organolithium. O'Brien and coworkers have solved this problem by the addition of stoichiometric amounts of an achiral diamine that undergoes ligand exchange with this complex, freeing the base.[18] Thus, deprotonation of *N*-Boc pyrrolidine (**12.27**) with ^{s}BuLi/(-)-sparteine in the presence of 1.2 equivalents of bispidine (**12.28**) followed by addition of Me$_3$SiCl yields the product (**12.29**) with high ee in good yield.

12.3   Protonation Reactions

The protonation of prochiral enolates can be achieved using a stoichiometric enantiomerically pure proton source to provide a ketone or ester product, often with good control of enantioselectivity.[19]

The catalytic version of this reaction is more complex, since the kinetics in the catalytic cycle must be such that the achiral acid does not directly protonate the enolate at a significant rate. A catalytic cycle is represented in Figure 12.1, showing how the enantiomerically pure acid (EPA-H) protonates the enolate. The

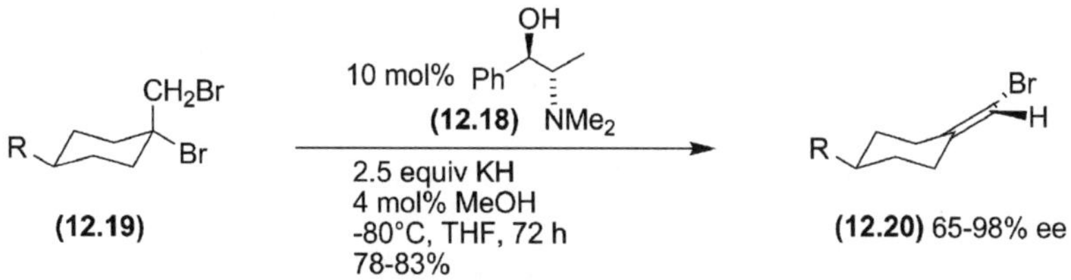

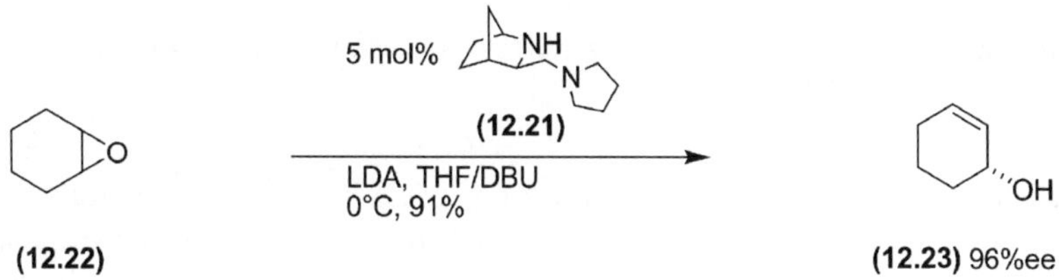

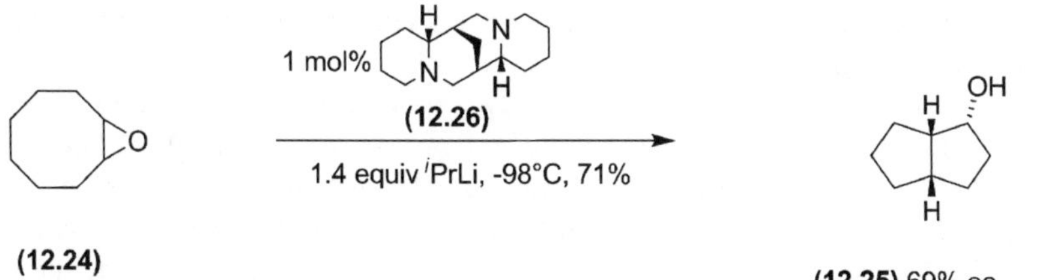

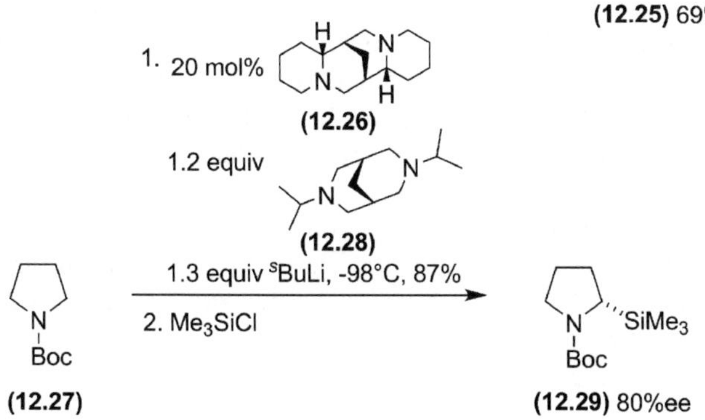

(12.19) → (12.20) 65-98% ee

Reagents: 10 mol% (12.18), 2.5 equiv KH, 4 mol% MeOH, -80°C, THF, 72 h, 78-83%

(12.22) → (12.23) 96%ee

Reagents: 5 mol% (12.21), LDA, THF/DBU, 0°C, 91%

(12.24) → (12.25) 69% ee

Reagents: 1 mol% (12.26), 1.4 equiv iPrLi, -98°C, 71%

(12.27) → (12.29) 80%ee

Reagents: 1. 20 mol% (12.26), 1.2 equiv (12.28), 1.3 equiv sBuLi, -98°C, 87%, 2. Me₃SiCl

achiral acid (AA-H) recharges the enantiomerically pure acid, but, ideally, doesn't protonate the enolate directly.

The achiral acid needs to be a kinetically slow acid, and is typically a hindered alcohol, an imide or a carbon acid (e.g. malonate). Enantiomerically pure imides (12.30),[20] diamine (12.31),[21] amino alcohols,[22] as well as a tetradentate amine have been used to protonate lithium enolates, including enolates (12.32) and

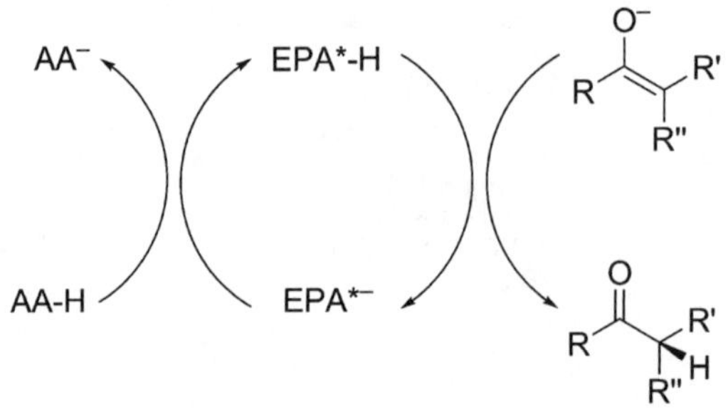

Figure 12.1 Catalytic cycle for protonation reactions. Abbreviations: AA = achiral acid; EPA = enantiomerically pure acid.

(**12.33**). In each of these cases, the stoichiometric achiral acid is kinetically slow, and is also added slowly to the reaction mixture.[23] The commercially available amino acid derivative (**12.34**) has also been used as an effective enantiomerically pure proton source in the protonation of tetralone enolates such as (**12.35**).[24] Samarium enolates have also been protonated enantioselectively (up to 93% ee) using catalytic amounts of a C_2-symmetric diol.[25]

The enantioselective protonation of silyl enol ethers, such as (**12.39**), by a catalyst has been achieved using 2 mol% of the proton source (**12.40**).[26] The acidity of (**12.40**) is enhanced by coordination to a Lewis acid. The silyloxy group is activated by fluoride ion and up to 99% ee in the asymmetric protonation of α-aryl substituted cyclic silyl enol ethers such as (**12.39**) has been obtained using a Lewis acidic BINAP•AgF complex.[27] In a similar vein, silyl enol ethers of tetralones and indanones undergo asymmetric protonation with moderate to good ee using catalytic quantities of hydrogen fluoride salts of cinchona alkaloids in the presence of acyl fluorides and ethanol, which act as a stoichiometric source of HF.[28]

An alternative approach to the protonation of silyl enol ethers involves the use of palladium catalysts which proceed via intermediate palladium enolates. The asymmetry can either be provided by ligands on the palladium[29] or from an enantiomerically pure acid.[30]

12.4 Alkylation and Allylation of Enolates

While the α-allylation of enolates occurs with high ee using palladium catalysts (see Section 10.2) there have been few reports on the enantioselective metal-catalysed enolate alkylation.[31] The best results to date have been achieved by Doyle and Jacobsen using the chromium(salen) complex (**12.42**) in the alkylation of cyclic tin enolates with a range of alkyl halides, including propargyl and benzylic halides

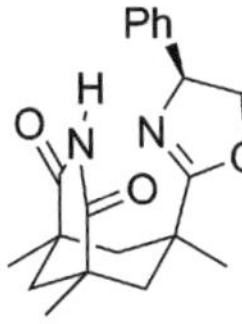

Ph
Ph
H
O
N
N
O
O

(12.30)

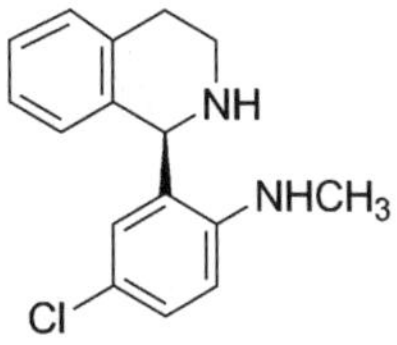

NH
NHCH3
Cl

(12.31)

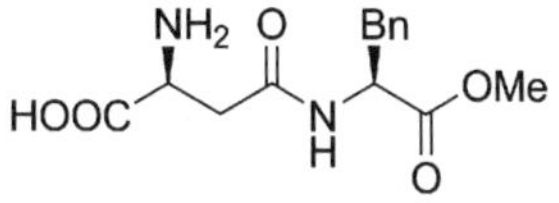

NH2
O
Bn
HOOC
N
H
OMe
O

(12.34)

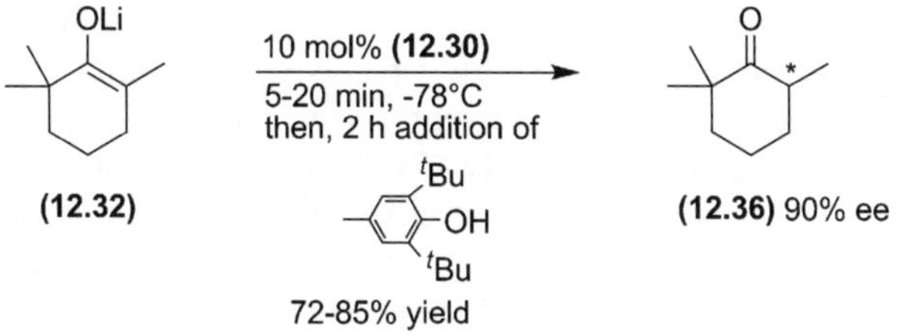

OLi
10 mol% (12.30)
5-20 min, -78°C
then, 2 h addition of
tBu
OH
tBu
72-85% yield
O
*

(12.32)

(12.36) 90% ee

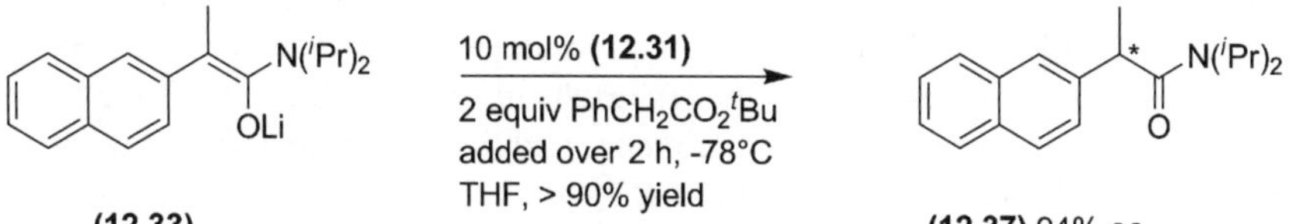

N(iPr)2
OLi
10 mol% (12.31)
2 equiv PhCH2CO2tBu
added over 2 h, -78°C
THF, > 90% yield
*
N(iPr)2
O

(12.33)

(12.37) 94% ee

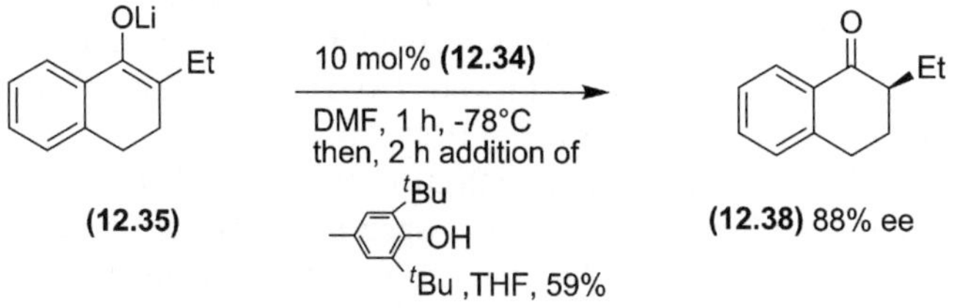

OLi
Et
10 mol% (12.34)
DMF, 1 h, -78°C
then, 2 h addition of
tBu
OH
tBu ,THF, 59%
O
Et

(12.35)

(12.38) 88% ee

and even methyl iodide,[32] and has also been recently applied to the alkylation of acyclic enolates.[33]

Phase transfer reactions have featured in several sections of this book, including epoxidation (Section 4.5), Darzens condensation (Section 7.5) and Wadsworth–Emmons reactions (Section 12.5). Another important aspect of phase-transfer catalysed reactions has been with alkylation reactions.[34] The asymmetric alkylation of glycinate Schiff base **(12.45)** using *N*-benzylcinchoninium halides as catalysts is particularly noteworthy, since the products are readily converted into amino acids.[35] Corey and coworkers have developed the original work, and have shown that the quaternary ammonium salt (**12.46a**) bearing an *N*-anthracenylmethyl moiety is remarkably capable as an asymmetric phase transfer catalyst.[36] Alkyl iodides and Michael acceptors have been used as the electrophile in these reactions.[37] Thus, alkyl iodide (**12.47**) affords the alkylation product (**12.48**) with excellent enantioselectivity. Other electrophiles also work well (92–99.5% ee). The anthracenyl unit has also been incorporated into cinchona-based phase transfer catalyst (**12.46b**) by Lygo and Wainwright for the enantioselective catalysis of alkylation reactions of imine ester (**12.45**).[38] Catalysts of this type have also been used in the enantioselective benzylation and allylation of oxazoline carboxylates such as (**12.49**).[39] Hydrolysis of the product yields alkylated serines. The use of β-chloro-α,β-enones as electrophiles results in the introduction of an alkenyl moiety via an addition-elimination sequence and this transformation occurs with high ee using derivatives of catalyst (**12.46b**).[40]

Maruoka and coworkers have obtained high ees in the alkylation of imine (**12.45**) using the C_2-symmetric ammonium salt (**12.51**),[41] and have applied this catalyst to the alkylation of the corresponding amide, Weinreb amide[42] and also cyclic β-keto esters.[43] Phase transfer reactions are not limited to quaternary ammonium salts for a successful outcome. Even the diol TADDOL (see Section 8.1) has been shown to be effective in the alkylation of related imines.[44]

L-proline and derivatives function as effective catalysts in the α-functionalisation of aldehydes with a range of heteroatomic species (see Section 5.3) and Vignola and List have attempted to apply this methodology to the asymmetric α-alkylation. While the intermolecular reaction has proved unsuccessful, high ees in the cyclisation of aldehydes such as (**12.53**) have been achieved using (*S*)-α-methylproline (**12.52**).[45]

MacMillan and coworkers have significantly expanded the scope of this enamine-mediated procedure by the addition of stoichiometric amounts of oxidant that leads to the *in situ* formation of a radical cation (**12.57**).[46] This intermediate then undergoes enantioselective radical-based addition with a range of unsaturated substrates (**12.58**). For example, α-allylation with allylsilanes such as (**12.60**) can be effected with high ee using CAN as oxidant in the presence of imidazolidinone (**12.61**) as catalyst, while an α-heteroarylation occurs using *N*-Boc pyrrole. Furthermore, an asymmetric α-enolation of a range of aldehydes can be achieved by addition of silyl enol ethers such as (**12.64**).

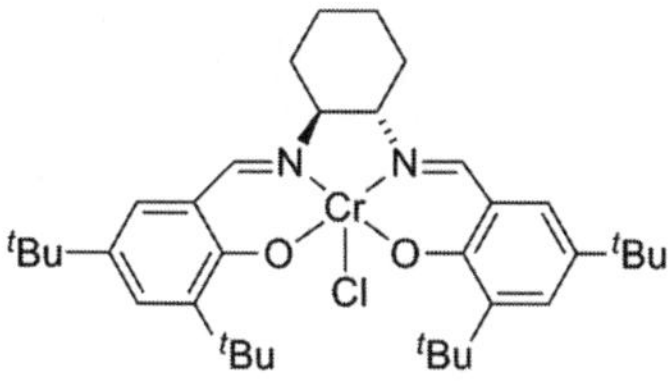

(12.42)

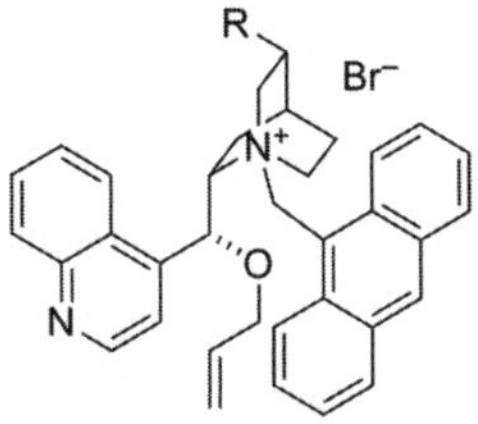

(12.46a) R = vinyl
(12.46b) R = ethyl

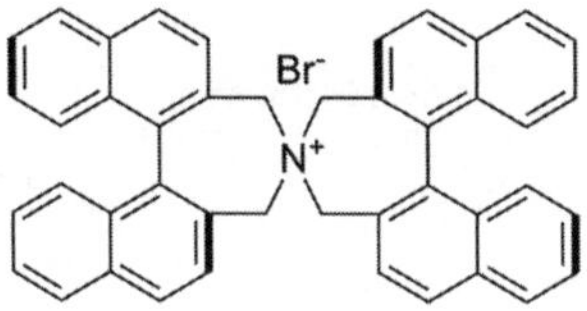

(12.51)

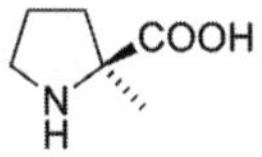

(12.52)

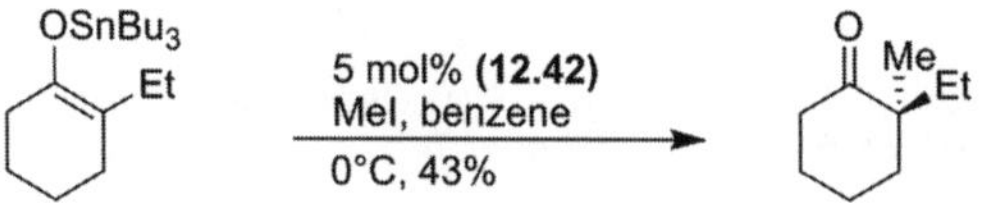

(12.43)

5 mol% **(12.42)**
MeI, benzene
0°C, 43%

(12.44) 90%ee

(12.45)

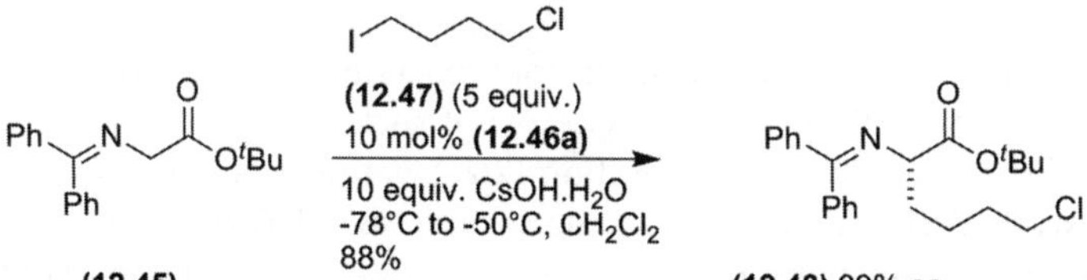

(12.47) (5 equiv.)
10 mol% **(12.46a)**
10 equiv. CsOH.H₂O
-78°C to -50°C, CH₂Cl₂
88%

(12.48) 99% ee

(12.49)

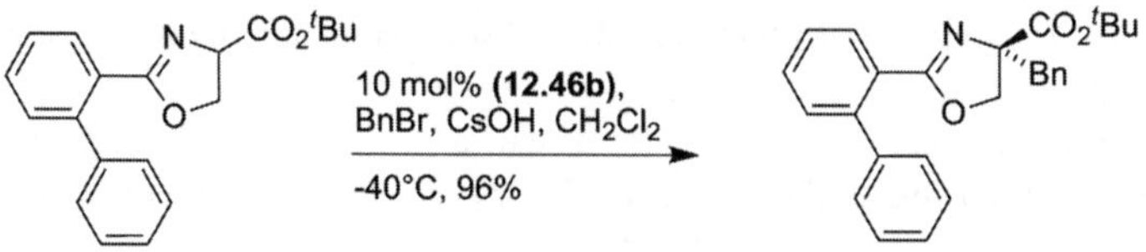

10 mol% **(12.46b)**,
BnBr, CsOH, CH₂Cl₂
-40°C, 96%

(12.50) 96% ee

(12.53)

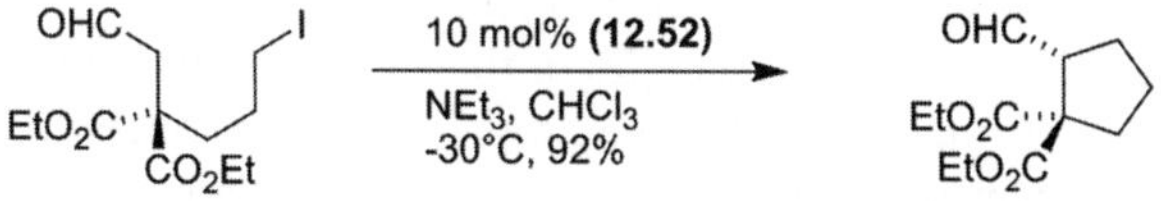

10 mol% **(12.52)**
NEt₃, CHCl₃
-30°C, 92%

(12.54)

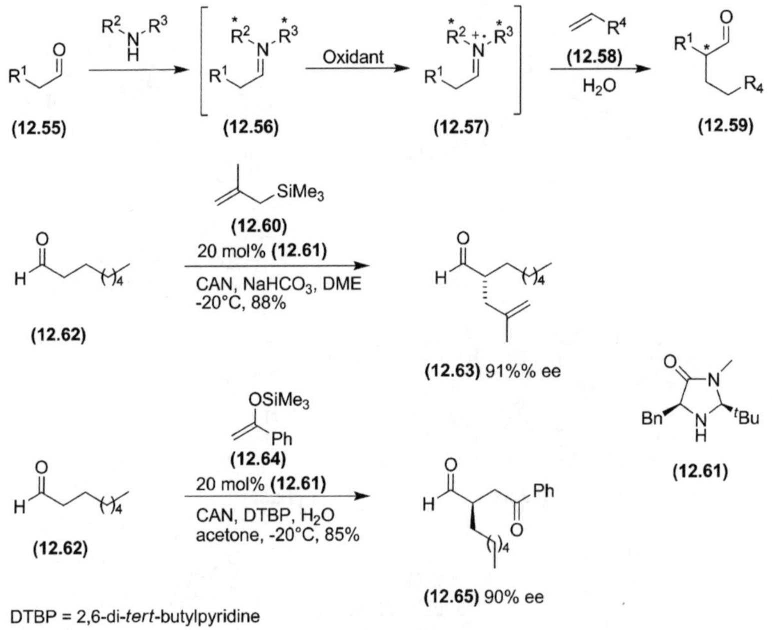

(12.55) **(12.56)** **(12.57)** **(12.59)**

(12.60)
20 mol% **(12.61)**

CAN, NaHCO$_3$, DME
-20°C, 88%

(12.62)

(12.63) 91%% ee

(12.61)

OSiMe$_3$

(12.64)
20 mol% **(12.61)**

CAN, DTBP, H$_2$O
acetone, -20°C, 85%

(12.62)

(12.65) 90% ee

DTBP = 2,6-di-*tert*-butylpyridine

12.5 Formation of Alkenes

The formation of alkenes is not an obvious reaction to explore for enantioselective catalysis. Nevertheless, some examples of such reactions have already been described in earlier sections (e.g. Section 12.2, Deprotonation Reactions).

The synthetically useful ring-closing metathesis[47] has been investigated using enantiomerically pure molybdenum complexes such as **(12.66)**, **(12.67)** and **(12.68)**. The kinetic resolution of acyclic dienes has been achieved with some success using such catalysts.[48,49] Thus, the racemic diene **(12.69)** undergoes a kinetic resolution such that the product **(12.70)** and recovered starting material can both be obtained with good enantioselectivity.

Desymmetrisations of *meso*-substrates can also be achieved using catalysts of this type.[49c, 50] For example, allyl ether **(12.71)** is converted into the heterocycle **(12.72)** with high ee using **(12.67b)**.[49c]

Enantiomerically pure ruthenium catalysts such as **(12.73)**[51] and **(12.74)**,[52] which are more air-stable and functional group tolerant than the molybdenum complexes, have also been shown to effect high ees in the desymmetrisations of *meso*-substrates. Furthermore, the stability of these catalysts allows recovery and recycling. High ees are not restricted to the simple ring-closing metathesis process and good levels of enantiocontrol have been achieved using both molybdenum and ruthenium catalysts in the ring-opening/cross-metathesis reaction of cyclopropenes and norbornenes such as **(12.75)**.[52, 53]

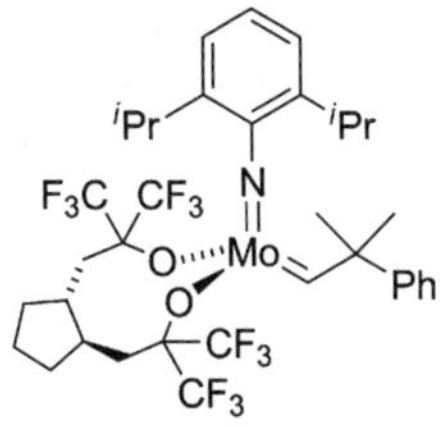

(12.66)

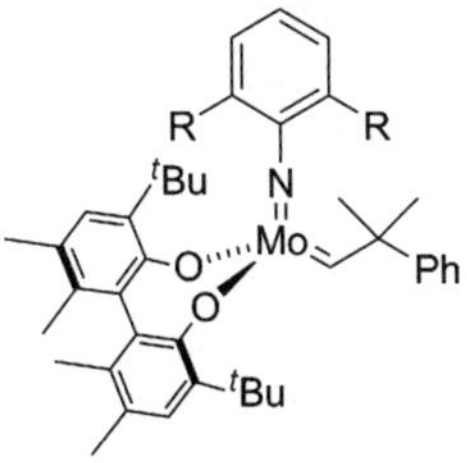

(12.67a) R = ^{i}Pr
(12.67b) R = Me

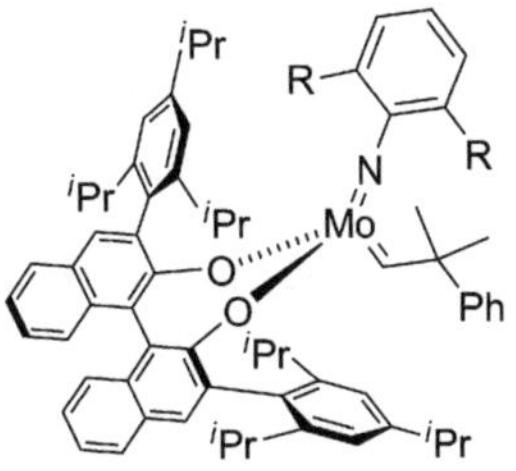

(12.68)

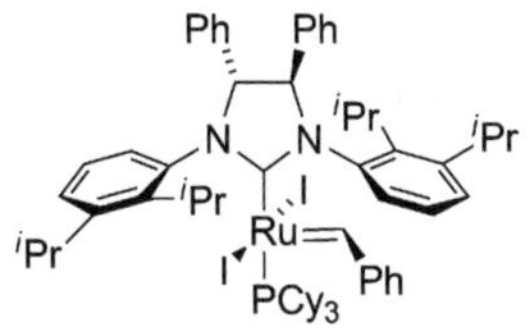

(12.73)

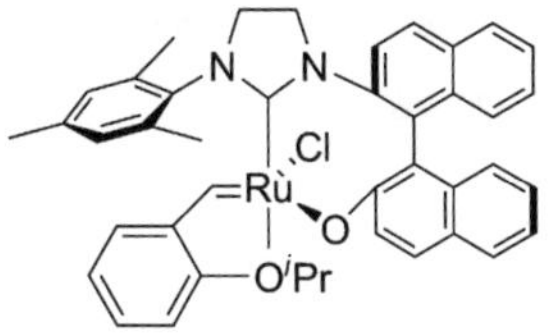

(12.74)

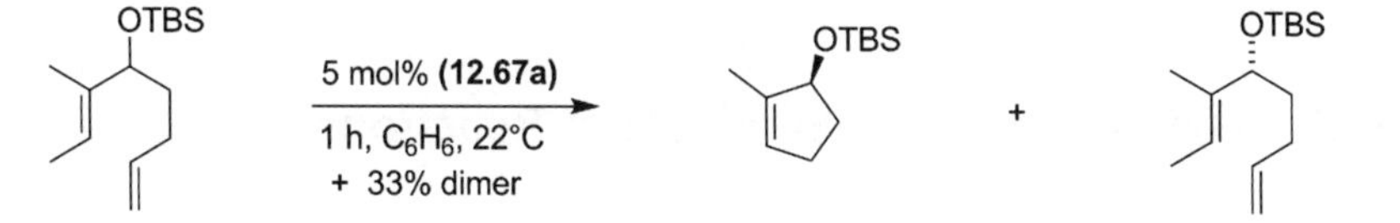

(±)-**(12.69)**

5 mol% **(12.67a)**

1 h, C$_6$H$_6$, 22°C
+ 33% dimer

(12.70) 93% ee
42%

+

(12.69) >99% ee
25%

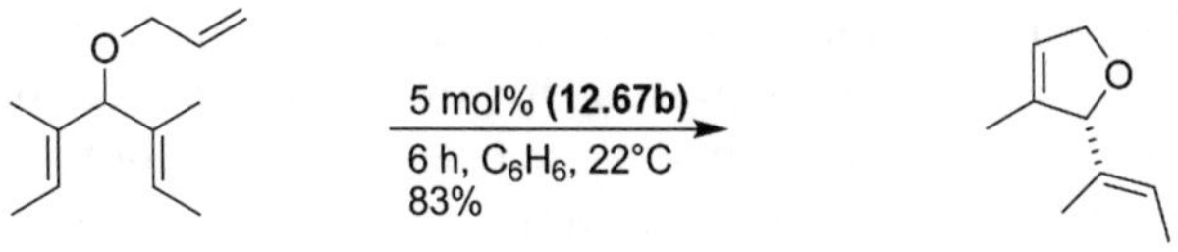

(12.71)

5 mol% **(12.67b)**

6 h, C$_6$H$_6$, 22°C
83%

(12.72) 99% ee

^{n}C$_5$H$_{11}$

5 mol% **(12.74)**

THF, 22°C, 77%

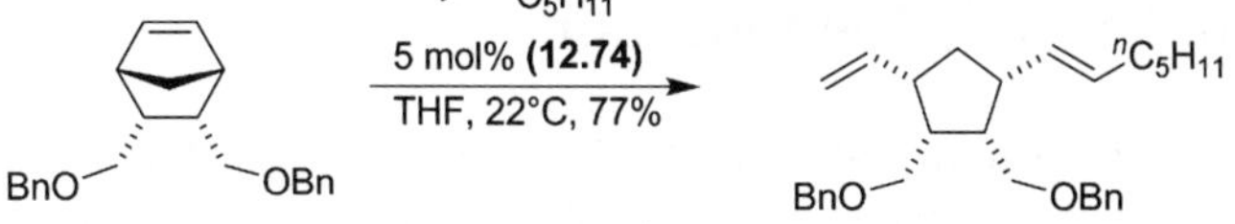

(12.75)

(12.76) 91% ee

As well as ring-closing metathesis reactions, the catalytic asymmetric Horner–Wadsworth–Emmons reaction has been achieved using phase-transfer catalysts including ammonium salt (12.77), with rubidium hydroxide as base. The achiral ketone (12.78) is converted into the alkene (12.79) with reasonable enantioselectivity. Currently, the long reaction time is a drawback.[54]

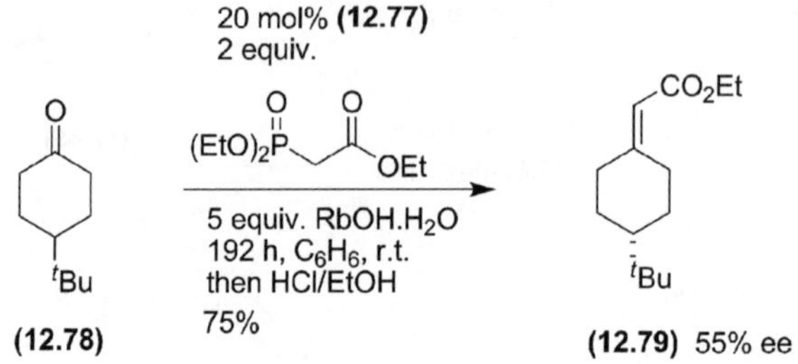

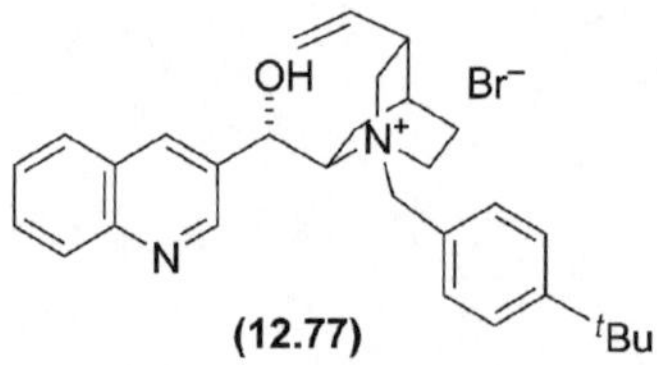

12.6   Oxyselenylation-Elimination Reactions

A catalytic oxyselenylation-elimination sequence has been devised that provides enantiomerically enriched allyl ethers.[55, 56, 57] For example,[58] the diselenide (12.80) is used to effect the transformation of *trans*-β-methyl-styrene (12.81) into the allyl ether (12.82). The catalytic pathway proceeds via initial oxyselenylation, followed by oxidation of the selenium, and elimination to provide the product along with a selenium reagent capable of repeating the catalytic cycle. Some of the best ees to date have been obtained by Tiecco and coworkers during the oxidation of β,γ-unsaturated esters such as (12.83) using the sulfur-containing diselenide (12.84).[59]

12.7   The Benzoin Condensation

Heterazolium salts such as thiazolium salts[60] (12.86) and triazolium salts[60d, 61,62] are able to catalyse the conversion of benzaldehyde (12.87) into benzoin (12.88). The mechanism involves deprotonation of the thiazolium salt to give the true catalytic species (12.89), which acts as a nucleophile towards benzaldehyde, and subsequently promotes the formation of benzoin.

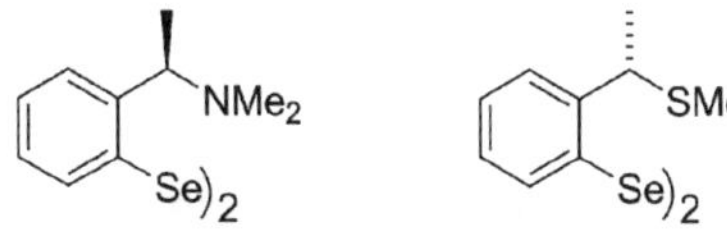

(12.80) **(12.84)**

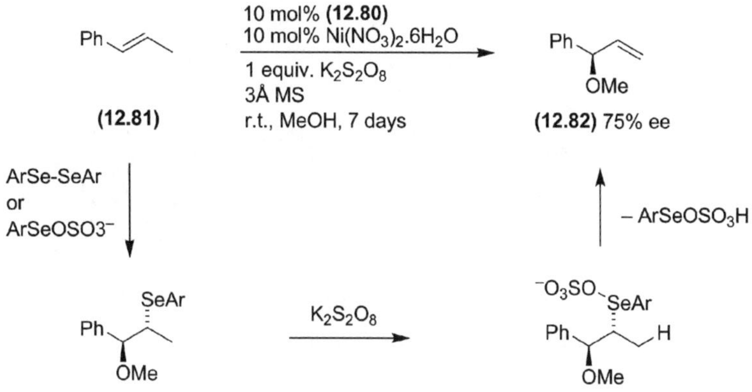

Ph

(12.81)

10 mol% **(12.80)**
10 mol% Ni(NO$_3$)$_2$.6H$_2$O

1 equiv. K$_2$S$_2$O$_8$
3Å MS
r.t., MeOH, 7 days

Ph

OMe

(12.82) 75% ee

ArSe-SeAr
or
ArSeOSO3$^-$

SeAr

Ph

OMe

K$_2$S$_2$O$_8$

$^-$O$_3$SO SeAr

Ph H

OMe

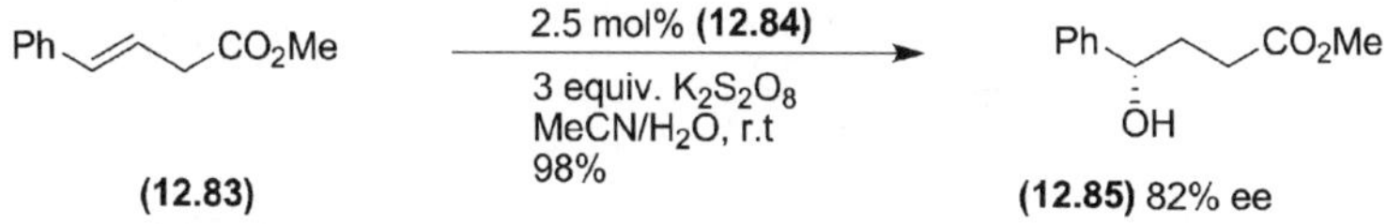

– ArSeOSO$_3$H

Ph CO$_2$Me

(12.83)

2.5 mol% **(12.84)**

3 equiv. K$_2$S$_2$O$_8$
MeCN/H$_2$O, r.t
98%

Ph CO$_2$Me

OH

(12.85) 82% ee

Several enantiomerically pure triazolium catalysts have been reported, generally with an enantiomerically pure moiety attached to the nitrogen of the triazolium ring with salts **(12.90)**[61] and **(12.91)**[62] providing some of the highest selectivities in the homo benzoin reaction. Some high ees have also been obtained in the intramolecular crossed benzoin reaction using enantiomerically pure triazolium salt precatalysts.[63] In one example, Enders and coworkers achieved up to 98% ee in the cyclisation of ketoaldehyde **(12.92)** using the polycyclic triazolium salt **(12.93)**.

12.8 Ester Formation and Hydrolysis

The enantioselective formation and hydrolysis of esters has been very heavily investigated using enzymatic methods, and has been well reviewed elsewhere.[64]

The use of nonenzymatic methods for the kinetic resolution of ester formation and hydrolysis is becoming increasingly important. A range of new catalysts has been shown to have good levels of enantioselectivity.[64b, 65]

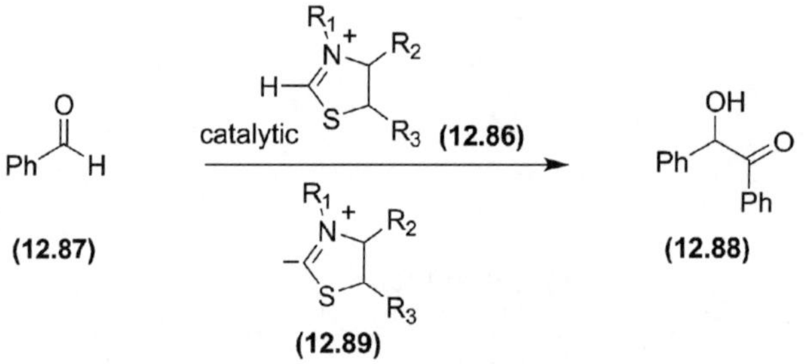

(12.87) **(12.88)**

(12.90) **(12.91)** **(12.93)**

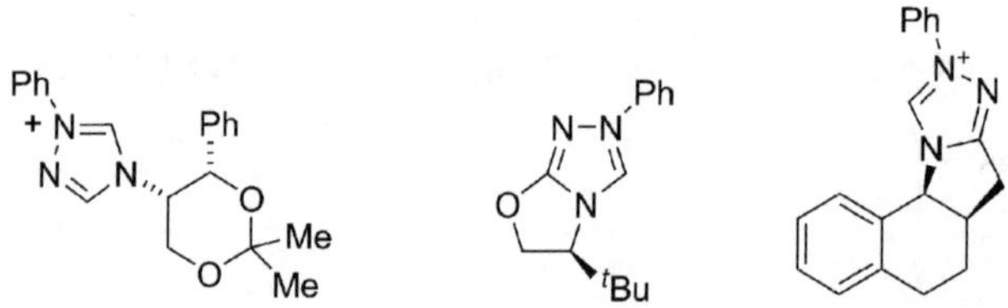

provides 95% ee   provides 95% ee

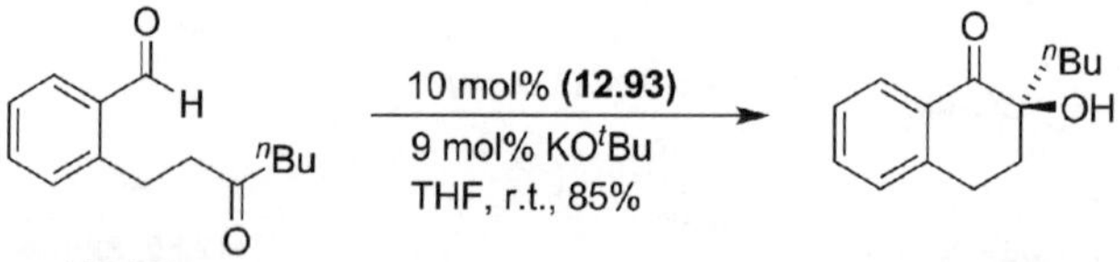

(12.92) **(12.94)** 98% ee

Fu and coworkers have developed planar nucleophilic catalysts, including complex (12.95), which is a versatile and efficient acylation catalyst.[66] This catalyst functions like the achiral catalyst DMAP (4-dimethylaminopyridine). The selectivity factor, S, is a good measure of the discrimination between enantiomers. It is defined as the rate of the faster-reacting enantiomer divided by the rate of the slower-reacting enantiomer (see Section 4.1).[66e, 66]

Racemic *sec*-alcohols (12.96) undergo acylation with acetic anhydride using the Fu catalyst (12.95) and achieving very good selectivity factors. Copper bisoxazoline catalysts also provide high selectivities in the asymmetric benzoylation of *trans*-1,2-diols.[68]

A range of metal-free organocatalysts have been used to good effect in the kinetic resolution of alcohols. A number of enantiomerically pure 4-aminopyridine-based catalysts effect kinetic resolution with high S values.[69] These include the conformationally restricted 2,3-dihydroimidazo[1,2-*a*]pyridine (12.98)[69c] and the atropisomeric catalyst (12.99).[69b, 69d] *N*-heterocyclic carbenes also function as nucleophilic

(±)-(12.96) (12.96) (12.97) (12.95)

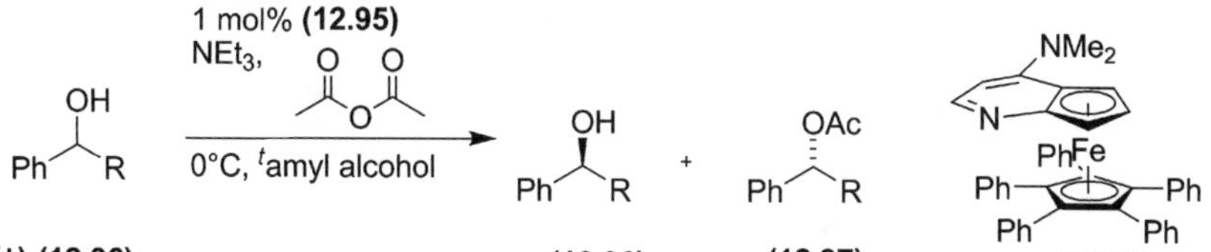

R = Me S = 43 99% ee at 55% conversion
CH_2Cl S = 32 98% ee at 56% conversion
^tBu S = 95 96% ee at 51% conversion

acylation catalysts and enantiopure carbenes such as (**12.100**) generated *in situ* have been used to good effect in the kinetic resolution of secondary alcohols.[70] Enantioselective acylations are also catalysed by phosphines and selectivity factors up to 380 have been achieved in the acylation of *sec*-alcohols using phosphines such as (**12.101**).[71] While most nonenzymatic catalysts are evaluated using secondary alcohols, the kinetic resolution of racemic primary alcohols has been achieved albeit with moderate ee and *S* levels using the proline-derived catalyst (**12.102**).[72] Miller and coworkers have shown that in addition to these small catalysts, the kinetic resolution of a wide range of secondary alcohols is also catalysed by octapeptides with a high degree of enantioselection at low loadings.[73]

The desymmetrisation of *meso*-1,2-diols has been achieved by Oriyama using enantiopure diamines including (**12.103**),[74,75] and also atropisomeric 4-aminopyridines.[68b] Metal-based catalysts have also been used to good effect in

(12.98)

(12.99)

(12.100)

(12.101)

(12.102)

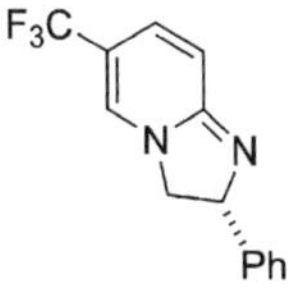

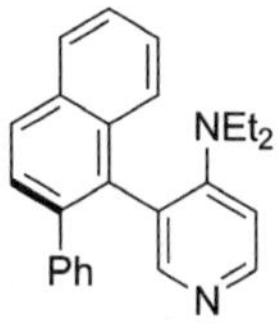

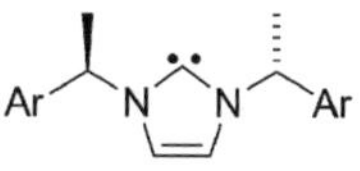

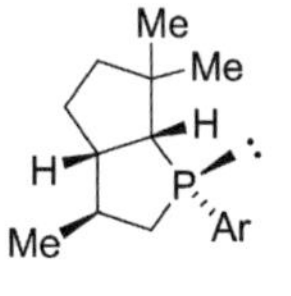

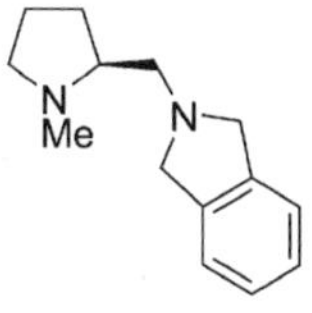

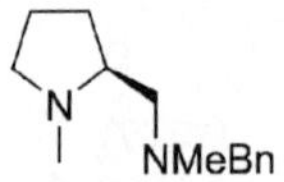

(12.103)

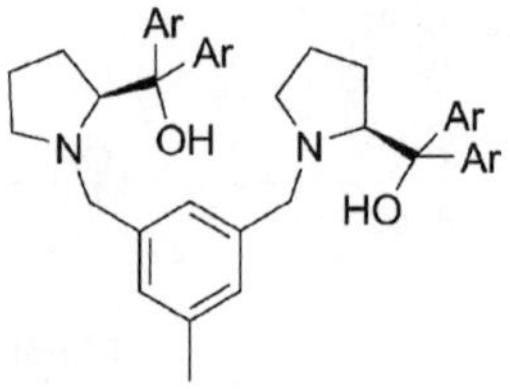

(12.109)

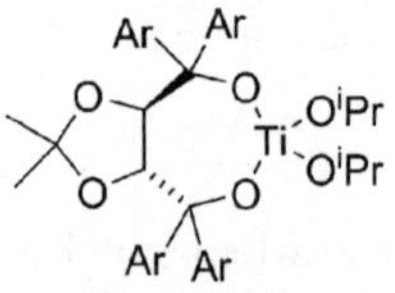

(12.111)

Ar = 2-naphthyl

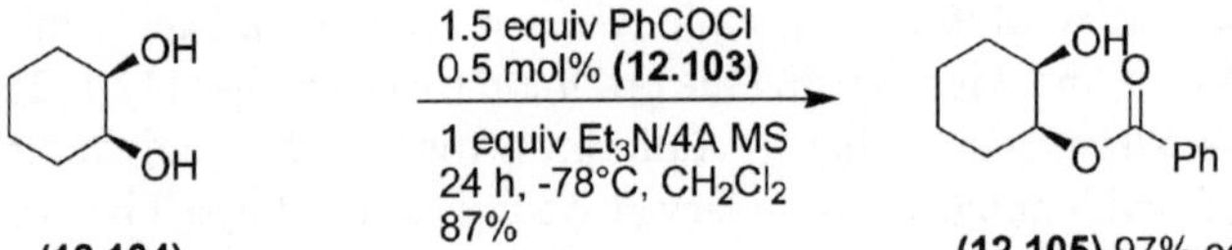

(12.104)

1.5 equiv PhCOCl
0.5 mol% **(12.103)**

1 equiv Et₃N/4A MS
24 h, -78°C, CH₂Cl₂
87%

(12.105) 97% ee

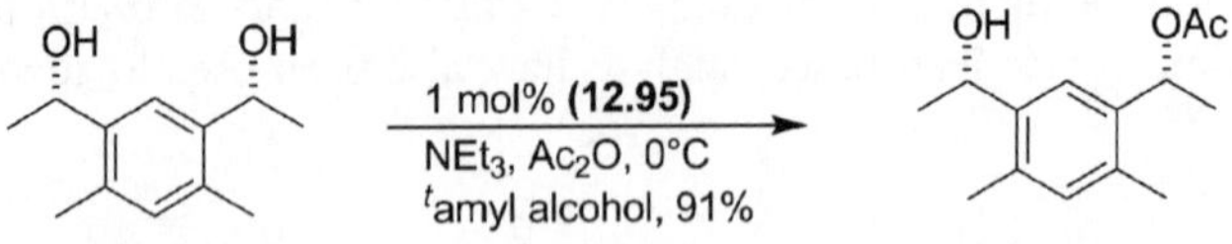

(12.106)

1 mol% **(12.95)**

NEt₃, Ac₂O, 0°C
ᵗamyl alcohol, 91%

(12.107) 99.7% ee

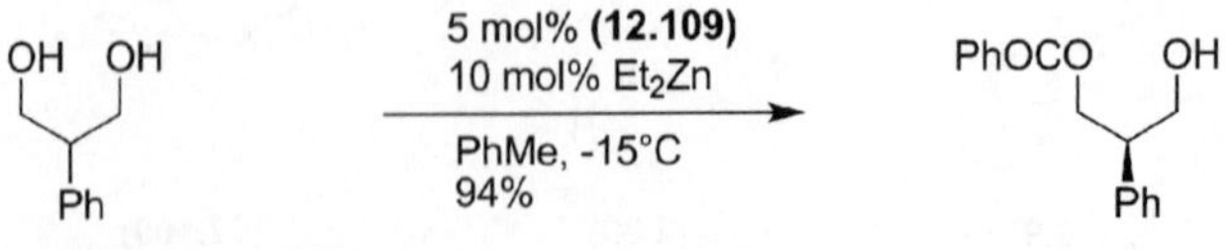

(12.108)

5 mol% **(12.109)**
10 mol% Et₂Zn

PhMe, -15°C
94%

(12.110) 91%ee

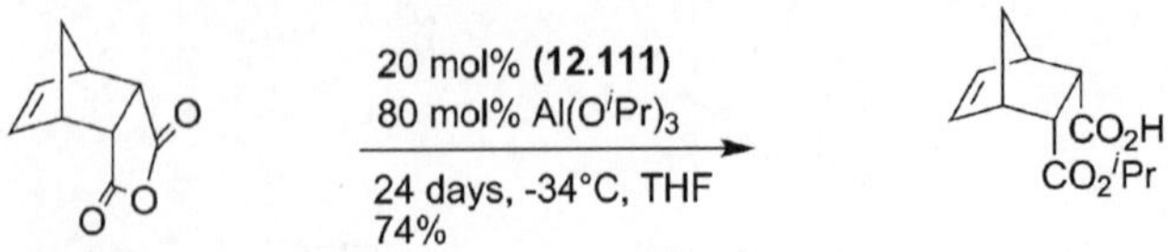

(12.112)

20 mol% **(12.111)**
80 mol% Al(OⁱPr)₃

24 days, -34°C, THF
74%

(12.113) 96% ee

the asymmetric desymmetrisation. For instance the *meso*-diol (**12.106**) is monoacylated with remarkable selectivity using Fu's catalyst (**12.95**).[66c] Copper bisoxazoline complexes have also been successfully applied to the asymmetric desymmetrisation of cyclic *meso*-*vic*-diols.[68a, 76] In addition *meso*-diols with primary hydroxyl group such as (**12.108**) have been desymmetrised by Trost and Mino using the dinuclear zinc catalyst derived from ligand (**12.109**).[77] Ti-TADDOLates, such as complex (**12.111**) have been used catalytically to open cyclic *meso*-anhydrides.[78] The *meso*-anhydride (**12.112**) is converted into the ring-opened product (**12.113**) with very good enantiocontrol, but the reaction time needs to be shortened in order for this process to become synthetically useful.

Fu has demonstrated that a dynamic kinetic resolution using the nonenzymatic catalyst (**12.95**) can be achieved.[79] The azlactone (**12.114**) is very prone to racemisation, whilst the ring-opened product (**12.115**) is stable under the reaction conditions. Thus the product is formed by methanolysis under dynamic resolution conditions (see Section 3.1), albeit with moderate enantioselectivity so far.

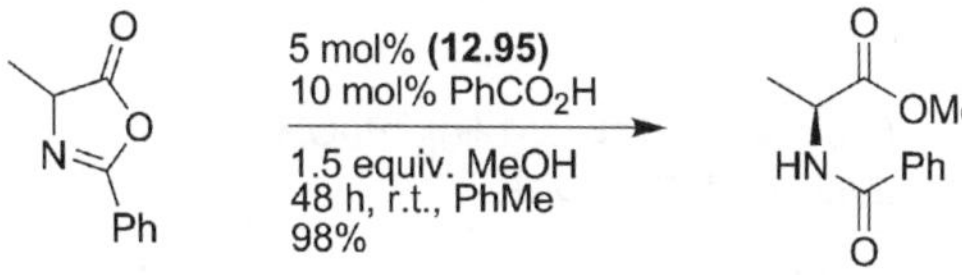

(12.114) **(12.115)** 54% ee

12.9 Ring-Opening of Epoxides

The nucleophilic ring opening of *meso*-epoxides leads to chiral products. Control over the position of attack will provide an enantiomerically enriched product, as shown in Figure 12.2, where cyclohexene oxide (**12.116**) undergoes ring-opening to produce either compound (**12.117**) or its enantiomer (*ent*-**12.117**).[70]

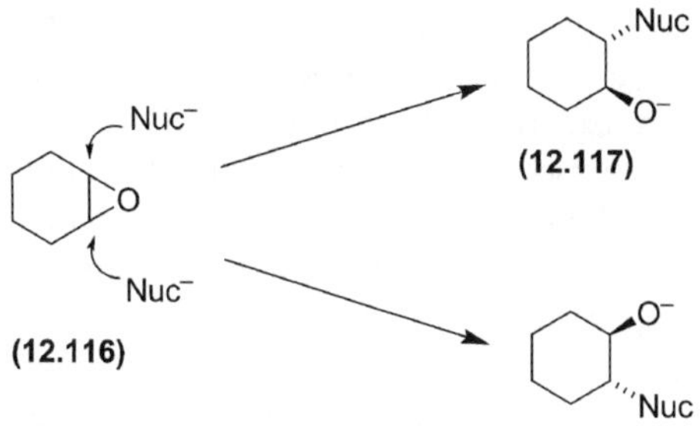

(12.116) **(12.117)** **(*ent*-12.117)**

Figure 12.2 Asymmetric ring-opening of expoxides.

Ring opening of *meso*-epoxides with azide can be a synthetically useful process.[81] Nugent has used ligand (**12.118**) associated with a zirconium catalyst,[82] whilst Jacobsen has used the chromium(salen) complex (**12.119**).[83] Representative reactions include the ring-opening of cyclohexene oxide (**12.116**) and cyclopentene oxide (**12.120**), which give good yields and enantioselectivities in the azide ring-opening reactions.

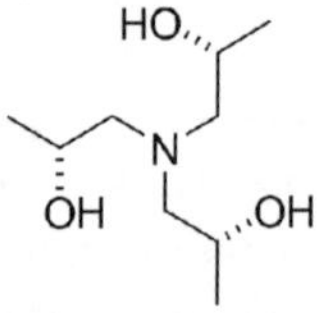

(12.118)

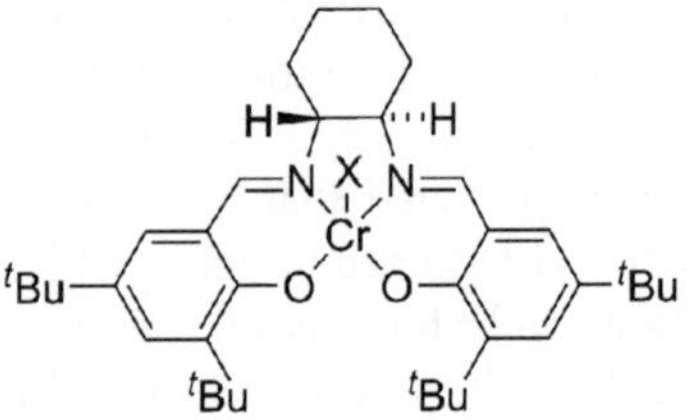

(12.119-X)
X = Cl or N$_3$

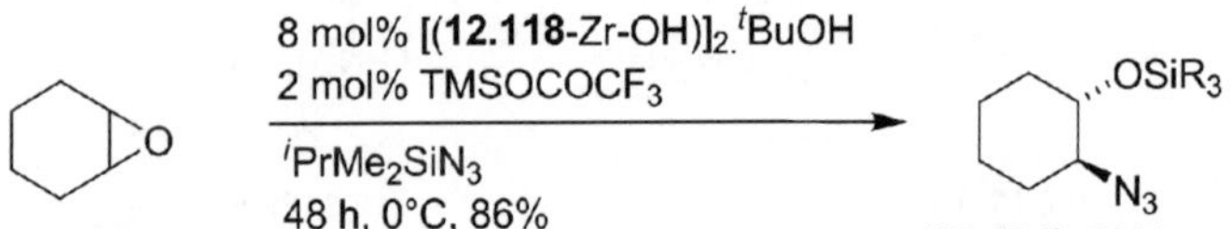

8 mol% [(**12.118**-Zr-OH)]$_2$.^{t}BuOH
2 mol% TMSOCOCF$_3$

^{i}PrMe$_2$SiN$_3$
48 h, 0°C, 86%

(12.116)

(12.121) 93% ee

2 mol% (**12.119**)

Me$_3$SiN$_3$
28 h, r.t., Et$_2$O, 80%

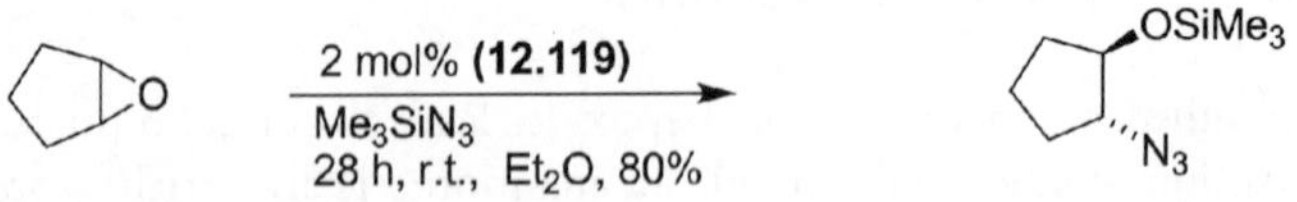

(12.120)

(12.122) 94% ee

The ring-opening of epichlorohydrin (**12.123**) with trimethylsilyl azide using catalyst (*ent*-**12.119**) merits special consideration.[83b, 84] The starting material is racemic, and on first consideration the product (**12.124**) should either be racemic or formed under kinetic resolution conditions. However, reversible ring-opening of the starting material by chloride provides a mechanism for its racemisation, thereby providing a dynamic kinetic resolution. The formation of symmetrical by-products lends support to the mechanism.

The asymmetric ring-opening of epoxides with amines is potentially problematic owing to competing reactions of the nucleophile with the Lewis acid catalyst employed. Nevertheless the catalytic asymmetric ring-opening of *meso*-epoxides with aromatic amines has been achieved with high ee. For example, aromatic *cis*-epoxides, such as *cis*-stilbene oxide (**12.125**), undergo ring-opening with up to

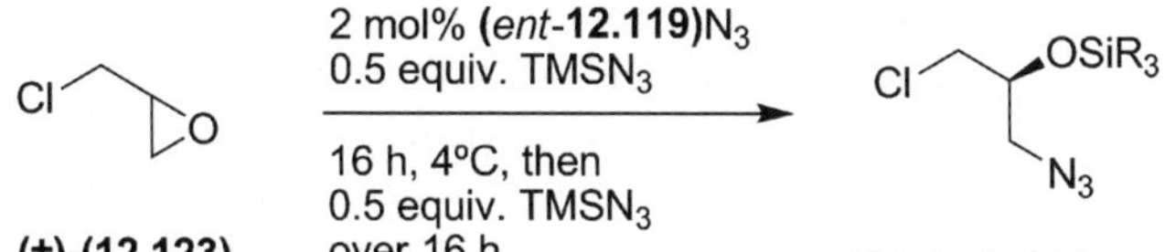

(±)-(12.123)

2 mol% (*ent*-**12.119**)N$_3$
0.5 equiv. TMSN$_3$

16 h, 4°C, then
0.5 equiv. TMSN$_3$
over 16 h

(12.124) 97% ee

+ 12% bis-chloro compound
+ 12 % bis-azido compound

96% ee in water, in the presence of the scandium complex generated with ligand (**12.126**), while ring-opening of cyclic epoxides such as (**12.120**) occurs with up to 93% ee using the samarium-BINOL catalyst (**12.128**).

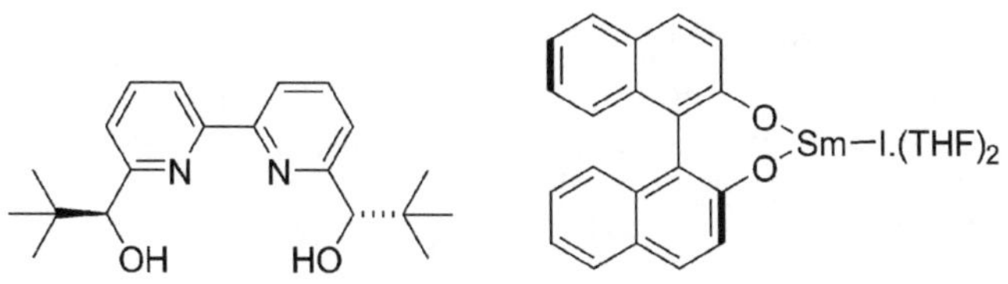

(12.126)

(12.128)

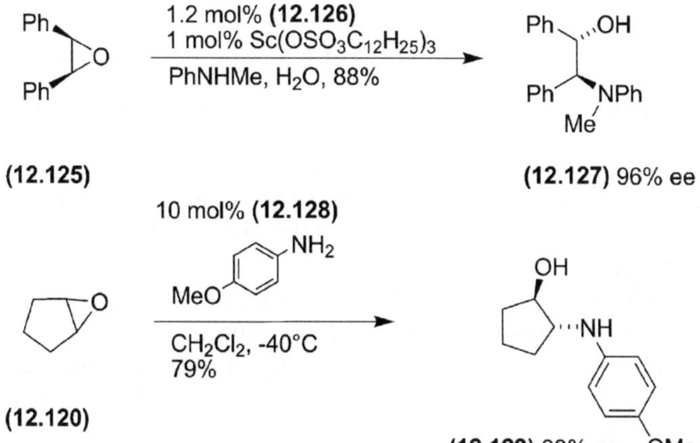

(12.125)

1.2 mol% (**12.126**)
1 mol% Sc(OSO$_3$C$_{12}$H$_{25}$)$_3$

PhNHMe, H$_2$O, 88%

(12.127) 96% ee

(12.120)

10 mol% (**12.128**)

CH$_2$Cl$_2$, -40°C
79%

(12.129) 93% ee

Snapper, Hoveyda and coworkers have used a ligand diversity approach for finding a catalyst for the addition of trimethylsilyl cyanide to epoxides.[86] The basic ligand structure (**12.130**) was optimised using combinatorial techniques, which allowed for the parallel synthesis of over 20 ligands a day. Variation of the first amino acid residue (AA1) identified *tert*-leucine as the best candidate for the conversion of cyclohexene oxide (**12.116**) into the ring-opened adduct (**12.131**) with a titanium catalyst. With AA1 fixed, the AA2 residue was varied, which identified L-threonine (*tert*-butyl ether) as the best candidate. Finally the Schiff base moiety was varied by the use of different aldehydes, and 3-fluorosalicylaldehyde was found to be the best choice. In this way, ligand (**12.132**) was identified through rapid screening, and the formation of ring-opened adduct (**12.131**) was found to occur with 86% ee and 65% yield using this ligand.

Epoxides are effectively ring-opened in the presence of lanthanide salts, and Schaus and Jacobsen have shown that the asymmetric desymmetrisation of cyclic *meso*-epoxides with TMSCN can be achieved with up to 92% ee using ytterbium bis-oxazoline catalysts.[87]

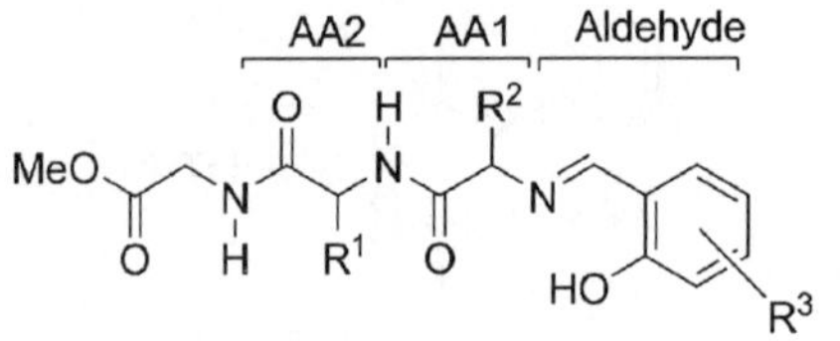

(12.130) 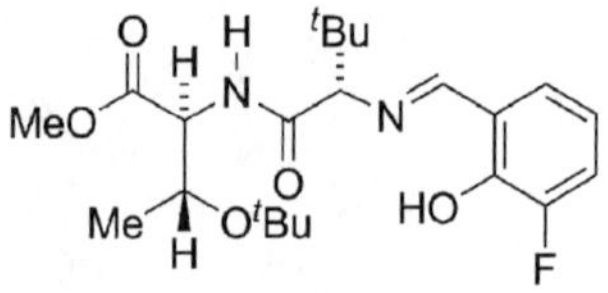   (12.132)

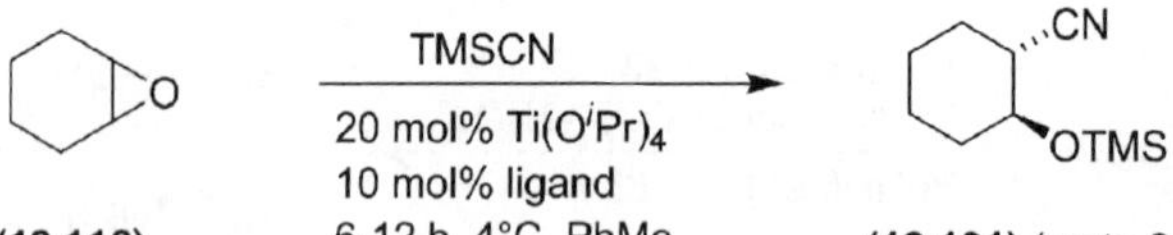

(12.116)   TMSCN

                20 mol% Ti(O^{i}Pr)$_4$
                10 mol% ligand
                6-12 h, 4°C, PhMe   (12.131) (up to 86% ee)

The use of oxygen nucleophiles for the asymmetric ring-opening of epoxides has been catalysed by the cobalt(salen) complex (**12.133**).[88, 89] Cyclohexene oxide (**12.116**) undergoes ring-opening with benzoic acid to give the mono-ester (**12.134**) with reasonable enantioselectivity. The kinetic resolution of propene oxide (**12.135**) by ring-opening with water is particularly noteworthy as an efficient route to either the epoxide or the diol (**12.136**), which are readily separated, and the catalyst can be recycled. Higher ees have been obtained using oligomeric cobalt(salen) complexes and up to 94% ee in the hydrolysis of epoxide (**12.116**) has been achieved using catalysts such as (**12.137**). These complexes also catalyse the kinetic resolution of terminal epoxides at much lower catalyst loadings than those used previously.

The higher selectivity of these catalysts may be attributed to simultaneous activation of both the nucleophile and electrophile within an asymmetric framework. The heterobimetallic gallium complex (**12.138**) also functions by binding to both

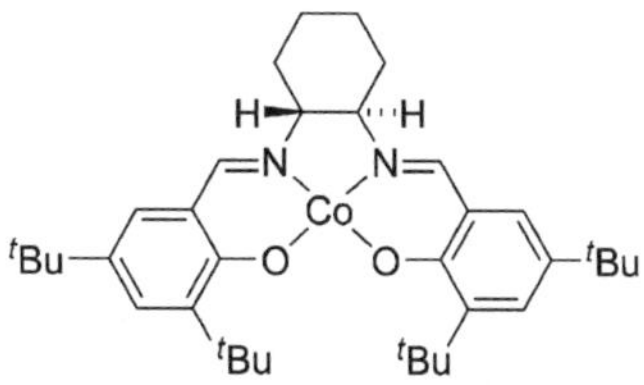

(12.133)

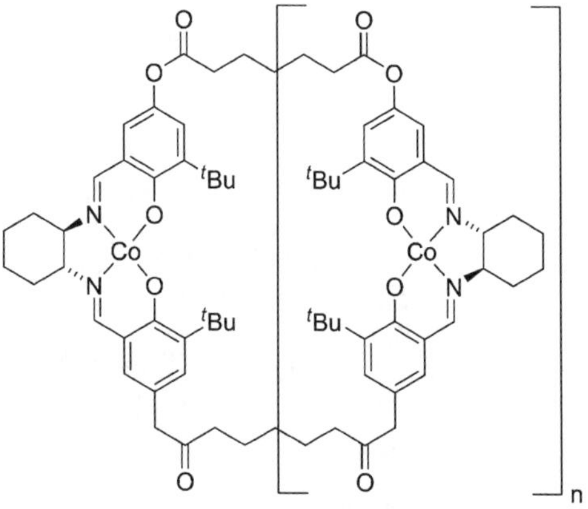

(12.137)   n = 1-5

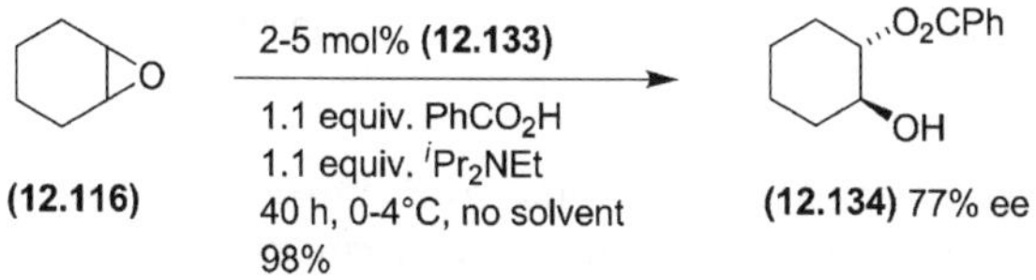

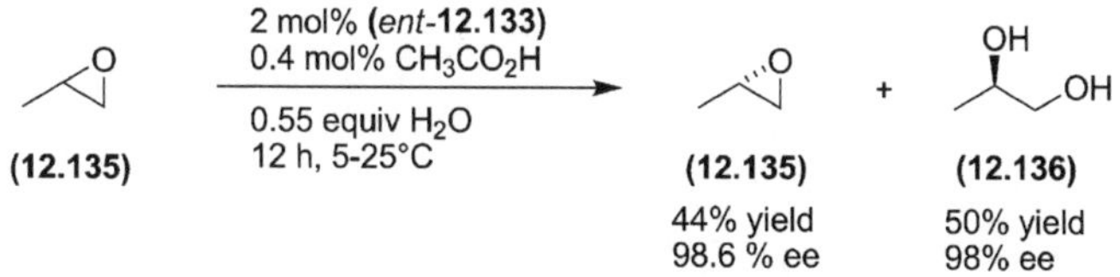

(12.116)

2-5 mol% **(12.133)**

1.1 equiv. PhCO$_2$H
1.1 equiv. ^{i}Pr$_2$NEt
40 h, 0-4°C, no solvent
98%

(12.134) 77% ee

(12.135)

2 mol% (*ent*-**12.133**)
0.4 mol% CH$_3$CO$_2$H

0.55 equiv H$_2$O
12 h, 5-25°C

(12.135)
44% yield
98.6 % ee

+

(12.136)
50% yield
98% ee

nucleophile and epoxide and has been used by Shibasaki and coworkers to effect between 66 and 96% ee in the ring-opening of cyclic *meso*-epoxides with phenols.[91] Scandium-bipyridine catalysts derived from ligand (**12.126**) are also effective in the epoxide ring-opening with oxygen-based nucleophiles.[92]

The formation of β-bromohydrins[93] and β-thioalcohols[94] from epoxides has also been reported. For example, Shibasaki has shown that the heterobimetallic gallium lithium BINOL complex (**12.138**) is a good catalyst for nucleophilic ring-opening of epoxides,[95] including the use of thiols. Cyclohexene epoxide (**12.116**) is converted into the sulfide (**12.141**) with very good enantioselectivity. Indium complexes of bipyridine (**12.126**) are very active catalysts in the thiolysis of *meso*-epoxides providing between 92 and 96% ee in the reaction of *cis*-stilbene oxides with a range of aromatic and aliphatic sulfides.[96]

In a recent example, Denmark has shown that an enantiomerically pure Lewis base (**12.142**) can be used in the ring-opening of epoxides.[97] The best substrate

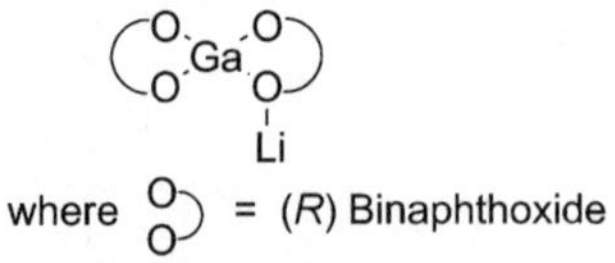
Ga
Li
where
= (R) Binaphthoxide

(12.138)

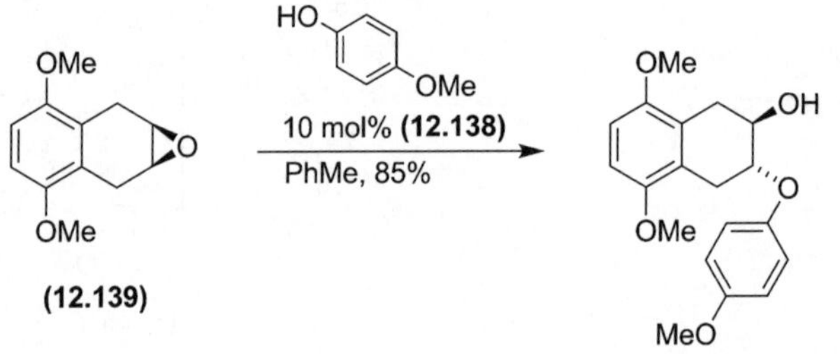
HO
OMe
10 mol% (12.138)
PhMe, 85%
OMe
OMe
OH
O
MeO

(12.139)

(12.140) 96% ee

reported was the *meso*-epoxide (**12.125**), which was converted into the chlorohydrin (**12.143**). The authors suggest that the phosphoramide displaces chloride to give a reactive silicon cation which activates the epoxide to nucleophilic attack by chloride.

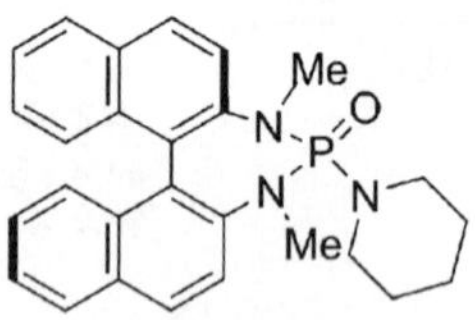
Me
N
P
O
N
N
Me

(12.142)

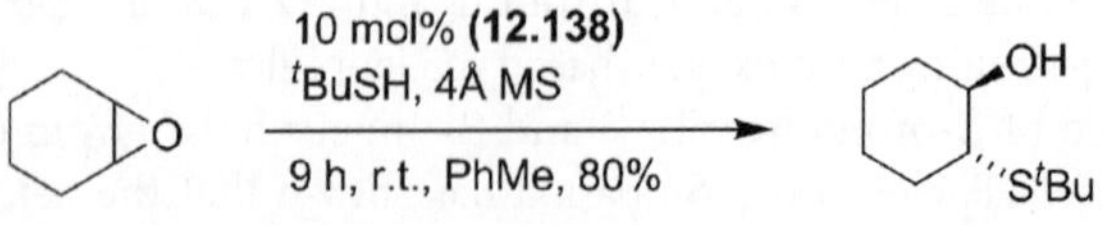
10 mol% (12.138)
^{t}BuSH, 4Å MS
9 h, r.t., PhMe, 80%
OH
S^{t}Bu

(12.116)

(12.141) 97% ee

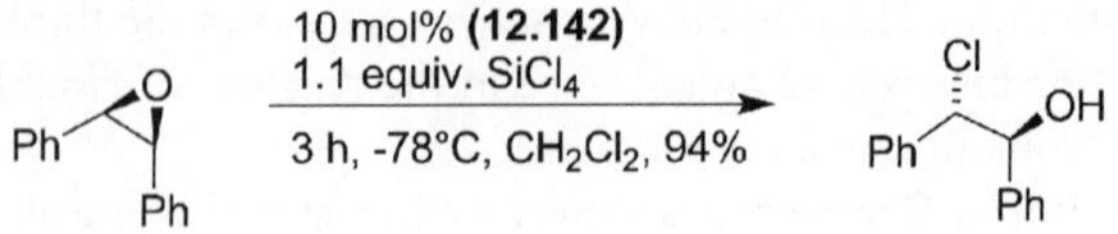
10 mol% (12.142)
1.1 equiv. SiCl$_4$
3 h, -78°C, CH$_2$Cl$_2$, 94%
Ph
Ph
Cl
OH
Ph
Ph

(12.125)

(12.143) 87% ee

References

1. K. Tani, T. Yamagata, S. Otsuka, S. Akutagawa, H. Kumobayashi, T. Taketomi, H. Takaya, A. Miyashita, and R. Noyori, *J. Chem. Soc., Chem. Commun.*, **1982**, 600.
2. (a) R. Noyori, *Asymmetric Catalysis in Organic Synthesis*, John Wiley & Sons, Inc., New York, **1994**, 95. (b) S. Akutagawa and K. Tani, in *Catalytic Asymmetric Synthesis*, ed. I. Ojima, VCH, New York, **1993**, Chapter 2. (c) S. Akutagawa in, *Comprehensive Asymmetric Catalysis*, Vol. *3*, ed. E. N. Jacobsen, A. Pfaltz and H. Yamamoto, Springer-Verlag, Berlin, **1999**, 1461.
3. (a) K. Tani, T. Yamagata, S. Akutagawa, H. Kumobayashi, T. Taketomi, H. Takaya, A. Miyashita, R. Noyori, and S. Otsuka, *J. Am. Chem. Soc.*, **1984**, *106*, 5208. (b) K. Tani, *Pure and Appl. Chem.*, **1985**, *57*, 1845.
4. M. Kitamura, K. Manabe, R. Noyori and H. Takaya, *Tetrahedron Lett.*, **1987**, *28*, 4719.
5. H. Frauenrath and M. Kaulard, *Synlett*, **1994**, 517.
6. H. Frauenrath, S. Reim and A. Wiesner, *Tetrahedron: Asymmetry*, **1998**, *9*, 1103.
7. H. Frauenrath, D. Brethauer, S. Reim, M. Maurer and G. Raabe, *Angew. Chem. Int. Ed.*, **2001**, *40*, 177.
8. Z. Chen and R. L. Halterman, *J. Am. Chem. Soc.*, **1992**, *114*, 2276.
9. (a) L. Abraham, R. Czerwonkla and M. Hiersemann, *Angew. Chem. Int. Ed.*, **2001**, *40*, 4700. (b) L. Abraham, M. Körner and M. Hiersemann, *Tetrahedron Lett.*, **2004**, *45*, 3647.
10. A. Pollex and M. Hiersemann, *Org. Lett.*, **2005**, *7*, 5705.
11. Q. Wang, A. Millet and M. Hiersmann, *Synlett*, **2007**, 1683.
12. (a) T. K. Hollis and L. E. Overman, *Tetrahedron Lett.*, **1997**, *38*, 8837. (b) M. Calter, T. K. Hollis, L. E. Overman, J. Ziller and G. G. Zipp, *J. Org. Chem.*, **1997**, *62*, 1449. (c) Y. Donde and L. E. Overman, *J. Am. Chem. Soc.*, **1999**, *121*, 2933. (d) L. E. Overman, C. E. Owen and M. M. Pavan, *Org. Lett.*, **2003**, *5*, 1809. (e) C. A. Anderson and L. E. Overman, *J. Am. Chem. Soc.*, **2003**, *125*, 12412. (f) S. F. Kirsch, L. E. Overman and M. P. Watson, *J. Org. Chem.*, **2004**, *69*, 8101. (g) D. F. Fischer, Z.-Q. Xin and R. Peters, *Angew. Chem. Int. Ed.*, **2007**, *46*, 7704.
13. T. Ooi, K. Ohmatsu and K. Maruoka, *J. Am. Chem. Soc.*, **2007**, *129*, 2410.
14. (a) B. M. Trost and T. Yasukata, *J. Am. Chem. Soc.*, **2001**, *123*, 7162. (b) B. M. Trost and J. Xie, *J. Am. Chem. Soc.*, **2006**, *128*, 6044.
15. M. Amadji, J. Vadecard, J.-C. Plaquevent, L. Duhamel and P. Duhamel, *J. Am. Chem. Soc.*, **1996**, *118*, 12483.
16. M. J. Södergren and P. G. Anderson, *J. Am. Chem. Soc.*, **1998**, *120*, 10760.
17. D. M. Hodgson, G. P. Lee, R. E. Marriott, A. J. Thompson, R. Wisedale and J. Witherington, *J. Chem. Soc., Perkin Trans. 1*, **1998**, 2151.
18. (a) M. J. McGrath and P. O'Brien, *J. Am. Chem. Soc.*, **2005**, *127*, 16378. (b) M. J. McGrath and P. O'Brien, *Synthesis*, **2005**, 2233.
19. For reviews, see; (a) C. Fehr, *Angew. Chem., Int. Ed. Engl.*, **1996**, *35*, 2566. (b) A. Yanigasawa and H. Yamamoto in, *Comprehensive Asymmetric Catalysis*, Vol. *3*, ed. E. N. Jacobsen, A. Pfaltz and H. Yamamoto, Springer-Verlag, Berlin, **1999**, 1295.
20. (a) A. Yanagisawa, T. Kikuchi, T. Watanabe, T. Kuribayashi and H. Yamamoto, *Synlett*, **1995**, 372. (b) A. Yanigasawa, T. Watanabe, T. Kikuchi and H. Yamamoto, *J. Org. Chem.*, **2000**, *65*, 2979.

21. E. Vedejs and A.W. Kruger, *J. Org. Chem.* **1998**, *63*, 2792.

22. C. Fehr and J. Galindo, *Angew. Chem., Int. Ed. Engl.*, **1994**, *33*, 1888. (b) J. Muzart, F. Hénin and S. J. Aboulhoda, *Tetrahedron: Asymmetry*, **1997**, *8*, 381. (c) K. Nishimura, M. Ono, Y. Nagaoka and K. Tomioka, *Angew. Chem. Int. Ed.*, **2001**, *40*, 440.

23. P. Riviere and K. Koga, *Tetrahedron Lett.*, **1997**, *38*, 7589.

24. K. Mitsuhashi, R. Ito, T. Arai and A. Yanagisawa, *Org. Lett.*, **2006**, *8*, 1721.

25. Y. Nakamura, S. Takeuchi, A. Ohira and Y. Ohgo, *Tetrahedron Lett.*, **1996**, *37*, 2805.

26. K. Ishihara, S. Nakamura, M. Kaneeda and H. Yamamoto, *J. Am. Chem. Soc.*, **1996**, *118*, 12854.

27. (a) A. Yanigasawa, T. Touge and T. Arai, *Angew. Chem. Int. Ed.*, **2005**, 1546. (b) A. Yanigasawa, T. Toue and T. Arai, *Pure Appl. Chem.*, **2006**, *78*, 519.

28. T. Poisson, V. Dalla, F. Marsais, G. Dupas, S. Oudeyer and V. Levacher, *Angew. Chem. Int. Ed.*, **2007**, *46*, 7090.

29. M. Sugiura and T. Nakai, *Angew. Chem., Int. Ed. Engl.*, **1997**, *36*, 2366.

30. S. J. Aboulhoda, I. Reiners, J. Wilken, F. Hénin, J. Martens and J. Muzart, *Tetrahedron: Asymmetry*, **1998**, *9*, 1847.

31. For a review covering catalytic asymmetric alkylations and allylations of enolates see: D. L. Hughes in, *Comprehensive Asymmetric Catalysis*, Vol. *3*, ed. E. N. Jacobsen, A. Pfaltz and H. Yamamoto, Springer-Verlag, Berlin, **1999**, 1273.

32. A. G. Doyle and E. N. Jacobsen, *J. Am. Chem. Soc.*, **2005**, *127*, 62.

33. A. G. Doyle and E. N. Jacobsen, *Angew. Chem. Int. Ed.*, **2007**, *46*, 3701.

34. For reviews covering phase-transfer catalysis see: (a) M. J. O'Donnell in *Catalytic Asymmetric Synthesis*, ed. I. Ojima, VCH, New York **1993**, 389. (b) T. Ooi and K. Maruoka, *Angew. Chem. Int. Ed.*, **2007**, *46*, 4222.

35. M. J. O'Donnell, W.D. Bennett and S. Wu, *J. Am. Chem. Soc.*, **1989**, *111*, 2353.

36. E. J. Corey, F. Xu and M. C. Noe, *J. Am. Chem. Soc.*, **1997**, *119*, 12414.

37. E. J. Corey, M. C. Noe and F. Xu, *Tetrahedron Lett.*, **1998**, *39*, 5347.

38. (a) B. Lygo and P.G. Wainwright, *Tetrahedron Lett.*, **1997**, *38*, 8595. (b) B. Lygo and B. I. Andrews, *Acc. Chem. Res.*, **2004**, *37*, 518.

39. Y.-J. Lee, J. Lee, M.-J. Kim, T.-S. Kim, H.-G. Park and S.-S. Jew, *Org. Lett.*, **2005**, *7*, 1557.

40. T. R. Poulsen, L. Bernardi, M. Bell and K. A. Jørgensen, *Angew. Chem. Int. Ed.*, **2006**, *45*, 6551.

41. T. Ooi, M. Kameda and K. Maruoka, *J. Am. Chem. Soc.*, **1999**, *121*, 6519.

42. T. Ooi, M. Takeuchi, D. Kato, Y. Uematsu, E. Tayama, D. Sakai and K. Maruoka, *J. Am. Chem. Soc.*, **2005**, *127*, 5073.

43. T. Ooi, T. Miki, M. Taniguchi, M. Shiraishi, M. Takeuchi and K. Maruoka, *Angew. Chem. Int. Ed.*, **2003**, *42*, 3796.

44. Y. N. Belokon, K. A. Kochetkov, T. D. Churkina, N.S. Ikonnikov, A.A. Chesnokov, O.V. Larionov, V.S. Parmár, R. Kumar and H.B. Kagan, *Tetrahedron: Asymmetry*, **1998**, *9*, 851.

45. N. Vignola and B. List, *J. Am. Chem. Soc.*, **2004**, *126*, 450.

46. (a) T. D. Beeson, A. Mastracchio, J.-B. Hong, K. Ashton and D. W. C. MacMillan, *Science*, **2007**, *316*, 582. (b) H.-Y. Jang, J.-B. Hong and D. W. C. MacMillan, *J. Am. Chem. Soc.*, **2007**, *129*, 7004.

47. For reviews on the RCM reaction in synthesis see; (a) M. Schuster and S. Blechert, *Angew. Chem., Int. Ed. Engl.*, **1997**, *36*, 2036. (a) S.K. Armstrong, *J. Chem. Soc., Perkin*

Trans. 1, **1998**, 371. (c) A. J. Phillips and A. D. Abell, *Aldrichimica Acta*, **1999**, *32*, 75. (d) R. H. Grubbs and T. M. Trnka in, *Ruthenium in Organic Synthesis*, ed. S.-I. Murahashi, Wiley-VCH, Weinheim, **2004**, 153. (e) J. B. Brenneman and S. F. Martin, *Curr. Org. Chem.*, **2005**, *9*, 1535. (f) A. Gradillas and J. Perez-Castells, *Angew. Chem. Int. Ed.*, **2006**, *45*, 6086. (g) S. K. Chattopadhyay, S. Karmakar, T. Biswas, K. C. Majumdar, H. Rahaman and B. Roy, *Tetrahedron*, **2007**, *63*, 3919. (h) A. H. Hoveyda and A. R. Zhugralin, *Nature*, **2007**, *450*, 243.

48. O. Fujimura and R. H. Grubbs, *J. Am. Chem. Soc.*, **1996**, *118*, 2499.

49. (a) J. B. Alexander, D. S. La, D. R. Cefalo, A. H. Hoveyda and R. R. Schrock, *J. Am. Chem. Soc.*, **1998**, *120*, 4041. (b) D.S. La, J.B. Alexander, D.R. Cefalo, D.D. Graf, A.H. Hoveyda and R.R. Schrock, *J. Am. Chem. Soc.*, **1998**, *120*, 9270. (c) S. S. Zu, D. R. Cefalo, D. S. La, J. Y. Jamieson, W.M. Davis, A. H. Hoveyda and R. S. Schrock, *J. Am. Chem. Soc.*, **1999**, *121*, 8251.

50. (a) S. J. Dolman, R. R. Schrock and A. H. Hoveyda, *Org. Lett.*, **2003**, *5*, 4899. (b) A. L. Lee, S. J. Malcolmson, A. Puglisis, R. R. Schrock and A. H. Hoveyda, *J. Am. Chem. Soc.*, **2006**, *128*, 5153.

51. (a) T. J. Siders, D. W. Ward and R. H. Grubbs, *Org. Lett.*, **2001**, *3*, 3225. (b) T. W. Funk, J. M. Berlin and R. H. Grubbs, *J. Am. Chem. Soc.*, **2006**, *128*, 1840. (c) C. Costabile and L. Cavallo, *J. Am. Chem. Soc.*, **2004**, *126*, 9592.

52. (a) J. J. Van Veldhuizen, D. G. Gillingham, S. B. Garber, O. Kataoka and A. H. Hoveyda, *J. Am. Chem. Soc.*, **2003**, *125*, 12502. (b) J. J. Van Veldhuizen, S. B. Garber, J. S. Kingsbury and A. H. Hoyveda, *J. Am. Chem. Soc.*, **2002**, *124*, 4954.

53. (a) W. C. P. Tsang, J. A. Jernalius, G. A. Cortez, G. S. Weatherhead, R. R. Schrock and A. H. Hoveyda, *J. Am. Chem. Soc.*, **2003**, *125*, 2591.

54. S. Arai, S. Hamaguchi and T. Shioiri, *Tetrahedron Lett.*, **1998**, *39*, 2997.

55. K. Fujita, M. Iwaoka and S. Tomoda, *Chem. Lett.*, **1994**, 923.

56. S.-I. Fukuzawa, K. Takahashi, H. Kato and H. Yamazaki, *J. Org. Chem.*, **1997**, *62*, 7711.

57. For a review see: A. L. Braga, D. S. Lüdtke and F. Vargas, *Curr. Org. Chem.*, **2006**, *10*, 1921.

58. T. Wirth, S. Häuptli and M. Leuenberger, *Tetrahedron: Asymmetry*, **1998**, *9*, 547.

59. M. Tiecco, L. Testaferri, C. Santi, C. Tomassini, F. Marini, L. Bagnoli and A. Temperini, *Chem. Eur. J.*, **2002**, *8*, 1118.

60. (a) J. C. Sheehan and T. Hara, *J. Org. Chem.*, **1974**, *39*, 1196. (b) R.L. Knight and F.J. Leeper, *Tetrahedron Lett.*, **1997**, *38*, 3611. (c) C. A. Dvorak and V.H. Rawal, *Tetrahedron Lett.*, **1998**, *39*, 2925. (d) R. L. Knight and F. J. Leeper, *J. Chem. Soc., Perkin Trans. 1*, **1998**, 1891.

61. (a) D. Enders, K. Breuer and J.H. Teles, *Helv. Chim. Acta*, **1996**, *79*, 1217.

62. D. Enders and U. Kallfass, *Angew. Chem., Int. Ed.*, **2002**, *41*, 1743.

63. (a) D. Enders, O. Niemeier and T. Balensiefer, *Angew. Chem. Int. Ed.*, **2006**, *45*, 1463. (b) H. Takikawa, Y. Hachisu, J. W. Bode and K. Suzuki, *Angew. Chem. Int. Ed.*, **2006**, *45*, 3492. (c) H. Takikawa and K. Suzuki, *Org. Lett.*, **2007**, *9*, 2713.

64. (a) C.-H. Wong and G.M. Whitesides, *Enzymes in Synthetic Organic Chemistry*, Tetrahedron Organic Chemistry Series, Volume *12*, Pergamon, Oxford, **1994**, Chapter 2. (b) H. Pellisier, *Tetrahedron*, **2003**, *59*, 8291. (c) A. Ghanem and H. Y. Aboul-Enein, *Tetrahedron:Asymmetry*, **2004**, *15*, 3331. (d) E. Garcia-Urdiales, I. Alfonso and V. Gotor, *Chem. Rev.*, **2005**, *105*, 313. (e) R. Chenevert, N. Pelchat and F. Jacques, *Curr. Org.*

Chem., **2006**, *10*, 1067. (f) E. Santaniello, S. Casati and P. Cuiffreda, *Curr. Org. Chem.*, **2006**, *10*, 1095. (g) A. Liljeblad and L. T. Kanerva, *Tetrahedron*, **2006**, *62*, 5831.

65. (a) P. Somfai, *Angew. Chem., Int. Ed. Engl.*, **1997**, *36*, 2731. (b) D. E. J. E. Robinson and S. D. Bull, *Tetrahedron: Asymmetry*, **2003**, *14*, 1407.

66. (a) J. C. Ruble and G. C. Fu, *J. Org. Chem.*, **1996**, *61*, 7230. (b) J.C. Ruble, H.A. Lathan and G.C. Fu, *J. Am. Chem. Soc.*, **1997**, *119*, 1492. (c) J.C. Ruble, J. Tweddell and G.C. Fu, *J. Org. Chem.*, **1998**, *63*, 2794. (d) G. C. Fu, *Acc. Chem. Res.*, **2004**, *37*, 542. (e) E. Vedejs and M. Jure, *Angew. Chem. Int. Ed.*, **2005**, *44*, 3974.

67. H.B. Kagan and J.C. Fiaud, *Top. Stereochem*, **1988**, *18*, 249.

68. (a) Y. Matsumura, T. Maki, S. Murakami and O. Onomura, *J. Am. Chem. Soc.*, **2003**, *125*, 2052. (b) A. Gissibl, M. G. Fin and O. Reiser, *Org. Lett.*, **2005**, *7*, 2325. (c) C. Mazet, S. Roseblade, V. Kohler and A. Pflatz, *Org. Lett.*, **2006**, *8*, 1879.

69. (a) G. Priem, B. Pelotier, S. J. F. Macdonald, M. S. Anson and I. B. Campbell, *J. Org. Chem.* **2003**, *68*, 3844. (b) A. C. Spivey, F. Zhu, M. B. Mitchell, S. G. Davey and R. L. Jarvest, *J. Org. Chem.*, **2003**, *68*, 7379. (c) V. B. Birman, E. W. Uffman, H. Jiang, X. Li and C. J. Kilbane, *J. Am. Chem. Soc.*, **2004**, *126*, 12226. (d) A. C. Spivey, S. Arseniyadis, T. Feckner, A. Maddaford and D. P. Leese, *Tetrahedron*, **2006**, *62*, 295.

70. (a) Y. Suzuki, K. Yamauchi, K. Muramatsu and M. Sato, *J. Chem. Soc., Chem. Commun.*, **2004**, 2770. (b) T. Kano, K. Sasaki and K. Maruoka, *Org. Lett.*, **2005**, *7*, 1347.

71. (a) E. Vedejs, O. Daugulis and S. T. Diver, *J. Org. Chem.*, **1996**, *61*, 430. (b) E. Vedejs and O. Daugulis, *J. Am. Chem. Soc.*, **2003**, *125*, 4166.

72. D. Terrado, H. H. Koutaka and T. Oriyama, *Tetrahedron: Asymmetry*, **2005**, *16*, 1157.

73. (a) S. J. Miller and G. T. Copeland, *J. Am. Chem. Soc.*, **2001**, *123*, 6496. (b) G. T. Copeland and S. J. Miller, *J. Am. Chem. Soc.*, **2001**, *123*, 6496. (c) M. B. Fierman, D. J. O'Leary, W. E. Steinmetz and S. J. Miller, *J. Am. Chem. Soc.*, **2004**, *126*, 6967.

74. T. Oriyama, K. Imai, T. Sano and T. Hosoya, *Tetrahedron Lett.* **1998**, *39*, 3529.

75. T. Oriyama, K. Imai, T. Hosoya and T. Sano, *Tetrahedron Lett.*, **1998**, *39*, 397.

76. K. Matsumoto, M. Mitsuda, N. Ushijima, Y. Demizu, O. Onomura and Y. Matsumura, *Tetrahedron Lett.*, **2006**, *47*, 8453.

77. B. M. Trost and T. Mino, *J. Am. Chem. Soc.*, **2003**, *125*, 2410.

78. G. Jaeschke and D. Seebach, *J. Org. Chem.*, **1998**, *63*, 1190.

79. J. Liang, J.C. Ruble and G.C. Fu, *J. Org. Chem.*, **1998**, *63*, 3154.

80. For a review dealing with catalytic epoxide desymmetrisation, see; (a) D. M. Hodgson, A. R. Gibbs and G. P. Lee, *Tetrahedron*, **1996**, *52*, 14361. (b) E. N. Jacobsen and M. C. Wu in, *Comprehensive Asymmetric Catalysis*, Vol. 3, ed. E. N. Jacobsen, A. Pfaltz and H. Yamamoto, Springer-Verlag, Berlin, **1999**, 1309. (c) C. Schneider, *Synthesis*, **2006**, 3919.

81. M. H. Wu and E.N. Jacobsen, *Tetrahedron Lett.*, **1997**, *48*, 1693.

82. W. A. Nugent, *J. Am. Chem. Soc.*, **1992**, *114*, 2768.

83. (a) L. E. Martínez, J. L. Leighton, D. H. Carsten and E. N. Jacobsen, *J. Am. Chem. Soc.*, **1995**, *117*, 5897. (b) E. N. Jacobsen, *Acc. Chem. Res.*, **2000**, *33*, 421.

84. S. E. Schaus and E. N. Jacobsen, *Tetrahedron Lett.*, **1996**, *37*, 7937.

85. (a) C. Schneider, A. R. Sreekanth and E. Mai, *Angew. Chem. Int. Ed.*, **2004**, *43*, 5691. (b) S. Azoulay, K. Manabe and S. Kobayashi, *Org. Lett.*, **2005**, *7*, 4593. (c) F. Carrée, R. Gil and J. Collin, *Org. Lett.*, **2005**, *7*, 1023.

86. B. M. Cole, K. D. Shimizu, C. A. Krueger, J. P. A. Harrity, M. L. Snapper and A.M. Hoveyda, *Angew. Chem., Int. Ed. Engl.*, **1996**, *35*, 1668.

87. S. E. Schaus and E. N. Jacobsen, *Org. Lett.*, **2000**, *2*, 1001.

88. E. N. Jacobsen, F. Kakiuchi, R.G. Konsler, J.F. Larrow and M. Tokunaga, *Tetrahedron Lett.*, **1997**, *38*, 773.

89. M.Tokunaga, J.F. Larrow, F. Kakiuchi and E.N. Jacobsen, *Science*, **1997**, *277*, 936.

90. (a) J. M. Ready and E. N. Jacobsen, *J. Am. Chem. Soc.*, **2001**, *123*, 2687. (b) J. M. Ready and E. N. Jacobsen, *Angew. Chem. Int. Ed.*, **2002**, *41*, 1374.

91. S. Matsunaga, J. Das, J. Roels, E. M. Vogl, N. Yamamoto, T. Iida, K. Yamaguchi and M. Shibasaki, *J. Am. Chem. Soc.*, **2000**, *122*, 2252.

92. A. Tschop, A. Marx, A. R. Sreekanth and C. Schneider, *Eur. J. Org. Chem.*, **2007**, 2318.

93. W.A. Nugent, *J. Am. Chem. Soc.*, **1998**, *120*, 7139.

94. T. Iida, N. Yamamoto, H. Sasai and M. Shibasaki, *J. Am. Chem. Soc.*, **1997**, *119*, 4783.

95. T. Iida, N. Yamamoto, S. Matsunaga, H.-G. Woo and M. Shibasaki, *Angew. Chem., Int. Ed. Engl.*, **1998**, *37*, 2223.

96. M. V. Nandakumar, A. Tschop, H. Krautschied and C. Schneider, *J. Chem. Soc., Chem. Commun.*, **2007**, 2756.

97. S. E. Denmark, P. A. Barsanti, K.-T. Wong and R. A. Stavenger, *J. Org. Chem.*, **1998**, *63*, 2428.

Index
